The GREAT BRIDGE

David McCullough

Simon & Schuster Paperbacks
New York London Toronto Sydney

SIMON & SCHUSTER PAPERBACKS
Rockefeller Center
1230 Avenue of the Americas
New York, New York 10020

For information about special discounts for bulk purchases,
please contact Simon & Schuster Special Sales:
1-800-456-6798 or business@simonandschuster.com

Manufactured in the United States of America

40 39 38 37 36 35 34

The Library of Congress has cataloged the hardcover edition as
follows:
McCullough, David G.
The great bridge.
New York: Simon and Schuster, 1972.
Bibliography: p.
Includes index.
1. Brooklyn Bridge (New York, N.Y.)
I. Title.
TG25.N53M32 624.5'5'097471 72-081823
ISBN-13: 978-0-671-21213-1
ISBN-10: 0-671-21213-3
ISBN-13: 978-0-671-45711-2 (Pbk)
ISBN-10: 0-671-45711-X (Pbk)

For my mother and father

Contents

AUTHOR'S NOTE

WHEN I began this book I was setting out to do something that had not been done before. I wanted to tell the story of the most famous bridge in the world and in the context of the age from which it sprang. The Brooklyn Bridge has been photographed, painted, engraved, embroidered, analyzed as a work of art and as a cultural symbol; it has been the subject of a dozen or more magazine articles and one famous epic poem; it has been talked about and praised more it would seem than anything ever built by Americans. But a book telling the full story of how it came to be, the engineering involved, the politics, the difficulties encountered, the heroism of its builders, the impact it had on the lives and imaginations of ordinary people, a book that would treat this important historical event as a rare human achievement, had not been written and such was my goal.

I was also greatly interested in the Roeblings, about whom quite a little had been written, but not for some time or from the kind of research I had in mind. Moreover, a good deal of legend about the Roeblings—father, son, and daughter-in-law—still persisted, along with considerable confusion. It seemed to me that the story of these remarkable people deserved serious study. It is an extraordinary story, to say the least, not only in human terms, but in what it reveals about America in the late nineteenth century, a time that has not been altogether appreciated for what it was.

And beyond that I had a particular interest in the city of Brooklyn itself, having spent part of my life there, when my wife and I were first married, in a house just down the street from where Washington and Emily Roebling once lived.

But early in my research another objective emerged. It became clear that this, to a large degree, was to be Washington Roebling's book. There was, for example, that day in the library at the Rensselaer Polytechnic Institute when I unlocked a large storage closet to see for the first time shelf after shelf of his notebooks, scrapbooks, photographs, letters, blueprints, old newspapers he had saved, even the front-door knocker to his house in Brooklyn. No one knew then what all was in the collection. There were boxes of his papers that had not been opened in years, bundles of letters that so far as I could tell had been examined by nobody. The excitement of the moment can be imagined. The contents of the collection, plus those in another large collection at Rutgers University, both of which are described in the Bibliography, were such that they often left me with the odd feeling of actually having known the Chief Engineer of the bridge. He was not only the book's principal character, he was the author's main personal contact with that distant day and age. So it has also been my aim to convey, with all the historical accuracy possible, just what manner of man this was who built the Brooklyn Bridge, who achieved so much against such staggering odds, and who asked so little.

I am not an engineer and the technical side of the research has often been slow going for me. But though I have written the book for the general reader, I have not bypassed the technical side. If I could make it clear enough that I could understand it, if it was interesting to me, then my hope was that it would be both clear and interesting to the reader.

During my years of research and writing I have been extremely fortunate in the assistance I have received from many people and I should like to express to them my abiding gratitude. For their kindnesses and help I wish to thank the librarians at both Rutgers and Rensselaer and in particular Miss Irene K. Lionikis of the Rutgers Library and Mrs. Orlyn LaBrake and Mrs. Adrienne Grenfell of the library at Rensselaer. Herbert R. Hands of the American Society of Civil Engineers, David Plowden, Dr. Milton

Mazer, Dr. Roy Korson, Professor of Pathology at the University of Vermont, W. H. Pearson, Sidney W. Davidson, J. Robert Maguire, Charlotte La Rue of the Museum of the City of New York, Regina M. Kellerman, William S. Goodwin, Allan R. Talbot, John Talbot, and Jack Schiff, the engineer in charge of New York's East River bridges, each contributed to the research. And Dr. Paul Gugliotta of New York, architect and engineer, said some things over lunch one day years ago that started me thinking about doing such a book and later very kindly walked the bridge with me and answered many questions.

I am especially indebted to Robert M. Vogel, Curator, Division of Mechanical and Civil Engineering at the Smithsonian Institution, to John A. Kouwenhoven, authority on New York City history and on James B. Eads, to Nomer Gray, bridge engineer, who has made his own extensive technical studies of the bridge, and to Charlton Ogburn, author and friend. Each of them read the manuscript and offered numerous critical suggestions, but any errors in fact or judgment that may appear in the book are entirely my own.

I would like to acknowledge, too, the contribution of three members of the Roebling family: Mr. Joseph M. Roebling of Trenton and Mr. F. W. Roebling, also of Trenton, who gave of their time to talk with me about their forebears, and Mrs. James L. Elston of Fayetteville, Arkansas, who let me borrow an old family scrapbook.

I am grateful for the research facilities and assistance offered by the staffs of the following: the Trenton Free Public Library; the Carnegie Library, Pittsburgh; the Brooklyn Public Library; the Long Island Historical Society, Brooklyn, and particularly to Mr. John H. Lindenbusch, its executive director; the Newport Historical Society, Newport, Rhode Island; the Library of Congress; the New York Historical Society; the New York Public Library; the Engineering Societies Library, New York; the Middlebury College Library, Middlebury, Vermont; the Baker Library, Dartmouth College; the Putnam County Historical Society and the Julia Butterfield Memorial Library at Cold Spring, New York; and the Butler County Library, Butler, Pennsylvania.

I wish also to acknowledge my indebtedness to two valued friends who are no longer living—to Conrad Richter, for his encouragement and example, and to Clarence A. Barnes, my father-in-law, who was born on Willow Street on Brooklyn Heights, when the bridge was still unfinished, and who could talk better than anyone I knew about times gone by.

Lastly I would like to express my thanks to Paul R. Reynolds, who provides steady encouragement and sound advice; to Peter Schwed, Publisher of Simon and Schuster, who had faith in the idea from the start; to Jo Anne Lessard, who typed the manuscript; to my children, for their confidence and optimism; and to my wife, Rosalee, who helped more than anyone.

—David McCullough

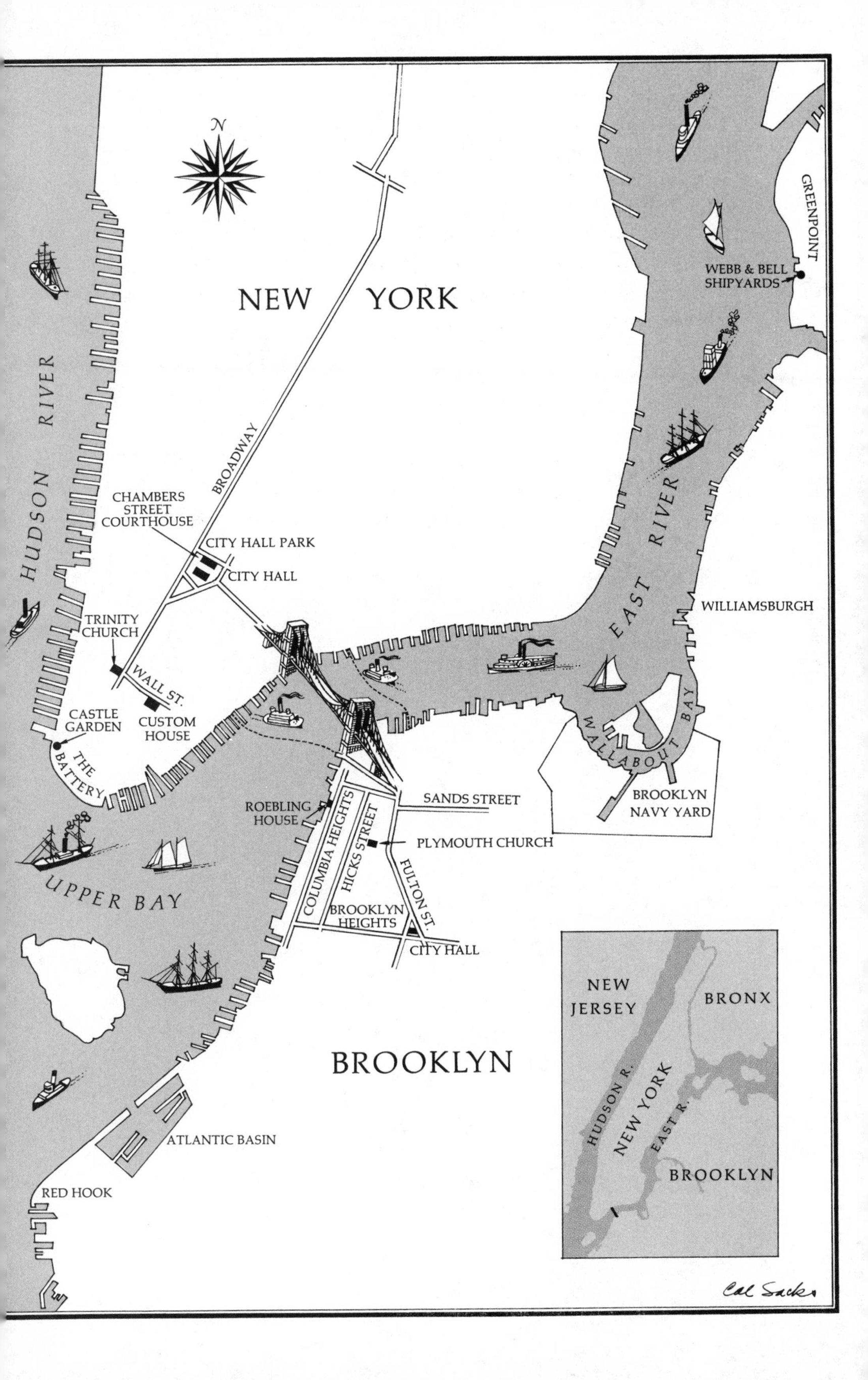
N
NEW YORK
BROADWAY
HUDSON RIVER
CHAMBERS STREET COURTHOUSE
CITY HALL PARK
CITY HALL
TRINITY CHURCH
WALL ST.
CASTLE GARDEN
CUSTOM HOUSE
THE BATTERY
GREENPOINT
WEBB & BELL SHIPYARDS
EAST RIVER
WILLIAMSBURGH
WALLABOUT BAY
BROOKLYN NAVY YARD
SANDS STREET
ROEBLING HOUSE
PLYMOUTH CHURCH
COLUMBIA HEIGHTS
HICKS STREET
FULTON ST.
BROOKLYN HEIGHTS
CITY HALL
UPPER BAY
BROOKLYN
ATLANTIC BASIN
RED HOOK
NEW JERSEY
BRONX
HUDSON R.
NEW YORK
EAST R.
BROOKLYN
Cal Sacks

It so happens that the work which is
likely to be our most durable monument,
and to convey some knowledge of us to the
most remote posterity, is a work of bare
utility; not a shrine, not a fortress,
not a palace, but a bridge.

—MONTGOMERY SCHUYLER

in *Harper's Weekly*, May 24, 1883

PART ONE

1
The Plan

The shapes arise!
—WALT WHITMAN

THEY MET at his request on at least six different occasions, beginning in February 1869. With everyone present, there were just nine in all—the seven distinguished consultants he had selected; his oldest son, Colonel Washington Roebling, who kept the minutes; and himself, the intense, enigmatic John Augustus Roebling, wealthy wire rope manufacturer of Trenton, New Jersey, and builder of unprecedented suspension bridges.

They met at the Brooklyn Gas Light Company on Fulton Street, where the new Bridge Company had been conducting its affairs until regular offices could be arranged for. They gathered about the big plans and drawings he had on display, listening attentively as he talked and asking a great many questions. They studied his preliminary surveys and the map upon which he had drawn a strong red line cutting across the East River, indicating exactly where he intended to put the crowning work of his career.

The consultants were his idea. In view of "the magnitude of the undertaking and the large interests connected therewith," he had written, it was "only right" that his plans be "subjected to the careful scrutiny" of a board of experts. He did not want their advice or opinions, only their sanction. If everything went as he wanted and expected, they would approve his plan without reservation. They would announce that in their considered professional opinion his bridge was perfectly possible. They would put an end to the rumors, silence the critics, satisfy every last stockholder that he knew what he was about, and he could at last get on with his work.

To achieve his purpose, to wind up with an endorsement no one could challenge, or at least no one who counted for anything professionally, he had picked men of impeccable reputation. None had a failure or black mark to his name. All were sound, practical builders themselves, men not given to offhand endorsements or to overstatement. With few exceptions, each had done his own share of pioneering at one time or other, and so theoretically ought still to be sympathetic to the untried. They were, in fact, about as eminent a body of civil engineers as could have been assembled then, and seen all together, with their display of white whiskers, their expansive shirt fronts and firm handshakes, they must have appeared amply qualified to pass judgment on just about anything. The fee for their services was to be a thousand dollars each, which was exactly a thousand dollars more than Roebling himself had received for all his own efforts thus far.

Chairman of the group was the sociable Horatio Allen, whose great girth, gleaming bald head, and Benjamin Franklin spectacles gave him the look of a character from Dickens. He fancied capes and silver-handled walking sticks and probably considered his professional standing second only to that of Roebling, which was hardly so. But like Roebling he had done well in manufacturing—in his case, with New York's Novelty Iron Works—and forty years before he had made some history driving the first locomotive in America, the *Stourbridge Lion*, all alone and before a big crowd, on a test run at Honesdale, Pennsylvania. He had also, in the time since, been one of the principal engineers for New York's Croton Aqueduct and so was sometimes referred to in biographical sketches as "the man who turned the water on."

Then there was Colonel Julius Adams of Brooklyn, a former Army engineer, who was usually described as an expert on sewer construction, and who, in truth, was not quite in the same league as the others. He had, however, a number of influential friends in Brooklyn and for years he had been dabbling with designs for an East River bridge of his own. For a while it had even looked as though he might be given the chance to build it. When Roebling's proposal was first made public, he had been among those to voice sharp skepticism. That he had been included as a consultant at this stage was taken by some as a sign that Roebling was not entirely the political innocent he was reputed to be.

William Jarvis McAlpine, of Stockbridge, Massachusetts, was the president of the American Society of Civil Engineers. Kindly, genial, widely respected, he had built the enormous dry dock at the Brooklyn Navy Yard, the Albany Water Works, and a fair number of bridges. He was also the proud possessor of what must have been the most elaborate jowl whiskers in the profession and he was the one man in the group, the two Roeblings included, who had had any firsthand experience working with compressed-air foundations, or caissons, as they were called, which, in this particular case, was regarded as an attribute of major proportions.

Probably the best-known figure among them, however, was Benjamin Henry Latrobe of Baltimore, who had the face of a bank clerk, but whose endorsement alone would perhaps have been enough to settle the whole issue. He was the son and namesake of the famous English-born architect picked by Jefferson to design or remodel much of Washington, and who rebuilt the Capitol after it was burned by the British during the War of 1812. He had laid out most of the B&O Railroad and had been in charge of building a number of exceptional bridges in Maryland and Virginia.

And finally there was John J. Serrell, the only builder of suspension bridges in the group except for the Roeblings; J. Dutton Steele, chief engineer of the Reading Railroad; and James Pugh Kirkwood, a rather mournful-looking Scotsman who was an authority on hydraulics, among other things, and who, in 1848, in northeastern Pennsylvania, had built the beautiful stone-arched Starrucca Viaduct, then the most costly railroad bridge in the world.

There is no way of knowing what thoughts passed through the minds of such men as they first looked over Roebling's drawings and listened to him talk. But it is also hard to imagine any of them remaining unimpressed for very long, for all their collective experience or their own considerable accomplishments or any professional jealousies there may have been. Nor does it seem likely that any of them failed to sense the historic nature of the moment. Roebling was the recognized giant of their profession, a lesser-Leonardo he would be called, and even on paper his bridge was clearly one of the monumental works of the age. To an engineer especially that would have been obvious.

A bridge over the East River, joining the cities of New York and Brooklyn, had been talked about for nearly as long as anyone could recall. According to the best history of Brooklyn ever written, a three-volume work by a medical doctor named Henry R. Stiles, Volume II of which appeared that same year of 1869, the idea for a bridge was exactly as old as the century, the first serious proposal having been recorded in Brooklyn in 1800. Stiles wrote that an old notation, found in a scrapbook, referred to an unnamed "gentleman of acknowledged abilities and good sense" who had a plan for a bridge that would take just two years to build. Probably the gentleman was Thomas Pope of New York, an altogether fascinating character, a carpenter and landscape gardener by trade, who had designed what he called his "Flying Pendent Lever Bridge," an *invention*, as he saw it, available in all sizes and suitable for any site. His bridge to Brooklyn was to soar some two hundred feet over the water, with a tremendous cantilever fashioned entirely of wood, like "a rainbow rising on the shore," he said in the little book he published in 1811. Thomas Pope's "Rainbow Bridge" was never attempted, however, and fortunately so, for it would not have worked. But his vision of a heroic, monumental East River bridge persisted. Year after year others were proposed. Chain bridges, wire bridges, a bridge a hundred feet wide, were recommended by one engineer or another. "New York and Brooklyn must be united," Horace Greeley declared in the *Tribune* in 1849, while in Brooklyn a street running down to the river was confidently christened Bridge Street.

But nothing was done. The chief problem always was the East River, which is no river at all technically speaking, but a tidal strait and one of the most turbulent and in that day, especially, one of the busiest stretches of navigable salt water anywhere on earth. "If there is to be a bridge," wrote one man, "it must take one grand flying leap from shore to shore over the masts of the ships. There can be no piers or drawbridge. There must be only one great arch all the way across. Surely this must be a wonderful bridge."

In April 1867 a charter authorizing a private company to build and operate an East River bridge had been voted through at Albany. The charter was a most interesting and important document, for several different reasons, as time would tell. But in the things it said and left unsaid concerning the actual structure to be built, it was notable at a glance. Not a word was mentioned, for example, about the sort of bridge it was to be or to suggest that its construction might involve any significant or foreseeable problems. The cities were not required to approve the plans or the location. The charter said only that it be a toll bridge. It was important that it have a "substantial railing" and that it be "kept fully lighted through all hours of the night." It was also to be completed by January 1, 1870.

A month after the charter became law, Roebling had been named engineer of the work. By whom or by what criteria remained a puzzle for anyone trying to follow the story in the papers. In September, that same year, 1867, at a private meeting held in Brooklyn, he presented his master plan in a long formal report. But such was "the anxiety manifested on the part of the press of the two cities to present his report to the public, that it was taken and published, as an entirety . . ." The bridge had no official name at this point, and in the time since, nobody seemed able to settle on one.

At an earlier stage it had been referred to occasionally as the Empire Bridge, but the organization incorporated to build it was called the New York Bridge Company, because the Brooklyn people behind the idea saw it as just that—a bridge to New York. Roebling, on the other hand, had referred to it as the East River Bridge in his proposal and the newspapers and magazines had picked up the name. But it was also commonly called the Roebling Bridge or the Brooklyn Bridge or simply the Great Bridge, which looked the most impressive in print and to many seemed the most fitting name of all, once they grasped what exactly Roebling was planning to do.

But it was the possible future impact of such a structure on their own lives that interested people most, naturally enough, and that the press in both cities devoted the most attention to. The *Times*, for example, described the bridge as a sort of grand long-needed pressure valve that would do much to alleviate New York's two most serious problems, crime and overcrowding.

In Brooklyn, where interest was the keenest, it was said the bridge would make Brooklyn important, that it would make Brooklyn prosper. Property values would soar. Roebling the alchemist would turn vacant lots and corn patches into pure gold. Everybody would benefit. Brooklyn was already expanding like a boomtown, and the bridge was going to double the pace, the way steam ferries had. Merchants could expect untold numbers of new customers as disaffected New Yorkers flocked across the river to make Brooklyn their home. Manufacturers would have closer ties with New York markets. Long Island farmers and Brooklyn brewers could get their wares over the river more readily. The mail would move faster. Roebling had even told his eager clients how, in the event of an enemy invasion of Long Island, troops could be rushed over the bridge from New York in unprecedented numbers. In such an emergency, the old Prussian had calculated, nearly half a million men, together with artillery and baggage trains, could go over the bridge in twenty-four hours.

Most appealing of all for the Brooklyn people who went to New York to earn a living every day was the prospect of a safe, reliable alternative to the East River ferries. Winds, storms, tides, blizzards, ice jams, fog, none of these, they were told, would have the slightest effect on Mr. Roebling's bridge. There would be no more shoving crowds at the ferryhouse loading gates. There would be no more endless delays. One Christmas night a gale had caused the river to be so low the ferries ran aground and thousands of people spent the night in the Fulton Ferry house. Many winters when the river froze solid, there had been no service at all for days on end.

Some of the Brooklyn business people and Kings County politicians were even claiming that the bridge would make Brooklyn the biggest city in America, a most heady prospect indeed and not an unreasonable one either. Congressman Demas Barnes contended Brooklyn would be the biggest city in the world, once New York was "full." New York, that "human hive" John Roebling called it, was running out of space, its boundaries being forever fixed by nature. Roebling and others envisioned a day when all Manhattan Island would be built over, leaving "no decent place" to make a home, neither he nor anyone else thus far having imagined a city growing vertically. "Brooklyn happens to be one of those things that can expand," wrote the editors of the new *Brooklyn Monthly*. "The more you put into it, the more it will hold."

And such highly regarded Brooklyn residents as Walt Whitman and James S. T. Stranahan, the man behind Brooklyn's new Prospect Park, looked to the day when the bridge would make Brooklyn and New York "emphatically one," which was also generally taken to be a very good thing, since the new Union Pacific Railroad was going to make New York "the commercial emporium of the world." This was no idle speculation, "but the natural and legitimate result of natural causes," according to John Roebling. His bridge was part of a larger mission. "As the great flow of civilization has ever been from East towards the West, with the same certainty will the greatest commercial emporium be located on this continent, which links East to the West, and whose mission it is in the history of mankind to blend the most ancient civilization with the most modern." The famous engineer, it had been noticed in Brooklyn, tended to cosmic concepts, but so much the better. If there were now forty million people crossing the East River every year, as was the claim, then, he said, in ten years' time there would be a hundred million.

"Lines of steamers, such as the world never saw before, are now plowing the Atlantic in regular straight line furrows," he had written in his proposal. "The same means of communication will unite the western coast of this continent to the eastern coast of Asia. New York will remain the center where these lines meet."

This, in other words, was to be something much more than a large bridge over an important river. It was to be one of history's great connecting works, symbolic of the new age, like the Atlantic cable, the Suez Canal, and the transcontinental railroad. "Lo, Soul, seest thou not God's purpose from the first?" wrote Walt Whitman at about this time. "The earth be spann'd, connected by network . . . The lands welded together." "The shapes arise!" wrote the Brooklyn poet.

> Singing my days,
> Singing the great achievements of the present
> Singing the strong, light works of engineers . . .

But it was Roebling himself, never one to be overly modest, who had set forth the most emphatic claim for the bridge itself and the one that would be quoted most often in time to come:

> The completed work, when constructed in accordance with my designs, will not only be the greatest bridge in existence, but it will be the greatest engineering work of the continent, and of the age. Its most conspicuous features, the great towers, will serve as landmarks to the adjoining cities, and they will be entitled to be ranked as national monuments. As a great work of art, and as a successful specimen of advanced bridge engineering, this structure will forever testify to the energy, enterprise and wealth of that community which shall secure its erection.

Roebling had written that in 1867, at the very start of his formal proposal, but in all the time since, for some mysterious reason, not a spade of dirt had been turned and numbers of people, some claiming to be experts, had begun saying they were not so sure about

Roebling's "advanced engineering," or whether it was worth the six to seven million dollars he had said it would cost, an estimate that did not include the price of the land required. Even if his figures were realistic, the bridge would also be about the most expensive ever built.

The editors of *Scientific American* said a tunnel would serve the purpose as well and cost less. A Navy engineer presented an alternative plan. He wanted to block off "the vexatious East River" with a dam several hundred feet wide on which he would build highways, stores, docks, and warehouses. By early 1869, when it looked as though the bridge might actually be started, the critics were sounding forth as never before. Warehouse owners along the river and others in the shipping business were calling it an obstruction to navigation and a public nuisance. The New York Polytechnic Society put on a series of lectures at Cooper Union devoted exclusively to the supposed engineering fallacies of the Roebling plan. Engineers expressed "grave apprehension." The bridge, it was stated on the best professional authority, was a monumental extravagance, "a wild experiment," nothing but an exercise in vanity. Even in Brooklyn the *Union* said another bridge and a tunnel besides would probably be built by the time everyone finished wrangling over details and questioned why, for so momentous a public work, only one engineer had been called on and no other plans ever considered.

So it had been to still such talk that Roebling had assembled his seven consultants and with total patience and candor went over everything with them point by point.

To begin with it was to be the largest suspension bridge in the world. It was to be half again the size of his bridge over the Ohio at Cincinnati, for example, and nearly twice the length of Telford's famous bridge over the Menai Strait, in Wales, the first suspension bridge of any real importance. It was to cross the East River with one uninterrupted central span, held aloft by huge cables slung from the tops of two colossal stone towers and secured on either shore to massive masonry piles called anchorages. These last structures alone, he said, would be a good seven stories tall, or taller than most buildings in New York at the time. They would each take up the better part of a city block and would be heavy enough to offset the immense pull of the cables, but hollow inside, to provide, Roebling suggested, room for cavernous treasury vaults, which he claimed would be the safest in America and ample enough to house three-quarters of all the investments and securities in the country.

The towers, the "most conspicuous features," would be identical and 268 feet high. They would stand on either side of the river, in the water but close to shore, their foundations out of sight beneath the riverbed. Their most distinguishing features would be twin Gothic arches—two in each tower—through which the roadways were to pass. These arches would rise more than a hundred feet, like majestic cathedral windows, or the portals of triumphal gateways. "In a work of such magnitude, and located as it is between two great cities, good architectural proportions should be observed," wrote the engineer. ". . . The impression of the whole will be that of massiveness and strength."

His towers would dwarf everything else in view. They would reign over the landscape like St. Peters in Rome or the Capitol dome in Washington, as one newspaper said. In fact, the towers would be higher than the Capitol dome if the dome's crowning statue of Freedom was not taken into account. So this in the year 1869—when the Washington Monument was still an ugly stone stump—meant they would be about the largest, most massive things ever built on the entire North American continent. On the New York skyline only the slim spire of Trinity Church at the head of Wall Street reached higher.

The towers were to serve two very fundamental purposes. They would bear the weight of four enormous cables and they would hold both the cables and the roadway of the bridge high enough so they would not interfere with traffic on the river. Were the two cities at higher elevations, were they set on cliffs, or palisades, such as those along the New Jersey side of the Hudson, for example, such lofty stonework would not be necessary. As it was, however, only very tall towers could make up for what nature had failed to provide, if there was to be the desired clearance for sailing ships. And as the mass of the anchorages had to be sufficient to offset the *pull* of the cables, where they were secured on land, so the mass of the towers, whatever their height, had to be sufficient to withstand the colossal *downward* pressure of the cables as they passed over the tops of the towers.

Below the water the towers were to be of limestone and each was to be set on a tremendous wooden foundation, but from the waterline up they were to be of granite. In plan each tower was essentially three shafts of solid masonry, connected below the roadway, or bridge floor, by hollow masonry walls, but left unconnected above the bridge floor until they joined high overhead to form the great

Gothic arches, which, in turn, were to be topped by a heavy cornice and three huge capstones. The total weight of each tower, Roebling estimated, would be 67,850 tons, but with the weight of the roadway and its iron superstructure added on they would each weigh 72,603 tons.

The suspended roadway's great "river span" was to be held between the towers by the four immense cables, two outer ones and two near the middle of the bridge floor. These cables would be as much as fifteen inches in diameter and each would hang over the river in what is known as a catenary curve, that perfect natural form taken by any rope or cable suspended from two points, which in this case were the summits of the two stone towers. At the bottom of the curve each cable would join with the river span, at the center of the span. But all along the cables, vertical "suspenders," wire ropes about as thick as a pick handle, would be strung like harp strings down to the bridge floor. And across those would run a pattern of diagonal, or inclined, stays, hundreds of heavy wire ropes that would radiate down from the towers and secure at various points along the bridge floor, both in the direction of the land and toward the center of the river span.

The wire rope for the suspenders and stays was to be of the kind manufactured by Roebling at his Trenton works. It was to be made in the same way as ordinary hemp rope, that is, with hundreds of fine wires twisted to form a rope. The cables, however, would be made of wire about as thick as a lead pencil, with thousands of wires to a cable, all "laid up" straight, parallel to one another, and then wrapped with an outer skin of soft wire, the way the base strings of a piano are wrapped.

But most important of all, Roebling was talking about making the cables of steel, "the metal of the future," instead of using iron wire, as had always been done before. There was not a bridge in the country then, not a building in New York or in any city as yet, built of steel, but Roebling was seriously considering its use and the idea was regarded by many engineers as among the most revolutionary and therefore questionable features of his entire plan.

The way he had designed it, the enormous structure was to be a grand harmony of opposite forces—the steel of the cables in tension, the granite of the towers in compression. "A force at rest is at rest because it is balanced by some other force or by its own reaction," he had once written in the pages of *Scientific American.* He considered mathematics a spiritual perception, as well as the highest science, and since all engineering questions were governed by "simple mathematical considerations," the suspension bridge was "a spiritual or ideal conception."

His new bridge was to be "a great avenue" between the cities, he said. Its over-all width was to be eighty feet, making it as spacious as Broadway itself, as he liked to tell people, and the river span would measure sixteen hundred feet, from tower to tower, making it the longest single span in the world. But of even greater import than length was the unprecedented load the bridge was designed to bear—18,700 tons.

The long river span was not to be perfectly horizontal, but would bow gracefully, gently upward. It would pass through the tower arches at an elevation of 119 feet, but at the center it would be 130 feet over the water. This, as Roebling pointed out, was thirty feet higher than the elevation fixed by the British Admiralty for Robert Stephenson's Britannia Bridge over the Menai Strait, built nearly twenty years earlier. Before long, sailing ships would be things of the past, he declared. His bridge therefore would be no obstruction to navigation, only possibly "an impediment to sailing." As it was, only the very largest sailing ships afloat would have to trim their topmasts to pass beneath the bridge.

But because of the great elevation of the river span and the relatively low-lying shores, the rest of the bridge, sloping down to ground level, would have to extend quite far inland on both sides to provide an easy grade. The bridge would have to descend back to earth rather gradually, as it were, and thus the better part of it would be over land, not water. Those inland sections of the bridge between the towers and the two anchorages were known as the land spans, and were also supported by the cables, by suspenders and diagonal stays. The ends of the bridge, from the anchorages down to ground level, were known as the approaches. In all, from one end to the other, the Great Bridge was to measure 5,862 feet, or more than a mile.

The red line Roebling had drawn on the map ran southeast from City Hall Park, in New York, crossing the river not quite at right angles, at that point where the river was returning to its essentially north-south course. At the Brooklyn Navy Yard—over to the right of the red line—the river turned sharply to the left, heading nearly due west, but then it quickly turned down the map again to merge with the harbor. And it was right there, where the river turned the second time, right about where the Fulton Ferry crossed, that Roebling had put his "Park Line" connecting New York, on the upper left of the map, with Brooklyn, on the lower right.

The precise terminating point on the New York side was at Chatham Street, opposite the park. This was *the* place for the bridge to come in, he said. For the next fifty years the park would remain "the great focus of travel, from which speedy communications will ramify in all directions." From there his red line crossed over North William Street, William, Rose, Vandewater, and half a dozen more streets, to the end of Pier 29, then over the river, straight through one of the Fulton Ferry slips, and into Brooklyn. Running parallel with Fulton Street, Brooklyn's main thoroughfare, the line cut across a patchwork of narrow cross streets—Water, Dock, Front, James—to Prospect, where it bent slightly toward Fulton, terminating finally in the block bounded by Prospect, Washington, Sands, and Fulton, or right about where St. Ann's Church stood.

Down the center of the bridge Roebling planned to run a double pair of tracks to carry specially built trains pulled by an endless cable, which would be powered by a giant stationary steam engine housed out of sight on the Brooklyn side. In time these trains would connect with a system of elevated railroads in both cities and become a lucrative source of revenue. He had worked it all out. His bridge trains would travel at speeds up to forty miles an hour. A one-way trip would take no more than five minutes. It was certain, he said, that forty million passengers a year could be accommodated by such a system, "without confusion and without crowding."

Carriages, riders on horseback, drays, farm wagons, commercial traffic of every kind, would cross on either side of the bridge trains, while directly overhead, eighteen feet above the tracks, he would build an elevated boardwalk for pedestrians, providing an uninterrupted view in every direction. This unique feature, he said, would become one of New York's most popular attractions. "This part I call the *elevated promenade*, because its principal use will be to allow people of leisure, and old and young invalids, to promenade over the bridge on fine days, in order to enjoy the beautiful views and the pure air." There was no bridge in the world with anything like it. And he added, "I need not state that in a crowded commercial city, such a promenade will be of incalculable value."

So the roadways and tracks at one level were for the everyday traffic of life, while the walkway above was for the spirit. The bridge, he had promised, was to serve the interests of the community as well as those of the New York Bridge Company. Receipts on all tolls and train fares would, he asserted, pay for the entire bridge in less than three years. To build such a bridge, he said, would take five years.

Horatio Allen and William McAlpine asked the most questions during the sessions Roebling held with the consultants. The length of the central span and the tower foundations were the chief concerns.

It had been said repeatedly by critics of the plan that a single span of such length was impossible, that the bridge trains would shake the structure to pieces and, more frequently, that no amount of calculations on paper could guarantee how it might hold up in heavy winds, but the odds were that the great river span would thrash and twist until it snapped in two and fell, the way the Wheeling Bridge had done (a spectacle some of his critics hoped to be on hand for, to judge by the tone of their attacks).

Roebling told his consultants that a span of sixteen hundred feet was not only possible with a suspension bridge, but if engineered properly, it could be double that. A big span was not a question of practicability, but cost. It was quite correct that wind could play havoc with suspension bridges of "ordinary design." But he had solved that problem long since, he assured them, in his earlier bridges, and this bridge, big as it was, would be quite as stable as the others. Like his earlier works, this was to be no "ordinary" bridge. For one thing it would be built *six times* as strong as it need be. The inclined stays, for example, would have a total strength of fifteen thousand tons, enough to hold up the floor by themselves. If all four cables were to fail, he said, the main span would not collapse. It would sag at the center, but it would not fall. His listeners were very much impressed.

There were questions about his intended use of steel and about the extraordinary weight of the bridge. Then at one long session they had discussed the foundations.

Roebling planned to sink two tremendous timber caissons deep into the riverbed and to construct his towers upon these. It was a technique with which he had had no previous experience, but the engineering had been worked out quite thoroughly, he said, in conjunction with his son, Colonel Roebling, who had spent nearly a year in Europe studying the successful use of similar foundations. McAlpine could vouch for the basic concept, since he had used it himself successfully, although on a vastly smaller scale, to sink one of the piers and the abutment for a drawbridge across the Harlem River. His caisson for the pier had been of iron and just six feet in diameter. Those Roebling was talking about would be of pine timbers and each one would cover an area of some seventeen thousand square feet, or an area big enough to accommodate four tennis courts with lots of room to spare. Nothing of the kind had ever been attempted before.

How deep did he think he would have to go to reach a firm footing, the engineers wished to know. Would he go to bedrock? And did he have any idea how far down that might be?

During the test borings on the Brooklyn side, the material encountered had been composed chiefly of compact sand and gravel, mixed with clay and interspersed with boulders of traprock, the latter of which, he allowed, had "detained this operation considerably." Gneiss had been struck at ninety-six feet. But below a depth of fifty to sixty feet, the material had been so very compact that the borehole had remained open for weeks without the customary tubing. So it was his judgment that there would be no need to go all the way to rock. A depth of fifty feet on the Brooklyn side ought to suffice and the whole operation would probably take a year.

About the prospects on the New York side, he was rather vague —but it looked, he said, as though bedrock was at 106 feet and there was a great deal of sand on the way down. Still there was a chance that rock might be found closer to the surface. An old well near Trinity Church showed gneiss at twenty-six feet, he noted, and in the well at City Hall the same rock was found at ninety feet. "The whole of Manhattan Island appears to rest upon a gneiss and granite formation," he said. The greatest depth to which similar caissons had been sunk before this was eighty-five feet. But he was willing to take his to a depth of 110 feet if that was what had to be done. His consultants said they did not think he would find that necessary.

Presently they took up the question of the timber foundations and their fate, once he left them buried forever beneath the towers, beneath the river, the rock, sand and muck of the riverbed. In his report, Roebling had explained at some length how the caissons would be packed with concrete once they were sunk to the desired position, and why, in their final resting place, well below the level where water or sea worms could reach them, they would last forever. But there were some among the consultants who wished to hear more on the subject and who had a number of questions.

That particular session on the foundations had taken place on March 9. Two days later, on the 11th, it was announced that the renowned engineers had approved the Roebling plan, "in every important particular." Their official report would come later, but in the meantime the public could rest assured that the plan was "entirely practicable."

Only Congressional authorization was needed now, since Congress had jurisdiction over all navigable waters and the bridge was to be a post road. Unlike the government in Albany, or those in either city, the government in Washington had some regulations it wished to see adhered to. Congressional legislation already drawn up stipulated that the bridge must in no way "obstruct, impair, or injuriously modify" navigation on the river. In particular, there was concern in Washington that it might interfere with traffic to and from the Navy Yard, and to be certain that every detail of the plan was fully understood, General A. A. Humphreys, Chief of the Army Engineers, decided to appoint his own review panel to give an opinion on it, irrespective of the conclusions reached by Roebling's consultants (This was to be the only public scrutiny of the design or the location.) So at about that point it had seemed the most sensible next step would be for everyone to go take a look at some of Roebling's existing works to see how he had previously handled somewhat analogous situations. Let his work speak for itself, he had decided.

The tour was arranged almost overnight and if there was any initial intention to restrict it to a relatively small body of professionals that idea was speedily overruled. A total of twenty-one gentlemen and one lady made up the "Bridge Party," as it was referred to in subsequent accounts. In addition to the two Roeblings, the seven consultants, and three Army engineers—General Horatio Wright, General John Newton, and Major W. R. King—several prominent Brooklyn businessmen were invited, most of whom were or were about to become stockholders in the New York Bridge Company. A Brooklyn Congressman named Slocum—General Henry W. Slocum—was included, as were Hugh McLaughlin, the Democratic "Boss" of Brooklyn, and William C. Kingsley, Brooklyn's leading contractor, who was known to be the driving political force behind the bridge and the largest individual stockholder. How many of the party were aware that the tall, powerful Kingsley would also be personally covering all expenses for the tour, in addition to the seven thousand dollars in consultants' fees, is not known.

Two young engineers, C. C. Martin and Samuel Probasco, both of whom had worked for Kingsley on different Brooklyn projects, were also to go, as was the wife of one of the consultants, Mrs. Julius Adams, who is described only as an "amiable lady" in existing accounts. Why she consented to join the group, or why she was invited in the first place, no one ever explained.

Nor is there anything in the record to indicate who determined the make-up of the group. Presumably it was taken to be a representative body, having an even balance of engineering talent, business acumen, and public spirit. In any event, the editor of the Brooklyn *Eagle*, Thomas Kinsella, was also included, so that in Brooklyn at least the expedition would receive proper notice, and young Colonel Roebling appears to have been the one delegated to make the necessary arrangements.

There were, however, two very important public figures who did not make the trip, both of whom had done much to bring the project along as far as it had come and who ought to be mentioned at this point in the story.

The first was State Senator Henry Cruse Murphy, lawyer, scholar, the most respected and respectable Democrat in Brooklyn, and in Albany the leading spokesman for Brooklyn's interests. Murphy had worked harder for the bridge than anyone in Brooklyn except Kingsley, the contractor. He was the one who had written the charter for the New York Bridge Company. He had seen it through the legislature and was currently serving as the company's president. Why he failed to make the trip is not known and probably not important. But he would have added a certain tone to the group certainly and John A. Roebling, in particular, would doubtless have enjoyed his company. (The idea of Roebling keeping company with the likes of Boss McLaughlin must have raised many an eyebrow on Brooklyn Heights.)

But the absence of the second missing party was quite intentional, one can be sure; it raised no questions and required no explanation, since there had been no mention as yet, scarcely even a whisper, that he had had anything whatever to do with the bridge. He was William Marcy Tweed of New York.

The itinerary called for stops at Pittsburgh, Cincinnati, and Niagara Falls, and the announced official purpose of the expedition was to inspect four of Roebling's bridges, each of which, in one way or other, illustrated how he intended to span the East River. But a week of traveling together was also supposed to give everybody a chance to get to know one another—nothing could so cement friendships as a long train ride, Thomas Kinsella would write—and particularly, it was presumed, everyone would get to know the key man in all this, John A. Roebling.

The great engineer was still largely a mystery to the people who had hired him. Except for the times when he had expounded on his plan at the meeting in 1867, his Brooklyn clients had seen very little of him. Their ordinary day-to-day dealings had been with his son. It had been young Roebling, not his father, who had set up the makings of an office and who had taken a house on the Heights. He had been the one on hand to answer their questions and keep things moving.

The father had wanted it that way. He had remained in Trenton, showing up in Brooklyn only now and then, and staying no longer than necessary. His time was always short it seemed and even when meeting with his board of consultants he had kept each session quite formal and to the point. He had no time for anything but business, and no small talk whatever.

On occasion the two of them, father and son, would be seen walking on Hicks Street, talking intently, or down by the slate-gray river pacing about the spot where the tower was to rise, the father pointing this way and that with his good hand. They resembled each other in height and build, even trimmed their whiskers the same way. But while the son was quite handsome in the conventional sense, with strong regular features, the father's face was a composite of hard angles and deep creases, of large ears and nose and deep-sunken eyes, all of which gave the appearance of having been hewn from some substance of greater durability than mortal flesh.

Most people, later, would talk about his eyes, his fierce pale-blue eyes. But just what sort of human being there might be behind them was a puzzle. He was a man of enormous dignity, plainly enough, full of purpose and iron determination, but accustomed to deference just as plainly, somebody to be admired from a distance. His look was all-knowing and not in the least friendly. Among those who were about to stake so very much on him and his bridge, or who already had, there was not one who could honestly say he knew the man.

And so on the evening of April 14, 1869, when General Grant

and his Julia were just taking up residence in the White House and the dogwood were beginning to bloom across the lowlands of New Jersey, the Bridge Party boarded a private palace car in Jersey City and started west. The only one missing from the group was the elder Roebling, who was to get on at Trenton.

2
Man of Iron

> We may affirm absolutely that nothing great in the world has been accomplished without passion.
>
> —G. W. F. HEGEL

ANYONE from Trenton who happened to be standing nearby on the depot platform that lovely April evening would have known who he was, and very possibly why he was waiting there. Trenton was still a small town, for all the changes there had been, and Old Man Roebling, as the men at the mill called him, was Trenton's first citizen. The whole town looked up to him and took pride in his accomplishments.

It was commonly said that he had done more in one life than any ten men. The town had seen him build the wire business from nothing, raise seven children, bury two others and one wife, then marry again when he was past sixty. He had survived hard times, fires, cholera epidemics, the hazards of bridgebuilding, accidents at the mill, and his own particular notions about maintaining good health, which to some may have seemed the surest sign of all that the man was indestructible.

John Roebling was a believer in hydropathy, the therapeutic use of water. Come headaches, constipation, the ague, he would sit in a scalding-hot tub for hours at a time, then jump out and wrap up in ice-cold, slopping-wet bed sheets and stay that way for another hour or two. He took Turkish baths, mineral baths. He drank vile concoctions of raw egg, charcoal, warm water, and turpentine, and there were dozens of people along Canal Street who had seen him come striding through his front gate, cross the canal bridge, and drink water "copiously"—gallons it seemed—from the old fountain beside the state prison. ("This water I relish much . . ." he would write in his notebook.) "A wet bandage around the neck every night, for years, will prevent colds . . ." he preached to his family. "*A full cold bath every day* is indispensable . . ." Illness he regarded as a moral offense and he fought it with the same severe intensity he directed to everything else he did in life.

The town knew all about him, or thought so. It was common knowledge, for example, that he was an inventor as well as an engineer, that he had designed every piece of machinery in the mill, that he was an artist, that he wrote prolifically for scientific periodicals, that he read Emerson and Channing and other freethinkers. At home he was writing his own "Theory of the Universe."

When he first came to Trenton, he had played both the piano and the flute, but then he caught his left hand in a rope machine and was left with three immovable fingers. Not long after his first wife died, he had taken up spiritualism. There had been talk ever since of after-dark gatherings, of table rappings and the like, inside the big Roebling house. The old man, on top of his other achievements, was now said to be on speaking terms with the dead.

The bridges were what he was best known for, of course, but only a few people in Trenton had actually seen any of them, except perhaps for a view in *Harper's Weekly* or one of the other picture magazines. Roebling the industrialist was the man Trenton people knew.

He was called a man of iron. Poised . . . confident . . . unyielding . . . imperious . . . severe . . . proud . . . are other words that would be used in Trenton to describe him. There had always been something distant about him; he kept apart and had no real friends in Trenton, but he had also been accepted on those terms long since and he in turn was always extremely courteous to everyone. "He was always the first to say good morning," a man from the mill would tell a reporter after Roebling's death. When he spoke they listened.

Roebling was sixty-three in 1869, but even when he was years younger, he had a special hold on men, it seems, with his commanding stares and wintry scowls, like an Old Testament prophet. His success in everything he turned his hand to was generally attributed to an inflexible will and extraordinary resourcefulness. "He was never known to give in or own himself beaten," one of his employees would recall and another would quote a saying of his they all knew by heart, "If one plan won't do, then another must." Charles B. Stuart, an engineer and author who knew Roebling, would later write: "One of his strongest moral traits was his power of will, not a will that was stubborn, but a certain spirit, tenacity of purpose, and confident reliance upon self . . . an instinctive faith in the resources of his art that no force of circumstance could divert him from carrying into effect a project once matured in his mind. . . ." It was a quality he had worked hard to instill in his children as well.

Time was something never to be squandered. If a man was five minutes late for an appointment with him, the appointment was canceled. Once, during the war, so the story went, he had been called to Washington by the War Department to give advice on something or other and was asked to wait outside the office of General John Charles Frémont, the illustrious "Pathfinder." Roebling took out a pencil, wrote a note on the back of his card, and had it sent in to the general. "Sir," the note said, "you are keeping me waiting. John Roebling has not the leisure to wait upon any man."

In all his working life John Roebling had never been known to take a day off.

He had settled in Trenton twenty years before, in 1849, when he was forty-three, or past the age, he knew, when most brilliant men do their best work. He had had no money to speak of then and not much of a reputation. All that had come in the years since. How much was generally known in Trenton of his life prior to that time can only be guessed at, but the story was well known among his family certainly, and, for the most part, in the engineering profession.

He had been born on June 12, 1806, in Germany, in the province of Saxony, in the ancient walled town of Mühlhausen, where for about a thousand years more or less not very much had ever happened. Bach had once played the organ in the church where he was baptized and in the spring of 1815, when Roebling was nine, five hundred of his townsmen had marched off to fight Napoleon at Waterloo, but other than that no one in Mühlhausen had ever done much out of the ordinary.

His father, Christoph Polycarpus Roebling, had a tobacco shop and the accepted picture of him is of an unassuming, rather comfortably fixed burgher of good family, who had no desire to be anything more than what he was and who smoked up about as much tobacco as he sold. Roebling's mother, however, was a fiercely energetic sort, with a mind of her own and some very fixed ideas about getting on in the world. It was their proud, determined, long-departed grandmother, Friederike Dorothea, John Roebling's children were raised to understand, who scraped and saved to send their father to the famous Polytechnic Institute in Berlin, and who later was the first to support his decision to leave Mühlhausen, something no Roebling had done before.

In Berlin, he had studied architecture, bridge construction, and hydraulics. He also studied philosophy under Hegel, who, according to one biographical memoir, "avowed that John Roebling was his favorite pupil." The renowned philosopher had been preaching a powerful doctrine of self-realization and the supremacy of reason to a generation of ardent young liberals hemmed in by an autocratic Prussian regime. The effect was pronounced, and not the least on Roebling. The contact with Hegel was a privilege and a calamity for Roebling, according to an old family friend in Trenton. Hegel had taught Roebling to think independently, he said, and to rely on the validity of his own conclusions, but the experience was a calamity "because it begat a pride and arrogance of opinion and a frigid intellectuality that came near putting the heart of him into cold storage." But according to family tradition, it was Hegel who started the young man thinking about America. "It is a land of hope for all who are wearied of the historic armory of old Europe," Hegel taught. There the future would be built. There in all that "immeasurable space" a man might determine his own destiny.

For three years after leaving Berlin, Roebling worked in an obligatory job building roads for the Prussian government. Once during a holiday in Bavaria, he had hiked to the old cathedral town of Bamberg, where he saw his first suspension bridge, a new iron chain bridge over the Regnitz and known locally as the "miracle bridge." He walked about it, made a number of sketches, and it is the traditional story that he decided then and there on his life's career.

In any event, not long afterward, in the spring of 1831, the year Hegel would die of cholera, Roebling returned to Mühlhausen and began organizing a party of pilgrims to leave for America, something that had to be done with caution just then since the government frowned on the immigration of anyone with technical training.

Talk of immigration was a common thing in Germany. Ever since the July Revolution of the previous year, there had been increasingly less personal freedom, less opportunity for anyone with ambition. Nothing could be accomplished, Roebling would write, "without first having an army of government councilors, ministers, and other functionaries deliberate about it for ten years, make numerous expensive journeys by post, and write so many long reports about it, that for the amount expended for all this, reckoning compound interest for ten years, the work could have been completed."

In the first week of May there had been the farewell visits with school friends and aged aunts, the last Sunday at church, the final evening walks through the ancient cobblestone streets. Then on the morning of the 11th, with his older brother Karl and a number of others, he had set off. His determination now was to become . . . an American farmer! Having had no previous experience in agriculture, having nothing in his background, training, or temperament that would indicate any interest in or bent for such work, he would become a man of the soil, in a distant land he knew only by reputation. The architect, the scholar, the musician, the philosopher, the engineer, the burning liberal idealist, the twenty-four-year-old bachelor, would now plant himself, willfully, somewhere in the American wilderness. His ambition was to establish his own community, which if not utopian in the religious sense—like Harmony, Pennsylvania, or some of the other earlier settlements founded by zealous Germans—would at least provide the honest German farmer, tradesman, or mechanic, men good with their hands and accustomed to work, a place where they could make the most of themselves, which to Roebling's particular way of thinking would be about the nearest thing possible to heaven on earth.

He never saw Mühlhausen or Germany again. In 1867, to prepare for the bridge at Brooklyn, he had sent his son Washington and his pretty, pregnant daughter-in-law back across the Atlantic. It was a journey he would have liked to have made himself no doubt. He could have returned in triumph. As it was, the young couple arrived at Mühlhausen to a rousing welcome, and in a small inn across the street from the old family home, his first grandson and namesake had been born. Later, he had sent the town a sizable gift of cash in gratitude.

In a bookshop in Mühlhausen in 1867, Washington Roebling found a rare printed edition of the journal his father had kept on route to America, which Washington carried with him on his own return voyage. *Diary of My Journey from Muehlhausen in Thuringia via Bremen to the United States of North America in the Year 1831* it is titled. It is an extraordinary little document, a recognized classic of its kind, describing days of howling winds and high seas, and a steamboat—the first Roebling had been—laboring mightily by, and later, like a specter, a derelict hulk of an abandoned sailing ship, a huge brig with all sails gone, drifting on the horizon; then days of no wind and bad drinking water, the burial at sea of a child, and at last, on a night in July, the smell of land in a warm westerly wind. "The odor was strikingly distant and . . . would also indicate that the entire American mainland is covered with an almost uninterrupted forest and a great abundance of plants, whereby the atmosphere is saturated with aromatic particles, which the winds blowing away from land carry away to a great distance. This scent of land produced a beneficial effect upon all the passengers."

His band of pilgrims consisted of fifty-three men, women, and children, most of whom had never laid eyes on salt water. Their ship was the *August Eduard*, a 230-ton American packet bound for Philadelphia, which, in all, took eleven weeks to make port, or longer than it had taken Columbus to make his first crossing.

Roebling himself was an immigrant of a kind the history books would pay little attention to, chiefly because they were so relatively few in number. He was seeking neither religious freedom nor release from the bondage of poverty. His quest was for something else. He came equipped with the finest education Europe could offer, he had a profession, and he was traveling first class, which meant he had one bed among four in a cabin he described as "very roomy" and "excellently lighted." Between them, he and his brother were also carrying something in the neighborhood of six thousand dollars in cash, a princely sum, and he had come on board with a whole trunkful of books—thick geographies, works of physics and chemistry, a German-French dictionary, Euclid's *Elements*, volumes of English literature and poetry, and one of English essays that opened with a favorite quote from Johnson: "No man was ever great by imitation."

What the American captain and his crew thought of this spare, incredibly energetic young German can be imagined. He started right off, for example, by instructing them on how to build a proper privy for the passengers in steerage, whose only facility was the usual sailor's seat perched precariously outside the stem of the ship, beside the bowsprit. Such an arrangement, Roebling announced, was altogether unacceptable for the women and children, or for anyone who might become sick or weakened by the voyage. He and the other cabin passengers, like the ship's officers, were entitled to use a relatively comfortable, enclosed affair that protected its occupant from sudden waves washing across the deck. The same or better should be made available for all on board, Roebling declared. He explained how it could be done and it was done. "If one earnestly desires it," he wrote, "everything will be brought to pass, even on board a ship . . ." The great thing, he believed, was getting people "to leave the accustomed rut."

His curiosity about all aspects of seamanship, navigation, ocean currents, rules for passengers, or the personal life history of the captain and each member of the crew seemed inexhaustible. He wanted to know the name of every sail, every stay, brace, bowline, halyard, every rope and how each one worked and he made diagrams to be sure he understood. He talked to the captain ("a very just, straightforward, and *sober* man") about astronomy, meteorology, philosophy, history, about Isaac Newton and the American coinage system. He was the first one on deck in the morning and generally the last to leave at night, and once, when nearly every passenger was miserably seasick and lay groaning in his berth, Roebling, his head spinning, his stomach churning, was resolutely walking the deck. The malady, he rationalized, "involves no danger at all," noting that "a cheerful carefree disposition and a manly, vigorous spirit will have great influence on the sickness."

For his son there must have been places in the old diary where the youthful and impressionable narrator seemed a little difficult to identify with the father he had known. One entry, for example, was taken up almost entirely with a long, vivid description of waves. Apparently his father had stood at the bowsprit watching them for hours on end and to no particular purpose. In the account of phosphorescence after dark, as the sea rebounded from the sides of the ship, it was as though the writer had been caught up in a spell:

> . . . then one perceives in the foam brightly shining stars, which appear as large as the fixed stars in the heavens. Along the entire side of the ship the foam has turned into fiery streaks. The spots of foam in the ocean, distant from the ship, which arise from the dash-

ing together of the waves, appear in the dark night to the astonished eye as just so many fiery masses. In front of the bowsprit, where the friction is greatest, the scintillation is often so bright, that the entire fore part of the ship is illuminated by it.

For the moment—except possibly for the word "friction"—it was as if nature was not something to be explained endlessly or to be "rendered subservient," as John Roebling would say in another time and place. And again, as the ship headed into Delaware Bay, there is a moment when the gifted young graduate of Berlin's Polytechnic Institute reflects with sadness on the Indians who once lived on shore—"quietly on the property inherited from their ancestors," long before "the sheltered loneliness of these wild surroundings was interrupted by the all-disturbing European."

From Philadelphia, Roebling and his followers headed west across Pennsylvania, having decided to settle on the other side of the Alleghenies. At Pittsburgh he and Karl purchased some seven thousand acres located to the north, in Butler County, not far from Harmony (the price was $1.37 an acre, with a thousand dollars down and the balance to be paid in two equal yearly installments "*without* interest," as he wrote home). And there he established his town, first laying out one broad Main Street exactly east-west, in the German fashion. He called the town Germania for a while, but then changed it to Saxonburg.

Roebling had concluded, his son Washington would write in jest, that western Pennsylvania was destined to be "the future center of the universe with the future Saxonburg as the head center, which then was a primeval forest where wild pigeons would not even light."

"My father would have made a good advertising agent," Washington would remark at another time. "He wrote at least a hundred letters to friends in and about Mühlhausen, extolling the virtues of the place—its fine climate—the freedom from restraint—the certainty of employment, etc. Many accepted and came. To each one was sent exact directions how to come, what to take—what to bring along, and what to leave behind. Most tools were to be left behind, because American tools were so much better, such as axes, hatchets, saws, grubbing hoes—nodody could cut down a tree with a German ax."

The beginning is hard, Roebling had warned. But there were "no unbearable taxes," no police commissioners. And finally: "If this region is built up by industrious Germans, then it can become an earthly paradise." But the soil turned out to be mostly clay, the winters were bleak and bitterly cold, and the roads to Pittsburgh or to Freeport, the nearest point on the Allegheny River, were "atrocious."

Among the early arrivals there were only two who knew a thing about farming. But according to one of the old histories of the town, they all "possessed to a remarkable degree the valuable attribute of industry, and, though many of their first attempts were ludicrous and miserable failures, they yet persevered until they became adepts at handling the ax and agricultural implements." Every newcomer was heartily welcomed and encouraged to stay. Presently more and more did come and settle and the surrounding country, only sparsely settled earlier by Scotch-Irish, began filling up with Germans. "They have made good farmers," an old Butler County history concludes, "succeeding, by patient industry and close economy, in gaining an independent condition where the people of almost any other nationality would have failed, in a majority of instances, to have secured more than a mere living."

The first building to go up in Saxonburg was a plain two-story house built by Roebling at the head of Main Street. It was clapboard on the outside, but brick behind that, and like everything he ever built, it was built to last. Five years later Saxonburg, if not exactly paradise, was at least a going concern, populated by a weaver, a grocer, a blacksmith, a cabinetmaker, about six carpenters, a tanner, a miller, a baker, a shoemaker, a Mecklenburg tailor, a Mühlhausen tailor, one artist, one brewer, a veteran of Waterloo, and an increasing number of plain farmers with names like Emmerich, Rudert, Goelbel, Heckert, Graff, Schwietering, Nagler, and Helmhold. And in May 1836, in his own front parlor, Roebling married Johanna Herting, the oldest daughter of the Mühlhausen tailor.

But in less than a year, with everything going about as well as he could have hoped, Roebling seems to have run up against the one problem he had not figured on. He had become bored. When he heard the state was in need of surveyors, he immediately wrote to Harrisburg. That was in 1837, the year he became a citizen, the year Karl died of sunstroke while working in a wheat field, the year Roebling became a father for the first time. In a letter to the chief engineer of the Sandy and Beaver Canal, he wrote, "I cannot reconcile myself to be altogether destitute of practical occupation . . ."

"So he took to engineering again, his true vocation," Washington Roebling wrote, "and let my mother do the farming again, *which she did very well* when he would let her." By the time the son was old enough to understand such things, the father's agrarian dream, if indeed that is what it was, was long since over.

Roebling built dams and locks on the Sandy and Beaver, between the Ohio and the lakes, then on the Allegheny feeder of the Pennsylvania Canal near Freeport. In 1839 he began surveying a prospective railroad route east of Pittsburgh that would later be adopted, in part, by the Pennsylvania Railroad. Living in tents, working in all kinds of weather through the roughest kind of wilderness, he and a few assistants covered more than 150 miles, plotting a line through the Alleghenies. His work was such that he was made Principal Assistant to the Chief Engineer of the state, a man named Charles L. Schlatter, and his report to Schlatter included not only full details on the grades, embankments, bridges, and tunnels required, but a number of prophetic observations about the locale around the village of Johnstown, where one of the nation's principal iron and steel industries would one day rise. "The iron ore on the Laurel Hill is only waiting for means of transportation to be conveyed to the rich coal basins below, where also limestone is to be had in quantity and, moreover, where an abundance of water power can be furnished by the never-failing waters of the beautiful mountain stream . . . and certainly capitalists could hardly find a more eligible situation for starting mammoth furnaces on the largest scale . . ."

At Johnstown he also became familiar with the workings of the newly built Portage Railroad, a system of long, inclined planes devised to haul canalboats up and over the Alleghenies, between Hollidaysburg at the foot of the eastern slope and Johnstown at the foot of the western slope. It was popularly thought to be one of the engineering marvels of the age and Roebling was fascinated by it. He also decided, after a good deal of study, that it could be greatly improved by dispensing with the immense hemp hawsers then in use. These were about nine inches around, more than a mile long in some cases and cost nearly three thousand dollars. They also wore out in relatively short time and had to be replaced or, as happened more than once, they snapped in two, sending their loads crashing down the mountainside. In one such accident two men had been crushed to death.

Roebling proposed to replace the hawsers with an iron rope just an inch thick, a product not made in the United States then, but which he had read about in a German periodical. Such a rope, he said, would be stronger, last longer, and be much easier to handle. Apparently he was the only one who took the idea seriously, but he was told to go ahead and try if he had such confidence in it—at his own risk and expense.

He began fashioning his new product at Saxonburg some time in the summer of 1841, using the old ropewalk system on a long level meadow behind the church he had built soon after finishing his house. The wire, purchased from a mill at Beaver Falls, northwest of Pittsburgh, was spliced inside a small building and wound onto reels for "running out." Separate strands of wire were laid up first, then twisted into the larger rope by means of a crude machine he had devised, which, like everything else in the process, was powered by hand.

A six-hundred-foot rope finished "in the best style," as he said, was tried out at Johnstown in September and it was a failure. Someone hired by the hemp rope interests had secretly cut it at a splice, with the result that it broke during the test. But the sabotage was discovered, Roebling was given a second chance, and his rope worked with such success that it was soon adopted for the entire Portage system. Orders began coming in from other canals with similar inclined planes. The rope was wanted for dredging equipment, for pile drivers, for use in coal mines. Roebling published an article on it in the *Railroad Journal*. "His ambition now became boundless," his son would write. Production in Saxonburg picked up sharply, as "farmers were metamorphosed into mechanics and an unlooked-for era of prosperity dawned."

"About eight men were needed for strand making," according to Washington Roebling, "but sixteen or eighteen were required for laying up the rope. These were recruited for a day or two from the village and adjacent farm—quite a task—in which I took my full share. The men were always glad to see me because it meant good pay and free meals for days. Work was from sunrise to sunset—three meals, with a snack of bread and butter in between—including whiskey. Meals were served at the house. My poor, overworked mother did the cooking—all done on an open hearth."

John Roebling could be sure, he was told in an admiring letter from Charles Schlatter, that before long he would be "at the head of the list of those benefactors to mankind who employ science to useful purpose."

In 1844, at age thirty-eight, he got his first real commission as an engineer. A prize of one hundred dollars had been offered in a notice in the Pittsburgh papers for "the best plan for a wooden or suspension aqueduct" to carry the Pennsylvania Canal across the Allegheny River in place of a ponderous, inadequate structure built years earlier by the state. Roebling worked out a plan for the world's first suspension aqueduct. He made a model and went to Pittsburgh to enter the competition, which he won, mainly because his bid was the lowest. He built the aqueduct in record time. He worked nine months nonstop and when he was finished, Pittsburgh, at a cost of $62,000, had a structure unlike any in existence.

From two iron cables seven inches in diameter, he had suspended a big timber flume, crossing the river with seven spans of about 160 feet each. The flume was sixteen and a half feet wide and eight and a half feet deep. It carried something over two thousand tons of water and a steady procession of canal barges that floated across high over the Allegheny, hauled by mules that walked a narrow plank towpath.* "As this work is the first of the kind ever attempted," wrote the *Railroad Journal*, "its construction speaks well for the enterprise of the city of Pittsburgh." But in 1861, after the canal had been put out of business by the Pennsylvania Railroad that Roebling had helped to lay out, the aqueduct was pulled down.

The winter he built the aqueduct had been the most trying, strenuous period in his life. Not only had he designed it himself, but he had directed and participated in every step in its construction, in freezing winds, sleet, snow, going back and forth over the spindly catwalk or swinging along one of the cable strands in a little boatswain's chair. The cables had been strung in place, wire by wire, in much the way his subsequent bridges would be. He had also devised a novel technique for anchoring the cables, attaching them to great chains of iron eyebars embedded in masonry, a plan not used in any prior suspension bridge and the one he would use on every bridge he built thereafter.

He had finished in exactly the time he had said he would and no one was more keenly aware of the real importance of what he had done than he. Judged against his later work, the bridge was crude, small, and uninspiring. And probably he knew the day it was fin-

* The Pittsburgh *Gazette* made much of the fact that the structure was strong enough to carry the water plus six heavily loaded barges all at the same time, the editors being unaware apparently that the boats merely displaced their own weight and so the total load remained the same, whether boats were crossing or not.

ished that its life-span would be brief. The significant thing was that he had demonstrated the immense weight that could be borne by a suspension bridge, not to mention his own skill and integrity as a builder.

In April of 1845, a month before the aqueduct was opened, more than half of Pittsburgh burned to the ground. "The progress of the fire as it lanced and leaped with its forked tongue from house to house, from block to block, and from square to square was awfully magnificent," wrote one observer. Among the victims was an old covered bridge over the Monongahela at Smithfield Street and as a result Roebling got the chance to build his first real bridge, which was also to be the first bridge on the tour he was about to lead.

In 1848 he began four more suspension aqueducts, these on the Delaware and Hudson Canal, linking the hard-coal fields of eastern Pennsylvania with the tidewater of the Hudson. In the meantime he wrote articles on his theories and in 1847 presented a twelve-thousand-word paper before the Pittsburgh Board of Trade (it was read at two sittings) calling for the immediate establishment of "The Great Central Railroad from Philadelphia to St. Louis." Like a magic wand, he said, the railroads were going to work a transformation over the land. A new nation was about to emerge and this would be the greatest of all railroads, "a future highway of immense traffic." It was another of his visionary proclamations. As it was, the Pennsylvania would not be completed to Pittsburgh for five more years, which was longer than John Roebling could wait.

It is not known when he first began thinking seriously about leaving Saxonburg, but by 1848, the year after his "Great Central Railroad" speech, with no such railroad in sight, he had concluded that Saxonburg would not become the center of the universe in all likelihood, and that in any event it was no location for a wire business. Having analyzed the problem as thoroughly as he was able, he decided to relocate in the old colonial town of Trenton, New Jersey, which then had a total population of perhaps six thousand people.

So he had departed from Saxonburg, leaving friends, relatives, everything they had struggled for so many years to build, and went east, against the human tide then pouring across Pennsylvania bound for the still-empty country beyond Ohio. His wife and children were to follow on their own. "He was disgusted with Saxonburg," Washington Roebling wrote, "and never revisited it. He was seized with a horror of everything Dutch and never alluded to it." In Saxonburg it would be said, "The dumb Dutch stayed behind."

It was a very changed man who was about to return now over that same route to Pittsburgh, to retrace his footsteps as it were, and review the best of his life's work. The bridges had made him famous in the time since, world-famous, and the wire business had made him rich. The John A. Roebling who stood on the station platform that April evening in 1869 was worth more than a million dollars, as his will would subsequently reveal. But other things had happened, private things, of which only his immediate family and one or two others knew anything, and these had affected him more than either notoriety or wealth, both of which, one would gather, he always had every expectation of attaining.

In the decade before the war, his most productive time as an engineer, he had grown increasingly distant and impersonal in manner whenever he was home, which was seldom. One April, while writing to tell him how green and lovely everything looked about the house, his young daughter Elvira suddenly realized that never in her memory had he been home during the springtime. The day-to-day running of the mill he had left largely to Charles Swan, a German from Pittsburgh who had worked on the Allegheny aqueduct and who had shown such promise that Roebling brought him to Trenton. Swan had the "happy faculty" of being able to get along with Roebling, "an important matter," as Washington commented knowingly. Swan also appears to have had no end of patience with his employer's mania for detail and his essential distrust of anyone's judgment other than his own. Time and again the two of them would ride down to the Trenton depot together, Roebling on his way to Niagara Falls or Cincinnati or some such place, and telling Swan as they went along how he was to have full authority to decide things. But it had never worked out that way. Swan heard regularly, almost daily, about what he was to do or not to do, and was expected to keep Roebling fully informed by return mail. Everything had to be done to the most exacting standards. If Roebling was dissatisfied with a clerk's handwriting, Swan would hear about it ("He must take pains to improve and examine attentively well written letters which you receive and which may serve him as patterns . . .") and a demonstration of the proper way to address a letter would be included. ("The direction should never be put up high in the upper part of the envelope, but rather below the center, else it looks uncommercial-like.") Appearances were exceedingly important.

The letters to Swan numbered in the hundreds as time passed and were always strictly business communications. Despite all the years Swan had been with him and all that Swan had come to mean to the family, never once did John Roebling write a line to suggest there could possibly be a bond of friendship between them. If he was meeting interesting people in his travels, there is no mention of it. If he had feelings for the places he went, he said nothing of them. If ever he had a sense of humor, there is not a trace of it.

His preoccupation with work became almost beyond reckoning. He was living in a time characterized by extraordinarily industrious men, when hard work took up most of everyone's life and was regarded as a matter of course; but even so, his immense reserves of nervous energy, his total devotion to the job at hand, whatever it might be, seemed superhuman to all who came in contact with him. If metaphysics was his only dissipation, as was said in Trenton, work seemed his one and only passion. Once, quite unwittingly, he revealed the extraordinary and rather ludicrous limits such preoccupation could reach. On New Year's Day, 1855, his wife had been delivered of still another child, but this apparently came as a great surprise to the bridgebuilder when the news reached him at Niagara Falls. "Your letters of the 2nd and 3rd came to hand," he wrote quite formally to Swan. "You say in your last that Mrs. Roebling and the child are pretty well. This takes me by surprise, not having been informed at all of the delivery of Mrs. R. Or what do you mean? Please answer by return mail." Swan was to waste no money on a telegram, in other words.

The war and Lincoln's murder had been terribly hard on Roebling. "I for my part wished the blacks all good fortune in their endeavors to be free," he had written when he first arrived in America. Slavery was "the greatest cancerous affliction" in an otherwise ideal land. When Lincoln called for volunteers after the attack on Sumter, Roebling had sat gravely silent at his end of the dinner table, then turned abruptly to his son Washington, "Don't you think you have stretched your legs under my mahogany long enough?" And the young man had enlisted the very next morning. "When a whole nation had been steeped for a whole century in sins of inequity, it may require a political tornado to purify its atmosphere," he wrote in his private notes. But as the years of the war dragged on he had worried incessantly about his son and the news of Lincoln's death fell on him like a massive personal tragedy. Bitterly he wrote, "We cannot close our eyes to the appalling fact that the prominent events of history are made up of a long series of individual and national crimes of all sorts, on enmity, cruelty, oppression, massacres, persecution, wars without end."

But the most shattering blow had been the death of Johanna Roebling in the final year of the war. In the years since Saxonburg they had seemed ill-matched. From her wedding day until the day she died, she served him faithfully and with love, but he had become increasingly preoccupied with his studies, his books, his work. She had had almost no education and understood very little about the things he considered so important. He was away most of the time, traveling always "in the first society." She went nowhere. Her world was scarcely broader than what she could see from her doorstep. Only in her last years would she feel enough at ease in English to get along in the most ordinary daily conversation.

"A purer-hearted woman or one gifted with warmer affections than my mother you will seldom meet," Washington Roebling had written in a letter to Emily Warren, who was shortly to become his wife. "It is therefore plain to you that before long my father outstripped Mother in the social race and she was no longer a companion to him in a certain sense of the word. A gifted woman like yourself would no doubt have suited him better from 40 to 50, but upon the whole he could not have had a better or truer helpmate for life. A man of strong passions and impulses he could only get along with a yielding and confiding woman."

That Johanna Roebling never understood, and therefore never fully appreciated, the range and fertility of her husband's mind or the extraordinary beauty of what he built seemed self-evident to almost everyone who did have a feeling for such things. But as his children knew full well, the failure of appreciation worked both ways, until it was too late. He was in Cincinnati when she died, but after the funeral the Man of Iron had taken down the family Bible and on a single blank page wrote the following:

> My dearly beloved wife, Johanna, after a protracted illness of 9 months, died in peace with herself and all the world, on Tuesday the 22nd November, 1864, at 12:30 P.M.
>
> Of those angels in human form, who are blessing the Earth by their unselfish love and devotion, this dear departed wife was one.—She never thought of herself, she only thought of others. No trace of

> ill will toward any person ever entered her unselfish bosom. And O! what a treasure of love she was towards her own children! No faults were ever discovered.—She only knew forbearance, patience and kindness. My only regret is that such a pure unselfishness was not sufficiently appreciated by myself.—
>
> In a higher sphere of life I hope I meet you again my Dear Johanna! And I also hope that my own love and devotion will then be more deserving of yours.

Always intensely philosophical, he now began filling hundreds upon hundreds of sheets of lined blue paper with his own private visions and speculations on man, matter, truth, and the nature of the universe. The words slanted across the paper as though in a tremendous hurry, heavy on the downstrokes, leaving no margin at all. Truth, he said, was "harmony between object and subject" and "the final idea, the absolute idea, which includes all other ideas." Truth was something that should appeal to every man "whose inner Self-consciousness is not yet worked out, whose spiritual manhood and mental integrity are yet asserting supremacy." He declared, "Existence has a cause." Life itself he saw in terms of a torrential, twisting stream "rolling along, ever driven by its own gravitating tendency towards the great Ocean of Universality."

The words sounded most impressive, but what he was getting at was sometimes very hard to tell, and apparently the few people he permitted to read his "Truth of Nature" and other essays found them extremely rough going. The afterhours philosopher seemed such a far cry from the clear, precise, no-nonsense person they knew. It was as though some impenetrable Teutonic mysticism had surfaced from a deep recess in his past. One friend of the family said he had never been invited to read any of John Roebling's philosophy, but from what he had heard, he prayed he never would be.

Still there were moments of great clarity. "We are born to work and study," he wrote at one point, which fitted him perfectly. "True life is not only active, but also creative," he asserted. And another time: "It is a want of my intellectual nature to bring in harmony all that surrounds me. Every new harmony is to me another messenger of peace, another pledge of my redemption."

Not for years had he taken an active interest in organized religion. Raised a Lutheran, he had joined the Presbyterians after arriving in Trenton, but for some time now the Roebling pew had been used as the visitors' pew. He made an appearance every so often, accompanied by one or more of his sons, and all eyes would be on them as they came down the aisle. But he held that spiritual communion with the Creator was more likely to be achieved through a vigorous life of the mind. "Human reason," he wrote, "is the work of God, and He gave it to us so that we can recognize Him."

He had been swept up by the teachings of Swedenborg, the brilliant Swedish physicist of the previous century, who rejected the dogma of original sin and eternal damnation and wrote of a spiritual evolution for the individual. And like Swedenborg he had embraced spiritualism.

For some twenty years and more, spiritualism had been gaining converts among educated people on both sides of the Atlantic. The Fox Sisters and their much-publicized "Rochester Rappings" had marked the start of it in America. And in the time since, it had become an intensely serious body of beliefs that had a strange, powerful appeal to a surprising number of intensely serious people. For those of a doubting analytical turn of mind, it seemed to offer proof of the existence of a spiritual realm. To practical men of learning, whose faith in traditional doctrine had been shaken by the revelations of science, it seemed at least an alternative. Why Roebling turned to it he never explained. But in the final years of his life he believed devoutly in a "Spirit Land" and in the possibility of mortal communication with its inhabitants. Specifically, he believed in the afterworld described by Andrew Jackson Davis, "The Poughkeepsie Seer," a pale, nearsighted son of an alcoholic shoemaker, who in Roebling's estimate was one of the great men of all time.

Davis had become a clairvoyant, healer, and overnight sensation in 1844, at age seventeen, when he took his first "psychic flight through space" while under hypnosis in Poughkeepsie, New York. For the next several years he traveled up and down the East delivering hundreds of lectures, taking his own attendant hypnotist along with him—to "magnetize" him for each performance—as well as a New Haven preacher who took down everything he uttered while under the spell, all of which was turned into books. (One such book ran to thirty-four editions.) His preachments were a strange mixture of occult mystery, science, or what passed for science, progressive social reform, intellectual skepticism, and a vaulting imagination. For Roebling the impact of all this was momentous. It was as though he had been struck by divine revelation. He wrote at length to Horace Greeley, proposing the establishment of an orphanage in which a thousand children would be "perfectly educated, physically and mentally" according to the Davis vision of the good life. An "earthly paradise" was still possible after all.

The hereafter as pictured by Davis was a complicated hierarchy of life spheres, successive states of consciousness, all worked out geometrically, that existed above, and concentric with, the earth's surface. Apparently, in terms of what Roebling knew of physics and astronomy, this made more sense than anything else he had heard of, and besides, there was the rich mystical language of Davis, which for Roebling seems to have reached farther even than reason could take him.

For the benefit of his family Roebling would expound on such things endlessly at the dinner table, using Davis or some philosophical discourse he had read as his text, his voice gaining strength as he went on and on, with no concern whatever that his small, respectful audience understood almost nothing he was saying, but just sat there, blinking like young owls in the sunshine, as Washington Roebling would say. Washington was old enough to remember when that "life force" his father liked to talk of had surged through the man with such vitality and there were scenes that would live on with the young man as long as he would: father at his drawing table at Saxonburg, before he needed spectacles to read, working long into the night, his books and things all about him; father in Pittsburgh before the war livid over some latest piece of political news and vowing to go straight home and fire every Democrat in the mill; father up and out of the house before breakfast, an old fur cap on his head, walking the fields with a stick and a dog, getting up an appetite as he said; father with strap in hand about to lay on terrible retribution for some childish misdeed, a burning, unforgettable fury in his eyes.

But in the years since Johanna's death he had seemed ever more engrossed in the spirit world and talk of sickness and death. His back bothered him. He suffered from indigestion. For those of his children still living at home it had been a disturbing, unpleasant time, and particularly for Edmund, the youngest, who had been his mother's favorite. When she was gone, his father, as always, had been too busy to give him any time. When his father married again, in 1867, the boy had been packed off to a boarding school, where, Washington would write, "he was subjected to evil influences of so galling and insidious a nature that he ran away—was caught, brought back, and nearly beaten to death by a brutal father, and sent back." The boy escaped a second time and vanished. For nearly a year the family agonized over his whereabouts. But then he was found, quite by accident, by a Trenton man who happened to be inspecting a prison in Philadelphia. "He had had himself entered as a common vagrant," Washington would explain, "to get away from his father, and was enjoying life for the first time."

The whole affair was kept very quiet. None of the family would ever speak of it. There was nothing said in the papers. Except for a private memorandum written by Washington years later, there would be no record of the incident. But in his philosophical notes, under the heading "Man. Conscience," John A. Roebling wrote the following at about the time Edmund was back home again.

> A man may be content with the success of an enterprise; he may have succeeded in overcoming obstacles; in vanquishing his adversaries and enemies; in achieving a great task; solving a great mental problem, or accomplishing work, which was previously pronounced impossible and impracticable. The hero is admired and proclaimed a public benefaction; observed of all observers, he feels himself elated, and in his own estimation a great man. Retiring for one calm moment within the recesses of his own inner self, he reviews his past deeds, his thoughts and motives of action. And before the stern judgment of his own conscience, he stands condemned, an untruth, a lie to himself. But nobody knows! Does he himself now know? Who can hide me from myself? . . .

Had their mother lived, Washington believed, none of this would have happened. And then one night she returned.

"The latest sensation we have had here are spiritual communications from Mother," Ferdinand Roebling, John Roebling's second son, wrote on November 12, 1867, to his brother Washington, then in Europe. A cousin, Edward Riedel, was the medium. He and Roebling's draftsman, young Wilhelm Hildenbrand, had been sitting in their room on a Saturday night when they heard three knocks under Riedel's chair. "He did not know what to make of it," Ferdinand said, "so they examined the room and the next room and porch and all around, the knocks still followed Ed, always under him, they then asked some questions." Was it a bad spirit? No answer. Was it a good spirit? Three knocks. After repeating these same two questions several times, they asked if perhaps they ought to give up and go to bed, and the response was three sharp knocks.

Roebling was told the next morning. That night they formed up in a circle in his office but got no response until Riedel, having lost hope apparently, went to his room and pulled off his boots. Suddenly he heard the knocks, coming from the kitchen. Roebling was called and they all quickly gathered there. "They then used the Alphabet and found out whose spirit it was. No answer could be given to anyone but Ed." Everyone was extremely excited, it seems, and Roebling especially, one would imagine. He suggested a few questions, but "none of any account," according to Ferdinand, and about the only important piece of information communicated by the spirit was that she would return two weeks hence—which she did, and this time Roebling was ready with a long list of questions carefully thought out in advance. If this was to be his first real chance to converse with "the other side," he would come to it as he had tried to come to every turning point in life, thoroughly prepared.

Having determined at the start of the séance that the spirit was indeed that of his wife, Roebling asked for Willie and Mary, their two dead children, for his own mother and father, and for Frederick Overman, a renowned German metallurgist from Philadelphia who appears to have been the one real friend he had ever had time for since leaving Saxonburg. Overman, dead for fifteen years by this time, once told Roebling that life was a result of movement and death only a change of movement and it had made a lasting impression on the bridgebuilder.

Then Roebling started down his prepared list.

"Do you remember, my dearest, the conversations I have had with you about the Spirit Land and Spheres? You remember the opinions and view taught by Andrew Davis on the subject?" Having been convinced he could talk again to his beloved wife after three years, his whole line of questioning was designed strictly to verify his own set of beliefs. Was Davis correct on the whole, he wanted to know. Yes. In detail? Yes. And so it went, through the rest of that session and in the eleven others that were to follow. On the night of January 25, 1868, for instance, the conversation had run this way:

"After your spiritual birth, did you feel like a new being, *young*, *energetic* and full of life?"

"Yes."

"Did you find that all disease had left you, and that your new spiritual body was free from pains, rejoicing in youth and vigor?"

No answer.

"Do you attend public lectures?"

"Yes."

"Who is your favorite lecturer? Will you spell out his name?"

"C-H-A-N-N-I-N-G."

"Are you taught religion?"

"No."

"Have you got a Bible there?"

"No."

"You are taught without books?"

"Yes."

"Is it taught by your teachers that inspired truth and the truth of nature cannot be at variance?"

"Yes."

"The Christian Bible, the Jewish Scriptures, the Turkish Koran, and all the other books, which lay claim to divine origin and divine inspiration, are all human compositions, therefore liable to error. Is this view correct?"

"Yes."

"Every man has a spark of divine principle within himself, which alone can save him and elevate him, is this so?"

"Yes."

He was getting all the answers he wanted. On the night of March 22, he had a total of sixty-one questions prepared for her. Their discourse should always be "serious," he said at the start; he should not waste time on trivial talk. Yes, she answered. It was truth he should be in search of, he said. Yes, she answered again.

Soon she was bringing a number of other spirits along with her, at his request. On the night of June 14, Willie and Mary, Grandmother Herting, Overman, Roebling's brother Karl, another brother, and three others besides Mrs. Roebling were present, making ten in all. Still Roebling took no time for personal exchanges with any of them, the line of questioning continued exactly as before. But the atmosphere in the dark room remained highly charged. At one point young Edmund exclaimed that he could see Grandmother Herting's face and that she was reaching out to touch him. Whether or not the new flesh-and-blood Mrs. Roebling was invited to sit in on any of these sessions is not known.

Had it not been for the bridge, such gatherings with the dead might have gotten to everyone in the household. But by this time the bridge had become the overriding passion of Roebling's life. It was the summer of 1867, the summer before the séances began, that he had drawn up his plans. In just three months, working at a fever pitch, he had produced the drawings, location plans, preliminary surveys, taken soundings, worked out his cost estimates, and written his proposal, nearly fourteen thousand words in all. Some would say later that it had been as though he knew how little time he had left, which seems unlikely, even though the subject of death, his own included, remained very much on his mind, as his oldest son would disclose.

Washington had been the one member of the family ever to go off and work with John Roebling at bridgebuilding. He knew the different man his father became then, out in the open air, a hundred men or more at his command, his bridge the talk of everyone who came to watch. More than anyone he could appreciate how long a shadow the old man cast. Moreover, he appears to have been the one person Roebling confided in, telling him things he had not said to anyone. He had unlimited confidence and pride in the young man and had agreed to begin the new bridge only with the understanding that the two of them would be working together.

And it would be Washington, later, in things he said and wrote, who would describe another change that had come over his father, something more than his remoteness or the ill temper of advancing age or his forays into the spirit world. It was a deep melancholic disillusionment growing out of what John Roebling thought he saw happening to the country since the war. The great dynamic of America, he had always said, was that every man had the opportunity to better himself, to fulfill himself. Now the great dynamic seemed more like common greed. It was not so much contempt for Germany that had brought him to America, he had told his children, but that in this new country a man was free to make the most of his abilities. If he had "personal energy and power of will," there were few limits to what a man might attain. Moreover, like numerous others of his day he had long equated works of monumental engineering—and his own work especially—with national grandeur. "The idea of an epoch always finds its appropriate and adequate form," his teacher Hegel had written. The steamboats, canals, highways, railroads, and bridges he himself had seen on first arriving in America were, he had written, the direct result of the "concerted action of an enlightened, self-governing people."

But now he had his doubts. Now he had seen men making the most of abilities he had no stomach for and self-government made a mockery. And lately he had seen his own work contribute to that kind of degradation. It had troubled him so deeply that he had talked seriously with his son of washing his hands of the entire affair in Brooklyn.

But now, dressed in a light topcoat and a soft felt hat, he stood waiting to join the Bridge Party. Whether any of his other sons, his wife, or perhaps the faithful Charles Swan had come to the depot with him, to keep him company or to listen to any last-minute instructions, is not known. Washington, however, had left Brooklyn with the others and would be at the door of the parlor car to greet him when the train stopped.

3
The Genuine Language of America

> He spoke our language imperfectly, because he had not the advantage of being born on our soil, but he spoke the genuine language of America at Cincinnati, Pittsburgh, and Niagara . . .
>
> —THOMAS KINSELLA, in *The Brooklyn Eagle*

CHAMPAGNE and sandwiches were served soon after Roebling came aboard. How late the little celebration lasted after that nobody said later. But at five the next morning, when he roused them all, there was no little grumbling. He was anxious, Roebling said, that nobody miss the sunrise over the Alleghenies.

By breakfast they were passing through Johnstown and he had everyone peering out at the steep, thickly wooded sides of Conemaugh Gap, a deep cleft in Laurel Hill that he and his railroad surveying party had first seen from a distant hill thirty years before. "There was our course!" he had written enthusiastically at the time.

The next town of any size was Greensburg, where the very first suspension bridge there is a record of was built over Jacobs Creek by a Scotch-Irish preacher, a Presbyterian named James Finley, in the year 1801, or before John Roebling was born. Finley had been a versatile and ingenious man. His "chain bridge" had a seventy-foot span, cost about six hundred dollars, and in the next ten years he built some forty more of them, including one over the Potomac above Washington. Perhaps Roebling told his traveling companions something about this, thereby getting a head start on their instructions in the history and theory of suspension bridges.

When the train pulled into Pittsburgh less than an hour later, he took them directly to their quarters at the Monongahela House, which stood at the end of his Smithfield Street Bridge. From the front door of the hotel, or possibly from their rooms, if they were on the river side, they had a perfect view of the pioneering work, now nearly twenty-five years old, that had started Roebling on his way. It had been built at a time when every floor beam had to be cut with a hand-pulled whipsaw, when screws were still turned on a lathe by hand, and steel, practically speaking, even in Pittsburgh, was regarded as a semiprecious metal. One of the Pittsburgh papers in 1846, the year the bridge was finished, had claimed "this admirable species" was "destined to supersede all others." For Roebling, from then on, it had been the only type he would care anything about building, and in its rather antique fashion, the bridge still illustrated several fundamental points about his own particular manner of building—all of which he no doubt explained as he and his entourage went out for a first look.

Here again, as at the aqueduct, he had fixed his cables to a chain of iron eyebars buried in masonry anchorages. Here, for the first time, he had used his system of inclined stays to add strength and rigidity. Only here, he explained, he had used iron rods rather than the iron rope used on all his later bridges. The bridge was fifteen hundred feet long (or not quite as long as the river span alone of the bridge he had drawn up for Brooklyn). It had eight spans of about 188 feet each and short cast-iron towers. The wind had no effect on it, he said, and the vibrations produced by seven-ton coal wagons and their teams were no greater than on a wooden truss bridge with spans the same length. The total cost had been $55,000—"a very small sum indeed for such an extensive work," according to the engineer.

But the real Roebling showpiece in Pittsburgh was across town at Sixth Street and there they all went first thing that afternoon. He had built the Smithfield Street Bridge largely to prove his engineering skill and the soundness of the suspension technique. He had been concerned with building an efficient structure at the least possible cost. But his Allegheny River Bridge, begun eleven years later, had been built with an ample budget. It had been his first real opportunity to display his gift for architectural design and he had

had a splendid time with it. Among people who knew bridges, it was considered one of the handsomest in the country.

It stood downstream from where his aqueduct had been and connected Pittsburgh with the small neighboring city of Allegheny. Its total length (1,030 feet) was less than the bridge over the Monongahela. It had four spans and was supported by four cables hung from six highly ornamental iron towers, each with iron latticework for bracing and iron spires for decoration on top. "The bridge will be beautiful," he had written when the towers were nearly finished. In truth it looked a little as though it had been designed to satisfy the aesthetic tastes of a Turkish sultan. This was also the first bridge he and his son had built together. "I am getting along well here," he had written home to Trenton in the spring of 1858. "Washington is about the work." As a matter of fact it was Washington who supervised most of the job thereafter and for whom numerous Pittsburghers had the most affectionate memories.

Once finished the Allegheny River Bridge was so sound that the owners—a private company—had not even bothered to take out insurance on it, and as a toll road, it had made money from the start—both points that must have been noted with interest by the delegation from the East. For about an hour they examined the bridge. There is no record of what was discussed during this time, but probably the cables were the main topic. These had been laid up, or "spun," in place, unlike those on the bridge just visited, where the cables were smaller and the spans between towers were much shorter. There the iron wires had been spun on land first, to form individual cable sections that were then hoisted into position. But here, one can picture him explaining, the cables had been spun on the bridge itself by a traveling wheel that went back and forth, stringing the wire over the towers, from shore to shore, making fourteen hundred trips in all, and this was the way that he meant to build his cables over the East River.

Thomas Kinsella, the editor of the *Eagle*, would report in an article written afterward that the floor trembled very little as trolleys to and from Allegheny went clattering by and everyone in the party thought Roebling's ornamental ironwork a feast for the eye. The remainder of the day was spent touring the ironworks of the young Carnegie brothers, where the manufacture and virtues of Bessemer steel were explained. Whether or not the wire in the new bridge would be of steel had still to be decided.

The itinerary called for a stay of several days in Pittsburgh, but so unpleasant was the air, in the opinion of several in the group, and so unsatisfactory the accommodations at the Monongahela House, that a decision was made to leave the next day. "If you ever visit Pittsburgh," wrote Thomas Kinsella for his Brooklyn readers, "and desire to stop at the best hotel . . . don't."

On the morning of April 16 they were again settled in their private car, "leaving Pittsburgh like a great sooty blotch behind." The sun out, they "swept across into Ohio" at the grand speed of fifty-four miles per hour, an experience everyone would have enjoyed had not the parlor car started rocking so that it greatly interfered with a poker game. At Cincinnati some time after dark they checked into the Burnet House, where they enjoyed a "very fair supper," after which, over cigars, the next morning's schedule was discussed. "Slocum, never lacking pluck, had the courage," Kinsella wrote, "to suggest that nine o'clock was, under the circumstances, a barbarous hour. He quickly won the majority over to his way of thinking, and the Untiring Old Man, Roebling, yielded an hour's grace, and it was tacitly accepted that no one would be greatly disappointed if the party should not leave the hotel before ten o'clock. As we retired the blessed spring rain was falling against the windowpanes, and after the day's fatigue sleep came as gentle as the dew." (All this still being written for home consumption, in the pages of the Brooklyn *Eagle*.)

The following morning one of the party, a man named Cary, reported sick. He had made the mistake, he said, of drinking some of the local water, a glass of which was described as eating and drinking combined. But the rest were in excellent spirits and the day was spectacular. It was Saturday and the streets were already crowded with people enjoying the sunshine as Roebling led his group out of the hotel.

The first view of the bridge proved to be a far more stirring experience than anyone from Brooklyn had been prepared for. It was built on a line running due south, reaching over the Ohio to Covington, Kentucky. But because of the way the streets were laid out along the river front, there was no way to see the bridge until nearly upon it. "It then broke upon us all at once," Kinsella wrote, "the stateliest and most splendid evidence of genius, enterprise and skill it has ever been my lot to see."

Eleven thousand people a day were crossing it, he and the others were told, as they stood gazing at the long, graceful arc of its river span (" . . . it was indeed a work to excite amazement and wonder."). For the next hour or more they walked back and forth from one end to the other (". . . it seemed as solid and as stable beneath our feet as the earth on either side of the river.").

This, they realized, was the nearest thing in existence to what Roebling planned to build over the East River. And if any of them was having trouble picturing the new bridge, he had now only to imagine something very like this one—only much bigger.* Here were the twin towers of stone standing foursquare and solid, a slender line of roadway stretched between them, slung on great cables and arcing the river with a single span. Here, as on the Pittsburgh bridges, were the inclined stays, slanting down from the towers like iron rays, angling across the suspenders that connected the cables to the roadway. The stays were the mark of a Roebling bridge, the traveling delegation had come to realize. But here the scale of the bridge was such that the combination of stays and suspenders looked like a gigantic web, or net, and the same effect at Brooklyn, it was understood, would be even greater.

Every diagonal stay, Roebling explained, formed the hypotenuse of a right triangle (the bridge floor and the tower forming the shorter sides) and thus provided tremendous stability, since, as he said, "The triangle is the only unchangeable figure known in geometry . . ." Altogether, cables, suspenders, stays, and bridge floor formed a kind of truss. The great horizontal stability of the bridge was due in large measure, he said, to such "bracing" of the cables. This was a proposition "readily comprehended by sailors, who are accustomed to stays on board ships."

The "Biggest Bridge in the World" had been opened to the public on December 1, 1866, to the tune of a thundering cannonade. By sundown 46,000 people had crossed it, with no ill effects to the bridge or to any of them. But the following day, an uncommonly mild winter Sunday, 120,000 people had turned out to personally examine the wondrous work. Then, on New Year's Day, 1867, the official opening, a big parade had marched over from the Covington side, led by Roebling and Amos Shinkle, the Cincinnati coal dealer who had been the principal organizer of the project, and who that

* The East River bridge was to be larger in every way. The river span of the Cincinnati Bridge was 1,057 feet, or 543 feet less than what Roebling had projected for the East River. The over-all length of the Cincinnati Bridge, 2,252 feet, was less than half the length, and its width, 36 feet, was also less than half that of the new bridge Roebling had planned.

sparkling spring morning in 1869 had come down to the bridge to greet Roebling and his new clients, some of whom had matters other than engineering on their minds. "Does the bridge pay, sir?" he was asked. "Yes, sir," answered Shinkle, "handsomely."

Roebling had first come to Cincinnati with plans for a bridge more than twenty years before, in 1846, and had felt very much at home in the brick city on the river, with its German theaters, its beer gardens and German newspapers. The Ohio was still the great dividing line between North and South then, between plantation slavery on the Kentucky side and in Cincinnati some of the strongest abolitionist sentiment in the country. (It was in Cincinnati then that stories told by slaves who had escaped over the river were making a deep impression on Harriet Beecher Stowe, the young wife of a local professor.) So there were reasons other than the mighty Ohio or the strenuous opposition of the steamboat interests for not building a bridge and it was nearly a decade later, and only when Amos Shinkle came on the scene, that anything began to happen.

But after Roebling had the work under way, he was hit by one of the worst winters on record in Ohio, the winter of 1856–57. In spring, when the ice broke up, the river flooded his foundations so badly that little could be accomplished for another six months. Roebling kept coming from Trenton to look things over, then headed home again. But it was a time of great productivity for him.

The year before, he had done a sketch for a bridge to Brooklyn, a multispan bridge to cross by way of Blackwell's Island, where the prison and poor asylum stood. In March of 1857 he wrote to Horace Greeley to propose "a wire suspension bridge crossing the East River by one single span at such an elevation as will not impede the navigation." His Cincinnati Bridge, scarcely even under way, was only a preliminary work, as he saw it. This East River bridge would be "without rival," its towers three hundred feet high. The letter appeared in the New York *Tribune* on March 27, 1857, and was Roebling's first public declaration of his plan.

That same March, in his Trenton study, he produced drawings for three different kinds of towers for the East River bridge—one an elaborate Egyptian doorway with a spread-eagle gargoyle for a corbel; another a notably plain Roman arch; the third, again a Roman arch but drawn with a bolder, heavier pediment and then a Gothic arch, a second thought apparently, sketched in tentatively in pencil, very lightly, like a ghost of things to come. None of these suited him, but still enormously excited about the idea, he wrote to

Abram Hewitt, head of Peter Cooper's Trenton Iron Works. It was Peter Cooper who had first urged Roebling to locate in Trenton and helped him pick a site for his wire mill, probably figuring the engineer to be a fine prospective customer. In the time since, Hewitt had become Cooper's son-in-law as well as his business partner. An energetic, self-assured young man, he was said to have a great future. Hewitt had Roebling's letter printed in the *Journal of Commerce*, but did no more than that, which must have been disappointing to Roebling. Indeed Hewitt's response would be barely worth mentioning were it not for the part he was subsequently to play.

The terrible Panic of 1857 burst upon the country that summer and Roebling had all he could do managing things in Trenton and at Cincinnati, where work on the bridge was shut down altogether, not to begin again until the early part of the Civil War, when a Confederate force under Kirby Smith, advancing into the Blue Grass Country to the south, threw all Cincinnati into a state of panic. Soldiers and citizens alike rushed to fortify the hills on the Covington side, discovering in the process how very advantageous a bridge would be, had there only been one. And beside the pontoon affair that was hastily assembled stood Roebling's half-finished towers to remind everyone what might have been.

There never was a siege of Cincinnati, but once the threat of one had passed, the fortunes of Roebling's troubled Cincinnati Bridge took an immediate turn for the better. Subscriptions to new stock poured into Amos Shinkle's office, the work commenced once more and with no opposition. For Roebling the bridge was a symbol of confidence rising above the "general national gloom." It proved, he said, that there were still men about "with unshaken moral courage and implicit trust in the future political integrity of the nation." When his Irish laborers, who shared no such feelings for the bridge or for the Union cause, walked off the job demanding higher pay, Roebling fired every man and hired only Germans as replacements. "The Germans about here are mostly loyal, the Irish alone are disloyal," he wrote. "No Democrat can be trusted, they are all disloyal and treacherous, more or less."

Two years later, the war nearly over, Washington Roebling was released from the Army and went almost directly to Cincinnati, where, by then, there was no longer any question about the relative importance of his father's bridge. "The size and magnitude of this work far surpass any expectations I had formed of it," the young man wrote to the rest of the family back in Trenton. "It is the highest thing in this country; the towers are so high a person's neck aches looking up at them. It will take me a week to get used to the dimensions of everything around here." From that point on, though his official title was Assistant Engineer, he had been in complete charge of the work. All the cable spinning, the most exciting, difficult part of the work, was done under his direction, his father having concluded that he would henceforth "leave bridgebuilding to younger folks."

The Cincinnati Bridge wound up taking a total of ten years to build and it cost just about twice what John Roebling had said it would. But no one had any complaints. It was unquestionably the finest as well as the largest bridge of its kind built until that time. Both structurally and architecturally it was a triumph.

Talking in retrospect, Amos Shinkle had nothing but praise for the manner in which it had been built. From an engineering standpoint everything had gone very smoothly. Only two lives had been lost during the entire time of its construction, a remarkable safety record, as the gentlemen from the East agreed. For the Roeblings, he had only the highest admiration, and especially for the redoubtable father, about whom, apparently, a few of the Brooklyn men had expressed some uneasiness. Getting along with him should prove no problem, Shinkle assured them. His advice was simple: "He is an extraordinary man and if you people in Brooklyn are wise you will interfere with his views just as little as possible. Give the old man his way and trust him."

At eight that evening the Bridge Party departed for Niagara Falls, by way of Cleveland, where they stopped for the night and where several of them decided things could be livened up a bit if the word was spread that they were a group of wealthy lunatics being conducted on an outing. The joke worked quite well, it seems, causing a considerable stir in the hotel dining room. But when Thomas Kinsella's account of such goings on appeared in the paper back in Brooklyn, it served mainly to substantiate what a number of people there had been saying right along, that the bridge was the scheme of madmen.

Spring was late arriving at Niagara Falls. The snowbanks had nearly all disappeared but the weather was sharp still and gigantic slabs of ice could be seen plunging down the river as Roebling led his group out to inspect what was generally conceded to be his masterpiece, a two-level suspension bridge over the great gorge.

This tour could have been arranged in the reverse order just as easily, with Niagara the first stop instead of the last, which would have made better sense in some ways, in that the Niagara Bridge was an earlier work than either the Cincinnati or Allegheny River bridges. But Roebling had saved Niagara for the last for good reasons.

At Pittsburgh he had been able to show as solid, dependable, and handsome a piece of workmanship as he had ever built. The Allegheny River Bridge was a bridge everyone liked. It had caused nobody any headaches when it was being put up or in the time since. At Cincinnati, the unprecedented length of the single river span was the most important thing on display, from the technical point of view. But the bridge had also a grandeur of a kind rarely seen and his new clients had come away from it with a keener appreciation of monumental scale as well as engineering genius.

But the Niagara Bridge, or International Suspension Bridge, or just Suspension Bridge as it was called locally, was neither terribly solid-appearing nor especially large. Indeed, every bridge they had inspected so far was longer. Also, unlike any of the others, it had been built with two levels—carriages and pedestrians traveled the lower level, while the Great Western Canada Railroad crossed on the one above—and the whole thing trembled quite noticeably when traffic was heavy. For some people the experience of crossing by carriage was positively terrifying. "You drive over to Suspension Bridge," wrote Mark Twain, "and divide your misery between the chances of smashing down two hundred feet into the river below, and the chances of having a railway-train overhead smashing down onto you. Either possibility is discomforting taken by itself, but, mixed together, they amount in the aggregate to positive unhappiness."

Its single span was 825 feet, which was nothing exceptional any longer. Its four stone towers stood only about half as high as those at Cincinnati. It had not been the first suspension bridge over the gorge and it did not stand alone, unrivaled, the way the Cincinnati Bridge did.

The first bridge had been built downstream at Lewiston, New York, in 1851, by Edward Serrell, brother of the Serrell traveling with the Bridge Party. It had been a very light suspension bridge and was badly shaken by a storm in 1855, after which, at Roebling's suggestion, it had been refitted with guy wires. But later when these wires were loosened by an ice jam, somebody neglected to tighten them. A spring storm tore the bridge floor to pieces and left the cables and suspenders dangling uselessly in mid-air. And there they were still, about as dramatic an example of what could happen to a poorly engineered suspension bridge as could be seen, short of an actual collapse.

The other bridge over the gorge was very much in view from where they stood, about two miles upstream, near the falls. It was a brand-new suspension span designed by a Canadian, Samuel Keefer. It had been opened just that January and was already an extremely popular attraction, being near the best hotels. But since it was only ten feet wide, or too narrow for carriages to pass one another, traffic had to cross it in turns, first one way, then the other, and long waiting lines built up. In years to come it would be remodeled extensively and become famous as the Honeymoon Bridge. But at this stage it was being publicized only as a greater span than the one at Cincinnati, which indeed it was in terms of length. It was more than two hundred feet longer, which meant it now held the world's record.

It would appear then that the gentlemen from Brooklyn might have been somewhat disappointed with this, the final Roebling bridge on the tour. But not so. Like nearly everyone who ever stood there at the brink of the gorge, with the bridge before them, they looked upon it with nothing less than awe.

The site alone was enough to take a person's breath away. Upstream were the falls, while directly beneath the bridge, deep in the abyss, was the first of a series of savage rapids that swept on downstream for a mile or better, ending in a tremendous whirlpool held in a looming rock basin. Past there the current veered to the right and disappeared through a narrow channel overhung by sharp cliffs and trees. It was an absolute no man's land below, but here above it had been conquered, bridged, beautifully.

Once, not many years before, an excursion boat, the *Maid of the Mist*, had gone shooting by below to the utter astonishment of those who happened to be on the bridge at the time. The boat had been built upstream, between the rapids and the falls, to take sight-seers for a rather terrifying close-up view of the falls. But the boat had never been a financial success. The owner, a Captain Joel Robinson, got into debt and when he heard that the sheriff was on his way to confiscate the boat, he decided his only chance was to escape down the rapids, something nobody had ever done before and lived to tell the story. Two men volunteered to go with him. The people on the bridge saw the boat make one long leap down the rapids. Her funnel was knocked flat by the blow, the whole boat was underwater from stem to stern. Then she was up again and skimming into the whirlpool, where "she rode with comparative ease upon the water, and took the sharp turn around into the river below without a struggle." Captain Robinson's wife later said her husband looked approximately twenty years older when he came in the door that evening.

But the bridge seemed to make the whole breath-taking panorama all the more terrifying, all the more magnificent. It was one of those occasions when the hand of man had enhanced that already wrought by the hand of God.

To begin with, the bridge seemed so serene and refined against such tumultuous doings of nature. Its essential components were four plain towers sixty feet high, four cables ten inches in diameter, their suspenders and stays, and a straightforward timber truss joining the two levels of the one span, which over such a gaping cavity in the earth looked ever so much longer than 820 feet. The bridge looked to be exactly what was called for, no more, no less. It was as though it was the only possible bridge for the place.

Actually, of course, one uninterrupted span was the only kind that would have worked there, since supporting piers in the gorge itself would have been out of the question. But this bridge was not simply for carriages and pedestrians, like the one upstream, indeed, like every other suspension bridge in the world at the time. It carried a railroad. That thought alone was enough to command the respect of anyone who knew a little about bridge engineering or recalled when it had been built. But even if a person were ignorant of such things, the sight of a moving train held aloft above the great gorge at Niagara by so delicate a contrivance was, in the 1860's, nothing short of miraculous. The bridge seemed to defy the most fundamental laws of nature. Something so slight just naturally ought to give way beneath anything so heavy. That it did not seemed pure magic. The reasons it did not were the very foundation of all Roebling's work. And if his band of clients, consultants, and other interested parties could return to Brooklyn understanding this bridge, they would understand what his work was all about.

In time to come, suspension bridges would not be used much for railroads, as Roebling expected they would. The Niagara Bridge was, in fact, the only noteworthy railroad bridge of its type ever built. The important thing, however, was that Roebling had demonstrated, at one of the most spectacular locations on earth, that the principles of suspension could be applied with perfect safety even to something so heavy as a locomotive and railroad cars, and this in turn had a profound effect on the whole evolution of bridge design, not to mention the acceptance of his own theories. At Niagara he had built the first truly modern suspension bridge.

The great appeal of the suspension bridge, apart from its beauty, was its economy. It required considerably less material than other kinds of bridges. But prior to the completion of Roebling's Niagara Bridge in 1855, suspension bridges had a dubious reputation. In all America then there were only two engineers who had any firm belief in them or who had built any of consequence. Roebling was one. The other was Charles Ellet, Jr.

The reason for so much distrust of suspension bridges was simply that so many of them had come crashing down over the years, and frequently with tragic consequences. In England in 1831 a suspension bridge had collapsed under the feet of marching troops. (The bridge was the work of Sir Samuel Brown, whose suspension bridges came down about as fast as he put them up, one after another—at Berwick, Brighton, Montrose, and Durham.) In France in 1850 another wire bridge had failed under almost identical circumstances, killing two hundred men. In America a number of small suspension bridges had collapsed under droves of cattle, including one at Covington, Kentucky, over the Licking River, just a few years before Roebling commenced his Cincinnati Bridge.

Nobody understood quite why these things happened. In actual fact the bridges had either been inadequately built to begin with or badly maintained, but whichever the cause, it had generally gone undetected and a large body of distrust had built up about the suspension method in general. In Europe especially, few engineers had confidence in such bridges for spans of any appreciable length or for heavy traffic, and this despite the fact that the earliest suspension bridges approaching the size of those by Roebling or Ellet were built in Europe from about 1820 on.

The basic idea was of course nearly as old as man. In China, South America, and other parts of the world, crude bridges had been slung from vines over rivers and ravines since before recorded history. There was, however, an obvious and important difference between such bridges and those that began to appear in the early part of the nineteenth century. The latter-day variety had a stiff, level floor that did not curve or sway with the ropes that held it, but was—or was supposed to be—as stable as any other kind of bridge floor. Moreover, these were no longer simple footbridges, but big enough to handle carriages and wagons.

A wire suspension bridge was built over the Schuylkill at Philadelphia as early as 1816, or fifteen years after the Reverend James Finley put his historic little chain bridge over Jacobs Creek. But it was the brilliant Scottish engineer Thomas Telford who completed the world's first great suspension bridge nine years later, in 1825, in Wales. It had two massive masonry towers and was hung on immense iron chains and it crossed the Menai Strait to the island of Anglesey, with a main span of nearly six hundred feet. It was the most famous bridge of its day and the prototype of all the great suspension bridges to came after it, including those by John A. Roebling.

Isambard Kingdom Brunel, the improbable little genius who would one day build the *Great Eastern*, the most colossal iron ship in history, also began building suspension bridges about this same time, as did the Swiss, the Germans, and the French. The Grand Pont, a suspension bridge built at Fribourg, Switzerland, in the 1830's, was not only very large for its time (more than eight hundred feet), but it would stand for a hundred years.

But in 1845, when a proposal was made to use a suspension bridge to carry a railroad over the Niagara Gorge, most of the experts declared the scheme quite impossible. Vibrations set up by so heavy a moving load as a train would, it was said, quickly destroy any wire-hung bridge. Still, the idea of a railroad crossing at Niagara made a great deal of sense to the American and Canadian railroad people and they were encouraged by four engineers who not only thought the thing could be done but were anxious for the chance to do it. Of the four, interestingly, three would eventually span the gorge with bridges of their own design—Serrell, Roebling, and Keefer. But as fate would have it, none of them got the first chance.

The man who did was Charles Ellet, who in 1845 was the best-known bridgebuilder in America. He was also the most flamboyant, the most interesting, and Roebling's one serious rival. Except for Roebling, he knew the most about suspension bridges and could turn on more fancy talk about them than anyone in the profession. Of all the American engineers of his time, Charles Ellet was the most impetuous and colorful, a genuine character of the sort who came and went with the nineteenth century.

Born in 1810, which made him four years younger than Roebling, Ellet had grown up in Pennsylvania, the son of a Quaker farmer. At seventeen he left home, worked on various canal jobs near home, taught himself French, and saved enough money to go to Paris to study at the Ecole Polytechnique. When he returned home after a year, he was the first native American with a European education in engineering. Almost immediately, he presented Congress with a plan for a thousand-foot suspension bridge over the Potomac at Washington and talked grandly of another over the Mississippi at St. Louis. Then he actually built one over the Schuylkill near Philadelphia in 1842, which was several years before Roebling had built anything. (Roebling had applied to build the same bridge himself, and when Ellet was chosen, Roebling wrote to commend him for so bold a plan. Thinking Ellet an older, more experienced man, Roebling applied for a job as his assistant. Ellet's reply was quite formal and vague, so Roebling wrote again, this time generously including drawings and notes on his own ideas; but nothing more ever came of it.) Five years after that Ellet had begun his greatest work, over the Ohio River at Wheeling, the first really long suspension bridge on earth. With a center span of 1,010 feet, it was only forty-seven feet shorter than Roebling's own bridge over the Ohio would be and Roebling's bridge would not be completed for another twenty years.

Ellet looked like an actor, with dark, brooding eyes and a lithe, athletic build. And all of his other talents aside, nobody made a better show of bridgebuilding. At Niagara he had a stage magnificently suited for the most thrilling performance of his career, and the last one, as it happens.

One of the first problems to be faced at Niagara was how to get a wire over the gorge and its violent river. Ellet solved that nicely by offering five dollars to the first American boy to fly a kite over to the Canadian side. The prize was won by young Homer Walsh, who would tell the story for the rest of his days. Once the kite string was across, a succession of heavier cords and ropes was pulled over, and in a short time the first length of wire went on its way. After that, when the initial cable had been completed, Ellet decided to demonstrate his faith in it in a fashion people would not forget. He had an iron basket made up big enough to hold him and attached it to the cable with pulleys. Then stepping inside, on a morning in March 1848, he pulled himself over the gorge and back again, all in no more than fifteen minutes' time, and to the great excitement of crowds gathered along both rims.

"The wind was high and the weather cold," he wrote, "but yet the trip was a very interesting one to me—perched up as I was two hundred and forty feet above the Rapids, and viewing from the center of the river one of the sublimest prospects which nature had prepared on this globe of ours."

Ellet appreciated the historic significance of his feat—he was the first man to cross the gorge—but he was not quite through. Several weeks later, after a plank catwalk had been strung across, he chose to demonstrate its strength in an even more memorable fashion. He leaped into a small carriage, gave his horse a slap of the reins, and went rolling headlong out onto the little bridge, which as yet had no guardrails and which swayed fearfully beneath horse, carriage, and Ellet, who drove standing up, like a charioteer. Everyone watched aghast, women fainted it is said, and Ellet and his bravado became a legend that would last longer at Niagara Falls than anything he built there.

In less than a year he had an angry falling out with the men who were paying for the job. He had finished the catwalk the summer of 1848 and opened it to the public. Very quickly it became a surprisingly lucrative property. Tolls collected came to five thousand dollars before a year had passed and a dispute arose as to whom the money belonged. Feelings between Ellet and his clients got so bad that Ellet drew up cannon at both ends of the bridge, proclaimed it was his, not theirs, and threatened to flatten anybody who came near. There had followed a few tense days at Niagara. Then, inexplicably, Ellet walked away from the greatest opportunity of his career, never to come back, leaving everything he had accomplished swinging uncertainly in the winds of the gorge.

Two years later Roebling commenced his own bridge at the same spot. In temperament and behavior he and Ellet were about as different as two men passionately committed to the same idea could possibly be. Where Ellet talked like a rain maker, Roebling was eloquent but precise, never promising more than he could deliver. Where Ellet was bold, impulsive, dramatic, Roebling was painstaking, methodical, working out every detail in advance. And once he had settled in his mind that he could do a thing, Roebling stuck to it. "Before entering upon any important work, he always demonstrated to the most minute detail its practicability . . . and when his own judgment was assured, no opposition, sarcasm, or pretended experience could divert him from consummating his designs, and in his own way."

Roebling started his bridge in 1851 and it took him four years. He worked carefully, steadily, and there were no hair-raising escapades anyone would remember later. For Roebling the excitement of the work, the drama of building a bridge, were chiefly matters of the intellect and spirit. Physical dangers were part of the job, inevitably, but to be taken as they came, or, better still, avoided entirely if a safer way could be figured. The bridge he built was a thorough demonstration of theories he had been perfecting and preaching for a decade and more. "The only real difficulty of the task," he wrote, "appears to be its novelty."

Put in its simplest form, Roebling's fundamental belief about suspension bridges was that the *stiffer* and *heavier* the roadway could be made, the more stable the bridge. To many this seemed contrary to common sense, since the weight of the roadway and its superstructure would seem to jeopardize those very elements that made a suspension bridge a suspension bridge—the cables.

Roebling was not the first to recognize the importance of a heavy, stiff roadway, just as he was not the first to use anchor stays or to spin his cables in place, all things he would be credited with initiating and reverently praised for by some of his more ardent admirers. James Finley had used stiffening beams and railings before Roebling was born and he knew the purpose they served. The scowling little Brunel, trudging about his bridges in a stovepipe hat, had directed that tension cables be attached to counteract the action of the wind. The French engineer Seguin wrote in 1824 that rigidity of the bridge floor was the surest means to prevent the "vacillations arising from moving loads of any considerable mass" and said the best way to achieve that rigidity was an arrangement of strong trusses.

There were others, too, including an English engineer named Rendel, who wrote the following before John A. Roebling had built a bridge:

> In the anxiety to obtain a light roadway, mathematicians, and even practical engineers, had overlooked the fact that when lightness induced flexibility and consequently motion, the force of the momentum was brought into action and its amount defied calculation. The

> author has long been convinced of the importance of giving to the roadway of suspension bridges the greatest possible amount of stiffness . . .

But unlike most every builder of suspension bridges then, and some much later, Roebling not only understood these ideas, he applied them, his system of inclined, or diagonal, stays being an excellent case in point. "I have always insisted that a suspension bridge built without stays is planned without any regard to stiffness, and consequently is defective in a most important point." And equally important, he did *not* apply some of the other theories in circulation at the time, many of them very bad theories, that were often taken seriously by the supposed experts. So if he cannot be honestly credited with originating all he preached, he at least was the one engineer who was practicing it properly.

In his original letter of proposal to the railroad men, Roebling had written that his bridge over the Niagara Gorge would stand up under a moving train because he would make it stiff enough to do so. He designed the two floors of the bridge and the open timber trusswork that was to bind them together as one enormous "hollow straight beam." The timber would be well seasoned, well painted, and the upper floor, where the trains would cross, would be caulked and painted as thoroughly as a ship's deck, and serve thereby, like the roof of an old-fashioned covered bridge, as a protective shelter for the lower floor and the trusswork.

To make the wire cables sufficiently strong to carry such a structure, as well as the trains, was, he said, "a matter of unerring calculation." So he had calculated (unerringly, he knew) and he had proceeded to build. Few other engineers gave him any hope of success. The most frequently quoted remark was one made by the great English engineer Robert Stephenson, builder of the famous Britannia railroad bridge, a tubular iron bridge, over the Menai Strait (the trains ran through a succession of enormous iron boxes set on stone piers). Stephenson wrote to Roebling from England: "If your bridge succeeds, then mine have been magnificent blunders."

Roebling had not a doubt in the world that Stephenson was wrong and said so. As far as he was concerned no Englishman, not even Telford, had ever built a suspension bridge worthy of the name; and to his way of thinking there was only one individual who had, or who really understood the subject, and that of course was John A. Roebling.

Then in May 1854 came the news from Wheeling that Ellet's Ohio River bridge had gone down. It had lasted just five years. As might be expected, the news created a great stir at Niagara Falls. Roebling especially was anxious to know exactly what had happened.

The details were provided in this vivid account published in the Wheeling *Intelligencer*:

> About 3 o'clock yesterday we walked toward the Suspension Bridge and went upon it, as we have frequently done, enjoying the cool breeze and the undulating motion of the bridge . . . We had been off the flooring only two minutes, and were on Main Street when we saw persons running toward the river bank; we followed just in time to see the whole structure heaving and dashing with tremendous force.
>
> For a few minutes we watched it with breathless anxiety, lunging like a ship in a storm; at one time it rose to nearly the height of the tower, then fell, and twisted and writhed, and was dashed almost bottom upward. At last there seemed to be a determined twist along the entire span, about one half of the flooring being nearly reversed, and down went the immense structure from its dizzy height to the stream below, with an appalling crash and roar.
>
> For a mechanical solution of the unexpected fall of this stupendous structure, we must await further developments. We witnessed the terrific scene. The great body of the flooring and the suspenders, forming something like a basket swung between the towers, was swayed to and fro like the motion of a pendulum. Each vibration giving it increased momentum, the cables, which sustained the whole structure, were unable to resist a force operating on them in so many different directions, and were literally twisted and wrenched from their fastenings . . .

From the description Roebling understood perfectly what had gone wrong. In a letter to the railroad officials describing his plans for the Niagara Bridge, Ellet had written ". . . there are no safer bridges than those on the suspension principle, if built understandingly, and none more dangerous if constructed with an imperfect knowledge of the principles of their equilibrium." Ellet's own knowledge had turned out to be imperfect, plainly enough. What Ellet had underestimated, Roebling knew, was the importance of building great rigidity into the bridge floor. A heavy floor would be less likely to move in a high wind, but weight alone was not enough. In fact, it was the weight of Ellet's bridge that had destroyed it, which Roebling later explained in his final report on the Niagara Bridge.

The Wheeling Bridge ". . . was destroyed by the momentum acquired by its own dead weight, when swayed up and down by the wind . . . A high wind, acting upon a suspended floor, devoid of inherent stiffness, will produce a series of undulations . . . [which] will increase to a certain extent by their own effect, until by a steady blow a momentum of force may be produced, that may prove stronger than the cables.* And although the weight of the floor is a very essential element of resistance to high winds, it should not be left to itself to work its own destruction. Weight should be simply an attending element to a still more important condition, viz: stiffness."

The best way to achieve such stiffness was with a strong truss, in this case of timber, composed of a combination of triangles. After the Wheeling news reached Niagara, Roebling straightaway reviewed all his plans and decided to build his trusswork stronger still. (To attain the necessary stability for his bridge over the East River, he had designed iron trusses twelve feet high to run the entire length of the suspended floor, from anchorage to anchorage.)

No lives were lost in the Wheeling catastrophe, but it aroused again all the old fears of suspension bridges, fears that would be a very long time dying. Later it would be said that Roebling went to Wheeling to rebuild the bridge properly. That was not what happened, however. Ellet rebuilt the bridge himself, just prior to the Civil War, using inclined stays the way Roebling would have. The bridge still stands.

When the war came, Ellet took command of a fleet of steam rams of his own design, on the Mississippi. Badly wounded at the battle of Memphis, he lingered on in great pain for several weeks, then died in June of 1862.

At Niagara, in June of 1854, Roebling had his cables finished and work had begun on the deck structure. "My bridge is the ad-

---

* The most famous latter-day example of this same phenomenon was the collapse of the Tacoma Narrows Bridge, over Puget Sound, in the state of Washington. On November 7, 1940, in a high wind, "Galloping Gertie," as the bridge became known, began heaving up and down so violently that it soon shook itself to pieces. The bridge lacked "aerodynamic stability" the experts concluded, for the simple reason that the necessary stiffness preached by Roebling had been overlooked by the designer. Eyewitness accounts of the disaster are strikingly reminiscent of the one from the Wheeling *Intelligencer*, written nearly ninety years before.

miration of everybody," he wrote; "the directors are delighted. The woodwork goes together in the best manner. The suspenders require scarcely any adjustment at all." In January, in a letter to Trenton, he said, "We had a tremendous gale for the last 12 hours; my bridge didn't move a muscle."

The bridge was completed in March 1855. Its span was nearly twice that of Stephenson's Britannia Bridge and was able to carry even heavier loading, and yet it had taken only one-sixth as much material in proportion to length. On Friday the 16th the first train rolled over. It was a triumphant moment for Roebling. The train, made up specially for the purpose, was as heavy a freight as could be assembled. The engine weighed twenty-eight tons and it pushed twenty double-loaded cars. As it started over from the Canadian side, it soon covered nearly the whole length of the upper floor. "No vibrations whatever," Roebling noted. "Less noise and movement than in a common truss bridge."

A few days later a passenger train made the crossing, going the other way. It was only three cars long, but people were packed into every available bit of space and some were perched on top as well. They went over in "fine style," Roebling wrote. After that, trains kept crossing at a rate of about one every hour, and to get an idea of the vibrations such traffic might be causing, he climbed to the top of one tower and sat there for some time. It shook less, he said, than his own house at Trenton whenever an express went by. "No one is afraid to cross," he wrote his family. "The passage of trains is a great sight, worth seeing it."

For those who stood with him now, the sight must have been no less stirring, even for the engineers—perhaps especially for the engineers. There was, after all, something quite special about a bridge, almost any bridge. Very few were ever outright ugly, and when built right, with everything in harmony, with everything superfluous done away with, with all elements doing exactly what they were supposed to, then a bridge was a thrilling thing to see, with its own kind of graceful majesty, something quite apart from the practicalities of engineering. This was that sort of bridge.

Of course for a nation so recently torn apart by civil war, a bridge was a particularly appealing symbol. But beyond that a bridge seemed such a magnificent example of man's capacity to master the forces of nature, and that, according to the preponderant wisdom of the day, was what the whole age was about. Building a bridge seemed such a clean, heroic thing for a man to do.

At a dinner given in Roebling's honor on the last night at Niagara Falls, General Henry Slocum was asked to give a toast, which he did, saying to great applause that he would gladly forfeit his war record for the bridge at Niagara—"to have been the engineer of that bridge." The general, whose political ambition was very large, sometimes said things he did not quite mean, but his toast would be repeated widely in Brooklyn, and it struck everyone present as a fitting tribute to end the tour on—his impressive war record being the most publicized in Brooklyn. (A general at thirty-three, he had commanded a wing of Sherman's army on the march through Georgia.) The war was past; the time had come to concentrate on "legitimate enterprise"—that seemed the true spirit of the day. A momentous new Age of Progress was dawning and for most of those who raised their glasses to toast John Roebling, as for most Americans, nothing was taken as such proof of that spirit as the works of engineers—". . . the great achievements of the present . . . the strong, light works of engineers . . ."

In Egypt the French had nearly finished the Suez Canal. In Europe the Mont Cenis Tunnel, then the longest on earth, was being blasted beneath the Alps. But nowhere was there so much happening as on the continent of North America. The Union Pacific was laying track at a rate of eight miles a day by this time. In Massachusetts a hole was being bored nearly five miles through the solid rock of Hoosac Mountain, just to slice a little time off the railroad run from Boston to Albany. Boston itself was being doubled in size by filling in Back Bay swamp. In New York Cornelius Vanderbilt was erecting a very grand new Grand Central Depot, the train-shed roof of which, an immense vault of glass and iron, would contain the largest interior space in the country. There was a new tunnel under the Chicago River, a first bridge over the Missouri at Kansas City, and at St. Louis a river captain named Eads had begun building a railroad bridge over the Mississippi.

That such outsize, unprecedented efforts frequently involved watered stock, political jobbery, kickbacks for contractors, and not a little human suffering was either not altogether apparent as yet or of minor concern. So much good was going to come out of so comparatively little evil, it was generally felt, that the evil seemed a reasonable price to pay, and probably inevitable in any event. What really counted was that things were being accomplished at last on a scale in keeping with the commonly held vision of the future. Man the killer, man the destroyer, would be man the builder for now—now and here, on the infinite, seemingly inexhaustible landscape of America. It was the time and place to be intensely, boldly constructive.

In less than a month, when a much publicized golden spike would be driven with humorous difficulty at Promontory, Utah, the completion of the transcontinental railroad would be hailed as "one of the victories of peace." In his way Slocum was saying the same thing. The real glory of American achievement lay ahead, as always. But the true heroes now would be those who made possible such victories of peace—the builders. One of the greatest of them, the architect Louis Sullivan, would later write of his own feelings as a boy at about this same time: "The chief engineers became his heroes; they loomed above other men . . . he dreamed to be a great engineer. The idea of spanning a void appealed to him as masterful in thought and deed. For he had begun to discern that among men of the past and of his day, there were those that stood forth solitary, each in a world of his own."

4
Father and Son

> Nothing lasts forever. The most unforeseen circumstances will swamp you and baffle the wisest calculations. Only *vitality* and plenty of it helps you.
> —WASHINGTON A. ROEBLING

THE BRIDGE tour ended on April 20. "The parties left for the east," reported the Niagara Falls *Gazette*, ". . . traveling in a special car well furnished with refreshments." All, apparently, were still on the friendliest of terms.

In the temporary offices overlooking Fulton Street, work picked up about where it had left off, and with renewed vigor. Directors and stockholders came and went. Orders were placed for drawing tables and filing cabinets and there was a steady tramp of feet on the stairs as inventors came to show patents of tools and machinery, as salesmen arrived with samples of granite or promises of speedy delivery and the best possible price for coal, lumber, sand, or nails. And whenever the senior Roebling was in town and conducting interviews for jobs, the applicants could be seen in the outer office, waiting their turn to go in and sit before the old Prussian and tell him about their special attributes.

A man named William Lane, carpenter, mason, and all-around mechanic, came highly recommended by the Army engineers, Generals Wright and Newton. ("Looks like an energetic good man," Roebling noted.) Charles Kinkel was a German who told Roebling he had had experience with foundations. John Morgan, an English draftsman, wanted employment badly, he said, while William McNamee looked like a "tolerably good man."

And so it went. J. W. Jenkins, a diver, said that if he was hired by the month he would work for twenty-five dollars a day, do whatever blasting was wanted, and provide his own tools. Otherwise his regular day rate was a hundred dollars. Two experienced surveyors, Rudolph Rosa and Colonel William Paine, each wanted ten dollars a day and both were hired on the spot.

Paine in particular seemed exactly the sort of man Roebling was looking for. Self-taught in engineering, he had surveyed the so-called Johnson Route of the Union Pacific, across the Sierra Nevada. During the war, wearing civilian clothes, he had slipped through Confederate lines and worked his way from Washington to Richmond, mapping the location of every destroyed bridge along the way. Lincoln had personally made him a captain of engineers for this and Paine was put on the staff of a major general, a position customarily held only by a West Pointer. By the time the war had ended, he was reputedly the leading topographical engineer in the Union Army. It was said he could prop a drawing board on the pommel of his saddle and as he rode along sketch a map of the surrounding terrain that would be accurate enough to go right to the engraver. He was also well read in chemistry, geology, the natural sciences, and he enjoyed literature, he told Roebling. He was from New Hampshire originally, and he was modest, but very firm and sure of himself. His most notable physical feature was a great sweeping handle-bar mustache. Finding him so early was taken as a good sign.

The one technical chore to be finished up before actual construction could begin was the final survey. A center line had to be located and the responsibility for this Roebling had turned over to his son, to whom Paine was assigned forthwith.

The two worked extremely well together and became a familiar sight in Brooklyn that spring. They took their sightings, hammered down their little iron pins, and worked their way steadily inland from the river, through a neighborhood of narrow shops and warehouses and a terrible tangle of waterfront traffic. As each iron pin went in, young Roebling recorded its location, making a small diagram in the black leather notebook he carried. To show where the center line crossed through the juncture of Fulton, Dock, and James Streets, for instance, he drew the basic outline of the intersection—no easy thing, since the streets, like most in the vicinity, did not join at right angles—then the center line, cutting across the intersection at its own angle. Along this he marked points A, B, C, and D. Point A was a notch he and Paine had cut into the belt course of a yellow house on the corner of Dock Street; B was a crow's-foot chiseled into the curbstone just up from the yellow house; C was an iron pin driven into the crosswalk on Fulton; D was another pin in the middle of the crosswalk, on the opposite side of the street.

St. Ann's Church was their most conspicuous landmark. The historic old building would have to come down eventually. It was one of those numerous pieces of property the value of which had not been included in the engineer's original cost estimate and there were those among its parishioners, as there were elsewhere in Brooklyn, who did not view the two intent surveyors with their brass instruments as necessarily the harbingers of progress.

Through most of this time John Roebling remained in Trenton. He would make an occasional visit of a few days, stopping at the Mansion House on Hicks Street, but the rest of the time he was content to leave things to his son, who by this time had apparently become greatly concerned over the possible verdict of the consultants, and who, like Charles Swan, was expected to keep Roebling regularly posted . . .

Brooklyn, May 21, 1869

DEAR FATHER,

Your Turkish Bath tickets came today.

Maj. King arrived yesterday. The Commission have made their report and sent it to Washington today. I think that if their report was at all favorable, they would not be so quiet about it.

Kingsley proposes to send Genl. Slocum to Washington next week in order to hurry up Humphreys . . .

Yours, Aff.
WASH.

Genl. Wright sent back the map for correction, to have the span put back to 1600 which I did.

Also to cables whether of steel and what diameter.

I said steel with 15″ diam.

WASH.

That he could write so matter-of-factly of Kingsley, the contractor, proposing to "send" Slocum, the United States Congressman, on such an errand suggests that the engineering department had no doubts about how things stood among the Kings County Democrats and that possibly the bridge tour had been something of an education for the Roeblings as well. The P.S. concerning steel for the cables also suggests that Roebling senior had at last made up his mind on this crucial matter and, moreover, that the son felt at perfect liberty to speak for the father on just about everything concerning the bridge.

In any event Slocum did go to Washington and was doubtless a good choice for the mission, since both A. A. Humphreys, Chief of the Army Engineers, and John A. Rawlins, the Secretary of War, happened also to be prominent Civil War figures and were both well known by Slocum. Rawlins was the one who counted. Grant's right-hand man through the war, he was generally respected in Washington for being both candid and decisive. That he was also dying of tuberculosis that spring was not generally known. Rawlins told Slocum not to concern himself, he could expect to see everything approved within a week. Kingsley was elated when Slocum returned with the news, as Washington Roebling dutifully informed his father. But by the end of the week nothing more had happened.

On June 12 John Roebling came on from Trenton to meet with his consultants still one final time, to receive their formal approval in writing. Washington Roebling ordered five hundred copies printed, then wrote to his father, who had immediately returned to Trenton, that General G. K. Warren, his wife Emily's famous brother, was going to Washington and would report back privately "how the matter stands down there."

On June 15 Slocum again saw Rawlins, this time at the sumptuous Brooklyn home of J. Carson Brevoort, where Rawlins was a guest briefly. Rawlins said Grant had told him to do whatever he liked about the bridge—the subject did not much interest the President, one would gather—and Rawlins guaranteed the whole business would be settled as soon as he got back to Washington. Taking no chances, Slocum once more was on his way to the capital, accompanied this time by Henry Murphy. "When Mr. Murphy returns we will have an authentic report," Washington Roebling wrote to his father, which might be taken to mean there was some skepticism between them concerning Henry Slocum or that they simply thought very well of Henry Murphy.

Rawlins was as good as his word. On June 21 General Humphreys informed Murphy by letter that Rawlins had approved both the plan and the location of the bridge so long as it conformed to certain basic conditions stipulated by the Army Engineers.

The center of the river span was "under no conditions of temperature or load" to be less than 135 feet "in the clear above the mean high water of spring tides." Nothing could be added that might project out from the towers and no guy wires were to be attached that might hang below the river span. (At both Niagara and Cincinnati Roebling had strung such wires below the bridge floor.) The river was to be kept perfectly clear, in other words, and the roadway of the bridge would have to be raised five feet higher than Roebling had intended.

Such seemingly small changes called for some rather serious revisions, however. The increased elevation would mean an increase in the grade of the approaches and land spans, which right away meant an additional cost of some $300,000, as near as the Roeblings could figure. To avoid all this it was decided to change the iron superstructure of the bridge floor. The stiffening trusswork would be built entirely above the roadway, instead of partly above and partly below as in the original design. And just to be certain that nothing projected beyond the pier lines, the length of the river span was extended from 1,600 to 1,616 feet. It was also decided to widen the bridge floor by five feet, to make room for two double roadways for vehicles, instead of single roadways as the elder Roebling had originally planned.

Such alterations were of little or no interest to the general public, but they were no trifling matter for the engineers and they would alter the looks of the finished structure.

Humphreys' letter was not a very long or impressive document, but it signaled the conclusion of bureaucratic red tape and was big news in Brooklyn. "THE ROEBLING PLANS FULLY ENDORSED" ran the headline in the *Eagle* the evening of June 25. The paper carried the complete text of the final report by the consulting engineers and concluded that now, with all obstacles at last out of the way, the work could commence.

Three nights later, at the Brooklyn Athenaeum, Congressman Demas Barnes delivered a lecture on the bridge before an audience "notable for its large representation of solid businessmen," who listened "with the most evident interest and attention." Barnes, who had made a fortune in patent medicines, had been the strongest voice for the bridge on the floor of the House. This night he began his talk with an impassioned description of Brooklyn and its future, from which he moved to the bridge itself, speaking with equal ardor. Then, for his grand finale, he proclaimed the following, summing up, it would seem, all that was so fervently felt, all the common expectations, concerning the Great Bridge:

> This bridge is to be built, appealing as it does to our pride, our gratitude and prosperity. When complete, let it illustrate the grandeur of our age; let it be the Mecca to which foreign peoples shall come. Let Brooklyn now take up the pen of progress. Babylon had her hanging gardens, Nineveh her towers, and Rome her Coliseum; let us have this great monument to progress.

But that same day, Monday, June 28, 1869, beside the Fulton Ferry slip, John A. Roebling had been involved in an accident, which, though extremely painful, seemed of no serious consequence.

The mental torture after the accident had been nearly as severe as the physical, according to his son, who had been with him almost constantly. "He felt at his age he could ill afford to lose any time: this circumstance, combined with the prospect of being crippled to some extent, had a most depressing influence on his spirits."

To have been struck down by such a foolish mishap did his spirits no good either. It was the sort of slip a new man might make, or one of the politicians or moneymen who invariably had to be conducted about bridge jobs. When he thought of the risks he had taken, the countless dangers he had exposed himself to over the years, to be felled this way was positively infuriating.

The afternoon of the accident had been clear and pleasantly warm in Brooklyn. He and Washington had been working since morning at the foot of Fulton Street, beside the ferry slip, where the Brooklyn tower was to go. He had come down to the waterfront to assist Washington and Colonel Paine in fixing the precise location of the tower. Paine had been over on the other side of the river, signaling to them.

At one point Roebling was standing as far out on the ferry slip as he could get, atop a cluster of piles. Seeing one of the boats approaching, he stepped back off the piles and onto a stringpiece, or beam, that was wide enough to get a footing and where, he supposed, he would be clear of the piles should they be forced against the beam by the docking boat. But there had been a knot—or something he had not noticed—sticking out from one of the piles and it had caught his right foot as the boat ground against the rack, crushing the tip of his boot and his toes.

The pain must have been excruciating, but he gave no sign of it. He went right on shouting directions until he toppled over, unable to stand any longer.

Washington rushed him to a doctor's office close by, where his father was no sooner in the door than he was telling the doctor what to do. He demanded a tub of cold water and plunged his foot into it to staunch the flow of blood. Other doctors were called in for an opinion and it was agreed that his toes would have to be amputated. To this Roebling promptly consented and requested that the operation be performed without anesthetic. When it was over, he insisted on binding the wound himself and in his own fashion. Then he was taken to his son's house on Hicks Street.

For several days there were no public announcements as to how he was getting along. But on July 8 the Brooklyn *Eagle* reported that he was busily engaged on his plans and drawings for the bridge, and that the injured foot had been so placed that a steady stream of cold water poured over it night and day. "The distinguished engineer has his notions about surgical treatment, and seems to be very stoic in regard to physical pain," the article said. "He thinks and talks of the bridge as incessantly as ever, and seems unwilling to have the conversation of his professional assistants diverted for a moment to his own accident." In another ten days, it was claimed, he would be out surveying again.

Dr. Brinkman, the family physician, came up from Philadelphia, and a Reverend John C. Brown from Trenton made a special trip. It seemed strange luck, the preacher told Roebling, that he should be laid up at the start of so great a work. "There is no such thing as chance," Roebling is supposed to have replied. "All is wisely ordered."

But in another week reports were he had taken a turn for the worse, though there was no mention of what was by then known inside the Hicks Street house.

Roebling, predictably, perhaps inevitably, had taken charge of his own case. He had fired one Brooklyn doctor, then another, much against his son's wishes, and though he seems to have tolerated the presence of Brinkman, he never paid any attention to him. Now things were not going at all well. Signs of tetanus had been detected. It would be commonly said later that had he obeyed the doctors, he would have recovered. "But Mr. Roebling was a man of indomitable will and perseverance," the *Eagle* would explain, "and the counsels of his friends were as naught."

For eight days, from July 13 on, Roebling suffered intensely. Medical experts would agree when it was all over that only a very tough and determined man could have endured what he did that long.

At first he had become extremely restless, complaining of savage headaches. But presently he began having trouble swallowing. After that there was no mistaking what was wrong with him. The muscles around his face, neck, and jaws grew rigid as iron. Within a day or so his eyebrows were permanently fixed in a raised position and his mouth was pulled back in a terrible grimace, the teeth all showing and locked tight. He was unable to eat anything solid, or to talk, but he kept scribbling notes to Washington and the others attending him, instructing them on his proper care.

Then the hideous seizures began, set off by the slightest disturbance. His room was kept dark, the long shades drawn against the July sun, and everyone who had reason to go in or out did so as softly as humanly possible. But then a window shade would rattle in the breeze or someone would inadvertently brush against the side of his bed, a door would squeak or there would be a noise from the street below, and he would go into a convulsion, the sight of which was something they would all live with the rest of their lives. All at once his whole body would lift off the bed and double backward with a fierce, awful jerk, his every muscle clenched in violent contraction. Sweat streamed from his body, but he made no sound, not even a groan, because during the spasm his whole chest wall was frozen hard.

He was being horribly destroyed before their eyes and there was not a thing any of them could do about it. Moreover, as nearly always happens with lockjaw, his mind remained as clear as ever, and this made the sight of his suffering all the more unbearable. They all knew the terrible, titanic battle going on behind those blazing eyes and the ghastly smile that stayed fixed like concrete on his ashen face throughout everything that was happening to him.

When the seizures passed, he generally slipped into a coma. But even toward the end, there were hours when he would lie there perfectly still in the darkened room staring straight up at the ceiling, one of his family sitting motionless beside him. During the final few days there were tears streaking down his face.

The watch went on hour upon hour. Downstairs, visitors came and went, talking in whispers. They were told their concern was deeply appreciated, that there was nothing they could do but pray, and they went away to tell others what they had heard about the particulars of his condition, which was very little.

But on the evening of July 21, quite contrary to all the professional forecasts, the patient took a turn for the better. With paper and pencil he began giving instructions to his nephew, Ed Riedel, on a special contrivance he wanted built to lift him up and move him about his bed. He made a sketch, explained how it should be done, and told the young man to get at it immediately. Through the rest of the night he kept issuing orders on a variety of matters, including the bridge, and a wave of hope swept through the house, until sometime after midnight, when it became clear from the things he was scribbling down that his mind was going. He thought he was back at the bridge office.

About three in the morning he had a convulsion so violent that he leaped clear from the bed and was caught in the arms of C. C. Martin, the assistant engineer, who with Washington and one or two others was standing watch at the time. Within minutes Roebling was dead.

Then in the gray light before dawn, Thursday, July 22, the undertaker arrived and an artist who had known Roebling in Cincinnati was called in to take a death mask.

The afternoon edition of the *Eagle* had the full story. Roebling was called a martyr, while in virtually the same breath the editors assured their readers that there was still great hope for the bridge. The implication was that the success of the bridge had been more or less assured now that it had claimed a life, like the bell in the old story that would not ring true until it had been cast of molten iron into which a man had fallen. Some people were saying the only safe bridge was one that had taken a life and stories were told of the lives sacrificed in the building of famous bridges of ancient times. The *Eagle*, for its part, said this:

> He who loses his life from injuries received in the pursuit of science or of duty, in acquiring engineering information or carrying out engineering details, is as truly and usefully a martyr as he who sacrifices his life for a theological opinion, and no less honor should be paid to his memory. Henceforth we look on the great project of the Brooklyn Bridge as being baptized and hallowed by the life blood of its distinguished and lamented author.

Flags were flown at half-staff all over Brooklyn, and when it came time to take the body down to the ferry, to start the trip to Trenton, there was slow going in the streets because of the crowds. As a subject of popular interest, Roebling seemed a more notable success dead than alive. His training, all his ambition and ability, his entire life's work had been building toward this greatest of bridges and he had not lived to do it—that was a tragedy people could readily understand regardless of how little previous interest they may have had in either the man or his work.

Word of Roebling's death reached Trenton early the same morning he died. Within hours the whole town knew about it, and though there had been ominous talk of his condition for days, no one seemed quite ready to accept the fact that the worst had happened. Talk of Roebling dead was one thing, but the idea of him laid out in a black suit of clothes like any other man, those pale eyes shut forever, was something else. Somehow, it was felt, he would figure a way.

But by nightfall Saturday, when the body arrived, the truth had long since sunk in. Nobody had any doubts that the extraordinary life of John A. Roebling was over and plans had been laid for the biggest funeral in Trenton's history.

The eulogies began that night at a special town meeting. Judge Scudder, General Rushing, and Charles Hewitt spoke, as did Reverend John Brown, who said that though Roebling was known the world over as a man of science, he ought to be remembered as a gentleman all the same. Then early the following morning, in twos and threes, some leading children, people began gathering outside the Roebling house.

Separated from the wireworks by a narrow strip of lawn, the house was a tall spacious affair, with some twenty-seven rooms, walls two feet thick, and few frills. Roebling had designed it himself before the war, in the Italian style and more for comfort than show, except for the glassy cupola on top. It had stood raw and pink-looking when it was first finished, taller even than the mill in those days, with nothing but bare fields to either side. But in the time since, Roebling had had it stuccoed over, the mill had more than doubled in size, and the trees he planted had closed in most of the property. In summer, only the windows of the cupola could be seen riding high above the treetops. They were the first windows in town to catch the morning sun.

The grounds themselves were neatly set off from the street by a tall iron fence. Flowers bloomed through the whole summer. Grapes hung from elaborate trellises. There were boxwood hedges, a handsome barn, an icehouse, and an especially fine orchard that he had been extremely proud of, adding to it year by year. As might be expected, everything was kept just so.

The house faced onto the street, a railroad track, and the old Delaware and Raritan Canal, which all ran side by side, parallel to the river. Past the canal and the state prison, the land sloped away toward the ironworks and the river. That part of town was all built up now, but behind the house, on the other side of the orchard, was a broad, flat wheat field that was just beginning to turn color.

By ten o'clock the small cluster of onlookers had grown big enough to fill the front lawn and most of the street. Carriages approaching the house had trouble getting through. But there was little commotion. The time passed about as quietly as on any Sunday morning, broken only by the sound of church bells from across town. Already the temperature was near eighty as the sun climbed into a cloudless bowl of summer sky. Nothing like this had ever happened in Trenton. Estimates were that perhaps two thousand people were gathered on the front lawn.

Inside the house the entire family was assembled—a rare thing for the Roeblings—surrounded by the books and paintings he had collected, the marble statuary and the steel engravings of his bridges. At eleven the doors were to be opened to the crowd outside, but for the time being, except for the servants, they had the house and its memories all to themselves.

With Washington Roebling, now head of the family, was his pretty and alert-looking wife, Emily, who had been a special favorite of her father-in-law's. He had admired her for her energy and intelligence, often showing her a degree of kindness seldom granted his own children. After her son had been born, she had written to Roebling in an affectionate letter from Germany, "The name of John A. Roebling must ever be identified with you and your works, but with a mother's pride and fond hopes for her first-born I trust my boy may not prove unworthy of the name . . ."

Then there was Ferdinand Roebling, slight, fine-featured, bespectacled, and now twenty-seven. This was the only one of his sons, John Roebling used to say, who had the makings of a merchant. His oldest boy, the bridgebuilder, he had ordered off to war, but Ferdinand had been kept at home, Ferdinand's services to the wire business being too valuable to spare, according to John Roebling.

Charles, younger still by seven years, was a strangely silent, thoughtful young man, whose chief interest was flower gardening and who was home for the summer from Troy, where he was a student at the Rensselaer Polytechnic Institute, like his oldest brother before him, and not particularly happy about it.

Edmund, or Eddie as he seems to have been called by most of them, was fifteen, very shy and uncertain-looking, and still a great worry apparently.

The sisters were Laura, Josephine, and Elvira. Laura was the oldest after Washington. She had dutifully played the organ at the German church every Sunday and married a "good German," a Mühlhausen man at that, her Mr. Methfessel, as she called him. They had a number of children and lived on Staten Island, where Mr. Methfessel had started a school and where they would have failed to make ends meet by this time had it not been for the checks she received regularly in the mail from her father.

Josephine was now the wife of Charles H. Jarvis, one of the finest American pianists of the time, and Elvira, the last of the three to be married, was Mrs. John Stewart. Elvira had always been the most playful and high-spirited member of the family, the least like her father in this respect and the one whose company he most enjoyed. Her wedding had taken place only a few weeks before, in the same large front parlor where his corpse was now on display. All that spring, as he went back and forth to Brooklyn, he kept bringing home expensive gifts for her, dresses he had picked out at A. T. Stewart's, hundreds of dollars' worth of silks, fancy carpets, Tiffany silver. A few days before the wedding he had insisted that she take a hundred dollars in cash, to have with her on her wedding trip. Then he had given her away to young Stewart in a room full of guests, several of whom would comment at the funeral on how exceptionally genial and good-spirited he had seemed then.

Very little is known about the new Mrs. Roebling, except that she was the former Lucia Cooper of Trenton and that their wedding had taken place in February of 1867. He had presented her with two gold bracelets and a painting by Rembrandt Peale and in his cashbook he entered $125 as the cost of the wedding trip, about which he made no other notation that is known of. Two years later she had still not been fully accepted by the family, nor would she be. Washington Roebling would write that after the summer of his father's death, he never saw her again.

And finally, there was Charles Swan, who, in a photograph taken some years later, sits in a stiff, upholstered chair, looking quite well upholstered himself as he focuses directly and amiably on the camera, every inch the solid, kindly, dependable man, it would appear, John Roebling's sons would say he was. That Roebling also appreciated Swan for what he was and all he had done seems clear, despite the cold formality of their working relationship. For when he sat down to write a new will in 1867, after he remarried, he included twenty thousand dollars for Swan and the clearly stated wish that his sons take Swan into the business as a full partner.

The contents of the will would not be made public for several days, but for the family its general outline had been known for some time. In addition to the money for Swan, Roebling had left some eighty thousand dollars for distant relatives and several charities. The bulk of his estate he had split eight ways, between his new wife and his seven children—except that he deducted from each child whatever money had been advanced to him during his lifetime. Year by year, in a private ledger, he had carefully itemized his expenditures for his children, down to the penny, and now in his last summing up he docked them each accordingly.

The wire business he left to his four sons, requesting that they keep the name John A. Roebling's Sons.

At eleven sharp the house was opened to the public and the crowd started moving for the front door. For the next two hours the town was permitted to pay its respects. The people passed through the dim front parlor in slow single file, in dark Sunday dress and funeral veils, the men with hats in hand, moving with almost no sound at all up to the rosewood casket with its huge silver handles, then on out through the back way, most of them glancing this way and that, trying to see as much as possible without appearing disrespectful.

Nearly everyone thought the Brooklyn undertaker had done extremely well. The body did look emaciated, it was agreed, and the massive brow stood out more even than in life, but when they considered the horrible way the old man had died, and the July heat, most of those who filed by thought he looked quite himself, perhaps even at peace, which was what seemed most unlike him of anything. For the majority of the people in line, it was a chance for a first real look at the man close up.

At one o'clock the front door was closed again. Then shortly after one, as the crowd regathered under the shade trees, the quiet was suddenly shattered by the shriek of a train whistle. People later described it as the most dramatic moment of the day.

Up the tracks crept a special train from Jersey City, five cars long. Like the one the night before that had brought Roebling's body home, it steamed slowly to the front gate and stopped. Then down stepped the delegation from Brooklyn and New York, some fifty or sixty men, most of them in high, shiny silk hats. They stood about in a cluster beside the train, squinting against the sunshine, until the last of them had gotten down. Then they started up the front walk in a body, the crowd making way for them. At the front door Washington Roebling stood waiting to usher them inside.

The services began at two, on schedule, the family sitting in the upstairs hallways, the guests crowded into neat rows of chairs set before the casket. The heat was terrific, fans were going at a great speed in gloved hands as four ministers—one Presbyterian, one Lutheran, two Episcopal—took turns with the services at the foot of the stairs. There was little out of the ordinary said. The Lutheran spoke in German.

Of the eight pallbearers who took the casket out the front door, four were Trenton men; the others were all associated with the New York Bridge Company—Julius Adams, Horatio Allen, Andrew H. Green, and Henry Cruse Murphy. That a large part of the funeral expenses would also be met by the New York Bridge Company, or more specifically by William Kingsley on behalf of the New York Bridge Company, was privileged information at this point.

The clergy and immediate family led the procession to Mercer Cemetery, riding in special carriages. The Brooklyn delegation came rolling along after, followed by the Board of Trade in still more carriages. Two hundred men from the Roeblings' rolling mill had marched up from South Trenton and they fell in behind, while the men from the wire mill walked two by two alongside the carriages. Estimates were that fifteen hundred people joined in the march, counting all the professional and trade associations, the orphans, and the singing societies. With everyone under way the whole procession stretched out more than a mile and a half and along the entire line of march—out Green Street, East Street, State and Clinton—sidewalks and doorsteps were thick with silent onlookers.

By four it was all over. John Augustus Roebling had been committed to eternity, beside Johanna Roebling and two of their children. Again the distinguished visitors from Brooklyn and New York were gathered beside their train, all looking a little worse for the dust and heat, each offering his own polite, soft-spoken farewell to the Roebling family, and to Colonel Roebling in particular. One by one they stepped forward to shake his hand and to wish him well—Henry Murphy, Henry Slocum, Horatio Allen, Julius Adams, Thomas Kinsella, Demas Barnes, Colonel Paine, C. C. Martin, William Kingsley. Then as he and the rest of the family turned and walked back through the gate and up the path to the big house, the Jersey City train rolled out of Trenton, gradually gathering speed as it broke into open country.

There is no record of what was talked about during the return trip to Jersey City and that is a great shame. There was quite a lot to be discussed, obviously enough, and nearly everyone who should have a say was present, and with nothing better to do. It was the sort of opportunity a politician seldom lets pass, and since the majority of them were politicians in one way or other, it is hard to imagine the time being wasted.

They had all been together on the ride down that morning, of course, but then, with Roebling not yet in his grave, any open talk about getting on with his work would have been considered out of line. Now the atmosphere was quite different, no doubt, and it seems reasonable to assume that as their train went steaming along through the late summer afternoon a number of highly interesting conversations were being conducted.

Years later it would be said that Roebling's death left everyone in a terrible quandary over who should take his place and that there were grave doubts about going ahead with the idea. "With its inspiration gone, the Brooklyn Bridge seemed impossible to build," one biographer would claim. But the truth is there was never any doubt at all.

As William Kingsley would reveal, Roebling had long since talked to him and to Henry Murphy about his son replacing him eventually. Kingsley even said Roebling had wanted his son in charge from the start, but that he, Kingsley, and the others would have none of that. Be that as it may, the very day of Roebling's death, Thomas Kinsella had stated in no uncertain terms on the editorial page of the *Eagle* that Washington Roebling would take up right where his father left off and that no man was better equipped for the job.

> Not long since, before the accident, which led to his death, Mr. Roebling remarked to us that he had enough of money and reputation. And he scarce knew why, at his age, he was undertaking to build another and still greater bridge. His son, he added, ought to build this Brooklyn bridge—was as competent as himself in all respects to design and supervise it; had thought and worked with him, and in short was as good an engineer as his father.

As a matter of plain fact those numerous different parties who wanted the bridge built for their numerous different reasons had been left with little choice but to go ahead with the young engineer. Moreover, to their way of thinking there was no good reason why they should not, and he himself, years afterward, would say there were three very good reasons why they should:

> First—I was the only living man who had the practical experience to build those great cables, far exceeding anything previously attempted, and make every wire bear its share.
>
> Second—Two years previous I had spent a year in Europe studying pneumatic foundations and the sinking of caissons under compressed air. When the borings on the N.Y. tower site developed the appalling depth of 106 feet below the water level all other engineers shrank back . . .
>
> Third—I had assisted my father in the preparation of the first designs—he of course being the mastermind. I was therefore familiar with his ideas and with the whole project—and no one else was.

He was also a very young man, which perhaps he ought to have added as reason four. He had that vitality his father prized so and that in his last years had come to be a thing only to hope for in the next life. "After your spiritual birth, did you feel like a new being," he had asked the spirit of his dead wife, "*young*, *energetic* and full of life?"

And beyond that it seems Washington Roebling had struck just about everyone with a say in the decision as quite a solid individual in his own right. The consulting engineers could vouch for his professional abilities. He had been a soldier and an exceptionally good one, which was also taken to be much in his favor. And on a strictly personal level he was simply a whole lot easier to talk to than his father had been and would probably be a whole lot easier to work with.

Indeed, the gentlemen from Brooklyn must have been most favorably impressed with Washington Roebling, considering what they were about to risk on him. It was true, just as he said, that he was the one man—the one and only man—in the country capable of building the unprecedented bridge his father had designed, but that of course meant that everything depended on him alone. It meant that unlike his father, he had no one standing by ready to take his place should anything happen to *him*, or between him and his employers.

Still, if the matter of a successor was self-evident and already settled, the death of John A. Roebling had raised other complications that remained quite unresolved. There was, for example, the vital question of public confidence in the work. The older Roebling's word had counted for something, among his peers as well as the general public, but even he had had to face a storm of protest. It had been necessary for him to resort to a committee of experts to testify to his judgment. How many more critics might surface now, now that he was no longer available to answer for his radical schemes? Public works of such magnitude demanded the smooth turning of many wheels, and wheels within wheels, a number of which were often carefully, cleverly concealed, and a collapse of public confidence could lead to all sorts of difficult, embarrassing complications.

John Roebling had known a great deal about the genesis of the bridge idea, about Brooklyn history and Brooklyn politics and who had the power. He also knew the role money played in getting things accomplished. Money had always been a secondary interest to him personally. He had made quite a lot of it, to be sure, but it had never been life's chief objective and he had little time for anyone who thought it was. Nonetheless, he knew the lengths some men would go for it and he himself had never been adverse to playing to that side of human nature if it suited his purposes. When he called for the building of the Great Central Railroad in Pittsburgh, for example, he described how the West was "ready to pour rich treasure into our laps," just as in Brooklyn he had pictured a toll bridge so lucrative that it would pay for itself—all six to seven million dollars—in just three years, which even some of his most ardent admirers in the Bridge Company recognized as foolishness. The bridge, after all, was to be built by a private corporation; it was a business proposition, just as the Allegheny River and Cincinnati bridges had been, and he took that as a matter of course.

But somewhere along the line he had found out more than he had known at the start, more perhaps than he had wanted to know about the ideas a few of his clients and their friends had for making money with his bridge—or at least so it appears from comments made considerably later by his son Washington. ". . . At the time of his death he was already arranging to retire and relinquish the work to me," Washington Roebling would write privately to a correspondent. ". . . You may not be aware that this bridge was started by the infamous 'Boss Tweed Ring' for the sole purpose of using it as a means to rob the cities. When this fact began to dawn on my father's mind he made up his mind to get out."

The statement is not quite accurate—the bridge had not been "started" by the Tweed Ring, nor is there any indication that either Washington Roebling or his father ever wrote or said anything to the same effect back at the beginning of the work. Nor is it known how much Washington Roebling himself knew at the time of his father's death, at the time he stepped into his father's place.

But for a number of those who were speeding toward Jersey City that late July afternoon in 1869, the full story was very clearly known; Brooklyn and its dreams of a bridge were essential elements in their own life stories and dreams.

It is intriguing to note what Thomas Kinsella said in the *Eagle* the day after the funeral. Possibly his remarks had nothing to do with his feelings about things said on the ride to Jersey City, but then again, possibly, that may have been exactly what he had in mind.

"The great boast of this land," he wrote, "is twofold—the political works of the [Founding] Fathers, and the material triumphs of science, of which Roebling was, with scarcely any exception, the greatest hero." But the politician of the present, he went on, was nothing more than "a thing of tricks and dodges." About all the modern-day politician could do was to undo "the grand creation of former days." The politician's words and deeds were as nothing, he said, when compared to the works of a man like John A. Roebling.

"One such life as Roebling's was worth more than those of a whole convention full of jabbering and wrangling politicians." Concerning politicians, Kinsella could speak with some authority, his Brooklyn readers knew, for he was one himself.

5
Brooklyn

A great future is opening before our city.
—From *The Brooklyn Eagle*, 1869

BROOKLYN in that high summer of 1869 was still a city quite unto itself, with its own paid fire department, police, schools, and a fierce local pride of a kind usually associated with smaller, less worldly places. It had sprung forth all at once, as suddenly as a mining town, on the western tip of Long Island, in King's County, New York, its population increasing a hundredfold in less than a lifetime. A man like Henry Cruse Murphy could readily recall when Fulton Street was lined with giant elms and an eccentric Hessian gunsmith named John Valentine Swertcope was free to go prowling about Washington's old fort on the Heights, popping away at songbirds. For nearly two hundred years, from the time it was first settled by Dutch farmers in the early seventeenth century, Brooklyn had changed hardly at all. At the start of the nineteenth century, when talk of a bridge first began, there had been fewer than five thousand people in the entire county, more than a thousand of whom were not there out of choice, being black slaves. Now there were close to 400,000 people who called Brooklyn home. In the words of one little guidebook, Brooklyn had been "transformed" in a generation "from insignificance into metropolitan importance."

Brooklyn's population was still less than half that of New York and among a good many New Yorkers it was regarded as a backwater, a familiar enough neighboring horizon, with its ships and church steeples, a place to go hear Beecher or to be buried at Greenwood perhaps, but a hinterland and scarcely worth mentioning in the same breath with New York. But Brooklyn, in fact, was the third-largest city in America and had been for some time. It was a major manufacturing center—for glass, steel, tinware, marble mantels, hats, buggy whips, chemicals, cordage, whiskey, beer, glue. It was a larger seaport than New York, a larger city than Boston, Chicago, St. Louis, San Francisco, and growing faster than any of them—faster even than New York, according to the *Eagle*—even without a bridge.

Already Brooklyn covered an area of twenty-five square miles, which made it larger than the island of Manhattan.

City Hall, with its attendant law offices and chophouses, was the political center of town. The white marble building with Greek columns and new cupola stood on a pie-shaped plot at the juncture of Court, Joralemon, and Fulton Streets, and there, most any day, could be found Brooklyn's own Commissioners of Water and Sewerage, the Street Commissioner, the City Auditor, the Comptroller, the Keeper of the City Hall, numerous frock-coated aldermen, and the Honorable Martin Kalbfleisch, Mayor, a vain, hard-drinking, foulmouthed little Democrat who would go down in history as "an enigma to the respectable and a delight to the reporter."

From City Hall, Brooklyn, to City Hall, New York, was less than two miles, but the pulsing salt river between them was a dividing line in more ways than one. The Brooklyn side was still strictly the domain of the Kings County Democrats.

Fulton Street, Old Ferry Road in earlier days, was the business district, Brooklyn's Broadway, it was said, but really more like its Main Street. From City Hall, Fulton Street sloped off a mile or so to the river, where it ended abruptly and its horsecars made their turnaround in front of the ferryhouse. The *Eagle* had its offices at the foot of Fulton Street, just up from the ferryhouse, as did the *Union*, the Republican paper. The banks and insurance offices were there, along with such up-to-date stores as Ovington Brothers China House or Frederick Loeser's (ladies' wear and "trimmings").

Unless a person lived on Long Island there was only one way to get to Brooklyn in 1869 and that was by ferry. Fulton Ferry was the one people meant when they talked of the Brooklyn ferry; it was the

"Gateway to Brooklyn" and the one Whitman immortalized in his poem. But it was only one of five different lines, all operated by the Union Ferry Company, each of which had its own slips and ferryhouse and was named after its destination on the New York side. (South Ferry ran from the foot of Atlantic Avenue in Brooklyn to South Street at the tip of Manhattan, for example. The Wall Street Ferry departed at the foot of Montague Street and Fulton Ferry ran from Fulton Street, Brooklyn, to Fulton Street, New York.) In all, thirteen boats were kept steaming back and forth, night and day, making something over a thousand crossings in twenty-four hours. They had names like *Mineola*, *Montauk*, *Clinton*, and *Winona* and were 150 and 170 feet or more in length, double-ended, and about six hundred tons on the average.

"What are these huge castles rushing madly across the East River?" wrote a visiting Englishman. "Let us cross in the *Montauk* from Fulton Ferry [to Brooklyn] and survey the freight. There are fourteen carriages, and the passengers are countless—at least 600. Onward she darts at headlong speed, until, apparently in perilous proximity to her wharf, a frightful collision appears inevitable. The impatient Yankees press—each to be the first to jump ashore. The loud 'twong' of a bell is suddenly heard; the powerful engine is quickly reversed, and the way of the vessel is so instantaneously stopped that the dense mass of passengers insensibly leans forward from the sudden check."

Once a ferry had landed and its passengers were ashore, the loading gates at the ferryhouse swung open and the waiting room emptied with a sudden rush of clerks and shop girls, day laborers with dinner pails, butchers, storekeepers, delivery boys, bankers, and business people. Outside, the street swarmed with more crowds coming and going, with vendors, dock workers, carriages, carts, farm wagons, and the clanging Fulton Street horsecars. It had been a long time now since the ferry captains and ticket boys knew their regular passengers by sight. One New Yorker who visited Brooklyn and went away quite impressed by the place also commented that he would just as soon stay in New York if living in Brooklyn meant riding on an East River ferry.

Upstream from the ferry slip and the spot where the colossal bridge tower was to rise were the Catherine Street Ferry; the Navy Yard, at Wallabout Bay; the Havemeyers and Elders sugar refinery, which on foggy mornings looked like a great Rhenish castle at the water's edge; the Roosevelt Street Ferry; and Brooklyn's famous old shipyards. Henry Steer's and Webb & Bell were builders of clipper ships before the war. At Samuel Sneeden's the Swedish genius Ericsson had built the *Monitor*.

Generally speaking, the East River was considered the best part of the harbor of New York. It had deeper water for wharves than along the Hudson, or North River,.as it was also known; it was less affected by prevailing winds, a little less troubled by ice. It was also the safest, most desirable place to build or repair ships and for this reason the Roebling bridge was still a bone of contention along the river. With the yards on the New York side taken into account, the shores of the East River represented one of the greatest concentrations of shipbuilding anywhere on earth.

Downstream from the ferry the waterfront ran beneath the brow of the Heights, on past Red Hook Point, clear around to Gowanus Bay. All told Brooklyn had nearly eight miles of piers, dry docks, grain elevators, and warehouses. The new Atlantic Basin, on Buttermilk Channel, was forty acres in area. More ships tied up in Brooklyn now than in New York and Hoboken combined. From the river the city looked as though it were enclosed behind a protective screen of ship masts and rigging. The sea lanes of the world ended at Brooklyn, an admirer of the city would write years later, but it was as true in 1869, and it was the sea, as much as anything, that gave the place its tone and distinction. Gulls wheeled and cried over the housetops. Sailors mingled with the evening crowds along Fulton Street. The salt air, reputedly, was "pure and bracing . . . wafted from a thousand miles seaward."

From half a dozen different high points, from Prospect Park, for example, or from Greenwood Cemetery, the world opened up in all directions and to the south was the Atlantic breaking on the shores of Coney Island. Brooklyn, it was claimed, offered "the most majestic views of land and ocean, with panoramic changes more varied and beautiful than any to be found within the boundaries of any city on this continent," and apparently that was no exaggeration.

Certainly the view from the Heights was as fine as anything on the eastern seaboard—a sparkling blue and green sweep of 180 degrees, taking in river, bay, Manhattan, the Jersey hills, Staten Island. There were ships everywhere one looked, making for port, heading out to sea. On any summer day in 1869, when the age of sail and the age of steam still overlapped, river and harbor were a ceaseless pageant. New York was the principal reason for most all of it, of course, but Brooklyn had the view.

Old engravings of New York harbor generally show the boats all out of scale, too big, that is, but the shape and nature of the various species represented are a great deal clearer that way and the over-all effect considerably more enjoyable. To judge from such views, there must have been few places on earth where a city dweller could drink in quite so much space and sky or see so much going on that was so everlastingly interesting to watch. The water is filled with schooners, packets, pleasure yachts, gleaming white excursion steamers the size of hotels, and giant iron-hulled, ocean-going sailer-steamers like the new *City of Brooklyn*, the latest and largest ship on the Inman Line. (At a banquet served on board the *City of Brooklyn* that spring, at the end of her maiden voyage, the spirit of good fellowship was such, reportedly, that Beecher broke bread with the Democrats.) Freight-car lighters, hay barges, sand barges, countless steam tugs move back and forth, up and down the river, and everywhere, cutting between them, sidling off crab-fashion against the tide, are the Brooklyn ferries.

It was a prospect to cleanse the spirit, no doubt, to put things back in proper balance at the end of a long business day. From such a vantage point, New York was clearly not all there was to life on earth. Even the ferries looked like nothing more than clever toys, perfect in every detail, down to the feathers of coal smoke trailing from each funnel. After dark, with their colored lights, they gave the river "a gala appearance."

A perfectly healthful place in all seasons and in all respects, Hezekiah Pierrepont had said of these gentle bluffs on the river, the heaped-up leavings of the last of the glaciers. The salt air filled the lungs, and to the rear stretched Long Island, a hundred miles of open country. An enterprising brewer with large cedar-dotted holdings on the Heights, Pierrepont, fifty years before, had advertised lots "for families who may desire to associate in forming *a select neighborhood and circle of society, for a summer's residence, or a whole year* . . ." Gentlemen whose business or profession required "their daily attendance in the city" could not do better, he said. His lots were 25 by 100 feet, many fronting on the river, others on "spacious streets 60 feet wide." By 1869 shade trees made green canopies over red brick sidewalks, upon which fronted some of the stateliest houses in America. As neighborhoods went, there was nothing in New York to compare to it. The Heights had become everything the brewer promised, all that the name implied.

Few in Manhattan could match Willow, Pierrepont, or Clinton

Streets, or Columbia Heights, the street running parallel to the river. Built of brick or brownstone, with rows of tall windows, the houses ran "plump out" to the sidewalk, almost without exception. Most of them were quite grand in dimension, beautifully detailed, with marble sills and cast-iron stair rails. Some, such as the Low place on Columbia Heights, were mansions by any man's standards. But there was little of the flamboyant display soon to characterize Fifth Avenue. No one house seemed designed for the express purpose of upstaging its neighbor. As the *Eagle* observed, "Almost everybody appears to have built his house like somebody else."

The Heights was the unchallenged social, cultural, and moral center of Brooklyn life, with the social and moral part of things taken the most seriously. It was also, one would gather, about as pleasant and lovely a place to live as there was to be found in urban America, then or since. It was not Brooklyn, but it was very often taken to be.

There on the Heights lived the oldest, wealthiest Brooklyn families—the Pierreponts, the Brevoorts, the Lows; A. S. Barnes, the book and hymnal publisher, with his family of ten children; Simeon Baldwin Chittenden, Moses Beach, Gordon Ford. They were second-generation New Englanders, in the main. They were the people who gave habitually to charity drives and figured on the boards of various Brooklyn institutions. Their names on a directors' list or an incorporating charter meant eminent respectability. They employed the best cooks, sent their sons to Yale or Columbia. On spring evenings along the shore drive to Fort Hamilton, they could be seen riding with "elegant equipages, well-dressed grooms, and spanking teams." As it happens, most of them were not around that summer of 1869. They had packed off weeks earlier, as was their custom, moving out en masse nearly—children, servants, steamer trunks, picnic hampers—to Oak Bluffs, Newport, Saratoga, or the White Mountains. All through July and August the *Eagle* carried regular columns to report their doings.

The heads of such families generally worked in New York, in banking, dry goods, "the China trade." Some owned the ships that tied up beneath their windows. A few were also men of learning. J. Carson Brevoort, to give just one example, had been educated in Europe and served for a time as private secretary to Washington Irving. He was a recognized authority on American history and entomology. "His knowledge of fish," it is reported on good authority, "was hardly exceeded by any naturalist and his collection of books and specimens was magnificent and valuable."

Among such men and in such a setting, New Yorkers of comparable station might well wish to make their home once the bridge was built and the inconvenience of the ferries was no longer an issue.

At Clinton and Pierrepont stood the Brooklyn Club, where, as one visitor noted happily, the members not only referred to one another's wives on a first-name basis, but their servants as well. Close by were the Mercantile Library, the Long Island Historical Society, and the Music Academy, where the Heights gathered for "uplifting" lectures, amateur theatrical productions and musicals, Johann Strauss on one occasion, and yearly charity balls. In the grand ballroom "one could not move a foot without appearing in mirrors."

And there, too, on the Heights, stood what was held to be not only the moral and spiritual center of Brooklyn and New York, but of all America. Brooklyn was "The City of Churches," Talmage and Storrs were among its pastors. But Plymouth Church, a big brick barn of a building on Orange Street, was its foremost institution, bar none, the thing Brooklyn was famous for from one end of the land to the other. For it was there, on an open platform, before a congregation of two thousand or more, that Beecher preached, weekly—except summers—taking the Rocky Mountains as his sounding board, as one man said.

From the photographs there are of Henry Ward Beecher and the volumes of printed sermons, it is a little hard to understand just what all the excitement was about. One eye droops quite noticeably, giving him an unbalanced, slightly unpleasant look. He wore his hair long and loose, as was the custom with many platform spellbinders of the time, and by 1869, at age fifty-six, he was beginning to look a little gray and too well fed. Still, by all accounts, he had a physical vitality, an exuberance that appealed enormously to both men and women. In an age that adored both oratory and showmanship, he was the supreme orator and apparently one of the great performers of all time.

A brilliant pantomimist and mimic, he could turn in an instant from radiant joy to real tears to thundering, righteous anger—whichever was called for. He used no notes and began his sermons very softly, as though holding a private conversation with the front pews. But then all at once the "full, round, sonorous" voice would fill the church. Mark Twain, who watched in awe from the gallery one Sunday, wrote, "He went marching up and down the stage, sawing his arms in the air, hurling sarcasms this way and that, discharging rockets of poetry, and exploding mines of eloquence, halting now and then to stamp his foot three times in succession to emphasize a point."

America had never produced anything quite like this man. Except possibly for Grant, no one alive was so highly regarded. His sermons were read avidly in the newspapers, and gotten up in book form they outsold the most popular fiction. Sunday-morning ferries to Brooklyn were known as "Beecher boats." The easiest way to find Plymouth Church from the ferry landing was to follow the crowd.

One could look through the hundreds, the thousands, of speeches and sermons delivered by Beecher during the years the bridge was being planned and built and doubtless find a Beecher quote on the subject. There was little in life he did not have something to say about, and particularly if it was a matter of popular interest. Perhaps, like the merchants on Fulton Street, he envisioned whole new elements of New York society suddenly discovering Brooklyn once the great span arched the river. On Sundays it might even become "Beecher's Bridge." Of greater interest, however, is what Brooklyn thought of him, what he meant to Brooklyn.

His name was in the papers almost daily, which suggests people never tired of reading about him—what he had to say on free trade or growing onions or the vagaries of the weather. He was regarded as a master of conversation, when in truth he seems to have been more a master of the monologue. His entrance into any Brooklyn auditorium or public gathering was the immediate signal for an ovation, and the plot he had picked out at Greenwood for himself and his stiff, severe-looking wife was a tourist attraction.

The rich and the famous paid him all kinds of homage—Lincoln once said he was the greatest man in America—and Plymouth Church paid him twenty thousand dollars a year, or the same as the President of the United States received. And nobody thought that out of line. It is perhaps impossible to imagine the hold he had on his time.

"Our institutions live in him," said the *Eagle*, "our thoughts as a nation breathe in him, our muscular Christianity finds in him the most vigorous champion. He is the Hercules of American Protestantism . . ."

The presence of such an individual would give a place a certain aura, needless to say, and was certainly a factor in determining the kind of people who had been choosing the Heights as a place to live over the twenty-odd years since Beecher first arrived there—not all of whom, it ought to be said, were excessively wealthy or prominent socially. Perhaps the most fitting description of people on the Heights at the time the bridge was about to be built is one written of the Plymouth Church congregation by a visiting reporter from Massachusetts. "A more intelligent body of people one would rarely find," he said. "A phrenologist would praise their intellectual developments, while there is a look of cheerful hearty satisfaction on most of their faces, as if they relished life and were seldom troubled with the blues . . . It is a well-to-do body also; not aristocratic or fashionable, though a score or more came in their carriages, but prudent and prosperous, as if they lived in good houses and both earned and enjoyed worldly comforts."

There was more to be said for Brooklyn. Gas rates were reasonable. Taxes were still lower than in New York. The schools were far superior. Local government was reputed to be honest, which it was not, but in contrast to the way things were done on the other side of the river, it looked pretty good. Streets were reasonably well lighted after dark and for a city of its size there was little crime. The drinking water was delicious.

The *Eagle*, Brooklyn's leading daily, was certainly another amenity, and Thomas Kinsella, its editor, who was soon to become a Congressman as well, was regarded as a perfect example of how far a deserving immigrant boy could rise in America.

Jobs happened also to be plentiful in Brooklyn just then. Charles Dickens, after a recent visit, dismissed Brooklyn as "a sort of sleeping place" for New York. But from the statistics available, it appears New York employed considerably less than half of Brooklyn's wage earners, perhaps even as few as one in three.

In any event, theirs was the fit place to live, most Brooklyn people felt. It was the more truly American city. New York, for all its enticements, was regarded as a monstrous, cold place, overcrowded, overpriced, bewildering—unwholesome. In Brooklyn a clerk could own a home. Of the five hundred miles of streets Brooklyn land speculators liked to exclaim over ("five miles for every one in New York") only half had anything built on them as yet.

Brooklyn had its shortcomings, of course. Even the best restaurants and shops were second-rate compared to those in New York. The air in the neighborhood of Peter Cooper's glue factory was not exactly that of the open sea and the average Brooklyn saloon, according to one source, smelled like a kennel. Nor was Brooklyn innocent of the filth and squalor so commonly attributed to "the modern Babel" across the way. The tenements on the "flats," south of the Navy Yard, where a large part of Brooklyn's Irish lived, were as foul as any in New York.

But Delmonico's, Barnum's museum, evenings at the theater, Wall Street, adventure, all the "delights" of New York were readily available over the river. That, after all, was one of the most appealing things about Brooklyn—it had New York at such easy reach, it offered the best of both worlds.

As for those unpleasant neighborhoods by the Navy Yard, well, most people never ventured down such streets or had any real idea of the life that went on there; and, naturally, what most people believed to be the truth was more important at the time than any latter-day objective appraisal of how things were. People then were still inclined to form opinions more from experience than information and it was the experience of most Brooklyn people that between their city and the other one, there was no comparison.

Moreover, the bridge was going to make things better still. Like most of their countrymen, Brooklyn people, the newcomers especially, were essentially expectant at heart, optimistic, looking forward, believing fervently in the future. How much was already known of the politics the bridge involved, one wonders, or of the various bargains that had been struck? How much was even suspected? How many people had speculated seriously on what the real cost of the bridge might be?

There is a story about where and when the bridge scheme was hatched in Brooklyn and a plaque that commemorates the event, in Owl's Head Park, near the place where Henry Cruse Murphy's house once stood.

The winter of 1866–67 was as severe as any on record. Ice conditions on the river were so bad on several occasions that a traveler by train from Albany could reach New York in less time than a commuter from Brooklyn could. Roebling's bridge had opened in Cincinnati with national acclaim and every Brooklyn paper was demanding a bridge to New York.

Then, on the night of December 21, 1866, young William Kingsley, convinced the time was ripe to get a bridge bill before the Albany legislature, decided to ride down to Bay Ridge to call on Henry Murphy. The night was bitterly cold, no night to be out, according to the story, and Bay Ridge was a good four-mile drive. Kingsley was accompanied by Judge Alexander McCue, a close friend of Murphy's, who went along, he said later, merely to give Kingsley what support he could, for Murphy then was known to be "far from persuaded of the practicability of the enterprise."

Kingsley had been conferring since summer with Julius Adams on the engineering involved, and in the library of Murphy's palatial home, beside a log fire, the conversation went on until past midnight. Murphy is said to have been highly skeptical at first, even hostile to the whole idea of a bridge. Kingsley is supposed to have responded with mounting enthusiasm for his subject, meeting Murphy's every argument with sharp, convincing rejoinders. Describing the scene later, McCue said of Kingsley, "His unexhausting and unresting mind, matchless in its clarity and invincible in its force, was my wonder and admiration." After a time Murphy was listening as though under a spell.

By the time they were at the door saying good night, Murphy had been converted. What exactly Kingsley said to him during the course of the evening was never revealed, however. All McCue said was that nobody could have withstood Kingsley's onslaught of facts and figures.

Possibly the story is true. Possibly Henry Murphy, like earlier patrician figures in Brooklyn, saw the bridge as a threat to a whole way of life. Old General Johnson, Brooklyn's "first and foremost citizen" when Murphy was a young man, had declared during his campaign for mayor in 1833 that Brooklyn and New York had nothing whatever in common, in "object, interest, or feeling," and that the river dividing them was a wonderful thing for Brooklyn. During the War of 1812 the general had been put in charge of Fort Greene, to stand in wait for an invasion of Long Island that never materialized. Later he had grown gravely concerned about an invasion of another kind. He liked Brooklyn the way it was, and said so. There were still people who felt that way, not many, but some, and perhaps Murphy was of the same mind.

But it does not seem very likely—if only in light of Murphy's total, unwavering devotion to the bridge from that night on. Furthermore, nearly ten years before, in 1857, the year of Roebling's letters in the *Tribune* and *Journal of Commerce*, Murphy said most emphatically that the East River would soon cease to divide Brooklyn and New York. Speaking at a farewell dinner in his honor at the Mansion House, just before leaving for the Netherlands, he had hailed the "spirit of advancement" stirring in Brooklyn and suggested that the new water works was only a sample of the monumental enterprise such a community was capable of. Even the Tammany guests had applauded.

But be that as it may, Kingsley was unquestionably the spearhead of the bridge idea in 1867, and he would have more at stake in the venture ahead than any other one man, with the single, notable exception of Washington Roebling. Kingsley and Murphy were the two most powerful, influential Democrats in Brooklyn—Boss McLaughlin not included—and the Brooklyn Democrats had just about all the political power and influence there was to have in Brooklyn. So whether it was that particular night by the log fire that the two of them struck their bargain is nowhere near so important to the story of the bridge as are the men themselves.

Had a political cartoonist of the time decided to do a simplified illustrated key to Kings County politics, circa 1867, he might have drawn a bird's-eye view of Brooklyn with Beecher commanding the Heights—just to orient people—and behind Beecher's back, Boss McLaughlin, looking a little dull-witted, holding City Hall in the palm of his hand. At the foot of Fulton Street, beside an office labeled KINGSLEY & KEENEY, CONTRACTORS, would be the strapping young Kingsley, energetically cranking a cement mixer that spews out something marked $$$. And down the bay, off to himself, gazing from a tower window at his "villa," his eyes on some distant horizon and looking very senatorial, would be Henry Murphy, noble as a Roman.

In their different ways both Kingsley and Murphy were very impressive men. Kingsley was all of thirty-four in 1867, which made him young enough to be Henry Murphy's son. Hard, resourceful, ambitious, he had established himself in business in Brooklyn the same year Murphy went to The Hague, 1857, a depression year. He came to town not knowing a soul, apparently. He had been a school teacher and also a construction boss on canals in Pennsylvania and on railroads in the Midwest. For a brief time he had worked on the Portage Railroad at Johnstown (where perhaps he heard tales of Roebling installing his iron rope a decade before). His prime attribute in that earlier time appears to have been an ability for snuffing out strikes.

Now he was Brooklyn's most prosperous contractor. He had paved streets, put down sewers, built the big storage reservoir at Hempstead, built much of Prospect Park, some of Central Park, branched out into the lumber business, the granite business, bought up real estate, and became "identified" with Brooklyn's gas company and banking interests. Just ten years after stepping off the Fulton Ferry, a total stranger, and with no money to speak of, he was worth close to a million dollars and was one of the best-known men in Brooklyn.

Boss McLaughlin had taken an almost immediate liking to him. McLaughlin himself had been nothing more than a waterfront gang leader until 1856, when, as a reward for services rendered locally in the campaign to put Buchanan in the White House, he had been appointed "Boss Laborer" at the Navy Yard. It was not long before he was "Boss" of all Brooklyn. He was, in fact, the first political manipulator to be called "Boss," a name he never cared for. Soft-spoken, dingy-looking, a man who played dominoes for off-hours excitement, he walked about Brooklyn with his shoulders thrown back, his great stomach thrust forward. His silk hat was always last year's and brushed the wrong way. And yet there was something "about the bearing of his round head, and the quiet keen look of his small blue eyes that betrays the leader." It must have been something of the same look that he himself spotted in Kingsley the first time the two laid eyes on one another.

McLaughlin had just begun to organize "The Brooklyn Ring." With Henry Murphy out of the country, he was moving fast. Kingsley got a few paving contracts to start, then work on the water works. Kingsley's interests in politics became very great. In no time the young man was reputed to be the most effective money raiser in the party and nobody, it was said, was closer to the Boss. For McLaughlin, plainly, he was a valuable find.

Well over six feet tall, powerfully built, with broad shoulders and a deep chest, Kingsley "cut a striking figure in the street." His face, smooth and honest-looking, was set off by a fine head of wavy dark-red hair and a neatly trimmed beard. He looked people right in the eye and was obviously many cuts above Boss Hugh McLaughlin. Even the New York *World*, which normally had no use for Kings County Democrats, credited him with "plausible" manners.

Kingsley was Irish, but like Tweed, he was a Protestant. He was also a natural politician but, having no gift for public speaking and no apparent yearning for public office, preferred working behind the scenes. Like McLaughlin he gave no signs of aspiring to more power than could be had right at home in Brooklyn. Between them—the boss politician and the boss contractor—they had worked out a very pleasant, profitable partnership. But unlike McLaughlin, Kingsley had what another generation would call "upward mobility." He had the potential of going very far.

Murphy was quite a different sort. He was Old Brooklyn, he was grace and learning. For a long time he had been considered the handsomest man in town. Where Kingsley and McLaughlin were men of great physical bulk and had made their way up in the world in part because of that, Murphy was small, spare, well knit, and clean-shaven, a refined-looking man with gray hair and sharp, intelligent eyes. Judge McCue called him "cautious and subtle." Henry Stiles, the Brooklyn historian, described him as "very earnest in manner, a little severe even." Everybody respected him, it appears, and it is not hard to see why. According to Stiles, "no public man has, probably, passed thus far through the trying ordeal of a legislative career, so entirely free from the taint of corruption." Once upon a time, as some of his other admirers liked to tell, Henry Murphy had nearly become President of the United States.

John Murphy, his father, a "thorough Jefferson Democrat," had been a Brooklyn judge, a man of some renown in his own time, who did well enough with one thing and another to send Henry to Columbia, where he was graduated in 1830. In the next few years, while reading law—at the same time young John Roebling was struggling to become a Pennsylvania farmer—Murphy made quite a name for himself. "His pen embellished and enriched" the pages of the *North American Review* and the *Atlantic Monthly*. He edited the old Brooklyn *Advocate* and helped organize the Young Men's Literary Association of Brooklyn. In 1835 he went into partnership with an attorney named John A. Lott and was soon joined by another named John Vanderbilt, both older men and already prominent in Brooklyn. "From the first his firm was in high favor with Brooklyn people," reads one biographical sketch, "especially wealthy and conservative old property-holders of Brooklyn, and it soon built up an extensive and lucrative practice." It also ran the Democratic Party on the Brooklyn side of the river.

In 1841 Murphy founded the *Eagle*, then more explicitly named the *Brooklyn Eagle and Kings County Democrat.* The next year he was elected mayor at age thirty-one and commenced his administration by cutting his own salary. He went to Congress presently, served there twice, and distinguished himself by sounding forth against slavery and fostering McAlpine's dry dock at the Navy Yard. By the time his political career had run its course, he would also serve six terms in the State Senate, try three times for the United States Senate, once for the governorship, and fail every time mainly because of the opposition of one man, William Tweed.

But his closest brush with real glory had come in 1852 at the Democratic National Convention in Baltimore. The convention had been deadlocked after forty-eight ballots, when the Virginia delegation put up a compromise candidate, an old-fashioned party regular from New Hampshire, Franklin Pierce. But the Virginians' support for Pierce had been anything but unanimous. Murphy had been the other choice and had lost by a single vote. Pierce had a military record, Murphy did not, and apparently that had been the deciding factor. Had Murphy been put up instead, his admirers held, he would have been nominated, elected, and done better as President than did the colorless New Hampshire man. (For one thing, coming as he did from Beecher's home town, it is doubtful Murphy would have underestimated the abolitionists.)

It was five years after that when Murphy was named the American Minister to The Hague, by Buchanan, and no sooner was he out of the country than Boss McLaughlin swung into action, taking complete control in Brooklyn at the same time Tweed was taking over in New York. "It was not a change for the better," the *Eagle* would write in retrospect, many years later, after Murphy was dead. "Deftness in speech was supplanted by deftness in manipulation of votes; dexterity in argument made way for the dexterity which makes one count for two for your side. The deterioration of our political methods began then . . ."

Recalled from The Hague by Lincoln in 1861, Murphy returned to face other problems as well. A man taken into his law firm had squandered money and left the firm unable to pay its bills. Murphy, who had been planning to retire once he was home again, made good on all the firm's commitments out of his own savings, which nearly ruined him. As a result he had not only gone back to practicing law, but began taking an interest in various local business propositions, such as the development of Coney Island, which was something he had not done before.

In the years since, it was commonly remarked, there had been a certain air of disappointment about Henry Murphy. Albany was now his only field of political influence and there he appears to have been a rather lonely, incongruous figure, with his literary tastes and perfect manners. But he was immensely influential all the same, and for Brooklyn, a most valuable asset.

Privately he turned more and more to his family, his books, his scholarly interest in Brooklyn history. At the moment, he was finishing up a translation of a journal kept by two Dutchmen during a trip to New Netherlands in 1679, something he had found in an Amsterdam bookshop.* This was his third such translation and the library where he, Kingsley, and McCue held their historic conversation housed what would one day be evaluated as among the two or three finest collections of early Americana in the entire country. "Mr. Murphy only failed as a politician," said one Brooklyn observer of the time; "in all else his life was a grand success."

In talent, disposition, age, background, physique, in just about every way, Kingsley and Murphy were as different as they could be, opposite and complementary, and they worked superbly together.

Murphy "threw himself" into the bridge enterprise. He drafted an incorporating charter and "with great energy . . . enlisted the interest of his friends," his prominent, respectable friends, to be more exact. The thirty-eight directors he rounded up included the mayors of both cities, such presentable Brooklyn Democrats as McCue and Isaac van Anden, owner of the *Eagle*, and for the Republicans and old Brooklyn, Simeon Chittenden, J. Carson Brevoort, and Henry E. Pierrepont (son of Hezekiah). He also talked up the bridge at every opportunity and took framed copies of a bridge Julius Adams had designed with him to Albany to pass about among his fellow legislators. To no one's surprise, the *Eagle* gave him full support. "Every Brooklynite, resident or capitalist, is interested in bridging the East River," wrote Thomas Kinsella.

The bill was submitted on January 25, 1867, and passed on

---

* The full title of the translated work, published in 1867, was as follows: *A Journal of a Voyage to New York and a Tour in Several of the American Colonies, in 1679–80. By Jasper Dankers and Peter Sluyter, of Wiewerd, in Friesland.* Murphy's other translations include: *The Representation of New Netherland* (1849), from the Dutch of Adriaen van der Donck, and *Voyages from Holland to America* (1853), from the Dutch of D. P. deVries. He also wrote *Henry Hudson in Holland* (1859) and *Voyage of Verrazzano* (1875), in which he took the mistaken view that Verrazzano's claims of discovery were unfounded.

April 16. Neither Murphy nor Kingsley, nor the name of any engineer, was listed as having any association with the proposed project.

The New York Bridge Company, a private corporation, was to have the power to purchase any real estate needed for the bridge and its approaches and to fix tolls. The legislation fixed the capital stock at five million dollars, with power to increase it, and gave the cities of Brooklyn and New York authority to subscribe to as much of the stock as determined by their respective Common Councils. The stock was to be valued at a hundred dollars a share. The company was to be run by a president who would be elected annually. In time this document of Henry Murphy's creation would be looked upon as little better than a license to steal, but at this stage, for some reason, nobody seems to have regarded it that way.

Kingsley was to line up private money and see about the engineer. Perhaps his initial impulse had been to go along with Julius Adams, as he had led Adams to believe he would. But that seems doubtful. Adams would have been a bad choice and it is known that Murphy never entertained the idea for a minute. Adams had no reputation and had never built a bridge of any consequence. Indeed, one might well wonder why Adams was ever brought into the picture in the first place were it not for some comments made later by Washington Roebling in his private notebooks, where, it happens, a very great deal about the bridge would be said that does not appear in the official records or the various old "histories."

Adams had been nothing more than a straw man all along, according to the young engineer. His role had been not so much to design a bridge as to concoct the lowest possible estimate for a bridge. That way the real engineer, if he seriously wanted the job, would have to pare his figures to the bone, which in turn would give the promoters of the scheme a more attractive price tag to talk about. The point was to work a little businesslike deception right at the start, before any real plans had even been drawn up. "Adams surpassed himself by an estimate of $2,000,000," Washington Roebling wrote. As a result John A. Roebling was forced to trim his own estimate by more than one million dollars, knowing perfectly well what was going on and that his figure was ridiculous.

Then Kingsley and Murphy did some more cutting of their own to arrive at the five-million-dollar figure used in the Albany bill.

In May 1867, a month after the bill was passed, a meeting of the New York Bridge Company was held in the Supreme Court chambers at the County Courthouse in Brooklyn. Henry Murphy was appointed to fill a vacancy caused by the death of one of the directors. At a meeting three days later, Henry Murphy was elected president, and a week after that John A. Roebling was appointed Chief Engineer with full authority to design any sort of bridge he wished. A man had been selected, rather than a particular plan—Roebling had no real plan at that point. Kingsley said later the very nature of the enterprise demanded someone of towering reputation. The name Roebling was "invaluable to this enterprise in its infancy," Kingsley would explain. There could not be a breath of doubt or suspicion concerning the integrity of the builder.

The explanation of the choice of Chief Engineer was worded thusly in the official company records: "Confidence on the part of the public and of those whose money was to be invested in the undertaking would best be insured by employing the Engineer who had achieved the most successful results, and who was thus most likely to accomplish this great enterprise." No other engineer was ever considered.

Roebling's salary was to be eight thousand dollars a year, but any work he did until the bridge actually got under way was strictly on speculation.

Things were moving very fast that spring of 1867. Roebling was told to proceed at once with his surveys and come up with a proposal. He was also led to understand that Kingsley, who was neither a director nor officer of the new company, would personally cover any expenses involved, although nothing was put in writing about it. Test borings were made and in two months' time, having scrapped all his earlier sketches, Roebling was back in Brooklyn with his plan. At the first meeting of the Bridge Company's newly formed Board of Directors, a Committee on Plans and Surveys recommended "the immediate commencement of the work."

That was in October. But more than a year would pass before another meeting was held. Nor would there be any noticeable progress made on even the most preliminary construction. Three thousand new buildings went up in Brooklyn in 1868—churches, stores, banks, more factories, an ice-skating rink, row on row of plain-fronted brick and brownstone houses that sold about as fast as they were finished—but to judge by actions, not words, the Great Bridge was no more than a great figment of the imagination, nothing but a lot of politicians' talk, by all appearances. The New York Bridge Company was nothing more than a name on paper as far as most people could tell. The famous engineer from Trenton was nowhere to be seen.

The delay, however, was not due to any indifference or lack of sincerity on the part of Kingsley or Murphy, or of Roebling certainly. The problem was in New York City. Back at the start of the century, when Thomas Pope's "Rainbow Bridge" had been a favorite topic of conversation in Brooklyn, it was said that the only thing needed to bring a bridge about was "a combination of opinion." Thus was the case now. Now especially that it had become something more than just a Brooklyn dream, the bridge could no longer be considered wholly a Brooklyn enterprise.

So it would be 1869 before the bridge was at last under way. And it would be the end of that summer of 1869, a full two years later, before things were finally settled behind the scenes. By then, of course, Roebling was dead. And only then did William Tweed emerge publicly as one of the leading spirits behind Brooklyn's bridge.

6
The Proper Person to See

Who owns the City of New York today?
The Devil!
—Henry Ward Beecher

ON THE MORNING of September 17, 1869, William Tweed took the Fulton Ferry to Brooklyn. The day was fine, the sun shining, a light, cool breeze blowing over the water. Tweed had a strong attachment to the river. Its sounds and smells, South Street and the ships docking there, were part of the New York he had known since childhood. Twice, when he was an alderman, he had voted to establish additional ferry lines to Brooklyn, in return for which he had been nicely reimbursed with several thousand dollars in cash. He was much younger then and that had seemed a handsome sum. But for his vote for the ferry to Brooklyn's predominantly Irish Williamsburg section, he had also received a promise of support for his Congressional ambitions. The Williamsburg people had proved as good as their word and in 1853 he had gone off to Washington, a Congressman at age twenty-nine. The life on the Potomac had had no appeal for him, however, and he gladly gave it up after a single term. There were better things to be doing at home.

Now, as the big boat swept out into the river, its bell clanging, there stretched behind him the largest, richest, noisiest city in

America, over which he and a few indispensable associates ruled supreme.

New York then, as later, was the country's financial capital. It was the seat of mighty newspapers—Greeley's *Tribune*, James Gordon Bennett's *Herald*, the *Sun*, the *World*, the *Evening Post*, the *Times*. It was the place where fashions were decided on and the center of fashionable society. It was home for the Astors, and it was home for P. T. Barnum, Samuel F. B. Morse, and Herman Melville. New York set the example, as somebody said. It had power, style, more Jews, more Irish, more priests and pickpockets and private art collections, more of just about everything than any city in the country. There were hotels with French waiters and five hundred rooms, theaters numbered in the twenties. As a contemporary book on the city stated on its opening page, Broadway, the principal street, was "paved, policed, and lighted for fifteen miles." Pretty girls and rich men were too plentiful to count.

From the Battery northward for five miles or so, nearly every last piece of available space had been built up, and on toward the outskirts, near the new park, which had been so optimistically named Central Park, the city had a "straggling, unfinished look." Downtown the noise was fearsome most of the day, the traffic terrible, the air thick with the smell of horse manure. Along Broadway the fast-moving throngs were more than the average person could take for very long and most women would have agreed with the Reverend Mr. Beecher's famous sister, Mrs. Stowe, never a timid spirit, who wrote a friend that New York "always kills me—dazzles, dizzies—astonishes, confounds, and overpowers poor little me."

Strangers to the city sensed an uncommon and offensive preoccupation with self-interest and the almighty dollar. "The light of Mammon gleams on nearly every face in Broadway and Wall Street," wrote a visitor from England. And two years earlier, after arriving from California, Mark Twain said of New York, "Every man seems to feel he has got the duties of two lifetimes to accomplish in one, and so rushes, rushes, and never has time to be companionable—never has time at his disposal to fool away on matters which do not involve dollars and duty and business."

Contrasts were everywhere, sharp and frequently appalling. The glistening new carriages that streamed through Central Park on Sundays reminded one French traveler of filigreed jewelry, but of the domain of the Irish and the Negro, between Broadway and the Hudson, the same man wrote, "Nothing could be more depressingly miserable than these wooden hovels, these long muddy streets, and this impoverished population." An English writer said the whole city, generally speaking, was "one of the worst lighted, worst paved, worst kept cities in the world."

Homeless children loose in the streets numbered in the thousands. The permanent floating population of homeless children, beggars, drifters, petty thieves, and prostitutes was said to be perhaps 100,000. Bawdyhouses advertised openly in the newspapers. Prostitutes were said to be "more numerous, more dangerous and more shameful than anywhere else on the continent." Probably that was so. The most influential woman in town was an abortionist, Madame Restell, who lived in a big showy house on Fifth Avenue, across from where the new St. Patrick's Cathedral was being built.

But nowhere else was quite so much happening every day or was there so much opportunity for the young, the talented, the ambitious, not to mention the lucky or the unscrupulous. Yesterday's ragpicker or coal heaver was today's millionaire (a new word). It not only happened in stories, it happened. A. T. Stewart had once been an ordinary shopkeeper, living over a store with his wife in a single room. Cornelius Vanderbilt began penniless, everyone knew.

The city was the undisputed center of the new America that had been emerging since the war. It was a place of a thousand and one overnight schemes, some brilliant, some preposterous, some plain evil, and all, it seemed, calling for enormous outlays of capital and pure nerve. It was the great gathering place for every imaginable kind of promoter, inventor, pitchman, entrepreneur, self-proclaimed visionary or ordinary crook, and nobody, it seemed, could remain bored there for too terribly long, whatever his condition.

Moreover, if only New York could have produced a Jim Fisk or a Tweed, as its critics liked to say, perhaps only New York could have produced a Peter Cooper or a Thomas Nast. And while the opening of the West was the great, popular human drama of the time, New York was the central attraction of a smaller, less celebrated, and colorful, but no less important and purposeful, migration in the opposite direction—eastward to the big cities. "The West is best for the person who is seeking a home," wrote a young Swedish immigrant upon arriving in Kansas in June of 1869, but added, "The East's large cities offer a rich field for clever money lovers," thereby crystallizing what to him and countless others like him seemed the two most compelling and divergent choices in American life. And as he and his kind streamed out onto the American plains, it was to New York that a Mark Twain headed from the gold fields of California or an only moderately rich young Andrew Carnegie moved from Pittsburgh, or that an even younger and penniless Thomas Edison of Ohio would set himself up in business that same June.

Tweed, as it happens, had been called upon to journey all of a few city blocks to find his promised land. He had been born just to the east of City Hall, toward the river, at No. 1 Cherry Street, a squat red brick house that stood beside No. 3 Cherry Street, which, ironically enough, had once, briefly, been the nation's first Presidential mansion, the home of George Washington when he took office for the first time and New York was the capital. Both of the houses, Tweed knew, would have to be demolished to make way for the anchorage of the new bridge.

Tweed's people were hard-working Scotch Protestants, and Cherry Street when he was a boy was still considered a respectable neighborhood. It was said he had been a good student, at mathematics especially, and he had started off quite conscientiously, bookkeeping in a mercantile store. But about the time he reached his full growth, he joined up with a volunteer fire company and discovered politics.

In 1851 the fireman became an alderman, being elected to that Common Council of New York destined to be known as "The Forty Thieves." From that time on, he would find honest work no longer necessary.

By 1869 Tweed was nearing his zenith. Like Beecher across the way, he was right in his prime. He was School Commissioner, Assistant Street Commissioner, President of the Board of Supervisors of the County of New York, Senator of the State of New York, Chairman of the Democratic Central Committee of New York County, and Grand Sachem of Tammany Hall, the title that pleased him most.

But Tweed was also one of New York's principal landowners. He was on speaking terms with the rich and powerful, some of whom he also did business with, and he was an extremely generous advertiser in the newspapers. He was in the printing business. One way or another, thousands were in his employ, including several hundred prominent Republicans. Judges rendered decisions according to his requests. Legislators passed or defeated laws as he determined. The Mayor, the charming A. Oakey Hall, did as Tweed wanted and the same could be said of the Governor, John T. Hoffman, who if not an outright stooge was at least "pliable." As Samuel Tilden would write, it was the first day of January 1869, when Oakey Hall took office, that the Tweed Ring "became completely organized and matured."

It was also in 1869 that Thomas Nast began his brave, brilliant attack on the Tweed Ring in the pages of *Harper's Weekly*, so vividly characterizing Tweed that the true shape and nature of the man would seem to recede and the cartoon figure would become the real Tweed in people's minds, then and in generations to come. He would be seen as a gross, half-comic character done in quick, sure black line, a figure of corruption incarnate, leering, lecherous, Falstaff with a stickpin.

The portrayal was deadly, as no one knew better than Tweed himself. "I don't care a straw for your newspaper articles," he is said to have exclaimed when under attack from the press, "my constituents don't know how to read, but they can't help seeing them damned pictures." But it was caricature, for all that; the man was something else again. He was no less corrupt certainly, but he was very much of flesh and blood, and apart from everything else that was said about him, it seems he was both extremely bright and enormously likable.

The Tweed on board the ferry to Brooklyn that September morning was in his mid-forties, married, the father of eight children. What hair he had left was a dark reddish-brown, which he wore in thick clumps about his ears. His mustache was approximately the same color, as was his short, stiff beard. The eyes were bright blue.

But it was the fantastic physical size of the man that set him off in any gathering. He stood five feet eleven, yet he seemed much bigger than that. He had an abnormally big head, big hands, big neck, great, thick sloping shoulders, and a vast, belligerent stomach across which was draped a heavy length of gold watch chain, a gift from his old fire-fighting pals of the Americus Engine Company Number 6. Estimates are that Tweed weighed somewhere near three hundred pounds.

James Bryce, the English historian, who watched Tweed in action with total fascination, considering him one of the phenomena of the American political system, wrote that Tweed's size was an important part of his professional equipment, since it made the role of "the genial good fellow" that much easier. And by almost every account, Tweed was a most amiable fellow indeed, buoyant, booming, with a fund of stories and a gift for making friends quickly and easily. "Tweed had an abounding vitality," Bryce wrote, echoing John Roebling's view of what counted for success, "free and easy manners, plenty of humor, though of a coarse kind, and a jovial swaggering way which won popularity for him among the lower and rougher sort of people."

Like numerous other swindlers in high places, before and since, Tweed was known to have his "good side"—a combination of small but popular virtues that were taken by thousands upon thousands of loyal followers as obvious proof that he could not possibly be the monster his enemies described. He drank but sparingly. He smoked not at all. He was a devoted, generous father. He was exceptionally charitable. (Come a hard winter, no family in his district went without coal.) And although he kept at least two mistresses, one of whom, a tiny blonde, did not reach his shoulder, he was considered quite charming in his way by the ladies. He *loved* to dance. He was never ever arrogant. And perhaps most important of all, as Bryce said, he was always loyal to his friends.

But all that aside, Tweed had a genius for organizing things, and more than any other man of his time, he knew how to make politics pay. Precisely how much he and his cohorts managed to steal from the people of New York will never be known for certain, but responsible estimates range from $75 million to $200 million. Tweed himself, personally, probably made off with $30 million before he was deposed.

When the ferry docked at Brooklyn, Tweed went up Fulton Street to the Gas Light Company offices. He had come to attend the very first meeting of the new Executive Committee of the Bridge Company, of which he was one of just six members.

The full committee was present according to the official synopsis of the meeting: "Messrs. H. C. Murphy, S. L. Husted, Wm. M. Tweed, J. S. T. Stranahan, Hugh Smith and H. W. Slocum." The committee had been formed at a directors' meeting the week before. From here on, it had been agreed, the committee was to make all arrangements for the work and have full control over the business of the company. It was to decide on all appointments, purchase all lands required, decide on all contracts for supplies, audit its own accounts, appoint its own attorney (Alexander McCue), and determine all salaries and all other "compensations."

Tweed's interest in the bridge was not new, but only on August 3, following the Roebling funeral, had his name been associated with the great work in any public way. An important directors' meeting had been held at the Gas Light Company. Colonel Roebling was officially appointed Chief Engineer and six new directors were named to fill the vacancies caused by one death and a total of eight resignations, all quite sudden it seems. Three of the new directors were Brooklyn men—Stranahan, General Slocum, and John W. Lewis. The three others were from New York—Tweed and two of his closest associates, Hugh Smith and Peter Sweeny.

Smith was a Republican, a bank president, and a dear friend of Boss Tweed's, who had made Smith Police Commissioner. Sweeny, also known as Peter "Brains" Sweeny, was a pudgy, black-haired, sinister-looking man who had been called the sneak thief of Tweed's entourage, but who preferred to describe himself humbly as "a sort of adviser" to the Boss. Like Tweed, Smith had been named a member of the powerful new Executive Committee and so had the pleasure of watching Tweed perform his first recorded service on behalf of the bridge.

As was frequently his way when commencing an association with a grand public endeavor, Tweed wanted an eminently respectable name also associated with it from the start. So as the Executive Committee got down to business, he urged that Horatio Allen be appointed consulting engineer at an annual salary the same as young Roebling was to get, eight thousand dollars, and that the appointment date from August 1, even though Allen had had no hand in any of the work during that time. A formal motion to that effect was presented, passed, and the genial Allen, who was present, immediately accepted. Then, after a few brief formalities were seen to, the meeting broke up and Tweed went back across the river.

There had been no fanfare of any kind and on the surface it might appear that little of importance had been accomplished. But Tweed doubtless felt he had put in quite a good morning.

The sudden death of the elder Roebling had created the single possible flaw in the venture, as Tweed saw it; so he had moved to solve that—or to appear to solve it, which was more important. Irrespective of the skills possessed by the young engineer, without the father in command, a failure of public confidence might develop and bring the bridge to a halt, something Tweed did not want. Tweed was anxious that this project continue for quite some time to come. But now young Roebling was backed up by an engineer of his father's vintage, whose professional standing was well known, whose character was unimpeachable, who was a founding member of the Union League Club, and so forth. If anyone should one day care to check the record to see who had put him there to safeguard this vast and important undertaking, he would find the name "Wm. M. Tweed." (Horatio Allen's professional contribution to the work in the next few years would add up to just about nothing.)

The Tammany chieftain had also made his first entrance onto the Brooklyn scene and not a stir had resulted. He had gone to Brooklyn in good faith and in broad daylight, he had been seen on Fulton Street, he had taken a place among such upright gentlemen as General Slocum and Henry Murphy, he had fixed his name to the greatest municipal enterprise of the day, and no one had raised a hand or a voice to stop him.

For nearly two years more Tweed would play a decisive part in the business of the Bridge Company, traveling to Brooklyn to attend stockholders' meetings or to serve on the Executive Committee when it served his purpose. He would make no secret of his interest in the bridge, nor did anyone else try to hide, or gloss over, the obvious fact that his was a very powerful voice in bridge affairs. But when it came time for Tweed to explain the earlier stages of organizing the Bridge Company—that business transacted *prior* to his open association with the project—everything he said would be angrily denounced by the two principals in his version of the story, William Kingsley and Henry Murphy.

Murphy stated there was no truth whatever in anything Tweed said. He denied absolutely playing the role Tweed credited him with, claiming he and Tweed had had no dealings concerning the bridge. And he stuck to that until his dying day. Which one of them, if either, was telling the whole truth will never be known.

But according to Tweed's version of the story, Murphy was a very different sort of man from the one Henry Stiles depicted and perfectly capable of Tweed's brand of politics if that was the only way to get what he was after. In late 1867 or early 1868 Murphy had been after a pledge from the City of New York to subscribe to a million and a half dollars' worth of bridge stock. (Brooklyn's subscription, exactly twice that amount, had been arranged quite smoothly. The three million dollars from Brooklyn was all set. The remaining shares, a half million dollars' worth, were to be sold to private citizens.)

Tweed's account of what went on behind the scenes is the only one available. It was presented a number of years later and is rather vague on such important details as to just when Murphy came to him for help—whether, for instance, it was before or after Murphy made his bitter fight for the Democratic nomination for governor in 1868, a fight Murphy lost to Tweed's hand-picked candidate, John Hoffman. The timing, needless to say, would be an interesting part of the story. In any event, it was Murphy, according to Tweed, and Murphy alone, who came to him in Albany to say the Bridge Company needed money from New York, which, as Murphy said, was a matter to be decided by the Common Council.

Tweed, at that time, was newly arrived at Albany, having recently become "a brother Senator" of Murphy's, as Tweed put it, representing New York's Fourth District. He had hired a suite of rooms on the second floor of the Delevan House, where amid potted plants and gleaming sideboards loaded with decanters of whiskey and Holland gin, and with the steady trill of his beloved canaries filling the air, Tweed transacted the people's business. If one wished something done at the old stone Capitol up the street, one went first to the Delevan House, to the second floor. Jim Fisk, Jay Gould, Vanderbilt, they all made an appearance sooner or later, as did countless lesser "lobbyists," one of whom later testified that the going price of a vote ranged anywhere from fifty to five hundred dollars. For Vanderbilt alone Tweed is said to have distributed $180,000.

If Tweed and Murphy did meet in private, as Tweed said they did—to reach an "understanding" about the bridge—it was doubtless there in Tweed's chambers and it must have been a memorable confrontation. The two of them were like the opposing sides of the same political coin, the one a great, florid mountain done up in a loud suit, the other small, neat, dignified, but tough as a nut and doubtless detesting every minute of the transaction. There they must have sat, face to face, a pair of Tweed's favorite enameled cuspidors, decorated with rosebuds, stationed conveniently close by.

According to Tweed he immediately reminded Murphy that he was no longer a member of the Common Council and therefore had nothing to do with its decisions. "But," said Murphy, "can't you influence them?" (Tweed, it seems, described this little exchange with a perfectly straight face.)

"I told him I hadn't done any lobbying business there, but might if necessary," Tweed continued.

"Shortly after he called again. In the meantime I had conversed with a gentleman occupying a position in the Board of Aldermen which entitled him to credence, and he told me the appropriation could be passed by paying for it." Tweed had asked how much it would cost, but when he tried to recall the answer after so many years, he was unable to say whether it had been $55,000 or $65,-000. (Considering the number of "understandings" Tweed took part in, his slip of memory is not surprising.) But a price was agreed on. "I informed Mr. Murphy of that fact. He told me to go ahead and make the negotiation. I did so, and the money was authorized to be appropriated or the bonds issued. I can't tell the manner in which it was done, but that was the result."

Tweed had no trouble recalling the gentleman of "credence" to whom the money was delivered, or who acted as the go-between.

"With whom did you have dealings?" he was asked during the subsequent investigations.

"Mr. Thomas Coman," Tweed answered. (Alderman Coman was a Tammany hack of long standing.)

"You gave him the sum of money to be paid in bulk to the members of the Board of Aldermen?"

"Yes, sir."

Just how the money was actually delivered to him, Tweed never said, but according to a story told later, one fine day all $55,000 or $65,000, whichever it was, came across from Brooklyn in a carpetbag, and there is reason to believe that the man carrying the bag was William C. Kingsley. For even if everything Tweed said was true, Murphy would never have involved himself directly in that part of the transaction; while Kingsley, on the other hand, was not unaccustomed to handling large sums in the line of duty; and it seems most unlikely that either of them would have entrusted such a mission to anybody else. The only thing of interest Kingsley seems to have put in writing on the subject back at the time was a comment to John A. Roebling in a letter dated April 16, 1868. There was among the New York aldermen, he told the engineer, "a strong combination made against the measure [the bridge] by a ring that want to be bought."

But however Tweed got the money, he did not turn it over to Alderman Coman until *after* he, Murphy, and Kingsley had reached still one further "understanding." Tweed was always very agreeable about passing money along to his political associates and generally he liked to take a little of it for himself, as he probably did in this case. But Tweed by now was no petty grafter. He too was a visionary, with his eye on the future, and bribing a few aldermen was simply not his line any longer, except as a necessary first step in a larger, grander scheme. Tweed was working up an arrangement whereby he and his Ring could get control of the entire bridge.

First of all Tweed wanted stock in the bridge and he wanted it at a bargain price, he wanted it as a gift actually. It was a courtesy he was accustomed to in such affairs. In his testimony he said Murphy told him there had been some difficulties selling bridge stock in Brooklyn, which was the case, and that additional private investors would be most welcome; whereupon Tweed had immediately suggested that he, Smith, and "Brains" Sweeny might like "to go in the direction of the bridge," as Tweed phrased it.

"What inducement was held out to you to become a stockholder in the Brooklyn Bridge?" Tweed would be asked during his soul-baring testimony.

"As the law then read," he answered, "five hundred thousand dollars subscribed by individual stockholders would control the entire bridge, the appropriations, expenditures of money and supplies, and everything."

Tweed was very familiar with the legislation Murphy had drawn up. According to the law the entire corporation, though representing four and a half million dollars of the people's money (from the two cities), was actually controlled by the private stockholders. So just as Tweed explained in his testimony, the man with ten shares of stock (a thousand dollars' worth) had as much say as the City of New York or the City of Brooklyn, with all their millions tied up in the venture.

"Now, how did you expect to be benefited by becoming one of the subscribers to this bridge?" Tweed's interrogator asked. Tweed answered with two sentences, the second of which is a classic sample of his gift for understatement.

"I expected," Tweed said, "that when the bridge was built by the citizens of New York and Brooklyn, and with their money, it would be a well-paying dividend stock. Then we expected to get employment for a great many laborers and an expenditure of the money for the different articles required to build the bridge."

It would not be until after the Tweed Ring collapsed and its incredible thievery was exposed that anyone would be able to appreciate just what Tweed might have in mind when he spoke lightly of "employment for a great many laborers" or "an expenditure of the money for the different articles required." The truth of the matter was that no politician alive had so keen or cultivated an appreciation for a large, costly, time-consuming public work.

For example, several years prior to the time Tweed developed an interest in bridgebuilding, he had commenced a new County Courthouse on Chambers Street, just across the park from City Hall, or almost directly in line with where the New York entrance to the bridge was to be. The architect's plans called for a three-story building, of iron and marble, in the style of a Palladian country house, and it was to cost, according to law, no more than a quarter of a million dollars. At the outset it had looked like a straightforward, relatively modest piece of business. But by 1868 it was still being built and rebuilt—and ever so slowly. The "city fathers" (Tweed's people) had authorized some additional three million dollars to keep the job going (such an edifice certainly ought to be in keeping with the greatness of New York itself, Tweed would say), and there seemed no end to the number of people needed to work on the structure, or to keep it running smoothly. It took, for example, thirty-two full-time employees just to maintain the heating apparatus. By the time it would be finished, in 1871, Tweed's courthouse would cost more than thirteen million dollars, or nearly twice the price paid for Alaska.

The act incorporating the New York Bridge Company had not stipulated a specific ceiling on how much could be spent on the bridge, or even a rough estimate of the ultimate cost, only what the capital stock would be. Roebling's estimate was a matter of public record, of course, but engineers' estimates seldom turned out to be accurate, and even so, as round numbers to work with, six to seven million must have struck Tweed as a much better start than he had had with the courthouse.

But what surely must have set Tweed and his closest associates to doing some very fancy reckoning was the prospect of such an immense, unprecedented piece of construction, where all manner of unexpected developments could call for vast outlays of public money. Three chairs and forty tables for the Chambers Street courthouse had been bought by the City of New York for $179,792. Windows had cost $8,000 apiece. One friend of the Ring, a man named Garvey who would become known as "The Prince of Plasterers," had been paid by 1869 half a million dollars for his plastering work inside the courthouse, plus a million more to repair what he had done. (That July Garvey's bill for plastering came to $153,755, and his total bill, for work that should have cost about $20,000, would be nearly three million.) Among the many checks made out for "articles required" for the courthouse, to cite one more example, was one for $41,190.95—for "Brooms, etc."

So for Tweed and his friends the bridge must have appeared as the most spectacular of dreams come true.

"You mean to say you expected to get a percentage out of the materials and labor upon the bridge?" Tweed was asked.

"Yes, sir."

"Was there an understanding with anybody that you should do so?"

"There was no direct understanding," Tweed said, only "a kind of implied understanding."

Tweed then went on to explain how William Kingsley, too, was to get a percentage of the money spent for materials, according to the arrangement, and that Kingsley was to be the general superintendent of the construction work, with a large say in contracts. A formal confirmation of this part of the bargain would not, however, be agreed to until after the bridge was under way, and Kingsley, in due time, would have a great deal more explaining to do than would Murphy.

Tweed said the only "understanding" he personally had with Kingsley was that "he was to pay the balance of my stock after I paid the installments of twenty per cent of my stock."

"Oh! He was?" exclaimed Tweed's interrogator. "After you paid the twenty per cent of your stock Mr. Kingsley was to pay the balance of it?"

"Yes, sir."

"Was he to do that for the others?"

"I think he was, but I don't know."

So the agreement reached was this: Tweed, Smith, and Sweeny were to receive a total of 1,260 shares in the bridge, valued at a hundred dollars a share. The split between them was to be even, 420 shares per man, or $42,000 worth of stock, for which they each were supposed to pay 20 per cent, or $8,400. But the way it finally worked out, they each got 560 shares, so that at a future date they could each turn 140 shares over to another of Tweed's confederates, who, Tweed decided, had to be in on the arrangement. He was Richard B. Connolly, "Slippery Dick" Connolly, as he was known, Comptroller of the City of New York and therefore a very useful man in any scheme involving the expenditure of public funds. This way all four of them, Tweed, Smith, Sweeny, and Connolly, would wind up with 420 shares, which meant that individually they would be among the largest private stockholders in the East River bridge. Kingsley would be the largest by far. He had arranged to have his construction firm, Kingsley & Keeney, purchase 1,600 shares. Murphy, by contrast, was in for only 100 shares. But in combination Tweed and his friends controlled a grand total of 1,680 shares —or $168,000 in stock. So, right at the start, they had almost as much stock as Murphy and Kingsley combined and Kingsley had paid for the lion's share and given it to them.

And that was about the size of the bargain, if Tweed's story is to be believed, which probably it should be, considering the circumstances under which it was presented.

His testimony was delivered under oath on September 18, 1877, exactly eight years and a day after his appearance at the first meeting of the Executive Committee. By then he was a very changed man. He was in jail, sick, disheartened, deserted by his friends. Furthermore, he had been led to believe that if he made a clean breast of things he would not only be released, but would be granted immunity from any further prosecution. With no one left of the old crowd to protect, with his own name long since synonymous with villainy, there was really very little reason for him to tell anything but the truth. He had nothing to lose and, it appeared to him, quite a lot to gain. So it seems reasonable that his account, except for incidental details, was close to what happened.

The stock arrangement was, of course, all in the records and quite as Tweed described it. The recorded breakdown on stock ownership, as of the autumn of 1869, as the bridge got under way, reads as follows:

Kingsley & Keeney	1,600 shares
J. S. T. Stranahan	100 "
H. W. Slocum	500 "

Hugh Smith	560 shares
W. M. Tweed	560 "
P. B. Sweeny	560 "
W. Hunter, Jr.	50 "
J. H. Prentice	50 "
J. W. Lewis	50 "
G. T. Jenks	50 "
H. C. Murphy	100 "
Alexander McCue	100 "
Martin Kalbfleisch	200 "
S. L. Husted	200 "
Isaac Van Anden	200 "
Samuel McLean	50 "
William Marshall	50 "
Arthur W. Benson	20 "

It is a list that reveals several very interesting points that Tweed neglected to raise in his testimony. It shows, for example, that of the thirty-eight directors listed in the incorporating charter of 1867, only nine (Prentice, Jenks, McCue, Husted, Kalbfleisch, Van Anden, McLean, Marshall, and Benson) thought enough of the venture or the people now involved with it to put any money into it, and all together they held fewer shares than Kingsley's construction company. There were, to be sure, several of the original directors who were not stockholders but who were serving still as directors, including some with the impressive Brooklyn names, such as Simeon Chittenden. But a number of other esteemed figures had departed entirely—Andrew H. Green of New York, as a notable example—and the control now, very obviously, rested with the gentlemen named in the top third of the list.

But the list showed something more as well. The *only* New York people on it were Tweed, Smith, and Sweeny. Not a single resident of New York listed in the original charter had subscribed for any stock. So if Tweed did not control a majority of all the stock, he was at least in absolute control of both the private and public commitment from the City of New York.

The part of Tweed's story that remains open to question, and must ever remain so, is the role he attributed to Murphy. By the time Tweed's disclosures were made, a sizable segment of the press, not to mention the public, was ready to believe every last word he said, both the press and the public having by then concluded that all public works, however noble, were rife with corruption, and that virtually every last politician, and Democrats in particular, were guilty until proved otherwise. Murphy and Kingsley both were very quick to deny playing any such part as Tweed described, although Murphy, during a rare interview in his Bay Ridge home, would admit to having at least heard of such dealings. He told reporters that *after* the bridge was under way he had heard rumors of money having changed hands and that he had been "greatly surprised" by the news. He also admitted that possibly Tweed could have been paid off by somebody without his, Murphy's, knowing about it (which would seem to leave Kingsley in a rather bad light), and in conclusion he offered the following thought on why somebody—in theory at any rate—might have struck such a bargain with the Tammany devil across the water:

"Mr. Tweed was a power in New York then," Murphy said, "and nothing could be done in the Common Council and hardly in Albany without his help. You know how he controlled everything at that time, and therefore he was the proper person to see, when anything was wanted in the way of legislation, to secure his influence for any measures that were to be passed."

And this, in retrospect, seems the most charitable and very likely the most plausible explanation for the entire arrangement with Tweed. It also suggests that just maybe Tweed was also in on passing the original charter, Chapter 399 of the Laws of 1867, and the wording of it as well (". . . when anything was wanted in the way of legislation . . . for any measures that were to be passed," Murphy said).

It is always possible, of course, that to settle some old score Tweed decided he would just smear Henry Murphy before his time in court was over. It is possible, but not likely. Tweed was not that sort, for one thing. And the hard truth seems to be that if a bridge was to be built, Tweed had to have a hand in it, a large hand. Otherwise there would be no bridge. It was that simple.

Judged by the standards of his day, Henry Cruse Murphy was a politician of exceptional quality, as straight as a string, as the saying went. He was also, plainly, an extremely attractive, accomplished human being. But if the only choice came down to compromising his principles or giving up all hope for a bridge in his lifetime, Murphy apparently was not about to let a squeamish conscience stand in the way.

Or possibly he may have figured that the Tweed Ring could not last the five years Roebling had said it would take to build the bridge. That way Tweed would not be around to collect, so to speak, and therefore any pact made with him now was far less chancy and, somehow, less corrupt. Or perhaps Murphy gave no time to any such thoughts. Perhaps, knowing what he did about the financial terms Tweed was accustomed to in his Albany transactions, Murphy concluded that he and the other Brooklyn people were getting off easy. Considering what Tweed might have held them up for, considering what this particular enterprise might be worth to the few who were in on it at the beginning—not to mention its immeasurable value to both communities and the prestige that would be attached to it—$55,000 or $65,000 for some alderman was a bargain price.

The true value of the stock was, of course, another matter and open to question. Its potential value was enormous, just as Tweed said and just as Kingsley, Murphy, Stranahan, Slocum, and the other Brooklyn stockholders fervently believed it would be. The potential power that went with it was also quite enormous obviously. So in this respect Tweed was being handed a very great deal indeed. But the out-of-pocket cost of the gift, for the time being, was relatively little. The real cost would be the size of Tweed's grip on the work, not the cash in the carpetbag, and only time would tell about that.

As it was, Kingsley & Keeney purchased almost three-quarters of all the private stock (that in their name, plus what Tweed's stock cost). With the respectable Brooklyn sources for capital that had been counted on suddenly dried up, there was really no choice for Kingsley.

Kingsley's role in all this would be disclosed soon enough, long before Tweed went on the stand, and it would be said then, in his defense, that his willingness to put up such vast sums to get the bridge started was the most impressive demonstration possible of his unshakable faith in it. The same, of course, could be said for his willingness to do business with Tweed; but of that his admirers would have nothing to say.

That Tweed had the final say on whether a bridge could be started and so had to be dealt with, there is no question. Moreover, there is, as it happens, a memorable example from this very time of the lengths a thoroughly honorable man would have to go if he wished to circumvent Tweed on such matters.

As Murphy and Tweed were making their arrangements in Albany, the editor and publisher of *Scientific American*, Alfred Ely Beach, was setting out to dig New York's first subway and without Tweed or anyone else knowing about it.

Beach was a most unusual character. At twenty-one he had invented a typewriter, at twenty-two he had become publisher of the New York *Sun*, the first penny paper, which his father, himself an inventor of sorts, had purchased some years earlier. But Beach had grown bored with popular journalism and turned the paper over to his older brother, Moses, soon to take up residence on the Heights and become a pillar at Plymouth Church. (It was Moses Beach who brought to Beecher an olive tree from the Mount of Olives, after a visit to Palestine with Mark Twain and others in 1867, a tour Twain would recount in *The Innocents Abroad.*) Alfred had more interest in the *Scientific American*, which he had acquired at age nineteen, as well as several other schemes, the most important of which was a pneumatic train.

The idea had occurred to him some ten years before. New Yorkers needed a better way to get about their city he was convinced. To go from his office near City Hall to his house on East 20th Street, for instance, took more than an hour during the evening rush hour. His solution was a whole system of high-speed, air-propelled trains underground.

Tweed, however, had already made public his intention of bestowing upon the city an elevated railroad, a grand, costly affair to be built on a great viaduct of stone. So any alternative means of rapid transportation was bound to be either killed off by Tweed or cost its proponents dearly in Tammany blackmail. Beach recognized all this and decided therefore to do what he wanted secretly.

In 1868 he managed to get past Tweed an inconsequential-appearing bill permitting him to establish an experimental pneumatic tube for moving mail. Then, toward the end of the year, with no more legal right than that, he went to work. He had rented Devlin's clothing store at Broadway and Murray, and there, in the cellar, he began digging a tunnel, nine feet in diameter, that was to run a block uptown, to Warren Street, directly beneath Broadway. The digging went on within a special device he had invented for the purpose, the Beach shield, as it would become known, which was shaped like a huge hogshead, open at both ends, and powered by hydraulic rams. The men doing the digging stood inside the shield, and as they progressed, the rams shoved the shield forward, like a gigantic cooky cutter. This way the workers were completely safe from cave-ins and there was no need to disturb any of the surface above ground. The only excavation necessary was for the tunnel itself. The dirt Beach had smuggled out in sacks after dark.

It was all quite ingenious and slightly fantastic, as was the vehicle he intended to put inside the tunnel. It would be a cylindrical car, large enough to carry twenty-two people, and it would be sent plummeting back and forth along its track by an enormous, reversible fan mounted at one end of the tunnel.

Beach put his son in charge of the actual construction, and the work proceeded with little difficulty, despite all the time and care taken to avoid any discovery or suspicion of what they were up to. The Broadway crowds that hurried by overhead had no notion of such industry beneath their feet. Nor did any of Tweed's people find Beach out until he intended them to. The complete project, which Beach considered nothing more than a demonstration model, would take a little more than a year to finish, largely because Beach took such pains with refinements. The one way to overcome Tweed's certain wrath, Beach reasoned, was to win instant popular acclaim. So his pneumatic tunnel would have to be more than just an engineering success. The car itself would have to be plushly upholstered, as elegant as a drawing room, and he made up plans for an elaborate entranceway and platform, with frescoed walls, a fountain, a tankful of goldfish, and a grand piano. In all he would spend $350,000, which was more even than William Kingsley did for his bridge.

When the time came to unveil it, in February of 1870, the tunnel would be an immediate sensation, but Tweed, in a towering rage over the deception, would have his governor, John T. Hoffman, veto Beach's charter and thereby bring down the curtain on the famous pneumatic tunnel, the first New York subway.

But if Beach had gotten as far as he had with his dream undetected and consequently unimpeded by pacts with Tweed, there was obviously no possibility of doing any such thing with a bridge over the East River.

The fall of 1869, when the real work on the bridge began, was an unsettling time. Much in American life was turning out to be something other than what it seemed and that left a lot of people wondering just what to believe in.

In early September, at Avondale, Pennsylvania, fire and explosions ripped through a coal mine killing more than a hundred men. Accounts of the suffering and the obvious deficiencies in the engineering of the mine shocked the entire country. "It is impossible to censure too severely the culpable carelessness of the mining company," wrote *Harper's Weekly*.

Then on the 24th of September, in New York, something close to a national disaster occurred when two notorious young stock gamblers, Jay Gould and Jim Fisk, both frequent companions of Boss Tweed, brought on a sudden financial panic, Wall Street's first "Black Friday."

Gould was rumored to have White House connections. He had assured Fisk that Grant would not sell government gold, so if the two of them were to work together, they could squeeze up the price of gold. They began buying that summer, nearly all on margin. By September 22 gold had risen to 137. On the morning of Friday the 24th, at ten o'clock, when trading opened, the price was 150. Word swept through the financial district that Gould and Fisk, with Grant somehow in league, had nearly succeeded in organizing a corner on gold and could thereby fix whatever price they wished.

The frenzy on the floor of the exchange was of a kind never seen before. In Brooklyn a National Guard detachment received orders to stand by ready "to quell the riot in Wall Street." About eleven the price was 155, then 160. Scores of brokers were being ruined. One man went home and shot himself. When the government decided to do something at last and dumped four million dollars in gold on the market, it was early afternoon and the quotation was nearing 165. Within half an hour the price fell to 135.

The effect on the country was terrible, apart even from the financial misery brought on. For it appeared that a couple of cheap New York gamblers could very nearly bring the whole financial system to its knees, and that the President himself had been mixed up in it, which was not the case, but which a great many people believed to be the case all the same. And even if Grant had not been involved, he had certainly been made to look a fool, the dupe of the most transparent kind of crooked maneuvering. The whole affair was a mystery, even to people who understood how such things worked.

Back in March, when Grant took office, the editors of *Harper's Weekly*, the first in their trade to see the Tweed Ring for what it was and to say so in print, were pointing to Grant as the towering symbol of political virtue. Now, only a little more than six months later, Grant's character was in question and that did not sit well with anyone.

Then in mid-October occurred one of those curious half-comic, slightly unbelievable and unexplainable little episodes that sometimes characterize an age as much as or more than the usual events of conventional history. A ten-and-a-half-foot stone giant had been unearthed on a farm outside the village of Cardiff, in upstate New York. It was described immediately as one of the major scientific discoveries of all time, but soon turned out to be one of the great hoaxes.

The Cardiff Giant, so called, was the creation of a Binghamton cigar manufacturer named Hull, who had gone to Iowa some time before, purchased a twelve-foot block of gypsum, shipped it to Chicago, and there, with the help of some hired sculptors, carved out what he hoped would be taken for the petrified remains of an earlier race of supermen. He had gone about his business with exceptional care. To achieve something that looked like the pores of human skin, he hammered the figure all over with darning needles. To get the aged skin tone he wanted, he drenched the whole thing in sulfuric acid.

The figure was shipped east in a crate marked "Unfinished Marble" and buried behind Stub Newell's barn in November 1868. But not until the following October, and exactly according to plan, was the giant "discovered" by men hired to dig a well. A tent was quickly erected over the site and a ticket booth was set up. The admission charge was fifty cents. People came by the hundreds at first, then by the thousands.

Some Syracuse businessmen soon paid Hull thirty thousand dollars for part interest in the giant and moved it to their city, where popular interest showed no decline even after the giant had been examined by scientists and declared "of very recent origin and a decided humbug." The scientists could say what they wanted. In New York, humbug capital of the world, P. T. Barnum, who had tried without success to buy the giant, had a copy made up and put on display with great success. Later on, when the "real" giant was brought to town and exhibited only a few blocks from Barnum's museum, both of them drew enormous crowds.

The giant also played a part in the elections that November. It was a small part, but not without interest to those amused by or gravely concerned over the extent to which sham and nonsense were taking over in American life. It was said the giant was causing such a fuss upstate that people were not taking the elections seriously, or that so much talk of giants and of hills haunted by "lost" races had clouded popular judgment in some counties.

In any event, when the votes were all in, the beefy colossus of Tammany Hall was bestriding the state as never before. The Democrats gained control of both branches of the state legislature for the first time in more than twenty years. Tweed was having things all his way.

Commenting on the mood of the American people at this time, Henry Adams wrote, "all were disgusted; but they had to content themselves by turning their backs and going to work harder than ever on their railroads and foundries." He might have added, their bridges.

7
The Chief Engineer

> On motion of Mr. Jenks, it was resolved that *Col. Washington A. Roebling* be appointed Chief Engineer; that the Executive Committee have power to fix his compensation, and that he have power to employ such assistance as he may deem proper, subject to the approval of the Executive Committee.
>
> —From the minutes of a meeting of the Directors of the New York Bridge Company, August 3, 1869

HE HAD KNOWN times like this during the war, when nothing much appeared to be happening, but every day counted, when a dozen plans had to be gotten up and decided on without delay, contingencies considered, countless little details seen to and orders given, any one of which might determine the whole course of events to follow. "There must be someone at hand to say 'yes' or 'no,' " he liked to comment, "and it often makes a great difference which word they use."

Until the first of the year, when the site for the Brooklyn tower would finally become the legal property of the Bridge Company, no work could be started of the sort the public had been anticipating. But behind the scenes innumerable matters had to be settled. And from early August, when he was named Chief Engineer, until March of 1870, when he would launch the first of the great caissons, Washington Roebling was an exceedingly busy young man.

Specifications had to be drawn up and agreed on—for such basic supplies as piling, lumber, cement, coal tar, blasting powder. All kinds of special equipment and machinery had to be ordered—air compressors, hoisting engines, stone derricks, clamshell scoops, rock drills, air locks for the caissons—most of which had to be designed from scratch and custom-built. The production capacities and workmanship of various manufacturers had to be evaluated, usually by personal inspection. Job applicants had to be interviewed, the beginnings of a labor force assembled.

Roebling had had a good deal of experience in this sort of planning, working with his father and in the Army, and the business of organizing big construction projects was William Kingsley's specialty. At a meeting of the Executive Committee on October 14, Kingsley would be appointed General Superintendent. Kingsley also, along with the Executive Committee, was to decide on the awarding of contracts. Roebling had been instructed by Henry Murphy that he was to have nothing to do with that. How Kingsley was to be reimbursed for his services was something of a mystery.

Roebling's own staff, as of the end of August, consisted of six: Colonel Paine; C. C. Martin, who had been Kingsley's choice; Sam Probasco, another Kingsley man; and three new men, all quite young, whom Roebling had hired soon after his father's death. They were Francis Collingwood, Jr., George McNulty, and Wilhelm Hildenbrand.

Collingwood had been a friend at Troy. He had been two years ahead of Roebling and finished first in his class, but in the time since, he had been working in the family jewelry business in Elmira and had not had much engineering experience. Had it been up to John A. Roebling, Collingwood probably would not have been hired, but the new Chief Engineer knew his man, he thought, and had written to Elmira to ask Collingwood to join him in Brooklyn. Collingwood agreed, on the condition that he would serve one month only (the jewelry business was prospering it seems). When he arrived in Brooklyn in mid-August, Roebling put him to work helping Paine with plans for the Brooklyn caisson.

McNulty, the youngest of them, was barely twenty, a New Yorker and a graduate of the University of Virginia. He had done a little surveying, but that was about the sum of his experience and he had been turned down when he first applied for a position. Then he offered to serve without pay and Roebling had been so impressed by his manner that he took him on, with pay, as an assistant to Martin.

Of the three new men, Hildenbrand was the only proved quantity. A strapping, smooth-faced young German who had arrived in the United States only a few years before, he was a draftsman of exceptional talent. Earlier he had done a number of finished drawings of the bridge for John A. Roebling, including a big panoramic rendering with clouds sweeping above the towers, which was the one three-dimensional view the engineers had to show how the finished bridge would look. More recently he had worked for Vanderbilt's architects on the new Grand Central Depot and was in fact the man who had designed, at age twenty-two, the great arched roof over the train shed. Hildenbrand would spend most of his time in the office, computing stresses in various parts of the bridge, and producing the finished drawings. But he would also go to Maine to supervise the cutting of the granite and his primary responsibility would be to keep the plans well in advance of the work. He was to be a most valuable man.

Charles Cyril Martin, C. C. Martin as he was known, was second-in-command after Roebling, with a salary of five thousand dollars a year. Older than Roebling by six years, he had a big, plain, manly face, handsome except for the ears, which were extremely large, and he wore his whiskers clipped trim, in the manner made popular by Grant—as Roebling would too now, after his father's death. Martin was another Rensselaer graduate. He had been a class ahead of Roebling, but already twenty-three when he first arrived at Troy, he had been regarded as an old-timer even then. By this time he had been married ten years, was the father of four children, and had worked for William Kingsley on three different reservoirs. He had put down Brooklyn's new water main and was head engineer in charge of building Prospect Park. (The actual design of the park had been worked out by the noted landscape architects Frederick Law Olmsted and Calvert Vaux, who had done Central Park, which they considered the lesser work of the two.) Martin had even named one of his sons Kingsley Martin. So if experience dealing with Kingsley or a knowledge of Brooklyn politics were to count for anything, Martin had added qualifications.

Martin was to concentrate on supplies and the hiring of the work force, Kingsley's chief interests. He was, for all practical purposes, to be the executive officer for the bridge. For help he had Probasco and young McNulty.

The average age of the engineering staff, Roebling included, was about thirty-one. Paine, who was forty-one, was the senior member.

The men were all new to the job and to one another. The job itself, they all recognized, was going to be unlike anything attempted before. A little later, a salty and resourceful character named E. F. Farrington, who had worked on the Cincinnati Bridge, would be named master mechanic. But for now the only one who had had any previous experience building a Roebling bridge—or a suspension bridge of any kind—was the Chief Engineer and not even he had had working experience with some of the problems presented by this particular bridge.

Still he would be the one with the final say. And however much staff help he got, every important decision would be his in the end, and at this stage there was seldom any sure way to know which decisions, of all that had to be made, might turn out to be the important ones in the long run. Roebling, too, would be the one to deal with the Board of Directors and the all-powerful Executive Committee, with public officials from both cities, with old Horatio Allen, the high-paid consultant who had never built a suspension bridge, who knew little about the subject, but who had a reputation to throw about if he chose to.

Roebling would be the one to answer the sort of criticism from other overnight experts that the newspapers and a few of the professional journals liked to give space to. His father had been dead only a short time when a crackpot named Samuel Barnes B. Nolan was crying out in the pages of *Scientific American* that the grade of the bridge was too steep for wagons in slippery weather, that the central span was going to sag, that the whole work would be a "dead failure." The claims were absurd, Roebling knew, but one such article could mean hours lost explaining why to some influential politician or overly conscientious board member.

To get what he wanted, and particularly if it was a departure from the original plan—or seemed to be—he would have to present his recommendations formally, in writing. As the work progressed, he would have to account for each and every step along the way, explaining his every decision in lengthy reports addressed to the Executive Committee. These would have to be quite explicate, thorough, candid, yet in language his nontechnical peers could understand. Every report would also become a public document, he knew, and so, come what may, each would also have to be convincing enough to maintain public confidence not only in the work itself but in the man in charge of it. If things went wrong, if materials proved shoddy, if equipment broke down, if the work fell behind schedule, if some part of the structure itself failed, if accounts were juggled or costs got out of hand, if there were mistakes in judgment by any of his subordinates, if there were accidents, he would be the one held accountable. In time, there would be nearly a thousand men under his command.

Most important of all, the plans he had to work with at this point were only of the most general sort, "the details not having been considered" by his father.

It was a responsibility of monumental proportions and there were older men in the profession, men with proved abilities, who might not have felt up to it. But there is nothing to indicate that Washington Roebling had even a moment's hesitation. He had never had the full charge of a bridge before, the absolute final say, that is, except for two military bridges.

As yet he had done nothing to earn the confidence of his subordinates or any of the private parties connected with the enterprise. He was where he was strictly on his father's say-so and because the men who wanted the bridge built had been left with no other choice. Moreover, as he doubtless sensed from the start, anything he did would be measured against what his father might have done. He would be forever compared to the old man and held accountable for things said or promised by him. And however well he might succeed, however much of himself he might put into the work, the odds were it would always be John A. Roebling's bridge. If there were failures, they would be all his.

The great question now, of course, was whether he would prove to be the man his father had been.

Washington Augustus Roebling, at age thirty-two, was much like his father in a great many ways, but also quite different. John Roebling was a European, a European intellectual to be more exact, a perfectionist at heart and by training, and an aristocrat if one accepts his own belief in an aristocracy of ability. He was painfully proper, vile-tempered, and widely regarded a genius.

His eldest son was an even-mannered, informal, kindly man and, as he himself would say, just a little lazy by his father's standards. Others in the family would say he was more like his mother. He had her patience with people, her calm, as they said. He was extremely bright, but not brilliant. He was not a genius. Nor did he have his father's creative vision, which was among the main differences between them. Still, as his wife, Emily, would write, he was a man of "very versatile attainments." He was a first-rate classical scholar, a good linguist, and a fine musician. He was also quite articulate when need be and a great deal more open-minded than his father had been. He was considerably more interested in his fellow man, in the flesh rather than the abstract, and though he never managed his father's commanding presence, he was really far better at working with people. He was, everything considered, much more of a human being.

Professionally, he was as good as they came. Quick at mathematics, a superb draftsman, extremely thorough about details, he had his father's passion for perfection and, like his father, he had a very great deal of physical courage. But he never considered himself a creative genius and nowhere along the line did he have any airs that might have given anyone that impression. His wife called him "rather indifferent to matters of courtesy." And while he had the Roebling pride, he had inherited almost none of his father's vanity, and this, in his view, greatly reduced his chances of ever attaining comparable fame. "History," he would write, "teaches us that no man can be great unless a certain amount of vanity enters into his composition . . . For a man to be important it is also necessary to have a good opinion of one's self, even if for no other purpose than to impress others."

At first glance he seemed a rather silent but pleasant person, relaxed, unassuming, attractively modest. Indeed, apart from his name, there appeared, on first meeting, to be nothing much out of the ordinary about him. He was not impressive, the way his father had been. His father, on top of everything else, *looked* like a great man.

But in this they were quite the same: they had an absolute, total confidence in their ability to do the job at hand.

Washington Roebling was a great believer in heredity and the part it played in determining one's "composition." He would write of what he called "a peculiarity of the Roebling mind," which he saw as a fixed determination to do things as one thought they ought to be done, no other way, asking little advice of anybody, and generally refusing it when offered. This "overweening self-reliance" was a family streak, he held, something much more serious than ordinary "Dutch" stubbornness and as much a handicap as a virtue. But he also wrote, "It might be argued if a man inherits everything he deserves no credit for it. That would be so in a life of universal monotony, but with each generation in turn totally different conditions and environments arise. These have to be met by the new individual who must develop his powers to adapt himself to them; to overcome them and use them as his tools."

As a "new individual," starting, say, from about age six on, he had neither grown up in an ancient walled city nor filled himself with philosophy nor dreamed of future liberation in some distant utopia. He had grown up an American and perhaps the most obvious, important difference between Washington Roebling and his famous father was just that. Furthermore, he had been through those two most characteristic and influential experiences for American men of his generation: he had spent his boyhood in the rural backwaters, where the frontier was recent history and there were still comparable privations; and he had been through the war. Unlike his father he had been both an American farm boy and a soldier and those two experiences had played a profound part in determining his own "composition," as he called it.

And there was something else: he had grown up with John A. Roebling as his father.

Washington Roebling's passport, dated May 27, 1867, and signed by William H. Seward, offers the following physical description. "Age, 30 years; Stature, 5 feet 9 inches; Forehead, broad; Eyes, light grey; Nose, short; Mouth, small; Chin, square; Hair, light; Complexion, fair."

A passport issued some years later has him an inch taller and according to some accounts his eyes were blue, but in any case a more memorable picture was put down in one of the wartime letters of a Union Army colonel named Theodore Lyman, who served on the staff of General George Meade.

> Roebling is a character . . . He is a light-haired, blue-eyed man, with a countenance as if all the world were an empty show. He stoops a good deal, when riding has the stirrups so long that the tips of his toes can just touch them; and, as he wears no boots, the bottoms of his pantaloons are always torn and ragged. He goes poking about in the most dangerous places, looking for the position of the enemy, and always with an air of entire indifference. His conversation is curt and not garnished, with polite turnings.
>
> What's that redoubt doing there? cries General Meade. "Don't know; didn't put it there," replies the laconic one.

Contrary to general belief, he had not been named for George Washington. It would be said that as an idealistic young immigrant John Roebling "reverently chose" for his first-born son "the name that had most inspired him in the history of the young republic." But the son himself told a different story. He had been named for Washington Gill from Richmond, Virginia, a surveyor his father had hired to help lay the railroad line over the Allegheny Mountain. "They were sitting on top of the mountain when the news of my arrival came," Roebling wrote, "and Mr. Gill begged that I be named after him. The Gill was dropped but Washington I have struggled with ever since."

He was born on May 26, 1837. If ever he considered his father's absence at the time a slight or a prophecy in any sense, he never said so. Nor was he troubled by the absence of clergy at his christening. "I was . . . baptized by the postmaster, Mr. Shilly, there being no preacher as yet—have received no ill effects therefrom." When he was six, his father would describe him as a "well-built, sturdy, quiet boy."

Across the street, catty-corner from the house, was the church his father had built, and beyond that were orchards planted by the first settlers, great stretches of open farmland, but still, also, big stands of virgin forest—black oak in the main—and woods of smaller second growth that were full of game. "As late as 1845 a black bear walked down Main Street," he wrote, adding, "he got away."

The social life was decidedly German and Monongahela rye was considered the staff of life. For entertainment, people put on plays or small parties and dances, at home. "Bernigau played the violin; Wickenhagen the violincello; Neher the cornet; Roebling the flute and clavier."

The native Pennsylvanians called them Latin farmers, meaning they knew more of Latin than farming. There had never been a dearth of interesting conversation, Roebling would recall. Next door lived Ferdinand Baehr, a wool carder from Mühlhausen, who had a splendid library and a brother-in-law named Eisenhardt who had been at Waterloo, in the regiment that held the Château Hougoumont against the French attack. Baehr took a great interest in Washington Roebling and the little boy became a daily visitor, listening to Eisenhardt tell over and over how the French bullets had rattled like hail against big oak doors that never failed.

A year or so before the Civil War, when he was living in Pittsburgh, working with his father on the Allegheny River Bridge, Roebling had gone back to Saxonburg to visit his grandmother Herting. His father had refused to go. For Washington it had been a terribly disappointing experience and he wished with all his heart he too had never returned. As a child, he said, it had seemed the finest place in the world.

> Being the "Roebling boy" I had the entree to all houses, to wonder over the many heirlooms the people had brought over—curious old clocks, old Bibles and books, quaint pictures, novel utensils of copper, brass or china, long German pipes. My grandmother Herting had a wooden travelling-box with a carved top inside of which a picture of the battle of Navarino was glued, showing the burning of the Turkish fleet; that was a treat. A similar picture depicted Marshal Blücher driving the French over the Katzbach.

There were farms not very distant where the old Indian trail to Venango could be plainly seen. Delawares, Shawnees, Senecas, and Muncies had used it for nobody knew how long. George Washington himself had traveled it by foot. An old blacksmith named Glover, the first known settler in Butler County, was still alive then and a subject of immense respect. He had been at Valley Forge. And probably the most famous person in the whole county was an old woman over in Buffalo township whose story was part of the pioneer folklore Washington Roebling had grown up with. Her name was Massy Harbison. In 1792 she had been captured by Senecas and Muncies, who murdered two of her children before her eyes, then set off on a terrifying forced march through the forest, driving her and her one remaining child, an infant, before them. But she had managed to escape and made an unbelievable run for her life, traveling for four days through the wilderness, still carrying her baby. When she reached Pittsburgh, scarcely half alive, it was recorded on good authority that more than 150 thorns were extracted from her feet.

Saxonburg was on the route of the annual flight of the passenger pigeons to Canada every spring, and the sight of them filling the sky was something he would talk about all his life. Fearful thunderstorms shook the little town, and once, in 1843, everyone poured out into the night to see the great comet, "with its head at the foot of Main Street and tail above the church."

Roebling was twelve when his mother, who was again pregnant, moved the family to Trenton, traveling without her husband, who was "tied down" to work in the East. The boy was put into the Trenton Academy, along with his brother Ferdinand, and seems to have gotten by well enough for the next five years. He took up the violin, developed an interest in astronomy and mineralogy, and decided on an engineering career, although it seems unlikely that he ever had much choice in the matter.

Once, in the winter of 1853, he went up to New York with his father. Work on the Niagara Bridge had closed down until spring and his father was seeing to other business. They went over to Brooklyn for some unknown reason and in the process spent several miserable hours on board an icebound ferry. For a man of his father's temperament, it was doubtless an infuriating experience, to be so at the mercy of such elementary forces. But as a result, the story goes, John A. Roebling, his son at his side, "then and there saw a bridge in his mind's eye."

At seventeen Washington Roebling was sent off to Troy to get his training, his father having concluded Troy was the place for him.

The Rensselaer Polytechnic Institute at Troy was a new kind of school, the first in America established for the specific purpose of providing an education in "Theoretical and Practical Science." It had been started some twenty years before by Stephen Van Rensselaer, the Hudson River patroon and politician, who held that the "aspiring energies of youth" had for too long been "chained down to a kind of literary bondage," and who made Amos Eaton, the distinguished geologist, its first head. By the time Washington Roebling came along, it was a small assembly of brick buildings set on a steep hill overlooking Troy and the Hudson River and one of the very few institutions in the country offering courses in civil engineering. There were just over a hundred students in all and the prescribed attire was a dark-green cloth cap and a velvet-collared frock coat to match.

There is a picture of Roebling taken at Troy. He was nineteen at the time, a very handsome, sturdy-looking youth, with his father's jaw and a rather intense stare. He himself thought he looked altogether too boyish and tried without success to grow a mustache.

"In regard to the mustache you covet so," wrote his sister Laura from Trenton, "I can only recommend something which will favor the growth of the desired article, namely, shave every day, and apply some Guano on the desired place and no doubt soon a luxuriant crop will spring up."

The work itself was extremely difficult. Once in a letter to Charles Swan he mentioned swimming the Hudson, but otherwise he seems to have done little else but study, which is not surprising, considering what was expected of him at home and what was required by the institution. His senior thesis was to be on "Design for a Suspension Aqueduct," but in three years' time he had also to master nearly a hundred different courses, including, among others, Analytical Geometry of Three Dimensions, Differential and Integral Calculus, Calculus of Variations, Qualitative and Quantitative Analysis, Determinative Mineralogy, Higher Geodesy (the mathematical science of the size and shape of the earth), Logical and Rhetorical Criticism, French Composition and Literature, Orthographic and Spherical Projections, Acoustics, Optics, Thermotics, Geology of Mining, Paleontology, Rational Mechanics of Solids and Fluids, Spherical Astronomy, Kinematics (the study of motion exclusive of the influences of mass and force), Machine Design, Hydraulic Motors, Steam Engines, Stability of Structures, Engineering and Architectural Design and Construction, and Intellectual and Ethical Philosophy.

A century later, D. B. Steinman, a noted bridgebuilder and professor of civil engineering, would write, "Under such a curriculum the average college boy of today would be left reeling and staggering. In that earlier era, before colleges embarked upon mass production, engineering education was a real test and training, an intensive intellectual discipline and professional equipment for a most exacting life work. Only the ablest and the most ambitious could stand the pace and survive the ordeal."

Roebling, however, would take a different view when he came to appraise the system long afterward. He saw no virtue whatever in what he called "that terrible treadmill of forcing an avalanche of figures and facts into young brains not qualified to assimilate them as yet." "I am still busy," he said, "trying to forget the heterogeneous mass of unusable knowledge that I could only memorize, not really digest." The strain was terrific. Of the sixty-five students who started out in his class, only twelve finished. And among those who did not finish there had been some rather severe breakdowns, it appears, and one suicide.

The suicide was Roebling's first real experience with tragedy. It is not entirely clear what happened, but in the late fall of 1856, during his final year at Troy, a classmate killed himself and apparently it was because of his feelings for Roebling. All there is of the incident in the written record is a letter Roebling wrote to Emily during the war, and two desperate notes written by the unfortunate young man shortly before he did away with himself, all three of which Emily saved.

"My candle is certainly bewitched," he wrote to her from Virginia, nearly ten years after the incident, "every five minutes it goes out, there must be something in the wick, unless it be the spirit of some just man made perfect, come to torment me while I am writing to my love. Are any of your old beaus dead? If I wasn't out of practice with spiritual writing I would soon find out.

"There is only one friend whose spirit I want to communicate with," he continued, "you have his picture with mine; he committed suicide because he loved me and I didn't sufficiently reciprocate his affection; I advised him to find someone like you for instance, but he always said no woman had sense enough to understand his love."

And that is all Roebling seems to have written on the subject, just one small paragraph in his neat copperplate hand that leaps out of the last page of a love letter written late at night in Virginia, after he had been "building bridges and swearing all day."

The first of the other two letters was in German and written on the evening of October 5, 1856, which was apparently after Roebling had rejected his friend's proposal for some sort of formally declared bond between them. The writer pleads for Roebling to understand the nature of his affections and his misery, and asked Roebling to "make allowances." "Our temperaments are so very different, that something which appears only natural to me may perhaps appear incomprehensible or ridiculous to you." And again he begged Roebling to declare his own affections for him, and for him alone. The letter is signed "Your friend," but the name has been erased, whether by Roebling or the young man who wrote it is not known. Roebling was still quite unwilling to agree to what was being asked of him.

The next letter was written on Thanksgiving Day. There is only a copy of the first part of it, written from memory by Roebling later that day. The young man, it seems, had taken to using chloroform as a narcotic and explained to Roebling how to bring him out of an overdose, should Roebling find him in that condition (". . . pour cold water over my head, then breath air from your mouth into my lungs and if there is no success get Dr. Bonetecon and tell him to cup me in the neck; as ultima ratio you may try Electro-Magnetism. . . ."). Then he wrote, "If your efforts should prove fruitless do this: Keep of my things whatever you like, it is all yours!"

Roebling noted on the letter, "At this moment he suddenly staggered in, asking why I did not stay with him. Accordingly I went to his room—he took the letter afterwards so that I had no opportunity to copy it. The rest was merely an inventory of his property, together with some parting words of love."

Just when the young man died, or how, remains unknown. But the whole pitiful affair was a dreadful experience for Roebling, an unpleasant memory ever after. It may also account in large measure for the bitter feelings he later expressed about the prescribed regimen at Rensselaer. Even the few who did graduate, he wrote, "left the school as mental wrecks," which was an exaggeration clearly, his own case being the most obvious evidence to the contrary.

It was the summer of 1857, the time of the great panic, the year his father first proposed an East River bridge, the same year William Kingsley moved to Brooklyn, and the same summer Henry Murphy sailed for the Netherlands, that Roebling finished college and returned home to Trenton, his boyhood over.

His first job was in the mill, and it seems he was actually running things there, entirely on his own, almost immediately, his father and Charles Swan both being away on other business. Then in the spring of 1858, he was on his way to Pittsburgh to work with his father on the Allegheny River Bridge. His salary was eight hundred dollars a year.

Pittsburgh was home for the next two years and he developed a great attachment to the place, writing his family that he regretted the day when he would have to leave. "Pittsburgh is getting along quite smart now," he informed Charles Swan. "I doubt if there is a lazy man in it, your humble servant perhaps excepted." Already he had ten times as many friends as in Trenton, he said. He kept a small notebook that he titled "Lists of Persons I have been introduced to in Pittsburgh, Pa." and scattered among the names of contractors and ironmongers were a Miss McClure, Miss Carr, Miss Mendenhall, Miss Blake, and a Miss Molly Smith of Chambersburg.

He lived in a boardinghouse on Penn Street, worked hard, played chess with the other boarders, went to the opera, argued with his father about invariably starting the work day at six thirty in the morning, recovered from a series of severe abdominal attacks (the cure was his father's vile concoction of raw eggs, warm water, and turpentine), and he wrote home at length, describing with spirit and humor, as his father never would, what was happening in the world about him, apart from work.

He had been to hear Edward Everett at a celebration of the hundredth anniversary of the taking of Fort Duquesne. He described the iron buildings going up everywhere ("There is a perfect mania here for improvements . . ."), a storm that nearly took part of the bridge away, and "dark, cloudy, smoky afternoons, when the sun doesn't shine and the gas is lit at 4¼." One letter he began this way:

> This is my first letter to you in 1860 and consequently I shall make it very short, because it is always a good rule to make a small beginning, but a big ending, and since I don't expect to be here towards the end of the year, you may always expect small letters. Cousin Henry from Cincinnati was here yesterday; he is a very fine young man now, a perfect beauty; he was raising a goatee, containing 11½ hairs, just about as I would have if I were to attempt it.

That fall of 1860 he finished up on the Allegheny River Bridge and returned to the wire mill. The following spring he was in uniform, marching up and down a dusty drill field in Trenton.

"My enlistment was rather sudden," he said later, recalling the night his father had driven him from the house. It would be said by others that the break with his father was so angry and unpleasant that the two neither saw each other nor communicated in any fashion for the next four years. But there is nothing to this. Roebling returned to Trenton several times during the war, and while his letters to his father were customarily quite formal, and answered in kind, he had more to say to him than to anyone else, until he met Emily.

He enlisted on April 16, 1861, as a private in the New Jersey State Militia. Two months later, fed up with garrison duty, he resigned to enlist in New York, again as a private. In January of 1865, the war nearly over, he resigned from the Army, a lieutenant colonel, age twenty-seven, and a veteran of Second Bull Run, South Mountain, Antietam, Chancellorsville, Gettysburg, the Wilderness, Spotsylvania, and the Crater at Petersburg.

The letters he wrote during the war years number in the hundreds and provide not only an extraordinary personal footnote to the history of the Campaign in Virginia, but reveal much about the young man himself. During the years of the bridge, with Emily with him constantly, with his father dead, his brothers and sisters scattered and living their own adult lives, there would be little call for personal letter writing.

Except for promotions and some moving about, his first year at war had been uneventful and disappointing. "Loafing in the camp seems to be the principal occupation," he wrote from Washington. Later, from Harpers Ferry, he told Elvira, "This is a mean little town about the size of Morrisville, presenting a deserted, sleepy appearance, like most Virginia towns. John Brown was hung in a cornfield next to us. The site of the gallows are marked by a cornstalk and pieces of the gallows sell at $1 per lb."

But even when the fighting began, he would have little to say about that side of soldiering. "This artillery business is very hard work," he wrote, and that was about as far as he would ever go.

He was made a sergeant after four months and spent his first winter at Budd's Ferry, Maryland, on the lower Potomac, where his battery was supposed to protect shipping from Confederate batteries over on the Virginia shore, but where nothing much happened. He was billeted in a tent housing "ten choice sports" and about the only memorable event had been "a musical soiree at the widow Mason's house, down on the river bank." The music consisted of singing, piano, guitar, and Roebling on the violin. A supper was served ("Very creditable for this part of Maryland") and a couple of Confederate shells landed in the yard but failed to go off. Years later one of the other musicians wrote that Roebling "could make a violin talk."

He was elected a lieutenant in February and the next month he was at Hampton Roads in time to witness the battle of the Brooklyn-built *Monitor* and the Confederate counterpart, the *Merrimac*. Then he was designing and building his own first bridge—substituting for his father. He had been transferred to McDowell's staff and was ordered to put a suspension bridge across the Rappahannock at Fredericksburg. "My father being too old to rough it, I was selected." He had no experienced help to work with, no proper tools, no material except for some reels of Roebling cable that had been sent down from Trenton. At times the enemy was only five miles away.

The bridge he designed was more than a thousand feet long, longer than the Niagara Bridge, in other words, but broken up into some fourteen short spans. He hired contraband Negroes, trained them as they went along, had his lumber sent on from Alexandria, smoked up a box of Plantation cigars, and had the bridge built in a month.

Almost immediately he was ordered to Front Royal to build another one, over the Shenandoah. With no boats available to cross the river to make his measurements, he jumped in and swam over with the tape in his mouth. But when he and his work party had the bridge about halfway up, the Confederates under Jackson drove them off. Another bridge at Waterloo shared the same fate, and in the meantime General Burnside, retreating from Fredericksburg, blew up Roebling's bridge there. It had lasted about as long as it had taken to build.

That had ended his bridge engineering for the time being. He was assigned next to a cavalry expedition, spent some ten days on the move, scarcely ever out of the saddle, and once, at about five in the morning, surprised Jeb Stuart at his breakfast and very nearly captured him.

Second Bull Run followed after that and Roebling was with McDowell, as a staff aide, through all of it. Less than a month later he came very close to being killed at the hideous bloodbath at Antietam. Then he was back at Harpers Ferry, building another suspension bridge and writing to tell his father how young General Slocum had come along and taken away fifty of his best men. But by December he had it finished. "The bridge has turned out more solid and substantial than I at first anticipated," he told Charles Swan; "it is very stiff, even without a truss railing, and has been pretty severely tested by cavalry and by heavy winds." It was the last bridge he would do on his own until he got to Brooklyn, but as he wrote later, "The Harpers Ferry bridge met the same fate as the others. When Lee came up for Gettysburg the suspenders were cut and [the] floor dropped into the river, but I rebuilt it completely and the army in parts marched over it. The following year [General Jubal] Early destroyed it absolutely."

Roebling rejoined the Army of the Potomac in February 1863 back at Fredericksburg, where he was quartered late one night in an old stone jail, from which he would emerge the following morning with a story that would be told in the family for years and years to come. The place had little or no light, it seems, and Roebling, all alone, groping his way about, discovered an old chest that aroused his curiosity. He lifted the lid and reaching inside, his hand touched a stone-cold face. The lid came back down with a bang. Deciding to investigate no further, he cleared a place on the floor, stretched out, and went to sleep. At daybreak he opened the chest to see what sort of corpse had been keeping him company through the night and found instead a stone statue of George Washington's mother that had been stored away for safekeeping.

It was shortly after that when he was reassigned to the staff of General G. K. Warren. Then came Chancellorsville, where Hooker was facing Lee with more than twice the men Lee had and seemed to have forgotten anything he ever knew about commanding an army. At one point Roebling found himself leaning against the same post as Hooker, just as it was about to be split in two by a cannon ball. For years afterward, he would speculate on how history might have been altered had he not shouted a warning when he did. "Fighting Joe" Hooker would have been fighting no more, Roebling reasoned, and with another man in charge his army might have won the battle.

In the weeks after Chancellorsville, Roebling began going up in a reconnaissance balloon every morning at daybreak to see what the enemy might be up to and it was he, on one such flight, who first discovered that Lee had started to move again, toward Pennsylvania and Gettysburg.

That was in early June. On the 24th he was handed orders from Warren to proceed at once to Washington, Baltimore, and Philadelphia if necessary, to find the best available maps of Maryland and the southern border of Pennsylvania. Warren happened also to be on his way to Baltimore to get married, so Roebling accompanied him that far, the two of them riding all night. Then he went on to Philadelphia and to Trenton, where his father, he knew, had one of the best and latest maps of Pennsylvania.

He startled everyone in the big house with his sudden appearance there after dark. His father especially was very much alarmed. He stayed all of an hour. By the evening of June 29 he was back in Baltimore with the maps, only to find that Warren had already left to rejoin his army. The whole city was in a state of panic. The following morning, every bell was ringing an alarm as he headed west on horseback, toward Frederick.

He did not find Warren until he reached Gettysburg, on the second day of the battle. After that, events followed swiftly.

Through the whole war Roebling said very little about battles in his letters and next to nothing about his own exploits. But he seems to have had a great gift for being on the spot when needed. Long after the war, at the request of a friend who was also at Gettysburg, he gave this account of what happened:

> At Meade's headquarters I found General Warren. After making myself familiar with the situation and looking around, Meade suddenly spoke up, and said Warren! I hear a little peppering going on in the direction of that little hill yonder. I wish you would ride over and see if anything serious is going on, and attend to it. (This is verbatim.)
>
> So we rode over . . . Arriving at the foot of the rugged little knob, I ran up to the top while Warren stopped to speak to General Weed. One glance sufficed to note the head of Hood's Texans coming up the rocky ravine which separates little and big Round Tops. I ran down, told General Warren, he came up with me and saw the necessity of immediate action.
>
> . . . Without waiting for General Sykes' approval, who was some distance ahead, Warren ordered these troops to face about and get into line, covering little Round Top and the adjacent ground. Firing began at once. It was deemed very important to get a section of artillery up there [on Little Round Top].
>
> Hazzlitt's battery was nearby, it started up the hill, but the horses could not pull it up, so all hands took hold of the wheels and tugged away. I strained at one hind wheel, and you, my dear Sir, at the other hind wheel, until we reached the summit, and some shots were fired. They had a great moral effect, as the enemy supposed the hill to be unoccupied.

The safeguarding of Little Round Top would be viewed by many historians as the turning point of the war. Warren, understandably enough, got nearly all the credit and would be remembered as one of the heroes of Gettysburg. But later on, speaking generally of his young aide-de-camp, Warren said, "Roebling was on my staff and I think performed more able and brave service than anyone I knew." Roebling himself would be characteristically laconic and self-effacing. "I was the first man on Little Round Top. There is no special credit attached to running up that little hill, but there was some in staying there without getting killed."

Roebling's morning on Little Round Top would be the thing people would talk most about when describing his war record. But he had also been near Sickles when that flamboyant figure lost his leg. He helped engineer the tunnel under the Confederate lines at

Petersburg, the daring scheme that so very nearly worked, but resulted in the disastrous Battle of the Crater. Once, before the great blast went off, before dawn, with the moon still up, he and Warren had crawled on their stomachs to the very edge of Lee's works.

Later, he would also write this memorable description of Abraham Lincoln:

> . . . I was in the Civil War for four years and saw Lincoln on two occasions—the first in May 1861, when he spoke a few words of welcome from the rear portico of the White House to the newly arriving soldiers, one of whom I was, and secondly about April 1, 1864, when he came down to Culpeper County to review the army previous to the Battle of the Wilderness. I was at that time major and aide-de-camp to General Warren, commanding the 5th Corps, and joined the cavalcade.
>
> The President was mounted on a hard-mouthed, fractious horse, and was evidently not a skilled horseman.
>
> Soon after the march began his stovepipe hat fell off; next his pantaloons, which were not fastened on the bottom, slipped up to his knees, showing his white homemade drawers, secured below with some strings of white tape, which presently unraveled and slipped up also, revealing a long hairy leg.
>
> While we were inclined to smile, we were at the same time very much chagrined to see our poor President compelled to endure such unmerited and humiliating torture. After repairs were made the review continued . . .

As the war dragged on, Roebling, like thousands of others on both sides, grew increasingly despondent, wondering if ever there would be an end to the killing days, as he said. "They must put fresh steam on the man factories up North," he told Emily, "the demand down here for killing purposes is far ahead of the supply; thank God however for this consolation that when the last man is killed the war will be over."

The real heroes, he said, were the privates in the line, but then added bitterly, "When I think sometimes what those men all do and endure day after day, with their lives constantly in danger, I can't but wonder that there should be men who are such fools, I can't call them anything else. And that is just the trouble we are laboring under now—the fools have all been killed and the rest think it is about played out to stand up and get shot." It was only a matter of time, he believed, before he too would be dead.

Off duty he played cards, picked fleas, smoked cigars, drank whiskey whenever he could get it, cursed the heat and tried to think of Troy, New York, in the winter of 1856, when the thermometer outside his bedroom window had marked 20 below. Like others about him, he developed increasing sympathy for the people he was working so hard to defeat. He wrote again to Emily, ". . . the conduct of the Southern people appears many times truly noble as exemplified for instance in the defense of Petersburg; old men with silver locks lay dead in the trenches side by side with mere boys of 13 and 14; it almost makes me sorry to have to fight against people who show such devotion for their homes and country."

Emily had come into his life on the evening of February 22, 1864, at the Second Corps Officers Ball, which had been held in a huge wooden hall especially built for the occasion under his supervision. "In point of attendance," he had written to Elvira, "nothing better could have been desired; at least 150 ladies graced the assemblage, from all quarters of the Union, and at least 300 gentlemen from General Meade down to myself." The occasion was a grand success.

> Our supper cost 1500 dollars and was furnished by parties in Washington. The most prominent ladies of Washington were present from Miss Hamlin, Kate Chase and the Misses Hale down. Last but not least was Miss Emily Warren, sister of the General, who came specially from West Point to attend the ball; it was the first time I ever saw her and I am very much of the opinion that she has captured your brother Washy's heart at last. It was a real attack in force. It came without warning or any previous realization on my part of such an occurrence taking place and it was therefore all the more successful and I assure you that it gives me the greatest pleasure to say that I have succumbed . . .

They wrote to each other almost daily after that and met again, two or three times, at General Warren's wife's home in Baltimore, and at camp, when Emily and young Mrs. Warren came to visit the general. At about the time Lincoln made his visit, Roebling was writing to his father to tell him he planned to be married, expecting all kinds of arguments in return. But the letter from Cincinnati was not what he expected and represents one of those rare instances when John A. Roebling revealed his affection for his oldest son, as well as his total confidence in his judgment.

My Dear Washington,

Your communication of the 25th came to hand last night, and I hasten to reply. The news of your engagement has not taken me by surprise, because I had previously received a hint from Elvira in that direction. I take it for granted, that love is the motive, which actuates you, because a matrimonial union without love is no better than suicide. I also take it for granted, that the lady of your choice is deserving of your attachment. These two points being settled, there stands nothing more in your way except the rebellion and the chances of war. These contingencies having all passed away, you and your young bride, as you know beforehand, will be welcome at the paternal house in Trenton. Our house will always be open to you and yours, and if there is not room enough, a new one can be built on the adjoining ground, or one can be rented.

As to your future support, you are fully aware, that the business at Trenton is now suffering for want of superintendence, and that no increase or enlargement can be thought of without additional help. Of course I do not want to engage strangers, and it is you therefore, who is expected to step in and help forwarding the interests of the family as well as yourself individually . . .

Should you be in want of money at any time, let me know.

I conclude with the request that you will assure your young bride of my most affectionate regards beforehand, and before I shall have the pleasure of making her personal acquaintance.

Your affectionate father,

She had gone to Trenton herself after that. His father had met her in New York and they had taken the train to Trenton together. "I like her very much and have not the least doubt that your union with her will be a happy one," John Roebling informed his son. And Washington wrote to her at Trenton, "I dare say you could not sleep the first night on account of the water in the raceway making such a terrific noise. . . . Be sure and tell me all about your impressions . . . what do you think of Tilton the Bridgetender or Mitchell the lockkeeper or Mrs. Reilly that keeps the Irish tavern across the canal?"

Thereafter his days seemed endless. Little was happening and the boredom was unlike anything he had ever known, as he tried to describe for her:

This day might be signalized as one of the most uneventful ones I ever passed. I wrote perhaps two hours, fooled around for two more, walked for one, and that besides eating and drinking was the end of it. The programme at night is still more stupid, as it is chiefly

> spent shivering, turning over fifty times and occasionally dreaming of you. My mind is no longer as imaginative as it was 10 years ago, many of my dreams at that period are still vivid in my recollection. Had a great time hunting for a button tonight, finding none after all my search, and as I write the string at the bottom of my drawers comes off; that will be another sewing job before I go to bed.

He worried about his mother, who was failing rapidly, and not the least of her troubles, his father wrote, was concern over his safety. He thought about the future, worried that he had forgotten everything he ever knew about engineering, puzzled over where he and Emily might settle once the war was over. He pondered the possibility of not following in his father's footsteps. Trenton had no appeal, despite his father's generous proposals. "The town is horribly dull," he told her, "and I always get tired of it after being there a week." When he daydreamed of home, it was nearly always of Saxonburg. "I have now more lasting memories of the first eight years of my life there, than I have of the intervening twenty," he wrote. But there could be no going back to Saxonburg.

He grew a beard, changed the way he combed his hair to suit her, adopted two stray dogs, a family of kittens, a lizard, and took to sleeping with the blanket pulled over his face to keep off the flies, his feet sticking out at the bottom, a pose that inspired a fellow officer to do a pencil sketch, which Roebling included in one of his letters to Emily. In another letter, he told her, "I have been solacing myself all evening by playing on a fiddle, had a great time getting it, borrowed a bow from another man and stole the rosin from a sick horse; it did me good and so I played until the tips of my fingers began to ache . . ."

In November of 1864 his mother died and he hurried back to Trenton for the funeral; ". . . the greatest giver of us all [is] gone," he told Emily. Then, at the end of the year, the war, for him, was over. He was commissioned lieutenant colonel, by brevet, as of December 6, for gallant conduct during the campaign before Richmond. By Christmas he was home to stay. In January, at Cold Spring, New York, he and Emily were married.

They lived in Trenton at first and he went back to the wire mill. But that lasted only a few months. In the early spring of 1865, just before the war ended, he left for Cincinnati to join his father on the Ohio River bridge and once he found a place to live, she came on.

By the time the bridge was completely finished, they had spent nearly two years in Cincinnati, and his father had been back in Trenton the better part of that time.

In early 1867, when it appeared that the East River bridge at last had some serious support, his father had written to say he wanted him to go to Europe the following summer to make a study of pneumatic caissons, that Emily ought to go too, and that he would pay all their expenses. "Your kind offer . . . I accept with pleasure," Washington Roebling wrote in answer; "Emily is especially delighted, to her the idea of going to Europe is something exceedingly grand."

They sailed at the end of June, when Emily was a few months' pregnant. They were in England for a number of weeks, then France later, and Germany, where the baby was born. In London they visited St. Paul's, Parliament, Westminster, the Zoological Gardens. From their window at the Royal Hotel, he had watched construction of the Blackfriars Bridge and made a drawing of it to send his father.

He made flying visits to Telford's bridge over the Menai Strait and Brunel's bridge at Clifton, for a long time considered the most beautiful suspension bridge in the world. He did not think Telford's towers very handsome, he wrote to John Roebling. The floor, very light in weight, had no strengthening truss. There was a stiff breeze blowing as he walked out on it and the vibrations were very strongly felt, he said. The towers for the Clifton Bridge he called remarkably ugly.

They visited Manchester and he spent several days looking over the noted steelworks of Richard Johnson & Nephew, where the wire for the Cincinnati Bridge had been made. They went to Birmingham, Sheffield, Newcastle, then on to Paris to see the Exposition, but left after a week because they were running low on money. One lady costs twice as much in Paris as two men, he explained to his father. The fair was considered the greatest international exhibition ever and among the American Commissioners was his Trenton neighbor and competitor in the wire business, Abram Hewitt. But Roebling thought little of the fair, nothing but "a great advertising show," he said.

At Essen, Germany, to his astonishment, he was given the grand tour of the Krupp works, as though he were visiting nobility. On first arriving in Essen he had been told by several townspeople that an outsider had virtually no chance of visiting the works. He had spent one long night drinking wine with some of the young engineers employed there, hoping to win a friend who would open a few doors, but he had received only long faces whenever he mentioned the subject. Finally, figuring he had nothing to lose, he went directly to the main office, expecting to get nowhere and finding instead that they were eagerly awaiting his arrival. The management, he discovered, was well aware of the forthcoming East River bridge, knew all about his trip to Europe, and had already made up a sample eyebar for his personal inspection.

In all three countries he talked to bridgebuilders, wire manufacturers, visited iron works, filled his notebooks and letters with tens of thousands of words and hundreds of tiny freehand drawings and diagrams. He had an almost uncanny gift for observation and could commit vast quantities of information to memory, yet he never gave any sign of making a special effort along those lines. He would walk about a construction site or through a mill listening politely to his host, paying, it would appear, only the most casual attention to what he was being shown. Then he would return to a hotel room and write down a full description of what he had seen, with the most extraordinary memory for detail and great critical analysis.

After his return from Europe, for instance, while stopping over in Pittsburgh, he was invited to take a walk through Andrew Carnegie's new Keystone Bridge works and that night, in private, at the Monongahela House, he had written his father a complete report of the entire Keystone operation, describing the different machines in use and how they worked, the production patterns followed, the personnel involved, the various products turned out and his opinion on their relative merits. The letter went on for pages, even in his minute hand, and it was all put down directly, with no apparent hesitation, no erasures or editing, and, apparently, with no special effort. At the close, after remarking that he found Carnegie and his brother "very pleasant people," he told his father, "I could fill another letter of 20 pages to describe all I saw—still I keep it all in mind."

The value of such a son was not lost on the father, and as the young man made his way across Europe, he sent off one letter after another, each taken up almost entirely with technical matters—on wiremaking, on the latest developments in metallurgy, Bessemer steel making in particular, and on caissons, which he spelled "cassoons" for some reason—and everything was set forth with the sort of clarity and thoroughness demanded by the exacting mind back in

Trenton. By the time he and Emily and the baby were on their way home across the Atlantic in March of 1868, he knew more on the subject of pneumatic caissons than any American engineer.

The success of the bridge at Brooklyn, he and his father knew, would depend on the caissons. Everything hung on their success. If they could be sunk beneath the river properly and to the required depths—and they would have to be bigger by far than any caissons ever constructed before—then there seemed little doubt that the bridge could be built. If not, there would be no bridge, or at least not the one John Roebling had described with such persuasive language in the fall of 1867, when, as it happens, his son was still in Europe trying to determine just how the thing could best be done.

After returning home, Roebling busied himself at the Trenton mill, waiting for the New York politicians to settle their affairs. In 1868 his father finished the basic plans for the caisson, while he made another long tour at his father's request, this time through the hard-coal regions of eastern Pennsylvania to see how wire rope could be used in mining operations. He and Emily were staying in the big house temporarily. His father had remarried. Ferdinand too was married by this time. Laura came visiting from Staten Island with her children. By fall the baby was walking.

When the word came from Brooklyn that the bridge was all set to go, he and Emily packed up and moved to the house on Hicks Street.

In the months immediately following his father's death, Roebling spent much of his time away from Brooklyn. He made repeated trips to Trenton, where he had his father's estate to settle, as well as the wire business to look after. There were long family gatherings in the big house. It was agreed that Ferdinand would take charge of the business eventually and that Charles would come in with him as soon as he finished at Troy. But for the time being Washington would make the major decisions. Nobody foresaw any problems with that, but not very long afterward, at a time when Washington was having troubles enough in Brooklyn, a full-scale family fight developed over what to do about Charles Swan. Ferdinand did not want Swan made a partner, as his father had requested in his will. Washington said he should be. Swan quit and went off somewhere, leaving no word behind; plainly furious. But things were patched up eventually and Swan returned to work, on generous financial terms, but not as a partner.

It was also agreed, after some difference of opinion, that Washington was to be Eddie's guardian. Roebling felt he had more than enough to cope with as things were, building his father's bridge and looking after his own small family, without trying to be the father his father had never been to his troubled little brother. But the others felt differently, it appears, and so Eddie was packed off to Brooklyn, where Emily had a room prepared for him.

For Roebling, there was still more traveling to be done. He spent a day with Horatio Allen in Port Chester looking at dredging gear. He went to Albany to look at the black limestone and granite going into the foundations of the new state capitol, which had been engineered by William Jarvis McAlpine. He stopped at Kingston to visit a limestone quarry and to talk to the people there. He filled his small black leather notebook with pages of names, addresses, and reminders. "Find out where they get the broken stone at the Post Office. . . . Find out all about calcium light. . . . Find out who makes derrick forgings." He went to Niantic, Connecticut, to look at granite and to Hallowell, Maine, to inspect the quarries of J. R. Bodwell. The infamous Tombs prison had been built of Bodwell's stone, as had the coping of the huge new reservoir in Central Park, some of which had been built by William C. Kingsley. The quality of the granite, Roebling wrote in his notebook, was "very fine, very durable . . . the whitest granite known."

His most vital concern, however, was the first giant caisson. In mid-August, immediately after Collingwood reported for work, Roebling had handed him rough drawings and a long written account of what was wanted, and Collingwood, Paine, and Hildenbrand had set to work on final plans, figured down to the last inch. The only one who could handle such an order, it was decided, was a shipbuilder, and on October 25, 1869, the contract was awarded to the firm of Webb & Bell, whose yards were up the river at Greenpoint. And if a date were to be picked to mark the beginning of the building of the bridge, it probably ought to be that one.

The caissons were the foundations of the great work, quite literally and figuratively, as everyone working with Roebling in Brooklyn was aware. There were also some among them who appreciated something that would be overlooked by most everyone as time passed—that it was Washington Roebling, more than his father, who was the closest thing to an expert on caissons and who had concentrated on how that part of the job ought to be handled. And perhaps this helped compensate some for the very obvious fact that he had never actually worked with a caisson before and that there were only two or three American engineers who had, McAlpine being one of them. In St. Louis Captain James B. Eads had just completed the first caisson for his bridge over the Mississippi. Still, to all but a handful of engineers, even the word "caisson" was unfamiliar.

With absolute, unqualified conviction John A. Roebling had proclaimed that his bridge would be the greatest in existence, the greatest engineering work of the age. All his son had to do now was build it.

PART TWO

8

All According to Plan

> The foundations for the support of these large masses of masonry must be unyielding.
>
> —JOHN A. ROEBLING

THE EASIEST way to explain how the caisson would work, Roebling and his assistants found, was to describe it as a huge diving bell that would be built of wood and iron, shaped like a gigantic box, with a heavy roof, strong sides, and no bottom. Filled with compressed air, it would be sent to the bottom of the river by building up layers of stone on its roof. The compressed air would keep the river out, help support the box against the pressure of water and mud, and make it possible for men to go down inside to dig out the riverbed. As they progressed and as more stone was added, the box would sink slowly, steadily, deeper and deeper, until it hit a firm footing. Then the excavation could stop, the interior of the box would be filled in solid with concrete, and that would be the foundation for the bridge tower.

The idea was quite simple really. Furthermore, it had been used successfully in Europe for more than a generation, first in France, where the word "*caisson*," meaning "chest," had originated, then later in England and Germany. An air lock that enabled workers to get in and out of a sealed chamber filled with compressed air had been devised as early as 1831 by Lord Thomas Cochrane, the controversial British admiral, and in 1851 a pneumatic caisson had been used on a bridge foundation for the first time, for piers in the Medway River at Rochester, England. Seven years after that Brunel had taken a caisson down more than seventy feet to build a pier for his last and greatest railroad bridge, the Royal Albert over the Tamar at Saltash, Cornwall.

But the caisson Roebling intended to sink beneath the East River on the Brooklyn side would be bigger by far than anything used in Europe or the few that had been used in the United States, and the one for the New York tower, it was then thought, would have to go thirty to forty feet deeper than even Brunel had gone.

The caisson the Roeblings had designed, and which in the late fall of 1869 began slowly to take shape at the Webb & Bell yards, was to be built like a fort and launched like a ship. A gigantic rectangular box, 168 feet long and 102 feet wide, it was to have nine and a half feet of headroom inside and overhead a roof of solid timbers five feet thick, bringing the total height of the box to fourteen and a half feet. In area it would be more than half the size of a city block, more than half the size of the new St. Patrick's Cathedral, for example. The sides of the box were to be V-shaped, being nine feet thick where they joined the roof and tapering to a bottom cutting edge of just eight inches. The inside slope of the V would be at an angle of about forty-five degrees and the entire cutting edge, or "shoe," would be shod with a heavy iron casting and sheathed the whole way around with boiler plate extending up three feet, inside and out. A heavy oak sill two feet square would rest on the casting.

Driftbolts, screw bolts, and wood-screw bolts would be used to secure the whole immense mass. The V-shaped sides would be fixed to the roof with heavy angle irons. At the corners, timber courses would be halved into each other and strapped together. The roof itself would be built of five solid courses of yellow pine "sticks"—timbers a foot square—laid up side by side and bolted both sideways and vertically.

To make the box airtight the seams would be caulked with oakum to a depth of six inches, inside and out, and between the fourth and fifth courses of roof timber, across the entire top of the structure and extending down all four sides, a vast sheet of tin would be put down. The tin on the outside would be further protected by a sheathing of yellow pine and the spaces between timbers would be filled with hot pitch.

Since Roebling had learned that air under pressure of forty or fifty pounds (or about what would be needed inside the work chamber) will readily penetrate ordinary wood, he had selected a yellow pine from Georgia and Florida that was so pitchy that the 12-by-12 sticks would not even float. And finally, just to be sure, he planned to coat the whole inside with a specially concocted, supposedly airtight varnish.

Once the caisson was launched, ten additional courses of timber would be added to the roof, thereby making it a solid timber platform fifteen feet thick, which would act like a colossal wooden beam for carrying and distributing the load of the tower.

As the work at Webb & Bell progressed on the raw and ungainly-looking structure, two air locks, two so-called supply shafts, and two water shafts were being built into the timber roof.

The air locks were made of half-inch boiler plate. They were cylindrical in shape, seven feet high by six and a half feet in diameter, or big enough to pack in about a dozen men, who would enter from the top, through an iron hatch about the size of a manhole cover. The hatch closed, an attendant stationed inside would open a valve near his feet, releasing compressed air from the work chamber below into the lock. As soon as a gauge on the wall of the lock showed the pressure there equal to that below, another hatch in the floor would be opened and the men, one by one, would climb down a ladder through an iron shaft, three a half feet in diameter, into the caisson itself.

When it was time to come out, the process would be reversed. They would climb up the shaft and into the air lock, the floor hatch would be secured, and another valve would be opened to release the air from the lock. When pressure in the lock decreased to normal atmospheric pressure, or 14.7 pounds per square inch, then the top hatch would be opened and the men would climb out. This way the loss of compressed air from the caisson each time a gang of men went in or out was no more than the volume of the lock.

The water shafts were a very simple but ingenious means by which the mud and rock excavated inside the work chamber could be hauled out with no loss of compressed air whatsoever, and with none of the time required to move through an air lock.

The water shafts were seven feet square, open at top and bottom, and they extended like twin wells through the roof of the caisson straight down into the work chamber to a depth nearly two feet below that of the caisson's bottom edge. These shafts were also built of boiler plate and once the caisson was in operation they would be filled with water to a level sufficient to "lock in" the compressed air below.

At the base of the shafts, at the two points where they extended deeper than the caisson itself, open pits would be dug in the river bottom, which would also fill with water. These would be the delivery ends of the shafts and the columns of water within the shafts would be kept suspended (kept from flooding in on the workers) by the pressure of the air within the chamber.

The water shafts, as one magazine of the time explained for its readers, were essentially huge barometers that measured the pressure of the air inside the caisson. The shaft itself was the barometer tube, filled with water instead of mercury, and the pool at the bottom was the cistern. Every pound of pressure in the caisson above normal atmospheric pressure (which of course was bearing down on top of the column of water) forced the water a little more than two feet higher in the shaft.

To get rid of the material they excavated, the men would shovel it into the pits, or pools, at the base of the shafts, where it would be hauled up and out by big clamshell dredge buckets dropped down from above, directly through the shafts of water. The theory was the buckets could work as fast as the men could feed them. It was a neat, efficient system, so long as the water in the shafts stayed at the proper level. But if the volume of water in one shaft became too great—too heavy, that is, for the compressed air below to support it—then the water in the pit would flood out into the work area. Or, if for some reason, the volume of water decreased to the point where its weight was no longer enough to counteract the pressure in the chamber, then there would be a terrific release of air, or blowout, from below.

The supply shafts were only twenty-one inches in diameter and simply the means by which Roebling intended to get the necessary cement, sand, and gravel into the caisson once the excavation was finished.

John Roebling's thought had been to make the interior of the caisson one big open space, with no divisions or supports to get in the way of the excavation work. But his son had to abandon that idea for several reasons. First of all, since the caisson would have to be launched like a ship—only in this case a ship built upside down—there would have to be supports of some kind between the launchways and the roof. Washington Roebling also anticipated that the cutting edge of the caisson would be striking on boulders on its way into the riverbed and when that happened he did not want the entire weight resting solely on those few points. But it was chiefly because of the particular nature of the East River that he decided to divide up the work chamber with a number of supporting walls.

The East River connects the Upper Bay of New York with Long Island Sound, and because it has two entrances—at the tip of Manhattan and at Hell Gate, the opening to Long Island Sound—and two distinct tidal movements arriving at these points at different times of the day, its currents are quite unlike those of any ordinary river. The water is full of whirlpools and eddies caused by a bottom of jagged peaks and huge potholes, some as much as fifty feet deep. And with the tides surging in and out of the narrow openings, the currents are swift, turbulent, and something very serious to contend with. Even with a favorable wind the great sailing ships of the day could make little headway against an outgoing East River tide and would often stand in considerable numbers in the bay, like small armadas, waiting for the tide to change.

"The extreme rise and fall," Roebling explained, "is seven and a half feet. If the inflated caisson is just barely touching the ground at high water, it will press upon the base with a force of 4,000 tons at low tide, all of which has to be met by the strength of the shoe and the frames." Not until the caisson was permanently "righted down" under several hundred tons of tower stone would this powerful, potentially destructive up-and-down action stop.

So he had Webb & Bell build in heavy truss frames of pine posts and stringers, with three-inch sheathing on each side and side braces to the roof every six feet. There were five of these inside partitions, each running the width of the caisson and dividing the interior into six separate chambers, each 28 by 102 feet. Once the caisson was in the water and resting on the bottom, doors would be cut in the partitions so the men could go back and forth from one chamber to another.

As the mammoth timber box grew on the ways, it looked like nothing ever seen before in an East River yard. Seven launchways were required (one for each of the outer edges, five under the interior partitions), and the total weight of the structure, by the time it was ready for launching, would be six million pounds, or three thousand tons, which was, for example, a thousand tons more than the *Challenge*, leviathan of the clipper ships built in East River yards. The caisson would contain some 110,000 cubic feet of timber and 230 tons of iron and it was being built to go down the ways in the usual fashion of a great ship, its long side toward the water. For the time being it stood fifty feet back from the ends of the ways, and as everyone who had had any experience with shipbuilding knew, the great danger of launching so large a mass was the chance that one end might get going faster than the other and the whole gigantic affair would wedge tight on the ways. It was also necessary of course that the caisson get up enough momentum coming down the ways to overcome the immense resistance offered by the water. So just getting the thing launched was an engineering problem of very major proportions. Indeed, Roebling said later that the problems of launching the caisson and of protecting it against sea worms caused him more anxiety than the prospect of sinking it.

As might be expected, all such questions and the steady progress of construction were of enormous interest to innumerable bystanders. Day after day people came down to the yards at the foot of Noble Street to take a look for themselves, even after the first snows arrived. Newspapermen and some of Tweed's people came over from New York, as well as a number of engineers, not the least of whom was Captain Eads from St. Louis, whose own caisson was being sunk beneath the Mississippi by this time and who happened to be in New York on some other matter.

So concerned were the Webb & Bell people over the problems involved, so different was this job from any they had ever handled in all their previous experience, that they had insisted on being paid in advance—$100,274.51.

Once the caisson was in the water, the plan was to tow it downriver. How seaworthy it would prove to be in the turbulent current was another open question. After giving the caisson a thorough inspection, James B. Eads told Roebling he could expect trouble and said it might topple over if he inflated it during the trip downstream, which was exactly Roebling's intention.

In the meantime, however, the waterfront had to be cleared at the point where the caisson would be docked and the riverbed had to be dredged deep enough for the huge structure to be floated into place. The clearing of the site began on Monday, January 3, 1870, and because the winter turned out to be abnormally mild, the work there, like the work at the shipyard, moved along faster than expected. Any other winter it would have been impossible to do much of anything.

Clearing the site took about a month. For daily commuters on the Fulton Ferry it all provided an interesting show and the first real sign they had had that the bridge was actually under way. About half of one big ferry pier had to be dismantled, fender sheathing torn out, massive stone-filled cribbing removed, and all without disrupting ferry service. An enormous steam crane, called the "Ox," was brought in on a barge to pull out the old piles, and as they came up one after another, there was much amazement over the toll the East River had exacted. Each one was infested with thousands of sea worms in the area between the low-water mark and the mud line. As Roebling noted, "A pile which was sixteen inches in diameter below the mud, perfectly sound and free from worms, would be found eaten away to a thin stem of three inches just above the mud, and all timber was affected alike." Then so that no one missed the point, he added, "This shows the necessity of going below the top of the riverbed with our timber foundation, and also proves its entire safety in that position."

Once the old dockwork was out of the way, a large basin was to be built to contain the caisson, open toward the river but bounded on three sides with new piling. Within this area the riverbed was to be dredged to a uniform depth of eighteen feet at high tide, or deep enough to keep the caisson afloat at all stages of the tide. The dredges made great headway at first, as long as there was only surface mud to contend with, but then they hit hardpan and boulders. "The character of this material was next to solid rock," Roebling wrote. The dredges could make but the slightest impression upon it. "Recourse was necessarily had to powder," and the blasting commenced at night, from about seven in the evening until daylight, when traffic was light on the river and few people were about the ferry slip. Holes were driven into the river bottom with steel-headed iron piles. Then blasting charges were packed into iron canisters and dropped into the holes by professional divers. When the divers were out of the way and the pile drivers hauled back to a safe distance, the charges were set off by electricity.

Three pile drivers were kept in action, and with a little practice the men had the work down to a neat system, setting off some thirty-five blasts every ten-hour shift. During the day the dredges moved in and cleared out the results of the night's work.

A number of the boulders encountered were too large to be

picked up by a dredge and had to be dragged clear—the divers assisting underwater. The whole process was about six times as expensive as normal dredging, but still quite effective, and it provided valuable knowledge of the ground the caisson would have to penetrate. On one side, for example, near the new piling, a dozen blows of the pile driver would sink an iron pile forty feet through soft clay, but in the center area it took a hundred blows to go three feet. Toward the ferry the clay gave way to boulders of all sizes, closely packed, with coarse sand in between, and at the open end of the basin, on the river side, all soft strata had been washed away, leaving hardpan.

As time passed, Roebling decided to concentrate the dredging along the lines of the caisson's edges and frames; the parts in between could be removed later, he said, from inside the caisson. He also had two holes blasted to an extra depth to accommodate the water shafts.

The work went slowly now, and while the blasting and dredging provided valuable knowledge of the riverbed, that knowledge itself was a most sobering reminder of the magnitude of what they were undertaking. To sink a wooden box as big as a fair-sized railroad station straight down through such material, and underwater, keeping the thing absolutely level the whole time, and bringing it to rest finally—perhaps fifty feet down—and at the exact spot it was meant to be, was a very tall order indeed. And added to that, along toward the end of January, reports began coming in from St. Louis of a strange malady among the men working inside the Eads caisson.

James Buchanan Eads was an authentic American genius and one of the looming figures of the nineteenth century. Slim, leathery, highly opinionated, disliked by many, he had survived an extraordinary life on the Mississippi that had included a lucrative underwater salvage business, a financially disastrous attempt at glass manufacturing, and the building of a fleet of ironclad gunboats during the Civil War. These slow, squat, ugly warships, built before the *Monitor* or the *Merrimac* and nicknamed "the Turtles," had played a decisive part in defeating the Confederates on the Mississippi, along with the rams built by Charles Ellet. Eads had not designed the ships himself, nor had he gone into battle with them as Ellet had with his rams, but he had organized everything, having timber cut in Minnesota and Michigan, iron armor rolled at St. Louis and Louisville, keeping four thousand men at work on a night-and-day basis, and financing much of the operation out of his own pocket. At the time Washington Roebling was distinguishing himself on Little Round Top, Eads's gunboats were assisting Grant in the successful siege of Vicksburg.

In early 1870 Eads was approaching fifty. He was the sort of person who liked to play chess with two or three others at a time, and in a recent weight-lifting contest among some of his blacksmiths, he had come in second.

During his years in the salvage business Eads had worked with diving bells up and down the Mississippi and was said to know more than any man alive about the river's treacherous currents and the character of its bottom. This had been an important factor when he presented St. Louis and New York financial backers with his radical proposal for a bridge over the Mississippi. But it was his unbridled self-confidence and his reputation as a man who could get things done that mattered most in the end. He managed to convince men who had worked with the country's foremost engineers that he, James B. Eads, was the one man fit to bridge the Mississippi at St. Louis, that the bridge he wanted to build was the only answer, and this despite the very well-known facts that he had had no formal training as an engineer and that he had never once built a bridge before. Both Charles Ellet and John A. Roebling had prepared plans for suspension bridges at St. Louis back in the 1850's. Later, the year before he died, Roebling had done an entirely new set of plans, combining both suspension cables and parabolic arches. But Ellet's and Roebling's ideas had been turned down. (The St. Louis people were fools, John Roebling wrote to his son.) Now Eads and his bridge were the talk of St. Louis.

The great need was for a bridge to carry a railroad and highway over the river without interfering with steamboat traffic. The Mississippi at St. Louis is about the same width as the East River. Instead of a heavy iron truss, the customary thing then for railroad crossings, or a suspension bridge, Eads had conceived a mammoth arched bridge, with arches of steel set on stone piers. He intended to span the river with just three of his steel arches, the biggest of which, the center span, would be longer than any arch of the time by several hundred feet. To avoid interfering with river traffic during construction, his assistant, an engineer named Henry Flad, had devised a cantilever system nobody had tried before. The halves of each arch would be built out toward one another from their respective stone foundations, like great jaws slowly closing over the river, which was the conventional way, except that here the temporary supports needed (until the jaws joined) would be supplied from *above*. The usual practice was to prop such arches up from below with temporary timber "falsework" that could be torn out once the bridge was finished. But since this would be impossible, obviously, if the river was to be kept clear, Eads would hang the arches from overhead cables attached to temporary wooden towers built above each of his stone piers.

So the design of the bridge, the material he intended to build it with, the way he planned to build it, just about everything about the bridge, was unorthodox and untried. And when he had first proposed it, Andrew Carnegie had decided that somebody who knew about things mechanical, as he said, had better look over the plans.

Carnegie's interest in the bridge was twofold. He had been approached by Eads's St. Louis backers to see if he might be interested in selling some of their bridge bonds. Also, it was a few years before this that he had organized his Keystone Bridge Company, one of the first to specialize in manufacturing iron railroad bridges. Carnegie enjoyed talking about his love of bridges. Like Thomas Pope and John Roebling he saw them, he said, as testimonials to the national spirit and professed great personal satisfaction in the part he played in building them.

The Keystone company was now being invited to come in on the St. Louis job as consultants and to handle the superstructure. So Carnegie, quite sensibly, asked for an opinion on the bridge from Keystone's chief engineer and president, J. H. Linville, whom Carnegie described with customary enthusiasm as "the one man in the United States who knew the subject best." This was an overstatement, but Linville was certainly among the finest men in engineering. He had been bridge engineer for the Pennsylvania Railroad before Carnegie hired him and the huge iron truss he had built over the Ohio at Steubenville in 1864 was considered the outstanding structure of its kind.

Linville asked that a set of Eads's plans be sent to him. He examined them carefully, then, a little like the paleontologists who had been asked to give an opinion on the Cardiff Giant, he solemnly declared the subject preposterous. "The bridge if built upon these plans will not stand up; it will not carry its own weight," he told Carnegie in private, and presently, in a formal statement, he called the bridge "entirely unsafe and impracticable" and said any association with it on his own part would imperil his reputation and was therefore out of the question.

Linville was quite wrong and Carnegie, who knew nothing about engineering, urged Linville to lead Eads "into the straight path." Eads, however, was not about to be dissuaded or to have any outsider, regardless of reputation or connections, begin doctoring his bridge. In the end he would convince even Linville that he knew what he was doing. The Keystone company went to work on the bridge; Carnegie went off to London to sell a block of bonds to the American financier Junius Morgan, father of J. P. Morgan; and by the summer of 1867 Eads was confidently proceeding with the preparatory work for the first abutment beside the St. Louis waterfront. In neighboring saloons it was said that the bridge would take seven million dollars to build—and seven million years.

As things turned out the final cost would come to something near ten million, and seven years would go by before the job was completed. Once in use the bridge would be acclaimed by everyone, and by engineers especially. As one engineering historian would write, the bridge was "an achievement out of all proportion to its size," something Washington Roebling thoroughly appreciated at the time Eads came over to visit the Webb & Bell yards.

Like every bridge engineer and every railroad official in the country, Roebling was keenly interested in the St. Louis bridge, but since Eads, along with everything else in his radical scheme, also planned to sink his piers by means of pneumatic caissons, Roebling perhaps more than anyone appreciated the full daring of the man and the tremendous importance of what he was attempting, not just to his own work at Brooklyn but to the whole future of bridge engineering.

When he first envisioned his bridge, Eads had originally planned to use coffer-dams to sink the two midriver stone piers upon which his great steel arches were to rest. But in April of 1869, he had returned from a trip to Europe, convinced he had a better answer. He had seen the French engineer Moreaux use a pneumatic caisson to sink piers for a bridge over the Allier River at Vichy and he came home full of faith in the technique and sure he could make it work at St. Louis, even though the Mississippi, as he knew better than anyone, was not the gentle Allier.

So through that summer of 1869 Eads and Roebling had been devising their own separate plans for the foundations of the two biggest, most important bridges of the age, each man working quite independently and with only his own judgment to go by. Eads, however, had his caisson in the water by mid-October, before the contract with Webb & Bell had even been signed, and by the time Eads arrived in Brooklyn, his caisson was already well on its way into the sandy bed of the Missssippi.

Of the two, Roebling was unquestionably the better educated on the development of caissons in Europe and the various ways they had been used. Eads had happened onto the technique almost by chance and took about the least time possible to educate himself. Roebling's father had incorporated caissons in his plans from the start, knew much on the subject, and Roebling himself had taken great pains in his studies, spending close to a year in Europe for that specific purpose. Furthermore, unlike Eads, Roebling was a trained, experienced bridge engineer and was fluent in both French and German. Eads, who spoke only English, had had a difficult time conversing with some of the European engineers he met.

Still and all, Roebling doubtless appreciated that Eads was a man with a most uncommon gift for solving problems, a man of extraordinary originality and determination, a man, in fact, very much like his own father. Roebling also knew that what Eads was up against at St. Louis was far closer to his own situation in Brooklyn than anything the Europeans or McAlpine or anyone else had ever had to cope with. And most important, Eads, unlike Roebling, now had some working experience with caissons.

The caisson Eads had in operation was only about one-third the size of what Roebling was having built in Brooklyn, still it was bigger than any used by the Europeans. More significantly, by January 1870, the Eads caisson was already as deep as Roebling expected he would have to go on the Brooklyn side, and it was still descending steadily through Mississippi sand and mud that offered almost no resistance. By the end of January the trouble had begun.

From the very first Eads's men had noticed certain peculiarities about working in the heavy atmosphere of the caisson. The most manly voices had a thin girlish sound, for example. It was impossible to whistle or to blow out a candle, as the men gladly demonstrated for the many visitors Eads liked to bring down. Some of the men mentioned a notable increase in their appetites. Others talked of trouble breathing or of a painful ringing in the ears. But by the time they were down forty feet there had been several clear cases of the mysterious sickness, a subject Eads and Roebling had both heard something about in Europe.

As early as 1664 an English doctor named Henshaw had published an essay proposing, ironically enough, that compressed air be used as a method of treating a variety of common disorders. In France and Germany, institutions sprang up offering the latest facilities for just such atmospheric treatment. Compressed-air "baths" were claimed to work miraculous cures and became something of a fashion, and particularly for curing indigestion. But the pressure in such baths was never much greater than normal.

The first civil engineer to work with compressed air of any substantial magnitude, however, was a Frenchman named Triger, who in 1839, or thirty years before Eads and Roebling built their caissons, used compressed air inside an iron tube to hold back water while sinking a mine shaft through quicksand. The technique had worked quite successfully, but before the job was completed, Triger observed a number of unexplainable reactions among his men and put down in his notes what are thought to be the earliest recorded cases of caisson disease, or "the bends." Two of his men, Triger wrote, had been hit quite mysteriously by sudden sharp pains in the arms and knees about half an hour after coming out into the open air.

Later in France there would be more serious cases. Men would be seized at home, long after coming out of compression. Sometimes the pain was accompanied by chills and vomiting. Other symptoms were recorded: a great dullness of mind, an incoherence of speech or stammering, nosebleeds, a distressing itching of the skin, tottering gait, an increased flow of urine, even pain in the teeth. One supposedly scientific study noted that Hungarians and French suffered least, while Italians, Germans, and Slavonians were said to have had by far the worst time. It was also known for a fact that one or two men had died of the experience.

The first signs among Eads's men had been occasional muscular paralysis in the legs. But there was no pain connected with it, the men said, and the sensation passed off in a day or so. But as the caisson went deeper more and more of them began having trouble. In some cases now the arms were affected, as well as the bowels and sphincter muscles. Men complained of severely painful joints and sudden, excruciating stomach cramps. Still, nine out of ten of those affected felt no pain whatever, they said, and so long as the phenomenon remained painless, it would not be taken very seriously. Indeed, according to one account, "A workman walking about with difficult step and a slight stoop was at first regarded as a fit object for jokes, and cases of paralysis and cramp soon became known popularly by the name of 'Grecian Bend.' "

To ward off trouble the men rubbed themselves with an "Abolition Oil" that was said to work like a charm. Some of them began wearing bands of zinc and silver about their wrists, arms, and ankles, and such were the claims of success that Eads decided to thus outfit every man on the force at the company's expense, only now the protective armor, as the men called it, was worn about the waist as well, and even under the soles of the feet. Still instances of the unaccountable malady continued to increase.

When one of his foremen got sick, Eads decided to shorten the shifts inside the caisson. The men would stay down for four hours only, then rest for eight hours before going back for another four. The caisson was at forty-two feet by then. By February 5, when it was at sixty-five feet, Eads again altered the schedule, to three two-hour shifts, with rests of two hours in between, none of which was very popular with the men, since with every change of the shift they had to make a long climb out of the caisson, up a spiral stairway. For those who felt no adverse reaction from the compressed air, the new routine was just one more big inconvenience, while for those who did, the climb was only added torture. As the official history of the bridge states dryly, "The fatigue of ascent added not a little to the distress and prostration of those affected with cramp." At seventy feet, on February 15, with the air pressure in the chamber at thirty-two pounds per square inch, or more than double that of normal atmospheric pressure, one man was in such pain that he was sent to the hospital.

Severe cases grew a lot more common after that. One man became unconscious and did not speak for three hours. Nobody considered the thing a joke any longer. But even so, as Eads would tell visitors, many of his men, the majority in fact, had been affected in no way at all. He had taken hundreds of visitors down into the caisson, even "delicate ladies," he said, without any of them experiencing ill effects. There was no doctor who could explain it satisfactorily for him. Some doctors said a slower transition from the abnormal to natural pressure would prove less injurious; others claimed the contrary, that the trouble came from passing too rapidly from natural into compressed air. But Eads argued that neither could be correct since none of his air-lock attendants had been hit. It was the amount of time spent under compression that caused the trouble, he maintained, plus the general physical condition of the individual.

He pointed out that most of the men who had been struck down were new hands, unaccustomed to the work, that they had been thinly clothed and poorly fed to begin with, or, in some cases, alcoholic. So as the caisson continued its descent, Eads ordered that only men in prime physical shape be hired for the work.

Then on Saturday, March 19, which happened to be the same morning the Brooklyn caisson was launched, Eads reported the first death. The man's name was James Riley. He had worked the first shift, just two hours in the chamber, came up feeling fine so far as anyone knew, then fifteen minutes later gasped for breath and fell over on his face. He was the first American to die of the mysterious disease. But at least fifteen more would die at St. Louis before Eads finished his bridge, and more would be crippled for life.

About three thousand people turned out to watch the launching of the Brooklyn caisson. The Kings County Democrats, to no one's surprise, took the opportunity to make it a day of speeches and band music. People had trouble thinking of a suitable way to describe the main attraction, but most eventually concurred that it looked "more like a huge war leviathan or battery for harbor defense than any other thing." And as the *Eagle* observed, a very large number of them had turned out chiefly because they doubted it could ever be launched.

The top, or deck, of the caisson was strewed with tackle and various odd-looking pieces of machinery. A number of lines were connected to a steamboat standing by in case of trouble going down the ways, and at the rear of each way, heavy wooden rams had been rigged, to be worked simultaneously, to get the huge structure started. Inside, a temporary airtight compartment had been built on the forward wall to buoy up that side as it hit the water, and a full complement of crabs, winches, and 150 wheelbarrows had been stowed away, battened down with strips of wood.

The launch took place at ten thirty and was in every respect a great success. As soon as the last block was split out, the giant mass began to move. It went down straight and even, with no need of assistance. It struck the river with just enough speed to overcome the resistance of the water and the air chamber worked to perfection, keeping the front side from sinking. The deck never even got wet.

A great roar went up from the crowd. An air pump on the deck was at once set in motion and in a few hours the water was all out of the work chambers, thus proving to Roebling's satisfaction that the thing was airtight. Later on the air inside was allowed to escape and the top of the caisson settled to within seventeen inches of the water, which, Roebling noted with pleasure, happened to agree exactly with his previous calculations.

But the difficult work of dredging the site for the caisson was running far behind schedule. It would be another month before everything was ready there and nothing much could be done to speed things up. So apparently Roebling decided this would be an excellent time for him to go to St. Louis and see how Eads was progressing. The Bridge Company agreed and funds were provided for Horatio Allen to go along too.

Eads had a regular routine for handling visitors and it appears that Roebling and Allen received the same treatment when they arrived in St. Louis in early April. Eads would go over the plans first, explaining things, then set out in a tender to the spot mid-river where a flotilla of barges and derricks hovered over his submerged caisson. The functions of the various workboats would be described, after which Eads would lead his guests down the narrow spiral stairway, through the air lock, and into the caisson proper.

Roebling, as he would write later, had the highest admiration and respect for Eads and "his remarkable inventive talent." Roebling also said later that Eads was extremely courteous to him during his two days in St. Louis and one man who was on hand at the time, a friend of Eads's, said Eads took special pains to explain each and every detail to the younger engineer. So if there was any friction between them at this point in the story, there is no evidence of it.

Roebling appears to have returned to Brooklyn confident he was proceeding along the best possible course, and although he must have heard a great deal about the caisson sickness in St. Louis, most of those he talked to, including Eads, were convinced that whether a man got hit or not was largely a matter of luck and to judge from things he said later Roebling had arrived at about the same conclusion. Certainly Eads then knew no more than Roebling did about how to prevent the trouble, or how to cure it, as must have been obvious to both of them. Men were still suffering, more of them were dying.

Eads would keep plunging ahead with his work, sure that solutions could be improvised somehow should the problem grow still worse. In his place Roebling probably would have done the same. The great tragedy was that both of them were almost totally ignorant of what others had already learned about the effects of compressed air. They were both unaware, for example, that the surest, fastest remedy for caisson sickness was already known.

Possibly things might have gone differently for each of them had they compared notes as time went on, or had they been in touch with the few others there were working on similar problems. But they were living in an age when communication among professional colleagues was, by later standards, frequently at the most superficial level. Engineering then, like nearly every other line of work, was intensely competitive. An organization such as the American Society of Civil Engineers was striving with some success to encourage an open exchange of professional information and there were several reputable journals publishing valuable technical material. Still there was as yet no strong tradition along these lines and in some quarters not even an inclination. The railroads, the biggest clients for engineering talent, as well as the training ground for a very large number of engineers, were not the sort of institutions to foster an open exchange of valuable ideas. Minding one's own business was considered among the basic rules of business. There were trade secrets in other words, and the sharp rivalry men had to live with frequently gave rise to the worst kinds of professional jealousies and animosity. Roebling's own father, for example, had once written to Charles Swan to warn him not to hire a certain man simply because he had once worked for Ellet. "I do not want any news carried between myself and Mr. Ellet," John Roebling had said.

There were exceptions, to be sure, but even then, often as not, it was because the party sharing his special knowledge stood to gain financially thereby. Carnegie had so agreeably granted Eads the benefit of Linville's experience only when a large contract for the Keystone Bridge Company was involved. Eads's own first instructions on caissons had been given by the French bridgebuilder

Moreaux largely because Moreaux happened to be chief engineer for a leading French ironworks that, like the Keystone company, wanted to do the superstructure for Eads's bridge.

Perhaps, after Roebling returned to Brooklyn, he and Eads simply felt they had little more to say to each other, or little to gain by saying more than they already had. Or possibly for all their courtesies, things did indeed go sour at the start, simply on personal grounds. Eads, after all, was an exceedingly proud person who knew most all the answers always and was forever on his guard with anyone who might try to prove otherwise. He viewed his bridge, and none other, as the single most important engineering event of the century. Roebling almost certainly felt the same about the bridge he was about to build but, unlike his father, never once would he say so. Quite possibly Eads considered Roebling a threat and he was not about to stand in the shadow of any man. Maybe he simply saw Roebling as a nuisance.

It is also understandable that a man who had achieved so much on his own, against all odds and despite the doubters, might be reluctant to go out of his way to help a young man who appeared to have been handed quite enough already, and who so far had done little to prove himself particularly worthy of all that. Furthermore, Eads at best was a difficult person.*

But on top of everything else there was the prevailing belief of the time that a stiff spirit of independence was in itself a very good thing. And both Eads and Roebling were exactly the sort of men others would have pointed to as shining examples.

So they would each go their own way, alone, set apart by half a continent and, in time, open hostility.

On May 3, in the early afternoon, the Brooklyn caisson made its maiden voyage, which, of course, was also its final voyage—four

* In his *Autobiography*, Carnegie would tell the story of a personable mechanic named Piper who was sent by the Keystone company to help on the St. Louis bridge. "At first he was so delighted with having received the largest contract that had yet been let, that he was all graciousness to Captain Eads. It was not even 'Captain' at first, but 'Colonel Eads, how do you do? Delighted to see you.' " But presently feelings between them became a little complicated. "We noticed the greeting became less cordial." Colonel Eads became Captain Eads, then Mr. Eads. "Before the troubles were over, the 'Colonel' had fallen to 'Jim Eads' and to tell the truth, long before the work was out of the shops, 'Jim' was now and then preceded with a big 'D'."

miles down the East River to the site beside the Fulton Ferry slip. The chambers were again fully inflated, the air pumps were kept running, and the gigantic box was now riding with its deck a full nine feet above the water. (This inflation was essential, since in one part of the river there would be only a foot of space between the river bottom and the lower edge of the caisson.) Half a dozen tugboats took it in tow and proceeded out into the current at about quarter to two, "creating a great sensation among all whose good fortune led them to view one of the wonders of the nineteenth century," which was so soon to be "hidden from the gaze of mortal eyes."

Roebling, Kingsley, Horatio Allen, Bell of Webb & Bell, and three or four others went along for the ride, standing forward on the long, flat deck. And any doubts Eads may have planted about the caisson staying afloat were quickly forgotten. In the words of one witness, it came down the river "as placidly as a swan upon the bosom of an inland lake."

They tied up a block above Fulton Street, and by the time the sun went down, several thousand people had given it their personal inspection. "Of course, everyone was anxious to be able to say in future years that they had been upon the monster," wrote the *Eagle*. The monster, it seems, appeared even more formidable than anyone had expected and especially on toward dusk.

The following morning, at the turn of the tide, the caisson was shifted into position inside the new basin, the whole operation taking little more than an hour. As the crowded ferries churned in and out of the slip next door, young men were seen climbing to the tops of the cabins for a better look.

For the next several weeks additional courses of timber were built on the roof, each course at right angles to the other, with spaces left between the timbers, which were filled in with concrete to add weight and to help preserve the wood. Additional sections for the water shafts, air locks, and supply shafts were also installed as the roof grew in size. And on May 10, Roebling, Colonel Paine, and Francis Collingwood made the first inspection below. The temporary air compartment put in for the launching was removed, two doorways were cut through each of the interior walls, and any loose rock or mud under the edges was shoved out. A few men complained of trouble breathing the heavy air and apparently there was a sharp change in temperature inside the air lock every time the pressure changed, about which something would have to be done.

The heat was up over 100 degrees. But otherwise everything was going as expected.

In his report to the directors of the Bridge Company, signed June 12, Roebling wrote:

> For three weeks past a gang of forty men have been at work in the caisson for eight hours every day, under the charge of Mr. Young, principally in leveling off and removing boulders which happened to lie under the frames and the edges. A deposit of dock mud, from two to three feet deep, has made this work exceptionally unpleasant. The dredges, which are now beginning to work, will remove it in short time. The removal of large stones from under the shoe, some of them 100 cubic feet, is a matter requiring considerable skill and perseverance.

During all this time the caisson was rising with every high tide, then resting on the bottom again at low tide, which, of course, meant that work within could be carried on only during the low-tide time of day, when the chambers were comparatively free from water.

As more timber courses were added on top and the over-all height of the caisson was increased by a full ten feet, its center of gravity was raised considerably, causing a condition of "unstable equilibrium"—that is, the caisson would no longer rise uniformly with the rise of the tide. One end would come up ahead of the other and this would cause what was known as a blowout, a phenomenon of imposing appearance, as Roebling said, and the subject of much excited talk in Brooklyn.

As the tide was rising, and the downward weight of the caisson was being overcome by the increased tension of the air inside, along with the buoyancy of the river outside, one end of the caisson would suddenly tip up six inches or more. For an instant the tension of the air inside exceeded the head of water outside, and there would be a huge rush of air from beneath the shoe, carrying with it a column of water weighing hundreds of tons to a height of maybe sixty feet. Fish would fall all over the top of the caisson and the men working there would scramble to gather them up.

For the men inside the caisson such occurrences were quite terrifying at first, but of little serious consequence. There would be a terrific roaring noise and a sudden blast of air, both of which were decidedly unsettling, but after it had happened two or three times the men grew accustomed to it and the loss of a few hundred tons of air from a volume so large (163,000 cubic feet) was nothing to worry about especially.

Seen from the shore or the ferry, however, the sudden appearance of a waterspout on the East River was a spectacle that would be talked about for years by all who saw it.

It took three courses of stone and most of June before the vast wooden box was bobbing up and down no longer and was grounded on the bottom to stay. The first stone to be placed on top, the cornerstone as it were, was a block of blue limestone from the Kingston quarry, three feet by eight, weighing 5,800 pounds. There was no particular ceremony that went with it and so far as is known nothing was carved on it.

A stoneyard, as it was called, had been established downriver, below the Atlantic Docks, near Red Hook, and four huge scows had been especially built to bring the stone up to the site. McNulty had been put in charge of laying the first courses and the work had gone much slower than normal since portable derricks had to be used to move massive blocks, weighing anywhere from two thousand to three thousand pounds apiece. But once the caisson was righted down, three permanent derricks were mounted directly on top of it. They had great wooden masts fifty feet high, like the masts of a ship, and booms that were capable of swinging to any point on the deck.

By now, too, six big air compressors, built by the Burleigh Rock Drill Company, of Fitchburg, Massachusetts, were in operation inside a long shed nearby in the yard. Each had a twenty-horsepower steam engine driving two single-acting air cylinders of fourteen-inch stroke and fifteen-inch diameter. Each engine had its own boiler and they were all so connected that the stopping or breaking down of one boiler or engine would not affect the others. All piping and connections were in good order and working properly. (A ten-inch main took the compressed air underground some 150 feet to the caisson, where two six-inch rubber hoses carried the air down the supply shafts to the work chambers.) Thomas Douglas, a mason who had done the finest stonework in Prospect Park, had been put in charge of the labor outside the caisson, while the foreman inside was a strapping man named Charles Young.

To date everything had gone exactly as planned. There had been no serious interruptions. Material had arrived on time. All necessary machinery had been purchased and installed. Proper offices had by now been established for the Bridge Company in the *Union* Building on Fulton Street, which was only a short walk from the Fulton Ferry. Everyone involved was to be congratulated, wrote General Superintendent William Kingsley in his own first official report.

The great caisson could now begin its descent.

9

Down in the Caisson

We have no precedent just like this bridge.

—WASHINGTON ROEBLING

IN ALL the thousands of years men had been building things, no one had ever attempted to sink into the earth so large a structure as the Brooklyn caisson and there were not very many places where the job would have been more difficult than the Brooklyn side of the East River.

Roebling and his assistants thought they had learned quite a lot about the ground they had to penetrate while dredging the site, but as he commented with his usual dispassion, "The material now became sufficiently exposed to enable us to arrive at the conclusion that it was of a very formidable nature, and could only be removed by slow, tedious, and persistent efforts." Compared to this everything before had been child's play. Now that which had looked so reasonable on paper was turning out to be quite a different matter in practice. Indeed, so bad was the first month of excavation inside the caisson, so painfully slow and discouraging, that it began to look as though the whole idea for the foundations had been a terrible mistake, that they would have to give up and try again some other way or some other place.

There was never any public awareness of such feelings, which was just as well. There was, for that matter, very little real awareness on the part of the public of what actually went on inside the caisson, the work being entirely concealed.

The best over-all view of the site was still from the deck of the ferry. So every day thousands of people on their way to and from New York got a splendid, close-up look at the three towering boom derricks swinging blocks of limestone into place and at the squads of men swarming about the masonry work or through the adjacent yards, every last man appearing to know just what was expected of him. There were half a dozen different steam engines sending up columns of black smoke and everywhere a bewildering clutter of tackle, hand tools, nail kegs, and tar barrels, stacks of lumber and great heaps of coal, sand, and stone. How anything orderly or rational might emerge from such seeming chaos was something for ordinary men to ponder in dismay.

Still, seen from above, the work did not appear all that different from other big construction projects. The activity around the gigantic new Post Office being built in New York, for example, was every bit as confusing and impressive to watch. All this was lit by the same light of day and the men appeared no different from other mortals, breathing the same good air. But down in the caisson, everyone had heard, things were different. That was the part of the work that had the most fascination and of course the fact that it was hidden away where no one could see it, except for a relative few, made the fascination that much greater.

The newspapers sent reporters down soon enough. By July better than two hundred workers were going down every day and naturally they had their own stories to tell. So as a result a picture began to emerge, of a strange and terrifying nether world at Brooklyn's doorstep, entered only by men of superhuman courage, or by fools, and as sometimes happens with ideas that grow in the imagination, it was not so very far from the truth.

Probably the most vivid description was one given by E. F. Farrington, Roebling's master mechanic, a plain, blunt, practical man ordinarily. There would be rumors later about who actually was doing Farrington's writing for him, or at least dressing up his literary style, but there is no doubting the authenticity of the image.

> Inside the caisson everything wore an unreal, weird appearance. There was a confused sensation in the head, like "the rush of many waters." The pulse was at first accelerated, then sometimes fell below

> the normal rate. The voice sounded faint unnatural, and it became a great effort to speak. What with the flaming lights, the deep shadows, the confusing noise of hammers, drills, and chains, the half-naked forms flitting about, with here and there a Sisyphus rolling his stone, one might, if of a poetic temperament, get a realizing sense of Dante's inferno. One thing to me was noticeable—time passed quickly in the caisson.

Even the air lock was an unnerving experience for most men the first time they went down. For some it was also an extremely painful experience. The little iron room was abundantly lighted by daylight through glass set in the ironwork overhead. But once the attendant had secured the hatch with a few turns of a windlass, the common sensation was that of being enclosed in an iron coffin. Then a brass valve was opened. "An unearthly and deafening screech, as from a steam whistle, is the immediate result," wrote one man, "and we instinctively stop our ears with our fingers to defend them from the terrible sound. As the sound diminished we are sensible of an oppressive fullness about the head, not unaccompanied with pain, somewhat such as might be expected were our heads about to explode." (For many the sensation did not pass and they were said to be "caught in the lock.") Then the sound stopped altogether, the floor hatch fell open by itself, and the attendant pointed to an iron ladder leading into the caisson. The immediate wish of most men at this point, whether they showed it or not, was to get back out into the open air just as fast as humanly possible. But once the ladder had been negotiated and three or four minutes had passed, most men also found they felt reasonably steady.

The initial view of the caisson interior was generally something of a shock, once the eyes had adjusted to the light. The six big chambers looked something like vast cellars from which a flood had only recently receded. Every post and partition, every outside wall, and the entire ceiling were covered with a slimy skim of mud. Every man in the place wore rubber boots and got about on planks laid from one section to another and between the planks the muck and water were sometimes a foot deep or more. Most days the work force would be concentrated in a few locations, leaving some of the huge chambers as dark and silent as subterranean caves.

Where there was light it came from calcium lamps, limelights as they were also called, which threw steaming, blue-white, luminous jets into the corners where the men worked, or from squat sperm candles that blazed like torches at the end of iron rods planted alongside the plank walkways. "The subject of illuminating a caisson in a satisfactory manner is rather a difficult problem to solve," Roebling remarked in his report to the directors of the Bridge Company. At first the candles had burned with such vigor in the compressed air and sent up such clouds of smoke that the air had become intolerable. This had been overcome somewhat by reducing the size of the wick and of the candle and by mixing alum with the tallow and drenching the wicks in vinegar. Even so Roebling worried about the quantities of floating carbon the men were breathing into their lungs.

Kerosene lamps had to be ruled out from the start. They smoked even worse than candles, and with fire a constant hazard in such a charged atmosphere, Roebling did not want the risk of spilled oil. So he had hit upon the idea of limelights, of the kind ordinarily used for stage lighting or nighttime political rallies. He had the gas—a combination of compressed oxygen and coal gas—piped into the caisson, put burners in every chamber, and found two lamps per chamber would do the job. One small explosion had singed the beard off an attendant, but other than that the system had worked most satisfactorily. Ordinary street gas would have been about five times less expensive, but when that had been tried, the heat inside the caisson had built up to the point where no one was able to take it.

The air as it was, besides being heavy and dank, was uncomfortably warm. On the way from the compressors it passed through a cooling spray of water. Even so, winter or summer, regardless of the time of day or the weather outside, the temperature inside stayed at 80 degrees or more and the air was so saturated with water that under the best conditions the chambers seemed continuously shrouded in mist. Visitors who did not have to exert themselves in any way soon found they were wringing-wet with perspiration.

Most of the people who visited the caisson—newspapermen, local politicians, an artist from *Harper's Weekly*, editors from some of the professional journals—came out with their clothes thoroughly mud spattered and quite relieved to have the experience behind them. Many of them also expressed open amazement that men could actually work in such a place day after day.

The first load of rock and mud was hauled out of the caisson by clamshell dredge buckets on July 5. Most of the effort inside was spent removing the sharp-edged boulders that threatened to damage the frames and shoe as the caisson began to come down on them with crushing force. Boulders under the water shafts were the most serious initial problem, for if the caisson were to settle suddenly, the shafts might be blocked shut or badly damaged. And there was no way to get the boulders out of there except to chip away laboriously hour by hour, by hand, with long steel bars and sledge hammers.

In the middle chambers the ground was nearly all traprock, packed like gravel and joined by what Roebling described as a natural cement made of decomposed fragments of green serpentine rock. Every boulder was coated with this unyielding substance, upon which a steel-pointed pick had virtually no effect. Only by driving in steel-pointed crowbars with heavy sledges were the men able to make the slightest headway.

In chambers No. 1 and 2, those nearest the ferry slip, there was clay and gravel between the rocks, which made the going easier, while in Nos. 5 and 6, those at the upstream end of the caisson, there was a gummy blue clay that extended down forty feet, just as indicated by earlier soundings. This made the digging there relatively easy, of course, but it also meant that the caisson would have to go down at least forty feet—or beyond the clay. As Roebling said, no better foundation could have been wished for than what they were finding in chambers No. 3 and 4, but only if it had extended all over. And with the nature of the material so vastly irregular from one chamber to the other, lowering the caisson uniformly seemed practically an impossibility.

Roebling kept careful track of the rock uncovered. Nine-tenths of it, he found, was of Hudson River Palisades origin, transported, like all of Long Island, millions of years before by the glaciers. This traprock, as it was commonly called, was basalt, an igneous rock, like granite, and nearly as hard. As the men dug into the caisson floor, the traprock emerged in chunks the size of paving blocks or in monstrous boulders, but when a shovel or pick first struck one of them, with a sharp metallic clink, there was no telling which size it would turn out to be.

Boulders of quartz and gneiss occurred here and there, but rarely. Two big boulders of red sandstone were also found. But a collection made by Roebling of all the different varieties of smaller rocks uncovered, most of which had been worn down to pebble size, presented a complete series of the rocks to be found for a hundred miles to the north and northeast of Brooklyn.

The idea of driving the cutting edge of the caisson through such material by building weight overhead had to be abandoned at the start. The pressure needed to do that would crush the cast-iron shoe and smash the bearing frames. So the cutting edge would simply not cut. Instead, every boulder, every rock of any size, had to be removed before the shoe or frames began bearing down on them. And all such work had to be done by probing underwater since there were trenches along the inside edge of the shoe, clear around, and these were nearly always brimful of water that seeped in from the outside. (This water flowed in turn into cross trenches at the foot of the frames, which supplied the big pools under the water shafts.)

Just finding the boulders under the shoe, let alone removing them, was an unbelievably tough and disagreeable task. The full perimeter of the cutting edge was 540 feet. This added to the five frames, each 102 feet long, brought the caisson's full bearing surface to 1,050 lineal feet, or a distance greater than the length of three football fields, every inch of which had to be probed beneath with a steel sounding bar twice daily with each shift. Whenever a new shift came down, the work accomplished in the preceding eight hours had to be carefully explained; and since most of the trouble spots discovered would be underwater, there was no way simply to point them out—the information had to be written down or memorized. "Moreover," as Roebling wrote, "a settling of the caisson of six inches or a foot would bring to light an entirely fresh crop of boulders in new positions, and very often half without and half within the caisson."

To keep weight off the shoe, and so off any such boulders protruding under the shoe, it was necessary that the frames, or chamber partitions, take up that part of the load not balanced off by the compressed air. And with the frames thus the prime structural supports of the whole enormous burden, there had to be a way to lower them as the caisson descended.

The system used at first seemed the simplest solution, but it did not work well at all. Small pillars of earth were left under the frames, each one about three feet square and from six to eight feet apart. These pillars were then to be dug away systematically and the caisson lowered in that fashion. But the earth pillars often concealed a boulder that had to be removed, or they would be eroded away by water, or still more often, the workers in adjacent chambers, not working in unison, would undermine them and destroy their usefulness.

The system next adopted worked extremely well and was used until the end. Beneath each partition, every eight feet or so, two wooden blocks, a foot square and two feet long, were placed, one on top of the other, with oak wedges jammed between them and the bottom edge of the partition. Whenever the shoe had been cleared of all obstructions to a depth of several inches the entire way around the caisson, the wedges were knocked loose with sledge hammers, one by one, frame by frame, until the whole caisson settled. New blocks were then put in beside the old ones, which, if the descent of the caisson had been sudden, were split in two or crushed to a pulp. "The noise made by splitting blocks and posts was rather ominous," Roebling commented dryly, "and inclined to make the reflecting mind nervous in view of the impending mass of thirty thousand tons overhead."

Collingwood and Paine were in charge of clearing boulders from beneath the shoe and seeing that the caisson settled properly. "Levels were taken every morning on the masonry above," Collingwood wrote later, "and a copy furnished the general foreman. . . . If the caisson were level, the usual method followed in lowering was to begin at the central frame, and loosen the wedges regularly from the center towards the ends. The two frames next to these were then treated in like manner, and finally the outer two. When no obstructions occurred, the blocks would all be gone over several times in the course of a day, and the caisson would settle easily, at the rate of three or four inches in 24 hours."

At first, however, things had not gone that way at all. Through July and on into early August, the rate of descent had been less than six inches a week, and the boulders, instead of diminishing in number, as had been expected, became more plentiful. It was a hopeless rate of progress Roebling reported to his directors. At this rate it would take nearly two years to sink just the one caisson.

Boulders within the work chambers were the lesser evil. Before they could be hauled up the water shaft, they had to be split into manageable pieces, never an easy job, but at least they could be dealt with under comparatively reasonable conditions. Boulders under the shoe, however, or those found beneath the frames, were each a major undertaking. The removal of a boulder from under the shoe, for example, went as follows.

The ground around the inner side of the boulder had to be dug away with pick and shovel, with the excavation filling with water as fast as the men worked. Then the boulder had to be drilled by hand, underwater, and a lewis inserted, a dovetailed iron eyebolt to which a hoisting rig of some kind could be attached. In the early days of the work, double sets of block and tackle were tried, with a gang of thirty or forty men hauling at the ropes, while others worked furiously with winches and crowbars. But very often the boulder refused to budge. So Roebling had hydraulic jacks lowered through the supply shafts. These were of a kind designed for pulling instead of lifting and had a capacity of two to three tons. The water chamber on such a jack was above, not below, the piston, and the piston rod had a big hook at the end instead of a lifting shoulder. This hook was attached to the iron eye in the boulder, while the opposite end of the jack was chained fast to the nearest substantial timber or, better still, to the ceiling. The jack pump was then set in motion and, as Roebling said, it would prove itself a "very effective instrument." There would be an immense momentary strain, then the boulder would give way and come sliding into the caisson, where it would be broken up.

When a boulder appeared to extend several feet outside the caisson, no attempt was made to haul it in. Rather the part inside was slowly split up until enough had been removed for the caisson edge to clear.

But whichever way they were handled, a few good-sized boulders beneath the shoe could hold up everything for three or four days. Such delays were maddening, and there were more and more of them as time passed and equipment began to break down or the water shafts failed to function as they were supposed to. The big clamshell buckets, armed with seven-inch teeth, were formidable-looking affairs and under normal conditions one of them could dredge up more than a thousand yards a day. This, in theory at least, meant that the equipment in use should have been able to haul out the whole volume of material that had to be removed for the tower foundation in about a month's time. But the buckets kept breaking down or getting caught under the bottom edges of the water shafts. As it was, the job would take five months, and these, as Roebling wrote, were "five months of incessant toil and worry, everlasting breaking down and repairing, and constant study to make improvements wherever possible." Bucket teeth that worked well for scooping would not last a day at grappling with stones. For every two buckets in working order, three were being repaired. "There was, indeed, one period," Roebling said, "when we were almost tempted to throw the buckets overboard . . ."

One of the greatest early disappointments was to find that the buckets were unable even to dig their own hole under the water shafts, as they were supposed to. Much of the time the buckets failed even to bite into the material dumped into the hole unless a couple of men were kept constantly stirring the pool, "to keep the stuff alive," as Roebling said. But even then the bottom of the pool kept filling in and had to be dug out by hand repeatedly.

Stone and clay would pack solid and actually fill the hole in a few hours, such was the incredible nature of the material being excavated. So it became necessary to feed all the stones into the pool at one time, separately, then the clay by itself. The kind of bucket in use could lift any stone it could catch hold of, but such a stone, or a chunk of split boulder, had to be placed just so in the hole for the bucket to get a proper hold, and the stones could only be taken out one at a time. Whenever a badly placed stone got wedged under a shaft, which happened fairly regularly, somebody had to dive under to see what could be done. "When the lungs are filled with compressed air," Roebling wrote, "a person can remain under water from three to four minutes." He knew this, it seems, from personal experience.

Any material fed into the pools from other parts of the caisson could be properly prepared, as it were, for the dredges to handle, but when the trouble was inside the water-shaft pools, as often happened, or when the pool had filled in, one to two days would be lost while the shaft was sealed off on top, with an iron cap, the water forced out by compressed air, the pool pumped dry, and the pit dug out by hand to a depth of six to eight feet. And the whole time this was going on (about two days on the average), the other shaft had to handle all the work. There were, in fact, so many occasions when the pools had to be cleared in just this fashion, so many repairs needed on the buckets, that most of the time only one water shaft was in operation.

Furthermore, whenever the work was held up a day or two, and the caisson stopped settling, its movement immediately afterward could be quite erratic, coming in sudden, unpredictable, uncontrollable starts. This, Collingwood explained, was due to the earth compacting around the caisson, as it does around a pile when driven slowly. "At such times it would seem impossible to get it started, and when once movement began, it was almost sure to split a set of blocks before it was arrested."

Once, after the caisson had been at rest for several days because of breakdowns in equipment, all the usual steps were taken to get it started again, but to no avail. The blocking was eased, the shoe was cleared of obstructions, and still the caisson just hung there, motionless, with nothing holding it. The men did not know quite what to make of this. The only real significance of the episode, however, was that it gave the engineers a chance to compute roughly how much side friction the caisson had to overcome during its descent. As Collingwood figured it and reported later to a meeting of the American Society of Civil Engineers, the average pressure in the chamber at the time was seventeen pounds per square inch, giving a lifting force from the compressed air only of 20,400 tons. The bearing surface (posts and frames) was carrying about 625 tons and estimates were that the whole outer edge was probably carrying about that much again, which gave a total upholding force of somewhere near 21,650 tons. But the total weight of the caisson then, including the stonework on top, was judged to be 27,500 tons. Therefore, when it failed to move, the weight being carried by side friction alone was 5,850 tons. So this meant that along with everything else that had to be overcome to get the caisson down even a single inch, there was about 900 pounds of friction working against every square foot of the exterior surface.

To add to the over-all physical discomfort of everyone involved, blowouts continued and with greater frequency than Roebling had figured on. After each initial rush of air out one side or other, a returning wave would follow, inflowing river water that would stand knee-deep over the work surface until the air pressure eventually forced it out. Blowouts were usually caused by changes in the tide, which in the early stages affected the balance of pressures inside and out and which apparently Roebling had anticipated. But even the wake of a passing steamer could cause enough of a change in the water level to bring on a blowout, and this came as quite a surprise.

To build up additional weight on the caisson, some of the excavated material was dumped on top, in the spaces not taken up by the masonry. The rest of the material was dumped into carts that ran on inclined tracks down to big scows tied up on the riverside. (Once the caisson was in position this side had been closed in with a cofferdam, as the others were, and docks and tracks and turnarounds for the stone carts had been built.)

Eventually, when the caisson got down about ten feet below the river bottom, water ceased to come in at all, so tightly was the ground packed about the outer sides. Now, much to their amazement, the boulder crews encountered a new phenomenon. As Collingwood wrote, "It was not an uncommon occurrence in removing a large boulder, that an opening would be made entirely outside the caisson, for three or four feet." Sometimes when this happened a man might crawl inside, beyond the limits of the caisson, that is, to dramatize the uncanny nature of such a space, not to mention his own nerve.

To step up the pace, Roebling organized a special force of forty men who worked at boulders exclusively, from eleven at night until six in the morning, when the regular shift came on. In time everybody grew more accustomed to the work. Roebling, in the words of William Kingsley, gave "the work his unremitting attention at all times," but especially at critical points was he "conspicuous for his presence and exertions." Like his father, he demanded much of every man under him, and even more of himself.

As the weeks passed he found that a slight lowering of the air pressure inside the chamber could work wonders whenever added weight was advantageous. The compressors would be slowed a little and the caisson would immediately bear down harder. It was a ready, effortless way to apply an additional twelve hundred tons or so any time that was needed, and for only as long as needed.

But when the caisson had reached a depth of some twenty feet, or approximately half the distance Roebling intended to sink it, the boulders became so large and numerous that there was no choice left but to begin blasting.

The idea of using powder on the boulders had, of course, been considered from the start. It would have saved all kinds of time and effort obviously, and as things grew increasingly difficult and frustrating inside the caisson, the men were more than ready to give it a try, whatever the supposed risks involved. But Roebling had held off. In such a dense atmosphere, he reasoned, a violent concussion might rupture the eardrums of every man inside. Smoke from the explosions might make the air even more noxious and certainly more unpleasant than it already was. The doors and valves of the air locks might be damaged.

His greatest fear, however, was the possible effect on the water shafts. The two immense columns of water that stood above the work chambers and every man in the caisson were held there in a critical balance only by the pressure inside the chambers. The margin of safety was just two feet of water—the distance from the surface of each pool and the bottom edge of each shaft. An explosion inside the caisson, Roebling explained, might suddenly depress the level of the pool and allow the air to escape underneath. A water shaft might blow out, in other words. All the compressed air would escape in one sudden blast and almost certainly with the following immediate consequences: with the work chambers instantly deflated, so to speak, the full weight of the caisson would come down all at once, smashing blocks and frames and outer edges. The impact might be so great as to crush every interior support and everyone inside; and in the early stages of the work, the river would have rushed in and drowned everyone. What the effect might be on top was anybody's guess, but it was realistic to assume that all that water bursting out of a shaft would be about the same as a major explosion.

Still, Roebling knew, such prospects, however sobering, were all hypothetical. There was no past experience to go by. So whether he was right or not remained to be seen. With luck, he might be wrong. He decided to find out.

He began by firing a revolver with successively heavier charges in various parts of the caisson. When it was clear this was perfectly safe and causing no adverse effects, he set off small charges of blasting powder, fired by a fuse, gradually working these up in magnitude until they were on the order of what was needed to get on with the work. The concussions bothered no one especially, nor did they have any noticeable effect on the air locks or water shafts. "The powder smoke was a decided nuisance," Roebling said. "It would fill the chambers for half an hour or more with a thick cloud, obscuring all the lights." But this he alleviated greatly by switching to fine rifle powder.

The results were spectacular. With a little practice the work moved ahead as never before. A long steel drill would be hammered into the rock to make a hole for the blasting charge and the charge would be tamped in and set off.* "As many as twenty blasts were

* It was about this same time, during construction of the Big Bend Tunnel, in West Virginia, that a Negro railroad worker named John Henry drove just such steel drills faster, it was said, than any man, for which he would be immortalized in what has been called America's greatest ballad. Henry supposedly met his death competing with a steam drill about 1870. No such steam drills were used in the bridge caissons.

fired in one watch," Roebling reported, "the men merely stepping into an adjacent chamber to escape the flying fragments." The hard crystalline traprock split more easily than the tough gneiss or rotten quartz boulders. Invariably the traprock broke neatly into three equal-sized boulders. The caisson now began descending twelve to eighteen inches a week, instead of six.

Care was taken to guard against fires igniting in the yellow-pine roof and the men did their best not to injure the shoe with the charges they set off beneath it. But the shoe by this time was in such shape that a little more damage hardly mattered. The armor plating was bent and torn, the shoe itself cracked or badly crushed in dozens of places.

One convenient method for disposing of a boulder lodged beneath the shoe was to drill straight through to the other side, plant a charge at the far end, then shoot the boulder bodily into the caisson. Some boulders encountered now were up to fourteen feet in length and five feet in diameter.

For the people of New York and Brooklyn all such activity was considered somehow removed from reality. The whole concept of an enormous wooden chamber descending below the river was a little difficult for many to understand and the men who went in and out of it seemed a breed apart. There was simply something quite unnatural about all this. "For night is turned into day and day into night in one of these bridge caissons," wrote the *Herald;* "and when the steam tugs, with their red and blue lights burning from their wooden turrets go creeping along the bosom of the river like monstrous fireflies, then do these submarine giants delve and dig and ditch and drill and blast . . . The work of the buried bridge-builder is like the onward flow of eternity; it does not cease for the sun at noonday or the silent stars at night. Gangs are relieved and replaced, and swart, perspiring companies of men follow each other up and down the iron locks, with a dim quiet purpose . . ."

The sheer physical exertion inside the caisson was as great as ever, the work every bit as unnerving as it had been. And the deeper they went, the more the men felt the discomforts of the compressed air.

The work went on around the clock, except for Sundays, with three shifts of eight hours each. The first shift went down at six in the morning, the second shift at three in the afternoon, the third, the special night gang, went down about eleven. Most men stayed in the caisson the full eight hours, taking their dinner pails down with them. Work in such an atmosphere brought on an uncommonly fierce appetite, they said, and the standard meal consisted of great slabs of bread and cheese or beef, washed down with beer.

The two day shifts were composed of 112 men each, while the night shift Roebling kept to roughly forty picked men. So the full force working inside the caisson came to about 264. Up on the surface there were two shifts to operate the dredging gear and two shifts to dispose of the material brought up from below. In addition, there were people to run the compressors and hoisting engines, blacksmiths, mechanics, men to look after the gas for the lighting below, a carpenter's force of some twenty-five men, and thirty men working on the masonry, bringing the total force aboveground to something like a hundred.

But the number of those inside the caisson who had been with the work from the start was quite small comparatively. According to the time books, a total of 2,500 different individuals worked in the Brooklyn caisson from start to finish. This means then that men were quitting in droves—at a rate of about a hundred a week on the average, or, to put it another way, every week about one man in three decided he had had enough of building the Great Bridge and walked off the job, never to return again.

There were notable exceptions, of course. One man named Mike Lynch went down with the very first shift to go into the caisson and would be the last man to come out. He not only never lost an hour's time during the ten months he worked in the caisson, but he made a day's extra pay in overtime. "He is strictly temperate and regular in all his habits," William Kingsley noted, "and is none the worse for his long service in compressed air."

That the turnover was so great is not surprising.

Amenities provided by the management were very few—about what was customary. Two unpainted frame sheds had been put up in the yard, with rows of pegs and hooks inside for the men to hang their clothes. (The temperature inside the caisson was such that most men went down wearing nothing but pants and a pair of company boots.) In front of the sheds were sets of washtubs, with hot and cold water. And that was about the sum of the comforts provided aboveground.

Inside the caisson itself there were generally a few dry spots where a man could eat his midday meal. And against one wall stood what was considered by all the world's most extraordinary toilet. It was described in one of Roebling's official reports as a pneumatic water closet and consisted of a wooden box with a lid and a large iron pipe that passed up through the timber roof. The box was kept about half full of water, and whenever its contents were to be discharged, a valve was opened and the pressure from within the caisson would blast everything instantly overhead in the form of a fine mist. This particular device was not installed until the work had progressed some little time, however, and until then the pools beneath the water shafts, or any convenient corner, had sufficed for the same purpose. When he came to describe the general working conditions, Roebling would note that the sense of smell was almost entirely lost in the "made air," as he called it. "This," he said, "is a wise provision of nature, because foul odors certainly have their home in a caisson."

For an ordinary laborer the pay was two dollars a day. But after the caisson reached a depth of twenty-eight feet, it was decided to revise that. The bad air, the increased unpleasantness over all, and the widespread feeling that the deeper down they went the more hazardous the work, all called for a commensurate hike in wages, the management decided. So from that point on the pay was $2.25 a day.

Men kept quitting just the same, but for every one who did, there were at least a dozen anxious to take his place, most of them Irish, German, or Italian immigrants who were desperate for work of any kind, and many of them, like those who had gone into the Eads caisson, were thinly clothed and undernourished.

But for all the talk and worry there had been over caisson sickness, and for all the growing fear of it as pressure inside the caisson increased steadily, only a few so far had experienced any ill effects.

One pound of air pressure equals two feet of tidewater, so for every two feet the caisson was lowered, one pound had to be added to the pressure. Gauges in the engine room indicated the height of the tide and the pressure of the air. The greatest the pressure would ever be in the Brooklyn caisson was twenty-three pounds per square inch above normal atmospheric pressure, or nearly ten pounds less than it had been inside the Eads caisson the day James Riley fell dead. In St. Louis several more had died miserably, but there had been only mild symptoms in Brooklyn. A little paralysis in the legs was all. Only three or four men had been bothered in the slightest and none of the engineering staff so far.

Like Eads, Roebling noted that the ones who did have trouble were all new to the job. His way of alleviating their discomfort was to send them right out of the caisson. Now that he had seen something of the problem first hand and had spent as much time under compression as anyone on the job, Roebling was convinced that Eads's system of shortening the hours was the best possible prevention and said he would follow that same system in the New York caisson. The thing to do, he said, was to "reduce the period that the human system is in contact with the exciting cause." The increased quantity of oxygen inhaled under pressure was what did the damage, he thought. "That the system struggles against this abnormal state of affairs," he said, "is shown by the fact that the number of inhalations per minute is involuntarily reduced from thirty to fifty per cent. It follows, therefore, that the shorter the period of exposure to compressed air the less the risk."

But any change in the schedule would wait until the New York side, since the Brooklyn caisson was not going deep enough to produce anything like what was happening in St. Louis, where Eads had had a special hospital ship fitted up and had hired a full-time physician who prescribed special diets and set down strict rules about rest. Eads's men by this time were permitted to work in the caisson only an hour at a time.

But the men in the Brooklyn caisson were having their troubles all the same. The work was a hazard to the health, it was agreed, and far more exhausting than anything any of them had ever done before. Collingwood said a full day inside would leave him feeling worn-out and in ill temper for days. And when the weather turned cold in the late fall, dozens of men began coming down with severe colds and bronchitis, caused by the abrupt drop in temperature inside the air lock. Every time they "locked out" at the end of the day, hot, tired, and dripping wet, the men would experience a sudden temperature drop, from at least 80 to 40 degrees. Roebling had steam coils installed in the air locks to keep the temperature the same as in the chambers below, but the men still had to face the chill open air once they emerged from the locks.

A hacking cough also became common among those who had been on the job any time. Candle smoke and the blasting were said to be the cause. Those who had been going down the longest could spit black and would still be able to do so several months after the work was finished.

But what plagued everyone most was the thought of all that

weight bearing down overhead and the river outside and the unspoken fear that sometime, sooner or later, something was going to go wrong and they would all be drowned like rats or suffocate or be crushed to death. And then just to confirm how very tenuous was the balance upon which they were all trusting their lives, there occurred what would afterward be called "The Great Blowout."

It happened at about six in the morning and on a Sunday, when only a few men were about, a fact the pious took to be more than a matter of coincidence. Eads had his men working seven days a week, it was noted, while Roebling kept the Sabbath. This was a sign, it was said, and thanks were given through Brooklyn and nowhere more fervently than in the Irish neighborhoods near the Navy Yard. Heaven only knew how many would have been left widows, people were saying, had it happened any other day of the week.

All at once in the very still early morning there had been a terrific roar. The few who actually saw the thing go off said it looked more like a volcano than anything else. It was as though the river had exploded, sending a column of water, mud, and stones five hundred feet into the air and showering yellow water and mud over ships and wharves and houses for blocks around. The column was seen from a mile off and the noise was so frightful that people began pouring out of their front doors and rushing pell-mell up Fulton Street. The whole neighborhood was on the run. Roebling described it as a stampede. "Even the toll-collectors at the ferry abandoned their tills," he said.

Nobody was inside the caisson at the time and only three men were on top of it. One of them, a yard watchman, said later that the current of air rushing *toward* the blowing water shaft was so powerful it knocked him off his feet, ruining his Sunday suit. He had been struck by a stone after that and could remember no more. One of the other men leaped into the river, while the third tried to bury himself in a coal pile.

Then in an instant it was all over and everything was as silent as before. Both doors of the air locks fell open and for the first and only time the submerged caisson was flooded with daylight. The quiet lasted but briefly. Within minutes there was another rush of people heading down Fulton to see what had happened.

Roebling, Collingwood, and one or two of the others from the work force were on the scene almost immediately. They turned hoses into the open water shaft, closed the air locks, and in about an hour had a head of water thirty-one feet high back in the shaft and fifteen pounds of pressure back in the work chambers. When it was time to go down to take a look at the damage, Roebling led the way. "The first entry into the caisson was made with considerable misgiving," he wrote. But incredibly none of the disastrous consequences he had feared had occurred, as he reported later to the Board of Directors:

> The total settling that took place amounted to ten inches in all. Every block under the frames and posts was absolutely crushed, the ground being too compact to yield; none of the frames, however, were injured or out of line. The brunt of the blow was, of course, taken by the shoe and sides of the caisson. One sharp boulder in No. 2 chamber had cut the armor plate, crushed through the shoe casting, and buried itself a foot deep into the heavy oak sill, at the same time forcing in the sides some six inches. In a number of places the sides were forced in to that amount, but in no instance were they forced outward. The marvel is that the airtightness was not impaired in the least.

His caisson had withstood the staggering blow of 17,675 tons dropped ten inches. By the way certain boltheads were sheared off, he could tell that the sides of the caisson—nine solid courses of timber—had been compressed two inches, such had been the impact. In the roof, however, there was not a sign of damage except for the slightest sag near the water shafts, where the support from the frames was the least.

With a little figuring Roebling concluded that once all the settling had stopped and before the compressed air was built up again inside the chambers, the caisson was carrying a total weight of twenty-three tons per square foot. This was an astonishing revelation. As nerve-racking as the whole episode had been, it had demonstrated just how large a margin of safety Roebling had built into the structure, since its ultimate load, once the bridge was built, would be only five tons per square foot. So he had built the caisson at least four times as strong as it needed to be.

Why the water shaft blew out was, needless to say, a question of the gravest concern and the answer takes a little explaining.

The problem was that the weight of the columns of water in the shafts was not always the same. Particles of rock and earth were constantly washing out of the clamshell buckets as they were hauled up through the water shafts, and as a result a fine silt was held in suspension and this, in a column of water seven feet square by, say, thirty-five feet high, could and did increase the specific gravity of that column to a remarkable degree. But when that shaft was not in use for some reason or other, the silt would settle, the water would become less thick and would weigh less. So great would be the difference in the weight between a nearly cleared column of water and one still in use that the difference in the levels of the two (both of which were, of course, being supported by the same air pressure) would be about ten feet.

Outside, on top of the caisson, such a disparity in the levels of water in the shafts could be an alarming sight for anyone on duty who did not understand the reason for it. The impression would be that the lower of the two shafts was getting close to the danger level—that it might blow out at any instant—and the immediate response would be to feed more water into that shaft as quickly as possible.

This is precisely what happened on more than one occasion and the effect inside, naturally, was that the pool beneath one shaft would begin overflowing rapidly, which led those in charge below to assume the air pressure in the chamber was giving out. And since there was no direct communication between those above and those below, somebody had to scramble up through the air lock and find out what had gone wrong. On one such occasion it was discovered that the water in one shaft was about twelve feet from the top while the water in the other shaft was twenty-one feet from the top and despite the fact that a heavy stream of water was pouring into the shallower shaft. The water was immediately shut off when it was known what was happening below and samples of water were taken from each shaft. Water from the lower shaft was found to weigh eighty-five pounds per cubic foot, which was twenty-one pounds more than the water from the other shaft.

On the Sunday of the blowout, apparently the sediment in one shaft had settled to such an extent that the water no longer weighed enough to contain the pressure inside. The normal precaution had been to keep a small stream of water playing into the shafts to make up for just such a likelihood during days when the work was halted and no dredging was going on. But this time that had not been done. Sounding very much like his father, Roebling said, "To say that this occurrence was an accident would certainly be wrong, because not one accident in a hundred deserves the name. In this case it was simply the legitimate result of carelessness, brought about by an overconfidence in supposing matters would take care of themselves."

Trusting matters to take care of themselves was something this extremely competent young man had seldom done in his life. He had had the contrary attitude drummed into him since childhood and from here on, more than ever, he would insist on the strictest attention to every detail and to safety precautions especially, and he would come down hard on anyone caught taking such matters lightly. The thing to fight against, he told the men, was the kind of carelessness that comes from familiarity with the job.

German-born John A. Roebling, designer of the Brooklyn Bridge, was one of the creative giants of the nineteenth century, a builder of record-breaking suspension bridges, inventor, spiritualist, philosopher, and manufacturer of the first wire rope made in America. This rare photograph, from a family scrapbook, was probably taken a few years before his tragic death in 1869.

Washington Roebling was nineteen and a student at Rensselaer Polytechnic Institute when this picture was taken. The oldest of John Roebling's four sons, he was the only one of them who was expected to become a bridgebuilder. As a boy of fifteen he had been with his father on an icebound Brooklyn ferry, when, it is said, his father first envisioned a monumental suspension bridge vaulting the East River with one uninterrupted span.

Emily Warren Roebling, wife of Washington Roebling, was once described as "strikingly English in style." Throughout the building of the bridge, she remained her husband's constant companion, private secretary, nurse, diplomatic emissary, unofficial aide-de-camp, and was the subject of much gossip. He never cared for this photograph, or for any other ever taken of her.

Colonel Washington A. Roebling, Civil War hero and long his father's principal assistant, was named Chief Engineer of the Great Bridge after his father's death and before any of the work had begun. He was then thirty-two years old. A markedly modest person, he had a gift for being on the spot when needed, great physical courage, and his father's iron will. Confusion over which Roebling built the bridge would last for years. Toward the end of his life, Washington Roebling would write, "Long ago I ceased my endeavor to clear up the respective identities of myself and my father. Most people think I died in 1869."

The elder Roebling began plans for an East River bridge as early as 1857. Among his designs for the colossal stone towers was this one, which, had it been built, would have looked quite akin to those famous Egyptian-style New York landmarks of the time—the Tombs prison and the Croton Reservoir on Fifth Avenue at 42nd Street.

COMERFORD & COLLINS

No 169 No 170

east of the two

The mark on the be

door is 11½ " fro

of the building

The centre lin

of the adjoining

There is a pin inside of curb

it is opposite the east c

There is a pin, on th

Cherry street, inside of curb

The pin opposite the spice

along side the flagging,

measured along the edg

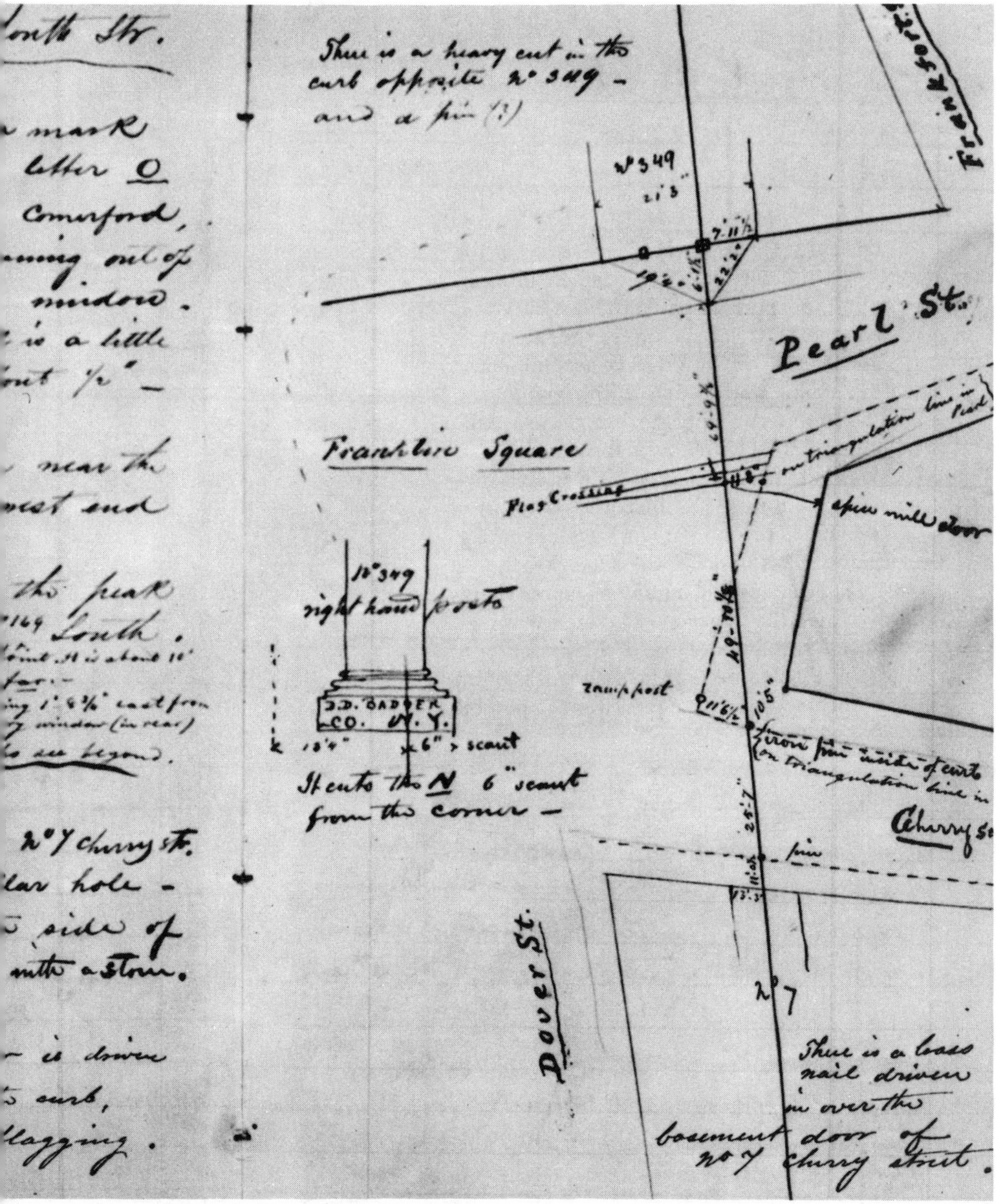

Two pages from one of Washington Roebling's field notebooks show where, in the spring of 1869, he and Colonel William Paine put the iron pins and other markings used to fix the center line of the bridge on the New York side of the river. On the right-hand page the center line can be seen passing through Cherry Street, at about that point where George Washington lived briefly after becoming President and where William Marcy Tweed spent his boyhood.

The bridge began with the sinking of the timber caissons, the immense pneumatic foundations upon which the towers were to stand. The caissons were the largest ever built (they could have accommodated four tennis courts each, with room to spare) and the whole technique was still in its infancy. At right is the Brooklyn caisson before launching. It weighed three thousand tons. The illustration at right below shows the men at work inside during the caisson's descent, the air locks, the supply shafts, and the ingenious water shafts, by which the material excavated was removed without any loss of air pressure in the work chambers. The diagram is of an air lock by which the men went in and out.

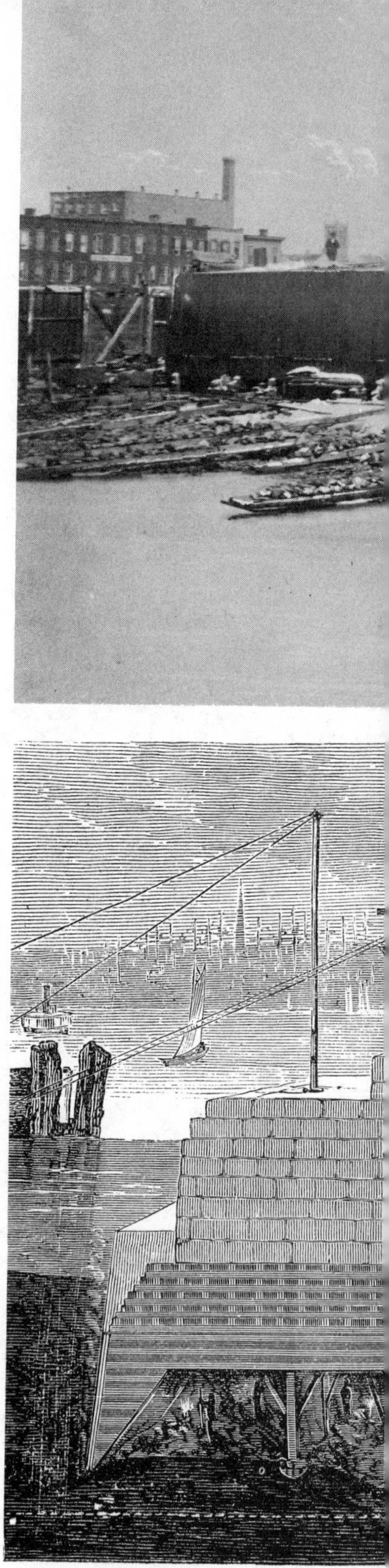

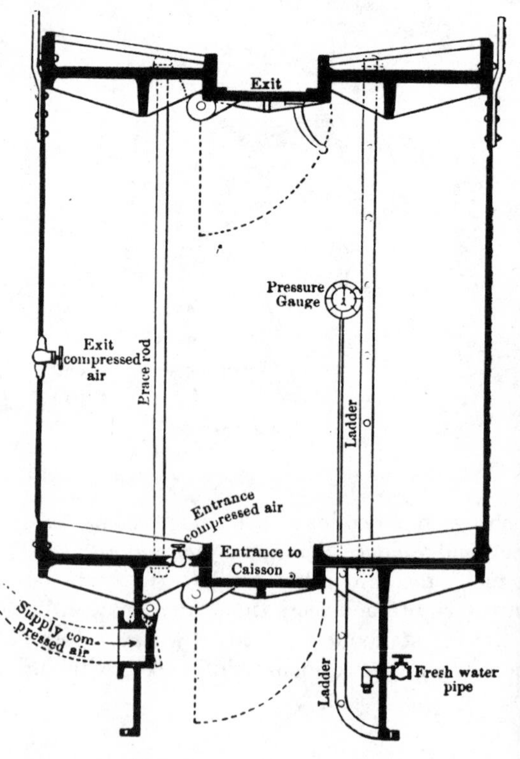

FOUNDATION LINE

Sketches by a *Scientific American* artist who ventured down into the heavy, damp air of the Brooklyn caisson show three formally attired visitors passing through an air lock (*far right*, *above*); the ladder that led from the air lock into the actual work area (*near right*); men with sledges and rods splitting one of the many boulders encountered (*middle*); and (*at far right*) the stirring up of a pool beneath one of the water shafts (this was done to ease the work of the clamshell scoops that dropped down the water shafts).

The caisson was being driven ever downward into the riverbed by the mounting burden of the tower. At left the Brooklyn tower begins to take form as huge granite blocks are laid up by three steam-powered boom derricks and by a swarm of laborers, here taking a moment off to pose for the photographer. The stone came from New York, Connecticut, and Maine.

The towers, when completed, reached an elevation of 276 feet above the East River. They dwarfed everything else on both skylines, as can be seen in this remarkable photograph. Taken from the top of the Brooklyn tower sometime in early 1876, it shows the nearly completed New York tower against all of lower Manhattan, from the Battery to Rutgers Street. The picture is composed of five different views that the photographer later spliced together to form a seven-foot panorama.

When the photograph at right was taken in September 1872, the Brooklyn tower had reached the level of the roadway at 119 feet, or less than half its ultimate height. The man standing at the upper left, looking out over the river, may be Washington Roebling, as is stated in some sources, but is probably his assistant, C. C. Martin, since Roebling had had his second attack of the bends by this time and was appearing at the work only infrequently and with the greatest difficulty. Martin and Roebling both trimmed their beards in the General Grant fashion and looked quite similar when seen from a distance.

The view from the completed Brooklyn tower was like nothing ever seen before. The illustration shows the East River (with a fair sampling of the variety of ships to be seen then), lower Manhattan up to where the New York tower is beginning to rise, the Hudson, or North, River beyond, and

New Jersey on the horizon. On the left, faintly indicated in the distance, is tiny Bedloe's Island, where the Statue of Liberty would be built after the completion of the bridge. The nearer, larger island is Governor's Island. The tallest building in New York was then Trinity Church.

William C. Kingsley, Brooklyn's leading contractor, was the driving political force behind the bridge. How much he himself profited from his official and unofficial connections with the great work remained a subject of heated debate.

State Senator Henry Cruse Murphy, Brooklyn's most respected and respectable Democrat, served as president of the Bridge Company. Most people took him at his word when he disavowed any personal, secret dealings with Boss Tweed.

Three highly influential trustees of the bridge were Congressman Demas Barnes (*top*), Mayor Seth Low (*middle*), and James S. T. Stranahan, "The Father of Prospect Park" (*bottom*). Each was a figure of major importance in Brooklyn, then the third-largest city in the country and a decidedly different sort of place from New York. For a very long time Brooklyn had been regarded by millions as virtually the moral center of America because it was there that Henry Ward Beecher (*below, right*), pastor of Plymouth Church, made his home. His congregation, he said, was all America and few men have ever known such adoration. At the time the bridge was starting to rise, he was Brooklyn's proudest symbol. But in 1872 the famous Beecher scandal broke in a New York paper, to be followed by a sensational trial in Brooklyn.

Abram S. Hewitt (*left*) was an ironmonger turned political reformer and the man delegated to root out all mismanagement and corruption inside the Bridge Company after the fall of the Tweed Ring in 1871. Few politicians of the day enjoyed a more impeccable reputation, a reputation that was undeserved according to private notes kept by the Chief Engineer, Washington Roebling.

Estimates are that William Marcy Tweed (*below*) weighed about three hundred pounds and that his Ring managed to steal as much as $200 million before it was put out of business. Tweed's hold on the bridge was considerable and went unchallenged until his "troubles" began. Before his death, Tweed confessed all, describing the details of his "understanding" with the Bridge Company.

10
Fire

> When the perfected East River bridge shall permanently and uninterruptedly connect the two cities, the daily thousands who cross it will consider it a sort of natural and inevitable phenomenon, such as the rising and setting of the sun, and they will unconsciously overlook the preliminary difficulties surmounted before the structure spanned the stream, and will perhaps undervalue the indomitable courage, the absolute faith, the consummate genius which assured the engineer's triumph.
>
> —THOMAS KINSELLA, in *The Brooklyn Eagle*

A SECOND contract was signed that October with Webb & Bell to build the New York caisson, according to plans drawn up during the summer. This time the caisson would be slightly larger and there were to be a number of important modifications. Roebling had decided to make the water shafts round instead of square, for example; he wanted the air locks positioned differently; and the entire interior would be lined with boiler plate, as fire protection, Roebling having realized by this time how very inadequate were the preventive measures he had specified in the Brooklyn caisson. With a caisson that might have to go twice the depth of the Brooklyn caisson, he wanted none of the anxieties there had been over fire.

In November actual construction of the New York caisson got under way at Webb & Bell yards on the other side of the river, while the Brooklyn caisson continued steadily downward. Things were proceeding very nicely now. In the middle of the month *Harper's Weekly* published a spectacular double-page view of lower New York and Brooklyn, as though seen from the air ("an unprecedented piece of engraving"), and there, bestriding the East River, bigger and grander by far than anything in sight, was the finished bridge. "This bold and peculiarly American design is not yet wholly beyond the scope of debate," wrote the editors, whose offices on Pearl Street looked out on the spot where the New York tower would stand. But the beautiful engraved view, placed in the magazine so it could be easily removed and saved by the reader, suggested quite clearly, if their words did not, that they had every confidence in successful results. And for the technical community, *Scientific American* announced about the same time that "the rapidity with which the work has proceeded is evidence that it is conducted by a man who is fully competent to conduct this greatest engineering feat of modern times," implying, it would seem, that there had been some doubts about that.

By the end of the month the Brooklyn caisson had reached a depth of almost forty-three feet, or very nearly as far as it would have to go. Another two feet, maybe even less, another ten days would do it, Roebling told the others. In the drafting room, plans were being drawn up for brick piers, ten feet square, that he had decided to build throughout the interior of the caisson, once it was at rest, to provide absolutely solid support for the mounting load overhead during the time it would take to fill the vast chambers with concrete. An order had already been placed for a quarter of a million bricks.

So the end of this first momentous step seemed very near indeed, when, on the morning of Friday, December 2, about the time most people were on their way to work, the news swept through Brooklyn that the caisson was on fire.

The fire had been discovered about nine thirty the previous evening. It had started several hours before that, as near as anyone could judge, in a seam where the frame between chambers No. 1 and No. 2 joined the roof. Every other seam in the roof had long since been pointed with cement, but this one for some unknown reason had been overlooked and the highly inflammable oakum caulking put in when the caisson was being built had been left exposed. In chamber No. 2, just below the place where the fire was discovered, a workman named McDonald had nailed a wooden box about head high, where he evidently stored his dinner pail, and during the change of shifts at three, in order to see into the box, he must have held his candle against the seam just long enough to start the oakum burning.

There had been four or five fires in the caisson before this one, in the summer when the men were still unfamiliar with working in compressed air. But none of those early fires had amounted to very much. They had been put out quickly by the men themselves, without any help from the fire department, or any public knowledge that they had ever happened. The pressure then was still comparatively mild and there was still plenty of river water in the chambers to drown fires with. But the need for extreme caution in such an oxygen-charged atmosphere had been pretty clearly demonstrated, and particularly if one imagined the same thing happening when the caisson got down to a depth where the river would be sealed out. Roebling wanted no chances taken. In the time since, he had had steam pipes introduced, as well as fire extinguishers and a couple of hoses that could throw an inch-and-a-half stream at sixty-five pounds of pressure. Two men had also been assigned to do nothing but watch over all the lights.

Because the fire began directly above the frame, it had remained undetected until the frame itself had burned through near the roof. But once it was discovered, there was a sudden, loud panic among the eighty men inside the caisson at the time. Tools were thrown down, wheelbarrows were overturned in a rush for the ladders. But that had ended quickly enough. Charles Young, the foreman, got the men under control. Nobody left the caisson and Young ordered some of the men to start packing wet clothes, rags, and mud into the cavity where the fire was burning to shut off the draft of compressed air as much as possible.

The charred opening of the cavity was only about as big as a fist, but inside it looked to be about seven feet long and maybe a foot or more wide and it was all a mass of flame. Men tried throwing buckets of water up into it. Then steam was turned into it for fifteen or twenty minutes. The fire extinguishers were tried after that: two big cylinders of carbon dioxide under several hundred pounds of pressure were discharged into the fire in an effort to smother it. But none of this had any noticeable effect. The instant the steam or the fire extinguishers were turned off the timbers would ignite again.

About ten o'clock, or half an hour after the fire was discovered,

Roebling was sent for, and by the time he had come down through the air lock, the hoses had been put into service and were enough to extinguish all the fire, or at least all anyone was able to see. Still, as Roebling noticed immediately, a violent draft of compressed air was rushing through the aperture. He had most of it stopped with cement, but kept the water steadily playing into it for the next two hours, the force of the water being greatly enhanced by the draft.

In the meantime, Farrington, the master mechanic, had also been sent for and Roebling put him to work drilling holes into the roof to see how far the fire had penetrated. Several holes bored up to two feet showed no signs of fire. Others were then bored a foot deeper and they too showed nothing. But this work went terribly slowly and the tension for everyone was agonizing. Time was lost lengthening out augers, as Roebling told the story later, and the draft carried the chips up into each hole as it was being drilled. There was also the anxiety of knowing that every new hole meant the introduction of another draft of compressed air into the yellow pine and even a small draft of such air, they all realized, would have about the same effect on smoldering wood as a huge bellows.

It was a strange, unnatural kind of fire they were fighting. There was no flame to be seen now and in the dense atmosphere of the caisson, charged already with lamp smoke and blasting powder, it was impossible to see or to smell whether the cavity was smoking. There was no telling either what damage was being done out of sight, or to what depths the compressed air might force the fire into the fifteen layers of timber overhead, in just the way a great weight might force a spike into wood.

Roebling worked feverishly with a few hand-picked men; the rest he had get back to their regular duties. He did not intend to lose a night's work unless he absolutely had to. His own efforts for the next several hours would be described as "almost superhuman" by those who were there.

The question of flooding the caisson came up. To put the fire out some less drastic, simpler way would be immensely preferable, Roebling said, but if they were to find that the fire was not out, as it appeared to be, then, he said, it was only a matter of time until the entire foundation would be destroyed. The fire would eat through the immense pine roof like a hidden cancer, destroying one course of timber after another, until the structure was so weakened that the vast weight overhead (now about 28,000 tons, Roebling calculated) would come crashing through.

The problem with flooding, however, was comparable to that of a blowout. Even if water could be substituted for the compressed air as rapidly as the compressed air was allowed to escape, the caisson would lose a considerable, perhaps vital, part of its support. During one of the earlier fires the river had been allowed to rush under the shoe as air was released and a uniform pressure had been maintained until the fire was out. Moreover, the load being borne then was quite light, comparatively speaking. But now the water would have to be poured down the water shafts from above. The supply might be limited, more than likely it would be variable, and so there was a chance the air might get out before the water reached the roof. In that event, of course, the caisson and all it carried would drop suddenly and the blow would probably be enough to destroy all supports.

To compound the problem, one water shaft was resting on several boulders at that particular moment. It had been capped above earlier, drained, and a gang of men were busy that same night digging the boulders out. But if the caisson were to be flooded immediately and settled abruptly as a result, even if only a foot or so, the water shaft would smash down on the boulders with such force as to wrench it permanently out of kilter. And this, quite likely, would leave the caisson leaking so badly that it could never be inflated again.

About three in the morning, water began to drip from the charred seam for the first time, suggesting that the compressed air had driven water into every possible crevice above, that the timber was now totally saturated for fifteen feet up, and that the fire was at last out.

Roebling was so exhausted he could barely stand. He had been in the caisson much of the previous day and had gone home that evening completely played out. The air was also bothering him a good deal. The men now urged him to leave. Then, about five o'clock in the morning, having decided that the fire was out, he had what appears to have been an almost total physical collapse. The accounts there are do not say exactly what his condition was, only that he had to be helped up through the air lock.

Apparently the sharp night air revived him some at first, but then suddenly he felt the beginnings of paralysis. In a matter of minutes he was unable to stand or walk. Nearby, Charles Young, the foreman, who had also been carried up through the lock about the same time, was in an equally bad state.

Roebling was driven directly home in a carriage just as it was turning light. For the next three hours he was rubbed vigorously all over with a solution of salt and whiskey. It was the best way to restore circulation they believed. He was conscious the whole time and he was in no pain apparently. After a bit, with a little help, he was able to get up and walk about, but he was very weak.

At eight, or thereabouts, a man was at the front door with a message. Fire had been discovered again deep in the caisson roof. Roebling dressed and returned immediately to the caisson.

He was down only a few minutes this time. The carpenters had drilled four feet into the overhead timbers and discovered that the whole fourth course of pine was a mass of living coals. The caisson would have to be flooded, Roebling said. It was a last-ditch decision and he made it on the spot, without any hesitation.

In his absence the men working on the boulders beneath the water shaft had succeeded in removing them, so that at least was one worry he could forget about. He ordered the men all up to the surface. An alarm was sounded in the yard and it was only a matter of minutes after he himself had come up out of the air lock that fire engines were clanging through Brooklyn toward the river with hundreds of people chasing after them.

The time was just about nine. Rumors were everywhere. People were saying there had been a terrible explosion beneath the river, that the caisson had been ripped apart, that half the men had been killed and the rest were still trapped below. At the ferry slip, along Fulton Street, Water and Dock, noisy crowds gathered, everybody trying to find out what was happening and nobody able to see very much. The best view, as usual, was from one of the ferries. But hundreds of people worked their way right into the bridge yards, mingling among swarms of firemen who were rolling out hoses from what appeared to be every last fire engine in Brooklyn. On the river, a New York fireboat and two tugs were being brought up alongside the caisson.

"The crowd dispersed, re-gathered, looked here and peered down there to discover the dread destroyer," the *Eagle* reported later in the day, "but to the general eye no fire was seen."

> Men, muddied by splashing liquid clay, dampened by the streams of bursting hose, made their difficult way over all obstacles, climbed upon the elevation whence the water shaft is accessible, and looked down, only to see the unrevealing surface of the column of muddy water, with which the shaft is filled. Others again, climbed upon the

platform about the air lock, up and down in which the huge rubber pipes go, and in pursuit of knowledge under difficulties, climbed down as far as they might.

Before the morning was out it seemed the whole of Brooklyn had been down to the river's edge. A double line of police was needed to keep back the crowd. "Everybody was there," the *Eagle* said, "and there was considerable lively calculations going on. Persons in every walk of life wandered about the spot, Senators, merchants, laborers. To most of them the whole thing was a mystery."

It was in truth the damnedest fire anyone could remember seeing. There was no flame, nothing to be seen burning, not even much smoke. In some respects the scene was more like that following a mine disaster, the trouble all concealed below, unexplained and out of reach, except here there was no anguish over human suffering, not even much indication that things were serious. Were it not for rumors and the anxious looks of a few officials, no one would know there was any particular trouble below. The idea that there was even a fire had to be taken pretty much on faith.

About the closest thing to real excitement the whole morning was the bursting of a hose that sent up a spectacular plume of spray "upon which the sunlight played," said the *Eagle*, "forming a beautiful, clearly marked rainbow, with a fainter one reflected on the mist and spray from the streams."

A grand total of thirty-eight streams were pouring into the water shafts by ten o'clock, eight from the fireboat, five from the tugs, the rest from fire engines, including two or three brought over on the ferry from New York. One tug alone was pumping eight thousand gallons a minute and estimates were that inside the caisson the water was rising across the entire interior about eighteen inches an hour. But even at this rate it would be another five hours anyway before the chambers were flooded to the top, and therefore it would be that long or longer until the water did any good, since the fire was all in the roof. So fire, fed by compressed air, would be eating through the timberwork with nothing to restrain it until at least three in the afternoon, which accounted for the "degree of anxiety" noted "on the faces of those familiar with the character of the works."

This same reporter singled out William Kingsley standing "conspicuously above all the others" and wrote, "He appeared calm and collected and preserved well his equanimity, but a few words of conversation with him showed him to be anxious for the work." Collingwood, who had not been in the caisson since the day before, was also interviewed and talked of the grave dangers involved with flooding it. Roebling, who remained on hand through the whole morning, would say only that he thought everything would turn out satisfactorily, but that naturally the work would now be delayed some.

By half past three the caisson was entirely filled with water. The compressed air had been replaced without any sudden loss of support. The total quantity of water required was 1,350,000 gallons, which if not quite enough to float a battleship of the day was fairly close to it.

A careful watch had been kept on the pressure gauges during the whole operation; when it appeared that the air was escaping too rapidly, the compressors were started up again. When the water got to within two feet of the roof, the valves used for releasing air were closed off and the balance of the air escaped slowly through leaks and two small pipes. At one point during this stage, pressure dropped suddenly and inexplicably from nineteen to ten pounds.

After the caisson was flooded, the water in the shafts was kept ten feet above tide level, where it stayed with only a little feeding, indicating how very watertight the caisson had become at forty-odd feet below the river.

Still, the prospects looked dim. That night George Templeton Strong, the noted diarist of the time, wrote, "Caisson of the East River bridge was severely damaged by fire yesterday. I don't believe any man now living will cross that bridge."

The caisson remained flooded for the next two days, during which time there was an inquiry conducted by the Brooklyn fire marshal. Some of the New York papers, on their editorial pages, questioned what sort of management had allowed such an accident to occur. The *Herald* said the damage done would cost $250,000. Incredibly, the *World* implied the fire had been an act of sabotage, that directly or indirectly, it had been the doing of someone connected with the ferry company. The *Eagle* ridiculed such speculation and worried about what effect the whole incident might have on the morale of the men who had to carry on with the work. But after the fire marshal's hearing on Saturday, everyone calmed down considerably and it was pretty well concluded that much too much had been made of very little. Collingwood and Farrington both said they did not think the damage would run to more than five hundred dollars. Collingwood thought they had been set back two days at most. C. C. Martin said, "All that the fire has done is to burn little spaces between the beams, very probably very small ones which will not in that mass of timber affect the stability of the structure in the slightest."

It was also reported by one of the assistant foremen that the man named McDonald, who supposedly started the whole thing, had not been seen or heard from since.

Roebling had testified separately earlier in the day. He was still feeling some paralysis, he said. He too thought the damage had been minor and reported that the caisson had settled only two inches during the flooding, which he said was less than the average daily rate of descent. He left no doubt at all that this highly precarious operation had been very successfully executed and would perhaps prove even beneficial to the caisson in the long run, since the timbers had been getting too dry.

Monday morning the air pressure was restored, the water pushed out in about six hours. It all ran out over the tops of the water shafts. When Roebling and the others went down inside, everything seemed to be in good order, beyond a few blocks crushed and some posts thrown over. The structure itself appeared tighter than before due to the swelling of the timbers.

The fire marshal went into the caisson a little later with C. C. Martin and reported that he watched Roebling and the others at work, checked things over, and said that if he had not been told differently, he would never have known there was any fire at all. There was not the least sign of fire below, he said, except through one small opening and he concurred that the damage had been very slight.

Work was resumed immediately. The brick piers, about a third of the way built by then, were finished in another two weeks and the caisson was lowered the final two feet to rest upon them.

The day before Christmas the men began filling the work chambers with concrete. To save time and cut the quantity of concrete needed by a third, the shoe of the caisson was allowed to sink into the ground three feet deeper than the average level of the caisson floor, which meant that headroom inside was reduced from nine and a half feet to six and a half feet.

The concrete consisted of one part Rosendale cement, two of sand, and four of a fine gravel from the Long Island beaches, where it had been washed perfectly clean by the surf. Outside the caisson the weather by now had turned so cold that the concrete had to be mixed below. So like the bricks for the piers, cement, sand, and gravel were all brought down through the supply shafts, which for some several weeks had been functioning quite flawlessly.

The shafts were iron tubes forty-five feet long and twenty-one inches in diameter, with doors at top and bottom. When the upper door was open, the lower door would be held shut by the pressure in the caisson and locked by two iron clamps worked by levers. Any material needed below would simply be dumped down the shaft and the upper door, which closed up, not down, would be pulled shut. Then compressed air would be allowed to enter the shaft from below, closing the upper door tighter still. As soon as the shaft was filled with compressed air, the lugs on the lower door would be removed, the door would fall open, the contents in the shaft would drop into the chamber. The system was fast, uncomplicated, and quite safe so long as the attendants responsible for it used their heads.

But it was only two weeks after the fire that again something went wrong. Every so often a load of bricks would get jammed in a supply shaft and the usual method of breaking the jam was to drop a weight down on a rope. But this time the men above decided instead to dump in a second load, then signaled for the men below to open the lower door while they neglected to close the upper door. The second load loosened the first, the two together landed on the lower door with a force greater than the air pressure against it from inside, and since the lugs on the door had been opened as directed from above, the door fell open.

Instantly there was an enormous, earsplitting rush of air out of the caisson. Stone and gravel shot from the shaft as if from a cannon. The men on the top dove for cover or fled as fast as their legs would carry them. Had any one of them had the least presence of mind, he could have closed the shaft instantly and had everything locked up tight quite simply by just reaching over and pulling at the rope connected to the upper door. It would have taken no effort whatever. The explosion of air from below would have slammed the door shut. But nobody did that.

Roebling was one of those trapped inside the caisson at the time. The noise, he said, was so deafening that no voice could be heard.

Water was pouring in from the water shafts. The lights went out. The air, he said, was full of a dense, impenetrable mist. Men were stumbling all over one another, running in terror, smashing into pillars, tripping and falling in the pitch-darkness, nobody sure where he was going.

In an instant the water was up to their knees. The river had broken in, they all thought.

"I was in a remote part of the caisson at the time," Roebling wrote, "half a minute elapsed before I realized what was occurring and had groped my way to the supply shaft, where the air was blowing out. Here I joined several firemen in scraping away the heaps of gravel and large stones lying under the shaft, which prevented the lower door from being closed." It took from two to three minutes for them to clear the door. Then they had it shut and everything was all over. Fifteen minutes later the pressure was restored.

Roebling had kept his head under the most nightmarish conditions, and when nobody else had. He had analyzed the situation in an instant and moved swiftly to put a stop to it. In the eyes of many, it was as commendable a demonstration of cool command as anything he had done on a Civil War battlefield.

Later, in his formal report to the directors, he wrote:

> The question naturally arises, what would have been the result if water had entered the caisson as rapidly as the air escaped? The experience here showed that the confusion, the darkness and other obstacles were sufficient to prevent the majority of the men from making their escape by the air-locks . . . Now it so happens that the supply shafts project two feet below the roof into the air chamber; as soon, therefore, as the water reaches the bottom of the shaft it will instantly rise in it, forming a column of balance and checking the further escape of air. The remaining two feet would form a breathing space sufficient for the men to live, and even if the rush of water were to reduce this space to one foot, there would be enough left to save all hands who retained sufficient presence of mind.

It is not known whether he had realized this *before* the supply shaft blew out.

Again, as after the "Great Blowout," an examination was made to determine what effect the impact of sudden weight had had upon various internal supports and particularly on the new brick pillars. By the time Roebling and the others got the supply shaft door closed, pressure in the chamber had dropped from seventeen to four pounds. He reckoned, therefore, that for several minutes the weight on the pillars was twelve tons to the square foot. Still, they showed no signs of strain, which was the clearest demonstration possible of their capacity to bear up under the load they were designed for and proof certainly that Roebling had been right to put them in. More important, the subsoil beneath the pillars, on which the bridge was to bear, had also withstood this same tremendous pressure.

But Roebling would be granted precious little time to take pride in the way things were being handled. Work on the masonry above had stopped because of the weather. Eleven courses of stone had been laid up within a wooden cofferdam, the top of the stonework being about even with the river at high tide. But the people manning the dump carts were still about, along with a number of others who looked after this or that piece of equipment, and they had begun noticing a strong smell of turpentine that seemed to be coming from air bubbles being forced up through the caisson. Large deposits of frothy reddish-brown pyrolignic acid, or "wood vinegar," as the men called it, had also been found, indicating, as Roebling said, "that a destructive distillation of wood had been going on."

Acting on what he called very unpleasant suspicions, Roebling quietly ordered Farrington to start drilling into the roof again. About two hundred borings were made to determine for everyone's satisfaction just how extensive was the internal damage from the fire.

Most of it seemed to be confined to the third and fourth courses of timber, as had been expected, but as nobody had imagined, it also extended out laterally in every direction, in some places as much as fifty feet, or about five times farther than anyone had judged earlier. Equally disturbing was the discovery that the compressed air was rushing out of every bore hole, which meant that any attempt to cut into the roof to make repairs would result in an enormous drain on pressure.

Roebling decided, however, that if the air chamber were filled in with concrete around the edges, the pressure might now be released entirely with no harm. He had decided, in other words, that he could trust the brick pillars to support everything. So if maintaining pressure was no longer the vital concern, then holes as large as need be could be cut overhead and the damaged areas seen to properly.

Still, as he wrote, "It was very desirable . . . to gain time and do as much as possible at once, while air pressure was yet on." It would be necessary therefore to plug the boreholes and at the same time compensate somehow for the honeycomb of charred pine they all pictured overhead.

Accordingly cement grout was injected into all two hundred boreholes. It was no easy task and it took quite some time. Roebling had a cylinder and piston fixed to a quarter-inch pipe. The cylinder was filled with liquid cement, placed under a borehole, the pipe inserted, then the cement forced up the hole by a screw jack. The technique worked well for the most part. The cement could be forced a good ten feet into the timber and appeared to spread out laterally to some distance. But the moment it met any resistance, all the water would be squeezed out, and to budge the charge another inch became impossible. So a thinner mixture was tried and it was found that the suction of compressed air alone, through the holes, was enough to draw this up the pipe and into the timberwork.

By the time they were finished, six hundred cubic feet of cement had been pumped into the caisson roof. The leaks had been stopped and a number of new boreholes in the area of the trouble failed to reveal a single place without its own vein of hardened cement. "We already flattered ourselves that this filling might answer every purpose . . ." Roebling wrote.

But just to be sure, he had a great hole, six feet across, cut into the roof through five layers of timber, directly over the spot where the fire had originated. And by opening up the roof this way they discovered that they had been exceedingly proficient in their work and that it had been a great mistake. The cement had indeed filled every crack and crevice, but most of the timber beneath the cement was covered with a layer of soft, brittle charcoal, anywhere from one to three inches thick.

It was a crushing revelation. It meant that every last bit of the cement put in so laboriously would have to come out, and any charcoal there was would have to be found and scraped away. There was no other alternative Roebling said. The caisson roof, the timber platform upon which the bridge tower would stand, had to be absolutely, permanently solid. He could take no chances on that. He could no more let it go this way then he could launch a ship with rotten timbers.

The immense, painstaking job of restoration that followed took a force of eighteen carpenters three full months to complete, working night and day. It was like gigantic dentistry, as someone said. To say that the work was extremely disagreeable, as Roebling did in his report to the Board of Directors, was to greatly understate the situation.

Not until the cement was all chopped out did anyone realize the full extent of the fire. Instead of one opening into the roof, five had to be cut, slowly, laboriously, each one three to four feet square and five feet deep or more. Above the original opening it was found that the fire had not only turned the third and fourth courses of timber to charcoal, but it had burned right through the sheet tin between courses four and five, destroyed the fifth course and made a start on the sixth. To judge by the traces it left, the fire had advanced mainly as a slow, intense charring that expanded equally in all directions. But in numerous places it had been strangely erratic, due no doubt to the multitude of leaks that fanned it. Roebling noted, for example, that a single 12-by-12 timber would be burned away for thirty feet, while one just like it directly alongside would be untouched. And since the courses had been laid up at right angles to one another, the fire had had opportunity to branch off in a zigzag pattern, jumping from one timber to another, heading off left or right, up or down.

Damaged timbers were carefully scraped clean and all jagged edges were squared off with chisels. Every foot of burned wood was cut out. New cement was rammed into the smaller places, while new timber was cut to size (in lengths of eight to ten feet usually), rammed into all the larger openings with a screw jack, and securely bolted.

The burned channels that wandered laterally between the big vertical openings were generally about two by three feet in dimension. All such channels had to be gouged out by hand, then filled in in the same manner. The men worked like coal miners along such veins, inching forward on their backs or sides, with barely enough room to move, digging out charcoal instead of coal, imprisoned in a mountain of wood instead of earth. For hours at a time a man would be confined to a single spot, unable to turn around, his only light a little bull's-eye lantern, and breathing candle smoke, cement dust, and powdered charcoal. Because the air pressure had been greatly reduced by the openings that had been cut, the ventilation was dreadful and the heat remained near 90 degrees.

"After everything was filled up solid," Roebling wrote, " a number of five-foot bolts were driven up from below so as to unite both the old and new timber into a compact body." He also had forty iron straps bolted against the roof from below, and inside the air chamber, directly under the line of the fire, he had great square blocks of traprock set in the concrete that was being put down over the rest of the chamber floor.

When the repairs were at last completed Roebling reminded everyone that there were still eleven perfectly sound layers of timber above the first four. And in his final report he stated, "From the faithful manner in which the work was done it is certain that the burnt district is fully as strong, if not stronger than the rest of the caisson." Most people believed him.

The fire and its aftermath had been a sobering experience. It had delayed the work two, possibly three months. With the payroll running about eleven thousand dollars every two weeks, this meant a loss of some fifty thousand dollars at least, on that score alone. The fire had done much to reinforce the arguments of the skeptics, of whom there were still plenty on both sides of the river. This said one New York paper was the "main mischief" of the whole unhappy affair. But it had also been a brutal physical and mental ordeal for many of the men, and for Roebling in particular, whose strength had never quite returned since the night he collapsed in the caisson. He was a changed man after that, his assistants would say later.

Had he decided early that night to flood the caisson, in spite of the boulders beneath the water shaft, then things might have gone differently. But if he ever speculated about that, it was only in private.

The last repairs were completed on March 6, 1871. Five days later the air chamber was completely filled with concrete. During the final few weeks of sinking the caisson, several fresh-water springs had been encountered, and now, much to everyone's astonishment, the water came right up through six feet of concrete in such quantity that it filled the water shafts clear to the top. The water was perfectly fresh, without a trace of salt, so it was all coming from directly beneath the caisson. The shafts were drained, therefore, and they too were filled with concrete. The air locks were removed and these empty spaces were also filled with concrete.

So by mid-March the Brooklyn caisson was permanently in place. The hardest, most treacherous and uncertain part of the work, the sinking of the caissons, was half done. No lives had been lost, no one had been seriously injured. Every man on the engineering staff had proved himself worthy of the faith Roebling had in him, and none had quit.

Work on the Brooklyn side from here on would be of the sort everyone had been anticipating. There would be something actually to see now, to watch grow and change from one day to another. The Brooklyn tower, it was commonly said, would be the greatest structure in the world except for the Pyramids. "America has seen nothing like it," Thomas Kinsella wrote on the editorial page of the *Eagle*. "Even Europe has no structure of such magnitude as this will be. The most famous cathedrals and castles of the historic Old World are but pygmies by the side of this great Brooklyn tower. And it is our own city which is to be forever famous for possessing this greatest architectural and engineering work of the continent, and of the age."

Such grandeur was still several years off, everyone knew, but it was not so very difficult to picture. "Think of Trinity Church as big at the top of the steeple as at the ground," said Kinsella, "and one solid mass all the way up, and we get some idea of what this great Brooklyn tower is to be . . . the fame of the Roeblings and the boast of Brooklyn forever will be that, where Nature gave no facilities for a suspension bridge, and seemed indeed to place a veto upon the idea in these low and shelving shores, the genius of the father designed, and the consummate inherited and acquired ability of the son executed, in spite of all obstacles, this most novel and unparalleled masterpiece. . . ."

But not until June 5, when the *Eagle* published Washington Roebling's annual report to the directors of the Bridge Company, did the people of Brooklyn and New York get a fair idea of what exactly had been accomplished to make the tower possible. Except for the flurry of excitement when the fire was discovered, almost nothing about the details of the work or the setbacks experienced inside the caisson had appeared in the newspapers. Roebling's report filled seven and a half columns. It was straightforward, unadorned, and it was read with enormous interest. People were utterly astonished to learn all that had taken place beneath their very noses. "We are not partial to long official reports," the *Eagle* said by way of introduction, "but this one is exceptional in the thrilling interest of the story." Roebling was praised for his modesty by the editors and lauded for his own personal heroism in such a way that it seemed they too were realizing for the very first time the extent of what had happened down at the end of Fulton Street. In 1870, when the caisson was making its slow, tedious descent, the English translation of Jules Verne's *Twenty Thousand Leagues Under the Sea* had appeared in America, with its adventures of the strange genius Captain Nemo. Now the *Eagle* wrote: "The adventures of Colonel Roebling and his twenty-five hundred men under the bed of the East River are as readable, as he tells them, as any story of romance which has issued from the imagination of the novelist."

What Colonel Roebling and his men might run up against on their next descent was now, naturally, a matter of much popular interest. On one side of the river a tower of imperishable granite would be rising straight into the sunlight, while on the other side, mortal men would be descending beneath the tides and into the earth. It was quite a picture to keep in mind for anyone crossing the river to clerk in a countinghouse or sit the long day at a sewing machine.

11
The Past Catches Up

The spectacle is appalling. We live in an atmosphere of hypocrisy throughout.

—WALT WHITMAN, *Democratic Vistas*

WHEN the New York caisson hit the water it made such a wave that several tugs standing close by were tossed about in a "very sportive manner" and two men in a rowboat, who had come in close for a better view, were immediately swamped and had to be rescued. Because of the comparatively shallow water in front of the Sixth Street launchways, the caisson had been built with a temporary floor, which was the reason for the wave. Otherwise the launching had gone off perfectly, as expected. Indeed, the engineers had been so confident of success that Emily Roebling and Mrs. William Kingsley had gone along with their husbands and several others on top of the caisson as it was sent hurtling down the ways.

At a large luncheon served in the Webb & Bell offices afterward everybody had been in high spirits. "We are now on foreign soil!" proclaimed John W. Hunter of Brooklyn, one of the stockholders. Everyone cheered. Then Henry Slocum got up and said he and Kinsella had agreed that Kinsella would do the speaking, while he did the thinking. But Kinsella interrupted immediately. Slocum was to do the *drinking* not the thinking, the editor said, and the laughter was very great according to later accounts.

After that Kinsella reminisced about the days when the bridge was no more than "the shadow in the brain of one man," as he put it. "When William C. Kingsley [loud applause] was the founder, he put up more of fortune and reputation than any man I ever knew to do in an enterprise at the time so shadowy." Then Slocum was on his feet again to propose a toast to Kingsley "as the man to whom we are more indebted than to all others." There was a standing ovation for Kingsley, who said only that he had never made a speech in his life and asked instead that everyone drink to the health of Colonel Roebling. According to one version of the scene, Roebling was "subsequently introduced and loudly cheered, but not threats nor blandishments could coerce a speech out of him."

The launching of the caisson and the celebrative luncheon afterward took place on May 8, 1871. There was still a great deal to be done, however, before the caisson would be ready for use. By the time it was completely fitted out, towed to position, and sufficiently loaded down so that the men could begin work inside, seven more months would have passed, the year would be nearly over, and by then if anyone were to use the word "shadowy" to describe the early business of building the bridge, it would be for quite different reasons. For by the time the New York caisson would start its descent, the Tweed Ring would have collapsed, something no one would have believed possible in May of 1871.

The year had begun splendidly for Tweed. On New Year's Day Oakey Hall was again sworn in as mayor of New York and John T. Hoffman as governor. If everything went according to Tweed's plan, 1872 would see Hoffman elected President of the United States, Hall governor, and Tweed a United States Senator, or at least so it seemed to a number of political observers.

The annual Americus Club Ball in early January had been a triumph. The club stood on a cliff overlooking Long Island Sound, at Greenwich, Connecticut, not far from Tweed's own summer place. It was as sumptuous as any club in the country and the pinnacle of Tammany social life. On the night of the ball the great halls of the clubhouse were described as "a labyrinth of festoons, flowers, fountains, flags, and fir trees." To the obvious delight of the club president, Tweed, a thousand canaries chirped in gilded cages hung everywhere about the hall, while on the stage before the dance floor there blazed an immense gaslight rendition of the Tammany banner, the familiar snarling tiger's head of Tweed's old Americus Six fire company. Vases of cut flowers lined the walls, and fronting the main entrance was a life-sized photograph of the great man himself, hand-colored with crayons. Some six thousand people were there, the papers said, making it "a gala pageant such as is rarely witnessed anywhere in the world."

But later in the month, on the night of the 24th, a sleighing accident had occurred north of Central Park, in which a man named James Watson had been badly injured. Twenty years before, Watson had been behind bars in the Ludlow Street Jail when the warden put him in charge of the prison records because of his exquisite penmanship. But Watson had advanced rapidly in the time since and was now County Auditor and trusted bookkeeper for the Tweed Ring. A week after the accident Watson was dead and so began the fall of the house of Tweed.

The newspapers made little of Watson's demise and no one thought much more about it. Such momentous events were taking place elsewhere in the world that public interest in the Ring and its doings had greatly subsided. The Franco-Prussian War, which had begun the summer before in Europe, was coming to a thunderous conclusion. On January 18, in a solemn ceremony at Versailles, the German Empire had been born. Incredibly, France was about to be overthrown. Popular support for the Germans was enormous in the United States and in New York especially. King William I of Germany was described by the papers as an affable, courteous old gentleman and a special favorite of children. Heroic Prussian infantry charged across the illustrated pages of *Harper's Weekly*. On January 27, newsboys in the streets were hollering the surrender of Paris. But the bloodshed had continued for months afterward, as civil war broke out within the city, and the New York papers were filled with grisly accounts of hostages murdered by the Communards.

Not until the last of May was Tweed back in the news to any great degree and even then the occasion was a happy one, the publicity just what Tweed wanted. One of Tweed's daughters, Miss Mary Amelia Tweed, was being married to Mr. Arthur Ambrose Maginnis of New Orleans and the father of the bride and his friends put on quite a show.

The wedding took place at seven o'clock on the evening of May 31, 1871, in Trinity Chapel, where, at the appointed hour, "a richly attired audience" watched Tweed, daughter on arm, come slowly, grandly down the aisle to the tune of Mendelssohn's march. The bride wore white corded silk, "*décolleté*, with demi-sleeves, and immense court train." There were orange blossoms in her hair and she carried a huge white bridal bouquet. But the diamonds were what everyone talked about later. "On the bride's bosom flashed a brooch of immense diamonds," said the *Sun*, "and long pendants, set with large solitaire diamonds, sparkled in her ears." On the white satin shoes she wore there glittered tiny diamond buttons.

The mother of the bride, attired in salmon-colored silk, also wore "splendid diamonds," and it was noted that "Mr. Tweed himself wore black evening dress, and a magnificent diamond flashed on his bosom."

After the service Tweed put on a reception at his Fifth Avenue mansion, where a blue silk awning and a Brussels carpet had been run out to the curbstone and a huge crowd had gathered in the street. The house was ablaze with lights. A fountain played at one corner. Inside, the rooms were a mass of flowers. "Imagine all this," one dazzled reporter wrote, "lighted up with the utmost brilliancy, and hundreds of ladies and gentlemen in all the gorgeousness of full dress and flashing with diamonds, listen to the delicious strains of the band and inhale in spirit the sweet perfume which filled the atmosphere, and some inadequate notion can be formed of the magnificence of the scene."

The dinner was catered by Delmonico's, but the main attraction was a display of wedding presents in a big room upstairs. The gifts lined all four walls and were said to surpass anything seen since the marriage of the daughter of the Khedive of Egypt two years before. There were forty sets of sterling silver. One piece of jewelry alone was known to have cost $45,000. Peter "Brains" Sweeny gave diamond bracelets "of fabulous magnificence," and Tweed's two other fellow stockholders in the Bridge Company, Smith and Connolly, proved equally generous. The *Herald* came up with an estimated cash value of everything on display: "Seven hundred thousand dollars!"

The wedding was the high-water mark of the Ring's opulence and for Tweed it was a great blunder. The public, dazzled and delighted at first, began to ask questions afterward. How could a man who had spent his whole life working in moderately paid positions with the city live in such style? Where did the likes of Police Superintendent Kelso get money enough to buy a wedding gift that, as the papers reported, was the "exact duplicate" of the one presented by Jim Fisk? People said the wedding was proof positive of the corruption the *Times* had been making such a commotion over and papers other than the *Times* took up the cry.

"What a testimony of the loyalty, the royalty, and the abounding East Indian resources of Tammany Hall," wrote the *Herald* with bitter sarcasm two days after the wedding. "Was there any Democracy to compare with thy Democracy, in glory, power, and equal rights, under the sun? Never! And it is just the beginning of the good time coming. Don't talk of Jeff Davis and his absurd Democracy; don't mention the Democracy of the Paris Commune, as representing true Democratic principles; but come to the fountainhead of Democracy, the old Wigwam, and you will get it there—if you get within the lucky circle of the 'magic' Ring."

So with summer coming on, the public was ripe for disclosures and as chance would have it things had been happening inside the "lucky circle" that would reveal for the first time and in plain figures examples of such outrageous plunder that Tweed's wedding expenditures would look rather modest by contrast. The pattern of events was like something from one of the popular melodramas of the time, for it was on the very day of the wedding that a man named Matthew J. O'Rourke quit his job as County Bookkeeper, taking with him a package of documents he had been quietly assembling.

O'Rourke was a newspaperman who had been hired two months earlier when the County Bookkeeper of longstanding was moved into the late James Watson's place as County Auditor. For some strange reason Tweed's people had not bothered to take O'Rourke into their confidence. Moreover, O'Rourke had long had a claim against the city which the Ring had ignored. So he had very carefully copied down a number of choice samples from the ill-fated Watson's old account books and these he delivered to the offices of *The New York Times*.

Watson had been at the nerve center of the Ring. Indeed, if there was "magic" to the Ring, Watson was the unseen assistant who made it work, and as with most grand deceptions the secret was extremely simple.

Watson merely required everyone who received a contract from the city to increase his bills before submitting them by 50 to 65 per cent. Watson paid the face amount of the bill, then the contractor returned the overcharge in cash, and Watson, like a dutiful paymaster, handed it out within the Ring. Among New York contractors it was commonly said, "You must do just as Jimmy tells you, and you will get your money."

Anyone who knew a little bookkeeping could look at Watson's voucher records and see what was going on. O'Rourke, for example, judged from what he saw that the Ring had made off with $75 million since 1869.

Watson had worked directly for Comptroller Connolly. Why Connolly had been so careless about whom he let see Watson's books is hard to fathom. But the books were left so unguarded that the *Times* soon received a second bundle of figures from a man named William Copeland, who had copied down still further revelations.*

Tweed found out that the *Times* had the material before any of it appeared in print and he reacted predictably. Figuring that George Jones, publisher of the *Times*, had his price, like any man, Tweed told Connolly to go see him. Connolly offered Jones five million dollars to kill the story. Jones declined, reportedly saying to Connolly, "I don't think the devil will ever make a higher bid for me than that." Connolly persisted. "Why with that sum you could go to Europe and live like a prince," he said. "Yes," answered Jones, "but I should know that I was a rascal."

To Tweed the time seemed also right for silencing Thomas Nast. A well-known banker was selected to call on Nast. The banker told Nast that his artistic talents were much admired by certain gentlemen, that these same gentlemen thought so highly of Nast that they would be willing to put up $100,000 to help him further develop his genius in Europe. Nast asked if he might get $500,000 to make the trip. "You can," the banker is supposed to have answered most enthusiastically. "Well, I don't think I'll do it," Nast said. "I made up my mind not long ago to put some of those fellows behind bars, and I am going to put them there." Whereupon Nast was told to be careful lest he put himself in a coffin first. (Later, after the *Times* opened fire on him, Tweed said that were he a younger man he would have gone and killed Jones personally.)

The *Times* began running transcripts from Watson's account books in early July, starting with some armory frauds O'Rourke

* Copeland, unlike O'Rourke, had not been acting alone, but was a spy for Sheriff Jimmy O'Brien, a political enemy of Tweed's, who got Copeland a job in Connolly's office and intended to use the material to blackmail Tweed. Tweed offered O'Brien $20,000 to keep him quiet and promised more. O'Brien, who wanted $350,000, took the money, then took Copeland's "research" to the *Times*.

had copied down. (Ten old stables rented by the Ring for practically nothing had been sublet to the county as armories for $85,000 a year; the buildings had not been used at all, still the county had paid out nearly half a million dollars just for repairs on them.) Later in the month a special supplement was gotten out by the *Times* documenting the outlandish courthouse swindles, among other things. Two hundred thousand copies were sold out at once. Horace Greeley wrote in the *Tribune* that if the *Times* had its facts straight, then Mayor Oakey Hall and Connolly ought to be breaking rocks in a state prison. In the meantime, however, on July 12, the Ring had been dealt still another blow and, quite unintentionally, by some of its most ardent supporters.

July 12 was the day for the annual Orangemen's Parade, in honor of the anniversary of the Battle of the Boyne. The event had caused a serious riot the year before, in Elm Park. Some two thousand Protestant Irish Orangemen gathered for a picnic had been set upon by about three times as many Irish Catholics armed with clubs and pistols. The police had arrived eventually, but not before several people had been killed on both sides and scores severely wounded. This year, fearing the same would happen again, Mayor Hall had forbidden the Orangemen to hold their parade. (Though himself a Protestant and of English ancestry, the Elegant Oakey was accustomed to reviewing St. Patrick's Day parades dressed in a green cassimere suit with shamrock buttons, a bright-green cravat, and green kid gloves.) Hall's decision provoked a storm of protest. The Ring was pandering to the Catholic rabble it was said. New York had come to a terrible pass if decent people were no longer able to parade peaceably without fear of mob violence.

As a result Governor Hoffman, probably acting with Tweed's consent, if not at his direction, came rushing down from Albany to issue a proclamation saying that anyone who wished to "assemble and march in peaceable procession" was at liberty to do so and would get full protection from the police and the military. It was an immensely popular move. Some 160 Orangemen decided to go ahead with their parade. Four regiments of National Guard were ordered to stand by, including the Ninth, in which Tweed's bosom friend Fisk was a colonel. Catholic priests called for peace and understanding, while it was rumored that in Brooklyn trained gangs of Irish thugs were getting ready for action.

The line of march was from the Gideon Lodge of Orangemen in Lamartine's Hall, at Eighth Avenue and 29th Street, downtown to

Cooper Union, at Eighth Street and the Bowery. Nothing much happened until the Orangemen reached 26th Street, where the police, marching on either side, had to force a path through a crowd blocking the way. At 25th Street the police were ordered to charge. By then stones and bricks were coming down from housetops. Near the corner of 24th Street a shot was fired. The police said later it came from an upstairs window, but others claimed a rifle went off accidentally among the Ninth Regiment, which had been drawn up at 25th Street, with the comic-looking Fisk prancing about on horseback.

Whichever the case, the bullet took off part of the head of a private in Company K, of the Ninth. Instantly, without orders, the soldiers opened fire into the crowds. The fusillade was very brief. When it was over, two soldiers, one policeman, and a total of forty-six bystanders, including a number of women and children, were dead.

(Fisk had dismounted in the midst of all this and disappeared into a saloon, where, it was later learned, he escaped out a back door, scaled several fences, got rid of his uniform in a house on 23rd Street, then fled as swiftly as possible to Long Branch, New Jersey, the fashionable seashore resort.)

Everyone was furious at the Ring, including most of the Irish Catholics who had long been the very lifeblood of Tammany. Among the Protestants there was angry talk of Irish despotism. "Write on the tombstone of Wednesday's victims: 'Murdered by the criminal management of Mayor A. Oakey Hall' " one man commented. Others felt the riot was symptomatic of a larger tragedy. "Behind the folly and wickedness of the Irish," said *The Nation*, "there lie American shortcoming, corruption, and indifference."

The problem was, however, that most of the people who could have organized any sort of movement against Tweed were out of town for the summer. The *Times* continued its assault day after day, but nothing happened. When Tweed was pressed to comment on what the *Times* was printing about him, he exclaimed defiantly, "Well, what are you going to do about it?"

As it was, nothing would be done about it until the end of the summer, when a mass meeting was called at Cooper Union on the night of Monday, September 4. Former Mayor William Havemeyer presided, thousands attended. In no time the rather refined and reserved-looking crowd was in a fine frenzy. "There is no power like the power of the people armed, aroused, and kindled with the enthusiasm of a righteous wrath!" said Judge James Emott from the speaker's platform. No more could there be any denying the frauds of the Ring, he said, and in obvious answer to Tweed's now famous retort, he asked, "Now, what are you going to do about these men?" "Hang them!" shouted voices from the audience.

A so-called Committee of Seventy was organized that night, and included among its membership some of the most distinguished names in New York—Judge Emott, Robert Roosevelt, Andrew Green, who had been a pallbearer at John A. Roebling's funeral, and Abram Hewitt. Also among them was Samuel J. Tilden, a wealthy friend and New York neighbor of Hewitt's. Tilden was a rather cold, calculating corporation lawyer and a Tammany Democrat of great ambition, who had had little derogatory to say about Tweed prior to this, but who now took charge of the committee and shortly became known as the leader of the entire reform movement.

The campaign to destroy the Ring was on in earnest. On September 12 someone broke into Comptroller Connolly's offices and stole all the vouchers for the year 1870. The theft was a great mystery according to city officials. But in *Harper's Weekly* Nast pictured Connolly, Sweeny, Tweed, and Hall looking highly ridiculous as they proclaimed their innocence and a few weeks later Nast had the same group cowering in the shadow of a gallows.

Secretly, Tweed began transferring all his real estate holdings to his son. Hall tried to get Connolly to resign, while Connolly, terrified the others were about to throw him to the lions, consulted with Tilden, who told him to take a leave of absence and appoint the upright Andrew Green as Acting Comptroller. This was done. Armed guards were stationed in the Comptroller's office to watch over what remained of the records. The reformers had achieved a beachhead.

Tweed, however, was anything but placid through all this. He had power aplenty still and he had no intention of giving up without a fight. The real test, he knew, the only one that mattered, would be in November, at the polls. His own term as State Senator would expire on December 31.

First he had himself unanimously re-elected chairman of the General Committee of Tammany Hall. Then he went off to the Democratic state convention at Rochester, taking along a gang of New York thugs to remind any wavering delegates where their sympathies lay. Bribes were handed out liberally. Tilden and some other reform delegates put up a fight of sorts, but Tweed was in control the whole time and everyone knew it. He got the nominations he wanted, including his own.

Back on the Lower East Side of New York the day after the convention ended, Tweed stood before some twenty thousand cheering followers, in Tweed Plaza, doffed a little Scotch cap, bowed low, and said the following:

> At home again amidst the haunts of my childhood and scenes where I had been always surrounded by friends, I feel I can safely place myself and my record, all I have performed as a public official plainly before your gaze. The manner in which I have been received tonight has sent a throb to my heart, but I would be unjust to myself and unjust to those who have seen fit to entrust me with office if at times like these, when to be a Democrat, when to hold a public office is to be aspersed and condemned without trial, traduced without stint, there was not felt to be engraven on my heart the proud satisfaction that as a public officer, I can go to the friends of my childhood, take them by the hand, take them in a friendly manner, and saying to them, "There is my record," and finding that it meets with their approval.

For most Americans the evils of the Tweed Ring were the natural outgrowth of the essential evil of big cities. New York being the biggest of big cities, it would quite naturally produce a Tweed. New York simply got what New York deserved was the feeling. The city was mostly foreign-born after all, better than half Roman Catholic. The golden age of representative government had lasted less than a hundred years learned men were saying gloomily. Jefferson had been right about what cities would do to American life. The future now belonged to the alien rabble and the likes of Tweed. "Perhaps the title 'Boss of New York' will grow into permanence and figure in history like that of the doge of Venice," wrote George Templeton Strong in his diary. Even Walt Whitman of Brooklyn, who celebrated the "power, fullness, and motion" of New York in his *Democratic Vistas* published that year, wrote savagely of the "deep disease" of America, which he diagnosed as "hollowness at heart."

But if in New York—even in New York—men of principle, men of integrity, education, and property, could rise up and triumph over a Tweed, then perhaps there was hope for the Republic. The idea was a tonic. Numerous other cities had their "mute, inglorious Tweeds," as E. L. Godkin wrote in *The Nation*, and other cities would have their own spirited crusades for the restoration of political virtue. Reform groups became the fashion. Committees were formed. Silent indifference to political immorality was no longer acceptable in polite society and Brooklyn was no exception.

As might be expected the impact of things happening so close by in New York was especially pronounced in Brooklyn. The talk that October was nearly all politics. The editors of the *Union*, in their offices directly below the Bridge Company, wrote that Brooklyn could well use the services of New York's Committee of Seventy and that there had been "fear and trembling" among certain Brooklyn officials ever since the assault on Tweed began.

That election frauds had made the Honorable Martin Kalbfleisch mayor of Brooklyn two years before was common knowledge. That there would be more of the same this time seemed inevitable. But even Kalbfleisch appeared to be caught up with reform spirit. He had denounced the Democrats and was running for re-election as an independent. So it was a three-way race. The Republicans had a candidate and Boss McLaughlin had put up a former mayor named Powell, a decent enough man, as were most McLaughlin mayors, with the notable exception of Kalbfleisch.

The Democrats appeared to be in trouble for the first time in years. The opposition was calling for the downfall of the Brooklyn Ring along with the Tweed Ring. Kalbfleisch would assuredly take away more Democratic votes than Republican. For McLaughlin, for Murphy, Kingsley, and Thomas Kinsella, these were exceedingly busy times, with much at stake, not the least of which was the total say on how the Bridge Company ought to be run. Nobody who understood the realities of Brooklyn politics seriously thought the Democrats might lose, but at the same time the Democrats were taking no chances.

Then on the 23rd of October there was a horrible accident at the bridge. Two men had been killed, both married men with families. A third was so dreadfully mangled that little hope was held out for him, and five others were badly injured. Until now the bridge had had a perfect safety record. In the two years since construction began there had not been a single serious accident, no injuries to speak of, no loss of life. This was extraordinary, in view of the known hazards, and it was generally taken as a sign of conscientious management.

The Brooklyn tower by then had reached a height of about seventy feet. The day of the accident an eight-ton block of granite was being hauled to the top of the tower by one of the three huge boom derricks mounted there, which were all held in position by wire rigging, like the masts of a ship. Suddenly a socket in one of the guy wires broke and two derricks fell from the tower.

One man, a rigger on the derricks, was struck by a great wooden boom that sheared off half his skull. He was thrown down on the top of the tower, stone dead. "If he had stood still he would not have been injured," C. C. Martin said later; "but when he heard everything crackling and crashing, he lost his presence of mind and ran out where the derrick was coming down just in time to be struck."

Another man had been standing on an elevated railroad track that ran along the side of the tower facing the river, about fifty feet above the dock where the stone scows tied up. His job was to shove a little flatcar under the stones as they were hoisted by an engine on the dock, then move them into position to be picked up by one of the derricks on the tower. When the derricks fell, the stone suspended from one of them crashed through the track about twenty feet from the spot where he was standing. It would have missed him, in other words. But seeing it coming, he had fled to the end of the track and leaped off. The fall broke both of his legs and no one knew how much else internally. He died later.

The other man killed instantly was also on top of the tower. He too saw the derricks falling and tried to get out of the way by jumping over a granite block sitting on the masonry. Just as he jumped, the mast of one derrick fell on him, crushing him, face downward, into a crevice in the masonry no more than eight inches deep.

Roebling, Paine, and Collingwood were over on the New York side at the time, but C. C. Martin, who was in his office in the yard below, heard the crash and rushed up the narrow flight of stairs built at one corner of the tower. Martin found several men trying frantically and futilely to move the fallen derrick. The man caught beneath was unconscious but still alive. Martin sent for jacks and levers and in another fifteen minutes had the derrick up and off. The man was pulled out from under; he breathed a few times, as though his awful agony had been eased, then died.

The man's name was Daugherty and he had been working near Thomas Douglas, the stonemason, who was foreman on the tower. Douglas had started off running in the same direction as Daugherty, but instead of trying to jump the stone block in the way, he had crouched down beside it. As a result the stone saved him. But when Daugherty was struck, his knees came down on Douglas' back, pinning Douglas so he could not move. In his agony Daugherty kicked Douglas in the kidneys so severely that Douglas would never recover. He lived on for nearly two years after that, working part of the time, but troubled terribly by the injury and gradually growing worse. The story goes that he wanted to live only long enough to know that the Brooklyn tower was finished. When he was told that it had been, he died almost immediately afterward. "He was a splendid man," Martin told reporters.

But the accident also left a lot of people wondering how well things were being managed by the Bridge Company and at a time when what the Bridge Company needed most was public confidence. The guy wire on the derrick broke because of a defective weld in one socket, as would be learned. The accident had not been caused by mismanagement or the use of shoddy equipment, but a great many people did not know that, or believe it if they did.

An investigation was called for and was conducted by the coroner, with the result that the Bridge Company was entirely exonerated and credited publicly with using the best-quality material and taking every precaution to guard against accidents. "This has been the case from the first and will continue to be," William Kingsley stated; "no labor or expense will be spared to insure the safety of the employees." The day before elections the Executive Committee would grant to each of the families of the deceased payments of $250, or a little better than three months's wages.

But a few days after the accident and two weeks before the elections several hundred reform-minded Brooklynites met inside the Brooklyn Skating Rink to organize their own Committee of Fifty, as they called it, or the Rink Committee, as it would become popularly known—"to investigate charges of fraud, extravagance, and corruption in several departments of the city government." Also included for investigation was the New York Bridge Company.

As expected, the Democrats won in Brooklyn, but by only a slim margin. They were immediately charged with bringing hired goons over from New York, with buying votes, with using repeaters, fictitious names, all the customary devices. Most of the charges were probably true. That there was fraud at the polls there is no doubt whatever. Later examinations would show that a minimum of six thousand illegal votes were cast. But only five men would be convicted in subsequent trials, all of them quite unknown and unimportant, largely because the Democrats had in their possession some damaging information concerning the prosecuting district attorney.

Powell was mayor now. McLaughlin, Kingsley, and the rest were all still in power. But across the river the election had destroyed the Tweed Ring.

The reformers had gone to work in New York as never before, and they had worked together, Republicans and Democrats. It was their one best and last chance, they believed. Young Men's Reform Associations were organized. Clergymen called upon their congregations to do their duty. Newspapers and picture magazines described it as probably the most important election ever. On Election Day A. T. Stewart closed his store so every clerk could vote. Four regiments of militia backed up the police to see that everything was conducted properly at the polls and for the most part it was. Among many older, wealthier citizens, however, it was thought that very little would come of all this.

The reform ticket won by an overwhelming majority. Only two Tammany aldermen managed to survive. Tilden was elected to the State Assembly. Of the five Senatorial districts in the city, candidates put up by the Committee of Seventy were victorious in all but one—Tweed's. But Tweed's majority was less than half what it had been two years earlier and it had cost quite a sum of cash distributed about the ward. A few days after the votes were counted it was announced that the annual Americus Club Ball would be postponed.

Almost immediately the New York *World* sent a reporter across the river to interview the man thought to be the brains of the Kings County machine, William C. Kingsley. The *World* had been notably patient with Tweed all these years and now seemed eager to demonstrate that New York had no franchise on corruption. Kingsley stated categorically that there was no such thing as a Brooklyn Ring. Hugh McLaughlin, he said, lived in the plainest fashion. Why McLaughlin did not even keep a horse, Kingsley said with emphasis, and was probably not worth $100,000. The contrast with Tweed went without saying.

Kingsley conceded that the Kings County Democrats had taken something of a beating at the polls largely because of Tweed's troubles. "That great tidal wave of corruption in sweeping over the land has damaged us here to a certain extent," he remarked. But just so nobody misjudged his own motives, he concluded the interview by saying that he would be richer by $250,000 had he never had anything to do with politics.

Sometime soon after that the Committee of Fifty began looking into the management of the Bridge Company. The committee had already searched first for foul play in the management of Prospect Park and failed to turn up much. One or two other investigations were proving equally inconsequential, and among Brooklyn Democrats, the very conscientious chairman of the committee, a man named Backhouse, became the butt of innumerable bad jokes. But with the bridge the investigators believed they were on to something, and chiefly as a result of their curiosity a number of matters that had only been talked about privately before got into print a good deal sooner than they would have otherwise.

Tweed's prominent role in the business of the bridge, not to mention that of his three Tammany compatriots, Smith, Connolly, and Sweeny, was, of course, an obvious embarrassment to the Bridge Company at this particular time and many Brooklyn people thought that role and its origins needed explaining. But it was Kingsley's rather ill-defined position in the management of bridge business and how precisely he was being rewarded for his "services" that now became subjects of the keenest interest.

Kingsley was a difficult man to figure. His name was a football in Brooklyn, as Beecher would say. On the one hand he was a force for progress, industrious, respected by the respectable (Murphy and Stranahan, most notably), a good citizen if ever there was one. His firm of Kingsley & Keeney was building the Brooklyn Theater, lately the most talked about new building in town. He was one of the commissioners for the new capitol in Albany. His personal success was obvious. He was a kind and devoted family man. He attended church regularly. He was generous with his money, openly compassionate in times of trouble. When news of the Chicago Fire reached Brooklyn that October, he had been among the very first to do something. Along with Kalbfleisch, Judge McCue, and Isaac Van Anden, he had donated $100,000 to aid the stricken city. This "princely contribution" was announced as only the prelude to Brooklyn's bounty (which it was) and the *Eagle*, of which Kingsley was part owner, led the campaign to raise more funds.

Still, there was the political side of the man, as everyone knew, and it was never quite clear just where his business interests stopped and his political interests began. When the Rink Committee people started poking into the financial side of the bridge, it began to look as though his ability to work with machine politicians and to profit thereby was not necessarily limited to the Brooklyn side of the river.

As it happens neither Tweed nor any of the three other directors from New York had been seen in Brooklyn since Tweed's "troubles" began in early summer. Only that June, Tweed and Hugh Smith had been appointed once more to the Executive Committee and nobody had raised any objections. But neither man, nor Connolly, nor Sweeny, had had a thing to do with the bridge, or a word to say about it, since then, a decision doubtless strongly encouraged in the offices of the Bridge Company.

But as the company's records would show, Tweed had actually appeared at only a very few committee sessions even before his troubles. Of some fifty-eight committee meetings held between September of 1869, when Tweed came over for the opening session, and July of 1871, when the *Times* opened its attack on him, Tweed had bothered to show up for just six meetings, including the first one—so very few, as a matter of fact, that anyone examining the records afterward would naturally take a special interest in those he did attend. Tweed's time was exceedingly valuable after all, as everyone could now appreciate, so whenever he did make the effort to come to Brooklyn he must have had some definite purpose in mind.

The records were only a "synopsis," the business of each meeting being described in about the briefest, most general language possible, but even so, by looking at those meetings that Tweed attended a pattern emerges. Nearly always some particularly interesting motion was taken up, put forward, often as not, by Tweed himself and, as would be noted, these same motions almost always appeared to be directly beneficial to William Kingsley.

It was on a motion from Tweed, for example, at the Executive Committee meeting of October 27, 1869, when the bridge was just getting under way, that the members resolved to pay Kingsley & Keeney $46,915.56 for lumber. Kingsley would later explain that the lumber had been bought from a Georgia dealer at the request of John A. Roebling, before the bridge began, in order to take advantage of an especially good price. Since the Bridge Company had no money then, Kingsley said, he had made the purchase in advance.

So the committee was merely paying back what he had coming to him. Be that as it may, the record showed that Tweed was the one who urged that Kingsley get his money, plus interest, and naturally that left some people wondering.

Much more important, however, was the motion passed at the meeting held on July 5, 1870, a meeting at which Tweed appeared after a notable absence of more than six months. This was the day the commission arrangement that Tweed would later divulge was made official policy by the Bridge Company. The resolution, as recorded then, reads as follows:

> *Resolved*, That fifteen per centum on the amount of expenditure for the construction of the foundation of the towers of the Bridge on both sides of the river, up to high-water mark, including payments for land, be paid to William C. Kingsley, the General Superintendent, for his services and advances on behalf of this Company, up to the completion of such foundations.

It was customary always to include in the record the name of the member who put forward a motion, but in this case that was not done. The arrangement was not announced in the papers, moreover, the business of the committee still being conducted in private. It had not been discussed at the annual Board of Directors' meeting a few weeks before or even mentioned, nor would it be when the directors convened again.

Also, quite interestingly, Tweed was back again at the bridge offices in early September to move that "such amounts as due Wm. C. Kingsley up to the present time . . . be paid, and that hereafter the amounts, becoming due under said resolution be paid monthly." By the end of the year, expenditures on the bridge totaled $1,179,521.40. This meant that a 15 per cent payment to Kingsley came to roughly $175,000. A check for this amount was made out to Kingsley January 3, 1871. If Kingsley's official connection with the bridge was to be dated from the time he was formally appointed General Superintendent in October 1869, slightly more than a year earlier, then his compensation to date worked out to be a very handsome salary—about seven times that of the President of the United States, as would be noted in the papers.

Why Tweed took Kingsley's interests so much to heart is not known for sure. Perhaps, as was speculated, the early "understanding" with the Ring people also included an agreement that Kingsley would turn over part of the 15 per cent in return for

Tweed's support in the Executive Committee or that Kingsley would merely apply part of the 15 per cent to what the four Tammany men owed him for the bridge stock he had handed over. That way nobody would suffer. Tweed, however, would testify that he himself had made no deals at all with Kingsley.

"Was he to divide the fifteen per cent with you?" Tweed would be asked.

"I had no understanding with him, sir," Tweed answered, but then added, "I don't know anything about the rest." The rest meaning Smith, Sweeny, and Connolly. What they may have worked out with Kingsley on their own, Tweed could not say.

But since such arrangements had been shown to be the common practice of the Tweed Ring, a great many people, rightly or not, would conclude that that was just what Kingsley had agreed to. Moreover, it would also be pointed out that buying bridge stock by installments, 10 per cent at a time, which was all that was required and what Kingsley and most stockholders were doing, then drawing 15 per cent on several million dollars in expenditures, was a most attractive proposition from a business point of view.

Kingsley's explanation of all this would present a different picture, understandably, but before he felt compelled to speak out publicly in his own defense, some noteworthy changes were made in the arrangement he had with the Bridge Company.

On November 13, just five days after the Brooklyn and New York elections—when it was very clear that Tweed was finished—the Executive Committee resolved that "the claim of William C. Kingsley under the resolution of July 5, 1870" was "hereby liquidated, with his consent." For his services in behalf of the bridge the General Superintendent was now to receive "an amount not exceeding $125,000, in full." No explanation for the change was included in the record and again nothing was said of it in the papers.

At the end of November Kingsley returned to the Bridge Company the fifty thousand dollars he had been overpaid, according to this new rule—returned it *voluntarily*, as his admirers would later stress, since there was nothing in the rule saying he had to do so. Again the record contains no explanation. The "construction" account is simply credited with fifty thousand dollars from Mr. Kingsley.

Presently, sometime in early December 1871, somebody went to the record book containing the old July 5 resolution and made an erasure, changing the "fifteen per centum" Kingsley was to get to read "five per centum." No mention was made of this in the records. Nothing was said to explain why it was done or by whom.

Kingsley's compensation as General Superintendent was to be the central issue to come out of the investigation, but serious questions would also arise concerning the company's methods for purchasing materials. Who exactly decided what was to be bought from whom? Kingsley's recommendations, so it appeared, always received the unqualified approval of the Executive Committee. Why were no bids advertised for? And more specifically, why had the Bridge Company in the past two years purchased more than $140,000 worth of lumber from the New York & Brooklyn Saw Mill & Lumber Company, or from that firm's treasurer, a Mr. Ammerman, when, as was no secret, the General Superintendent was part owner of the firm? With how many other firms doing business with the bridge did Mr. Kingsley have "connections"? The impression was that the General Superintendent was deciding what to buy to build this bridge he had so eagerly wanted the two cities to pay for and that he was deciding to buy quite a lot from himself.

The Rink Committee would not be ready to report its conclusions until early spring the following year, 1872. By then Tweed had fallen on terrible times.

The elections had been over only a few days when Peter "Brains" Sweeny resigned his office as President of the Park Board and departed as swiftly as possible for France. Hugh Smith also fled the country, while Connolly by now was doing all he could to ingratiate himself with the New York Committee of Seventy. Only Tweed remained and his arrest seemed a matter of days.

Then it happened. On December 15 Tweed had been indicted by a Grand Jury on 120 counts. The following evening he was arrested by an old confederate, Sheriff Matthew T. Brennan, who came personally to Tweed's office to conduct the historic ceremony. Touching Tweed lightly on the shoulder, Brennan said, giggling, "You're the man I'm after, I guess."

To nobody's surprise Tweed got out on bail almost immediately. His first night as a "prisoner" he spent in the Metropolitan Hotel, which he owned. But he was forced to resign from the Public Works Department and on December 29 at Tammany Hall he was voted out of power as Grand Sachem.

Because the courts seemed unable to decide whether the county or the state had the right to bring Tweed to justice, it would be another year before he would go on trial. But his world was crumbling all around. On January 6 Jim Fisk was shot down in the Grand Central Hotel by Edward Stokes. Fisk had once swindled the wealthy Stokes out of the ownership of a Brooklyn oil refinery and Stokes for his part had been pursuing Fisk's mistress, Josie Mansfield, with considerable success.

Tweed was among those at Fisk's bedside as he lay dying. When Fisk expired at last, Tweed was beside himself. Of all the innumerable characters he had been in league with over the years, Fisk had been the most fun. The papers made much of the broken Tweed and his grief. Ten thousand people turned out to view the last remains of Fisk, lying in state, as it were, in his Erie Railroad office, but none would be so overcome as the former strong man of Tammany. The *Eagle* described the scene:

> Most of all affected was William M. Tweed. He cried like a child. His sobs were heard all over the house and while his hand, trembling as if with ague, continually stroked the broad and placid forehead of his dear friend, his hot tears dropped like a rain shower into the coffin.

For a very large part of the public, Fisk's murder seemed the just vengeance of the Almighty. From Plymouth Church came the voice of Beecher describing Fisk as "that supreme mountebank of fortune," "absolutely without moral sense," "absolutely devoid of shame," "abominable in his lusts" and proclaiming that "by the hand of a fellow culprit, God's providence struck him to the ground."

Tweed's arrest, the Fisk murder, Stokes on trial, smallpox in Brooklyn, the annual Assembly Ball at the Music Academy ("In the United States there is no pleasanter place than is our city socially . . ." said the *Union*); Talmage preaching out against Fielding and Dumas as obscene literature, Bill Cody playing in Brooklyn, the new Metropolitan Museum of Art opening in New York; talk of the upcoming Presidential elections, two thousand skaters on the lake in Prospect Park, and the East River spanned by gigantic cakes of ice—there was ever so much to read and talk about through that winter of 1871–72. But only rarely would there be anything in the papers about the actual building of the bridge. There was plenty of talk certainly about what the Rink Committee might turn up. But when Roebling announced in mid-December that the New York caisson was ready to begin its descent, only a line or two appeared in the papers. Not until later would it be considered a matter of the greatest interest, or would some observers find a certain irony in the idea that the New York foundation for the bridge was being sunk in a sewer.

12
How Natural, Right, and Proper

> Although the bridge from every element of its use and from the source of its finances, is considered a public enterprise, yet it is entirely a *private corporation* in which the public has no voice . . .
>
> —DEMAS BARNES

THE SENSATIONAL Stokes trial dragged on for months in early 1872, keeping the excesses of Fisk's (and Tweed's) unsavory world very much in the public mind at the time when the Rink Committee began presenting its findings. Predictably, Tweed's former association with the bridge was made much of. But Kingsley's "compensation," the possible reasons for it, and his relationship to the Saw Mill & Lumber Company were the heart of the committee's case.

The *Eagle* immediately jumped to the defense of the Bridge Company. "It is true that Tweed, Connolly, and Sweeny are among the subscribers to this stock . . . but what corporation would have refused the cooperation of these men one year ago?" Kingsley was the man who got the bridge going, assumed the responsibility, and so "would have been justified in asking for any contracts the Bridge Company had to give, on the condition that he would do as well by the Company as any other contractor." Furthermore, if the people of Brooklyn were not able to trust the likes of Murphy, Stranahan, Demas Barnes, or Henry Slocum to see that an enterprise was run legitimately, then whom were they to trust?

Kingsley himself replied right after that, by letter to the *Eagle* and the *Union*, rather than through an interview, since, he said, he wanted his answer worded properly.

The letter, however, was an angry tirade and raised new doubts about the man. The Bridge Company was composed of Brooklyn's first citizens, all men to be trusted, he said, echoing the *Eagle*'s theme and citing the same names over again. To suggest that the company was being mismanaged, to malign the reputation of anyone connected with the bridge, was to be against the bridge, and to be against the bridge was to be against Brooklyn and against progress. The Committee of Fifty was made up of "vagabonds and scoundrels," plus "a parcel of old fogies, who have accumulated wealth, though by their connection with the old farm titles of Brooklyn . . . and who have been made rich by the very progress they have persistently resisted." These were the people, he said, who had been against the water works, the park, "every railroad track." The landed gentry were out to block the people's bridge was the theme.

As for his dealings with Tweed, Kingsley answered with a question of his own, again sounding quite a lot like Thomas Kinsella or whoever wrote the *Eagle* editorial, except that Kingsley was perfectly frank about how far back that relationship actually dated. "Will any man whose memory goes back five years dare hazard his veracity by saying that the Company were unwise at that time in soliciting the co-operation of Tweed, Sweeny, and Connolly . . .?" Besides, Kingsley said that in his dealings with Tweed and the other Tammany people, he had found "their characters were then at least infinitely better" than those of the men on the Committee of Fifty.

Then he said something he regretted later. He said he had paid for everything in getting the bridge under way and was "out of pocket more than a quarter of a million before a blow was struck." (It was the same figure, interestingly, he had said he would be richer by had he avoided politics.) And as a kind of blanket justification for all those things he had done that seemed even the slightest bit suspect, he said, "I wanted to build the bridge."

"Every man must have somewhere an objective in life with which a greater [good] than money allies him," he said. "My objective is to build the bridge."

The committee quickly responded with its own letter to the *Eagle*, saying it was ridiculous to tell the public to relax and be confident just because men of money and position were in charge. The committee also wanted to know, if the Bridge Company was employing a Chief Engineer, not to mention a consulting engineer and four assistant engineers, why William Kingsley's services were so all-important and what exactly was he doing? In addition, the committee was now most interested to know how Kingsley had managed to spend a quarter of a million dollars. This new figure of Kingsley's was exactly *twice* as much as anyone had ever implied he had spent. Moreover, the committee had been led to understand, by officials at the bridge offices, that the $125,000 Kingsley received from the company was mainly to cover those earlier expenses, all of which could be itemized supposedly. If that was so, then for what purposes had the rest of the money been used? And did not the people of Brooklyn deserve an honest accounting of such transactions? Perhaps he had been extremely liberal in Albany to get the sort of legislation that would allow private stockholders such free rein with the people's money?

Kingsley answered with a second letter, but briefly this time and not at all belligerently. He said the money had been spent on lumber, for office rent and office equipment, for the bridge tour to Pittsburgh, Cincinnati, and Niagara Falls, for consultants' fees, for printing documents. But he neither retracted the quarter of a million figure nor attempted to explain it. The issue was left hanging and the Committee of Fifty, which had never managed to deliver the sort of sensational scandal the public had grown accustomed to, gradually faded from the picture leaving a number of questions unanswered.

The directors of the Bridge Company were not, however, about to let matters go at that. Nor was the public or the New York press. Irate letters to the editor demanded further investigation. Why, it was asked, had Roebling taken no stock? Why had so many of the prominent people listed as directors in the original charter withdrawn from the whole scheme, purchasing no stock? The *World* took to calling Kingsley a "cormorant," said he was shrewder even than Tweed, and claimed to have solid proof that Kingsley had once received several hundred thousand dollars in kickbacks from a Brooklyn paving company. The *Tribune*, meanwhile, reported that there were at least eight different factories, stone quarries, and lumber mills furnishing materials for the bridge in which Kingsley had a direct interest.

Something would have to be done, it was pretty generally agreed among the directors, and it was now that James Stranahan moved swiftly to restore public confidence.

The directors met next on June 4, 1872. Murphy was again elected president (nothing derogatory had been said about him as yet) and Andrew Green, the Comptroller of New York, was named to the Finance Committee. Stranahan then had four other esteemed New Yorkers appointed directors, as replacements for the old Tammany quartet. The new men were William H. Vanderbilt, Lloyd Aspinwall, William H. Appleton, and most important, Abram S. Hewitt.

Hewitt by this time had a great reputation as a reformer. He had been asked by William Havemeyer, soon to become mayor again, to become a director of the bridge, in order, as Hewitt explained it, "to investigate the expenditures, and to report as to the propriety of going on with the work." Havemeyer, who had never much cared for the bridge, saw it as a perfect "illustration of the dishonesty which enters into public undertakings." He thought a city government ought to be conducted with all the efficiency of a business, so he had sent an efficient, successful industrialist to clear things up in Brooklyn. Stranahan welcomed the idea, or so he said, and Hewitt was promptly named to fill Tweed's vacancy on the Executive Committee. It was a little like putting the teacher's pet in the seat long occupied by the class troublemaker, now expelled, as a way of restoring order in that corner of the classroom.

Hewitt meant business it seemed. At his first Executive Committee meeting less than a week later, with Stranahan and Murphy present, he personally approved the payment of bills amounting to some $71,000. At a meeting a week later he moved that the Chief Engineer "be requested to examine and report, at the earliest possible date, whether the prices paid for stone, lumber and other materials, and for labor have been reasonable and just." He also wanted Roebling to report whether the cost of the bridge thus far exceeded what his father had estimated, and if so why. His motions were immediately agreed to, as was a suggestion from Henry Murphy that for the next order of granite the Chief Engineer be directed to advertise for sealed bids.

On July 1, the Board of Directors, which had met only nine times since the bridge began, convened once more to hear the special report of the Chief Engineer, which had been turned in on June 28, or just four days after it was requested. Murphy was out of town for some reason, so Abram Hewitt was called to the chair. To begin with, the records of the three Executive Committee meetings held in June—the three Hewitt had attended—were read before the board, something that had never been done before. That over and the real business about to start, Demas Barnes proposed another radical departure, that the reporters waiting downstairs be allowed to come up and sit in on the meeting. But Henry Slocum moved as a substitute that after adjournment all the papers be provided copies of the proceedings by the secretary, a full-bearded, pious-looking gentleman with the memorable name of Orestes Penthilus Quintard. Slocum's motion carried. Then Roebling's report was read.

The opening statement probably left everyone breathing a little easier. Roebling began by saying that at the very start of the work he had been told explicitly by Henry Murphy that he was to have nothing whatever to do with making contracts or purchasing supplies. Still he was "personally cognizant," Roebling said, of nearly every such transaction that had taken place. "I know that all contracts have been made in a judicious manner, and have resulted in the best interests of the Company. They have, in most instances, been given to the lowest bidder, and where they have been awarded to another bidder, it has been at a figure as low as the lowest bidder."

"It has been alleged," he said, "that supplies have been furnished by members of the Company, at prices prejudicial to the interests of the Bridge. In all such cases I know that the supplies have been furnished after a reasonable competition, and at rates lower than those of any other bidder.

"I can further say that every dollar's worth purchased for the Bridge has been expended in a legitimate manner, and for the proper purpose for which it was designed, and nothing whatever has, to my knowledge, been diverted into any outside channel. I am in daily attendance at the Bridge, give it my whole time and constant superintendence, and am therefore in a position to give an honest judgment on this question."

Nobody, Roebling said, had been employed because of political influence. Wages were about what was customary. Wages might properly be higher considering the danger of the work. There was not a man on the job, he said, "who does not earn every cent he gets." How much Kingsley was on the job, he did not say.

Roebling did not deliver the report himself, he was not even present. The New York caisson had been sunk by this time and as was known by most of those in the room, the work had not gone at all well. The Chief Engineer was worn out and had worries enough of his own.

Attached to this initial statement was a long accounting of expenses to date and a detailed explanation of why the bridge was costing more than had been anticipated—and could not therefore be completed for the sum John Roebling had set. The presentation was so very thorough, so concise and solid, that it was obvious that more than four days had gone into its preparation. Clearly Roebling had been ready in advance for just such an accounting. Quite likely he had even welcomed the opportunity.

So far costs were running more than a million dollars above the original estimate. This, he explained, was due primarily to several large items that had not been taken into account by his father—the increased size and elevation of the bridge (the changes would cost about $113,000), the land that had to be purchased for the approaches (for which some $330,000 had already been spent), the troubles with the New York caisson (about $375,000, everything considered), and then a number of other expenditures that he lumped together, describing them as "outside of ordinary construction contingencies." These included the following: $20,000 for a consulting engineer for two and a half years (Horatio Allen); $7,000 for the board of consulting engineers (the original seven); $8,000 for his father's funeral expenses, travel, legal fees, donations, doctors, etc.; nearly $6,000 for taxes and interest; and lastly $125,000 for the General Superintendent. All told they came to $165,771.65. Such items could not have been anticipated, Roebling said, making it clear, in a respectful way, that neither he nor his father had been prepared for how much William Kingsley was to cost.

The way things were going, Roebling concluded, the bridge would wind up costing $9.5 million, or nearly $3 million more than his father's original figure.

The "integrity and fidelity" of the Chief Engineer had never been questioned by any of the stockholders, as Hewitt would write later, and so the report was taken as a most encouraging document, as far as it went. But in the opinion of several of the directors it did not go far enough. It was agreed, therefore, that a special Committee of Investigation be formed. Demas Barnes was the one who put forward the idea. When Hewitt asked him who would be acceptable to serve on such a committee, Barnes said he, Hewitt, would be, and F. A. Schroeder; "whereupon the Chair appointed as said committee Messrs. Barnes, Hewitt and Schroeder." Barnes was to be the chairman.

The *Eagle* angrily charged Barnes with concocting a political scheme, the purpose of which was to check Brooklyn's growth and prosperity. The editorial read as though it had been written by none other than William Kingsley.

In Congress three years before, Barnes had worked harder than anyone for the bridge. He had been among its most rhapsodic spokesmen and the one who called it a monument to progress. But Barnes had made trouble before. Ostensibly a Democrat, he had an unorthodox and aggravating habit of siding with the Republicans whenever he thought they were on the right side of an issue. He had never been one to "go along." Now the *Eagle* was calling him a notorious demagogue, an ass, and a quack. Defeat the bridge, the paper warned ominously, and watch what happens to the value of Brooklyn property.

Accusations of a bridge scandal were thus written off by Brooklyn's most influential paper as purely politics, nothing more than the work of destructive little men of mean ambition. At the same time, by way of contrast Thomas Kinsella began giving more space to the bridge itself and to its Chief Engineer. Kinsella was a tough, expansive, and undisciplined man who had had his own personal experience with scandal. The father of ten children, he had been serving on the Brooklyn school board when it became public knowledge that he was having an affair with the wife of the superintendent of schools. Another man might have packed up and left Brooklyn under the circumstances, but Kinsella had stayed on, faced down his accusers, and despite the gossip maintained his grip on the *Eagle*. Kinsella was, of course, one of the earliest, most enthusiastic backers of the bridge. A number of New York editors considered him little better than a paid propagandist for the project, diligently serving the interests of the Kingsley-McLaughlin machine. But for all his obvious partisan feelings politically, Kinsella, as he would prove later, was no mere stooge, and beyond that, he had an unshakable belief in the great work itself and did not intend to see it destroyed by scandal any more than he had been.

Moreover, it seems Kinsella had a very genuine, unbounded admiration for Washington Roebling, just as he had had for Roebling's father. Another of Roebling's reports on the progress of construction had been released earlier that same June of 1872 and Kinsella had published the entire thing, saying it set the kind of confident tone everyone ought to use when talking of the bridge. Most of the report concerned the sinking of the New York caisson. It told quite a different story from that of the Brooklyn caisson and it did not tell the whole story, as Kinsella was quite aware, but the spirit of its plain, confident language and the extraordinary achievements described were about as sharply contrasted with the other things being said about the bridge as could possibly be—as Kinsella fully appreciated. The strongest possible defense for the bridge, he had decided apparently, was just such a factual, unadorned accounting of what was being accomplished by brave men every day. And no more exemplary specimen of such men could be found than young Roebling. So while Demas Barnes was lumped with that "class of croakers that exist in every community," the loyal readers of the *Eagle* were asked to consider "The Engineer":

> . . . He is the thinker who acts. He contributes to his country's sum of achievements as much as and less expensively than the soldier. His ends, in the elevation of the race and in increasing the aggregate of its capacity and performance, are kindred to the statesman's. And if there be those who think that the work of the Engineer is only hard and material, that there is no charm of art in its processes, let them read the story of the building of the Bridge.

The Bridge Company's new Committee of Investigation would spend six months at its studies, submitting its conclusions more than a year after Kingsley had first come under fire.

In the months following that July directors' meeting, however, the heat was on as never before, with the New York papers making much of a bridge scandal. The bridge was entirely the doing of the Brooklyn Ring, it was charged, none of whom had paid a cent for their stock. The stock was merely a sham to hide "the too palpable intention of defrauding the corporations of New York and Brooklyn." The superintendent of the work was himself "The Monarch of the Ring." His duties appeared only to be selling material from his own mills to the Bridge Company at an enormous profit and then pocketing a percentage of the expenditures. If a superintendent was really necessary, then any one of the finest men in the business could be employed for no more than ten thousand dollars a year.

Not even the word of the Chief Engineer was above suspicion. It was charged that he too might have his motives for concealing the truth and that he was, in any case, scarcely more than a hired hand who stood to lose his job if he said anything other than what he was supposed to say. "He is too good a son of his father not to wish to identify his own name and fame with the building of the structure his sire designed," wrote the *World;* "and he could hardly be blamed for not quarreling with the powerful superintendent and Executive Committee of the company by which he was employed."

Scientific American commented that the bridge would end up costing forty million dollars unless something were done. A bridge over the river was a bad idea anyway, said the editors, who claimed they had been for a tunnel all along.

Early in November, Abram Hewitt announced that "the agreement with Mr. Kingsley, General Superintendent, was at an end." Kingsley still had his job, but his pay had been stopped.

But by then, all of a sudden the ins and outs of bridge business, talk of kickbacks, investigations, and the rest, had become very bland fare in Brooklyn. For it was in early November that the Henry Ward Beecher scandal broke wide open. Victoria Woodhull, a notorious lady stockbroker and publisher, spiritualist, feminist, magnetic healer, free lover, and all-around adventuress, had branded Beecher an adulterer in the pages of her newspaper, *Woodhull & Claflin's Weekly.** She had said as much about Beecher earlier that fall, speaking in a trance, as was her platform style, before a convention of spiritualists in Boston. But when no respectable paper had been willing to print the story, she had decided to publish it herself.

The charge was not new. There had been whispered stories about Beecher for some little time in Brooklyn. Now, however, it was in print. It was said he had been carrying on an affair with one of his young parishioners, Elizabeth Tilton, the wife of Theodore Tilton, a prominent liberal editor and poet and a former protégé of Beecher's on the *Independent*, a religious paper. Mrs. Woodhull said she thought Beecher perfectly within his rights to have done what she accused him of. Beecher's "immense physical potency," as she called it, was, in her view, "one of the grandest and noblest of the endowments of this great and representative man." Beecher's only sins, she held, were concealing his acts and not joining her to expound the glories of free love. Her fervent hope was that her arti-

* Claflin was Tennessee Claflin—sometimes spelled Tennie C.—Mrs. Woodhull's younger sister and the consort to old Commodore Vanderbilt.

cle would "burst like a bombshell into the ranks of the moralistic social camp." That it did.

More than a hundred thousand copies of her paper were sold out immediately. Secondhand copies were soon selling for as much as forty dollars. The story was a sensation on both sides of the river, with people talking of little else.

In no time Mrs. Woodhull and her sister were arrested on charges of sending obscene literature through the mails and were locked up in the Ludlow Street Jail, where they would be held for six months. The Sunday crowds at Plymouth Church grew steadily larger. Beecher, against the advice of family and friends, refused to say anything one way or the other on the subject. Mrs. Woodhull was known as a habitual liar, among other things. His best policy, he believed, was to hold his tongue and wait for the storm to pass. But it was not to be that way. And Brooklyn was never to be quite the same again.

The majority report of the Committee of Investigation was presented December 16. It ran to six printed pages and was signed by Hewitt and Schroeder. Its conclusions, in brief, were that everything was on the up-and-up inside the Bridge Company except for the expenditure of $125,000 for the superintendent's services, which was politely termed a "misapplication of money."

After examining the purchases made under Kingsley's supervision, and particularly those from his Saw Mill & Lumber Company, Hewitt and Schroeder concluded that though public competition was not in all cases required ("as is customary in enterprises where public moneys are disbursed"), the Bridge Company did not appear to have suffered any thereby. But they added:

> Your Committee believe, however, that this practice is objectionable, and that no purchase should ever be made, except upon public tender, with adequate notice, nor from parties who may be in any wise associated in the management of the work, unless such parties should be the lowest bidders upon fair and open competition, and under no conditions should contracts be given to parties identified in interest with the officers of the Company, who after first making or approving specifications are called upon to judge and certify as to compliance with such specifications in the execution of the contract.

All of which was a long way of saying that Kingsley would have to mind his ways from here on out, but which also suggested, as doubtless several people immediately realized, that another major conflict of interest lay directly ahead, when the bridge would be further along. For it was common knowledge that the foremost manufacturer of steel wire was John A. Roebling's Sons of Trenton.

The two investigators said they did not know "how far in reality Mr. Kingsley was interested in any stock besides that which appears in his name," but their impression was pretty far. This, they agreed, was not as it should be and they said that "a mere partnership between the public and private individuals who have the expenditure of the money is not sufficient to protect the public interests." So they concluded with two specific proposals.

First, that Alexander McCue, as counsel for the Bridge Company, draw up for approval by the Board of Directors an amendment to the original charter authorizing the cities of Brooklyn and New York a vote on their stock. The cities would also decide on the choice of directors, "taking care, so far as may be practicable, that the private stockholders are represented in the Board in proportion to the stock held by them." In short, the two cities would henceforth control the bridge and the power enjoyed by the private stockholders—as granted by Murphy's charter—would no longer be absolute.

The second proposal was to continue to employ a General Superintendent (no name specified), but at an annual salary not to exceed that paid the Chief Engineer, which by this time had been raised from eight to ten thousand dollars.

Demas Barnes, however, insisted on submitting his views separately. The criticism in the majority report he found too mild and generalized; the recommendations made fell far short of the mark, he said. So as a minority of one Chairman Barnes presented his own report, which was twice the length of the other one and infuriated nearly everybody who had had anything to do with the business management of the bridge. It was a remarkable document.

Barnes's basic contention was that since the bridge was being built with public money, its managers ought to be accountable to the public, which they were not. He said that the original charter provided the people of Brooklyn and New York with no adequate protection against fraud; that the business side of the bridge had been carried on in much too much secrecy; and that certain members of the company were benefiting personally more than was proper. He reported that the Finance Committee kept no records, that there was no record of a minority motion ever voiced at a meeting of the Executive Committee, or of any substantive debate on any question. He noted that Kingsley, the largest stockholder, was not a director or on any of the committees, so he could not be held responsible for anything decided on by the board or by the committees. And yet: "While no restriction appears limiting his discretion, there is no act or recommendation of his which has not the unqualified approval of the Committee." No part of the material purchased thus far had been advertised for. There was no indication of who opened bids, if indeed any of them had ever been submitted sealed.

Kingsley's arrangement for a percentage of expenditures Barnes found to be the most reprehensible irregularity and he was puzzled why the various alterations in the agreement made the previous November had been done without any explanation. And it was Barnes who totaled up all the money paid out to Kingsley's Saw Mill & Lumber Company or to its treasurer, A. Ammerman.

Barnes also reported that checks seemed to be handled in a rather peculiar fashion. When most organizations deposited checks in the bank, he said, it was customary to record the name of the drawer on the check in the margin of the checkbook. But, he went on, "The following credits are given for money received April 12th, 1870: Five parties, $23,000. The amount did not reach the bank until June 3rd, and only one check was deposited instead of five. The name of the drawer of the check deposited has been erased and the word 'check' inserted. The same circumstances (erasures, etc.) occur several times." The procedure struck Barnes as most mysterious.

But this and one or two other things in his report struck some of the other directors as nit-picking, and for all the rather fishy-looking inconsistencies in the record books, Barnes had in fact found precious little proof of foul play except, again, for Kingsley's $125,000.

Still the report wound up as a rigorous indictment of the private corporation as a means of building such an important public work and a strong endorsement for an immediate change in that policy:

> The Company is under no financial restriction, and is accountable to no authority. It may do its business in private; its members may furnish any part of the supplies; it may pay such prices for the supplies, labor, superintendence, etc., as it chooses. The company violated no law in agreeing to pay fifteen percent on its total disbursements . . . Neither would it have done so had it paid fifty per cent

or one hundred per cent. There are men in this city who will pay for all the private stock, and give one million dollars for the privilege of completing the Bridge under the existing charter.

Kingsley, he said, might even be commended for taking only what he did, considering the opportunities available. The miracle was, in Barnes's view, that *more* money had not been stolen. He left little doubt that he thought Kingsley was profiting handsomely, but he had no solid facts and figures and so never condemned the man outright. All Barnes could conclude in print was that Kingsley was doing no more, in fact a good deal less, than the law permitted him or any other big stockholder to do. The man was not crooked, or at least Barnes could not prove him so, but the law was and something could be done about that.

Unlike Hewitt and Schroeder, Barnes was not willing to leave any changes in the charter to the Board of Directors. He wanted such changes agreed to and approved by an impartial conference of responsible citizens, none of whom had had any previous connection with the bridge.

He also wanted further investigations conducted and he made no mention of keeping Kingsley on. "The people need the Bridge," he concluded; "make them realize that when they have expended ten or fifteen million dollars on a bridge [he was not ready to accept Roebling's latest figure apparently] that they will have ten or fifteen million dollars' worth of bridge, and they will be ready to furnish the money."

In the months to come, Barnes would argue for a long list of commendable amendments to the charter. He wanted all meetings of the board open to the public, all supplies in amounts over a thousand dollars advertised for, all bids to be opened by the secretary of the company and in the presence of the board. He wanted the records of all committees open for inspection by any director at all times.

When the Executive Committee filed its report, in answer to the Committee of Investigation, its defense would be that the stockholders were doing only what was within their rights. Barnes would be chastised for quoting from the records only those references that supported his own arguments and for filling his report with "disingenuous insinuations against the Executive Committee." The arrangement with Kingsley was defended on the grounds that it covered only the building of the foundations, that such work was highly precarious, and as a man to take charge of such an unprecedented effort Kingsley had been the ideal choice. That Kingsley expected to make money out of the arrangement was not denied. "It is, no doubt, true that Mr. Kingsley, in connecting himself with this great work, looked to some pecuniary advantage. It is hardly to be supposed that anyone would spend so much time and labor and incur such pecuniary liability as he had done without some expectation of remuneration." The Executive Committee claimed Kingsley's 15 per cent was no more than the federal government allowed contractors furnishing stone for the new Post Office in New York. (The situation was not quite the same clearly, but that was overlooked.) And finally, argued the committee, if Kingsley had not put up the money he did, when he did, the bridge would not have been built, and nobody could say the job had not been handled superbly in all the time he had had anything to do with it.

What exactly his duties were, how much of his time he was giving to the bridge and how much to his other business and political interests, were never explained.

Why he had agreed to the ceiling of $125,000 and had returned the $50,000 when he did was explained, however. He had agreed "on the condition that he should be relieved of a certain portion of the stock of the company he was carrying, which we agreed to do in our individual capacities, with the assistance of two other members of the Board, at our request, by purchasing from him and paying him for $130,000 of the stock at its par value, taking the consequences of our act upon ourselves."

In other words, by returning $50,000, Kingsley got back $130,000 that he was in the hole for. So, in actual fact, instead of returning $50,000, he was really having $80,000 returned to him, and if that plus $125,000 he had been paid for his services did not add up to the $250,000 he claimed to have spent—but was never able to itemize to anyone's satisfaction—it was a tidy sum nonetheless.

As for the purchases made from his Saw Mill & Lumber Company, the defense was that since the Bridge Company got everything at a good price, then no damage had been done. That the Saw Mill & Lumber Company and Kingsley may also have benefited by the business was apparently not considered a pertinent issue.

The Executive Committee was able to report, however, that the money paid to Mr. Ammerman, treasurer of the Saw Mill & Lumber Company, should not be viewed as payments to that firm. Mr. Ammerman, it was explained patiently, had been paid some

$84,000 not as a representative of the Saw Mill & Lumber Company, but as the representative of mill owners in Georgia (the same Georgia firm perhaps from which Kingsley had bought the $46,000 worth of lumber back in 1869).

But the real issue, in the view of the Executive Committee (then composed of Murphy, Slocum, Stranahan, and Husted), was not Kingsley's innocence, but whether or not the Bridge Company had been run according to the accepted practices and ethical standards of a *private* corporation, and in their unanimous opinion it most certainly had. The purchasing of materials, for example, was, they said, conducted as it would be by the officials of a railroad company or a private citizen building a house; such people, it was reasoned, ask bids for the work from whichever builders and dealers they choose, whichever "they think best for their interest, without deeming it necessary or advisable to make public advertisement of them."

"This Company was chartered as a private company," they reminded everyone, "and although the cities of New York and Brooklyn subscribed to its capital stock, that fact did not change its character. There were scores of railroad corporations in this State to whose stock towns and cities have subscribed under authority of law, and which retain their private character. We refer to this fact merely for the purpose of showing how natural, right and proper in itself it was for the Executive Committee, in the absence of all legal provisions to the contrary, to exercise their discretion in this respect in the same manner as other private corporations."

So for all the talk of the bridge as a noble public work, the men who had the power saw it as a business proposition.

This particular document was signed on the last day of December 1872. On January 11, 1873, the Board of Directors convened. A committee of five was named to consider how the charter ought to be amended. The five were Hewitt and Schroeder, William Marshall, Judge McCue, and James Stranahan. Murphy, the man who wrote the original document, was not chosen to participate, nor was Demas Barnes, its severest, most conscientious critic.

Then it was resolved by the board that the Executive Committee be directed to appoint "Mr. Wm. C. Kingsley . . . General Superintendent at a salary of $10,000." This was done by the Executive Committee later the same day on a motion by Hewitt, which included the provision that Kingsley's pay be effective as of the previous July 12, when his "former appointment" had terminated with the completion of the tower foundations up to high-water mark. The motion was carried. The management would remain precisely the same as it had been all along.

Kingsley, who was present for this little ceremony, accepted his "new" appointment gratefully, then remarked that his physician had advised "some recreation from his labors was absolutely necessary" and with that in view he was forced to ask for a six-week leave of absence. This request was granted unanimously.

Hewitt had been convinced somehow. He was the key man now, clearly enough, the one known voice of reform on the Executive Committee, the one new face, the one and only representative for the city of New York. And apparently his own examination of the Bridge Company, plus explanations offered by the others on the Executive Committee—not just in their formal rebuttal to Barnes, but in personal conversation—had been enough to satisfy Hewitt that there need be no shake-up in the over-all method of operation or in the principal members of the cast.

Hewitt still wanted the charter changed, but he was in no hurry about it and he had concluded that for the time being there was no reason why things could not continue as before. So as far as the bridge management was concerned the crisis had passed, the case was closed. Nothing was ever said to that effect, but that was the situation. No one on the Executive Committee would oppose changing the charter, then or later, but there would be no great rush to see it done either.

For anyone else who had troubled to follow the situation as it had evolved over the past year and a half, and who had been able to keep it all straight, things did not seem all that tidy, however. Quite a little had been left unsaid, several very important questions had been left unanswered.

Kingsley had still not explained how he spent a quarter of a million dollars prior to the start of construction. (Barnes had been able to justify slightly over $59,000, but that included the early lumber purchase of $46,000, for which Kingsley & Keeney had already been reimbursed. The Executive Committee in its answer to Barnes had not bothered with an estimate of Kingsley's early expenses or even a suggestion of what they might have involved. Kingsley was simply credited with putting up a large sum of money "at risk," for which he rightfully deserved to make a profit.)

Nobody said why Kingsley's agreement had been changed so suddenly after the Tweed Ring's defeat at the polls or why an erasure had been made in the records instead of simply amending the agreement in the usual fashion.

Kingsley's duties were still unexplained. In all that had been said in the man's defense or to justify the peculiar privileges he enjoyed, no one had bothered to mention what his responsibilities were or how much of his time he was expected, or requested, to devote to them. His presence was vital according to the members of the Executive Committee, but just why that was so none of them ever said. Moreover, it seemed curious that whenever the management of the Bridge Company wished to present an authoritative, reliable opinion on the status of the work, Kingsley was never quoted—it was always Roebling. The bridge was always progressing under Roebling's direction, never under Kingsley's, according to Bridge Company pronouncements. Unfortunately Roebling himself had not been very clear about what Kingsley contributed to the job; in the two annual reports he had prepared thus far, Roebling cited various members of his staff by name, going out of his way to credit them for this accomplishment or that, which, as he said, often involved "a certain amount of risk to life and health." But Kingsley he mentioned only in passing—as having made some contracts.

If Kingsley had spent $100,000, $200,000, $250,000, whatever, before the bridge began, then where did he spend it? In Washington? Albany? Or if he had spent no such sums, but more like the sixty thousand that seemed a reasonable figure for the sort of initial expenses such a work customarily entailed, then why was he entitled to such a disproportionately high compensation for his services?

How much had his Saw Mill & Lumber Company profited from the bridge business he had arranged for it? In how many other firms doing business with the bridge did he have an interest? How much of the stock owned by the members of the Executive Committee had he made it possible for them to own? Why had he been so very eager for this bridge in the first place and how many other bargains had been agreed to privately with politicians and contractors? These were the obvious questions and there were still no answers for them. Nor would there ever be from those in a position to know.

Hewitt would staunchly contend that nothing improper had occurred. The others would simply not say anything more, except to deny the charges that came along. How much Roebling knew of what went on behind the closed doors of the Bridge Company was very hard to surmise. But the impression was that anything undeniably dishonest could have been kept from him and would not have been an especially difficult thing to do. Roebling was not a stockholder and so was not eligible to attend directors' meetings or meetings of the all-powerful Executive Committee, except when specifically invited. And the record revealed that he had been invited very seldom and only when engineering matters were to be discussed.

Why he owned no stock was never explained. It had been his father's practice to invest in the bridges he built, whenever possible. And surely if Kingsley and the other Brooklyn people were as anxious for private capital as it appeared they were in the earlier days, then the wealthy John Roebling should have been a prime prospect. But the elder Roebling had not invested in the bridge, nor would his son.

Perhaps Kingsley and the others discouraged John Roebling from coming in with them, preferring to keep the imperious old man and his unbendable integrity at a safe distance. More likely, both Roeblings, knowing something of the other stockholders, and suspecting more, decided they wanted nothing to do with that side of the bridge. The answer will never be known.

Washington Roebling was a very sick man by the time the reports from the Committee of Investigation were issued. One day he would say something of what he knew about the alleged bridge scandals, but that would be many years later.

For the moment, even to those willing to give Kingsley the benefit of the doubt, it seemed obvious that if he had not gotten rich on the bridge, he certainly had been nicely set up to do so. Furthermore, he had managed it all in such a way that it had cost him virtually nothing. Neither he nor any of the others had lost a cent for their efforts.

Indeed, it seemed terribly naïve, quite unrealistic really, considering Kingsley's basic nature and the ground rules of the political and business circles he customarily operated within, to conclude anything other than this: Kingsley had passed out money liberally in New York, perhaps in Washington, but most certainly in Albany. In Washington there had been comparatively little resistance and Demas Barnes, never a friend of the Brooklyn Ring, was probably not a likely candidate for bribery in any event, nor a safe confederate in any such dealings. But in Albany, support could be extremely dear, as everyone who read a newspaper knew perfectly well by this time, and especially for a charter that was virtually a license to steal, if anyone wanted to view it that way. It was even suggested by one New York paper that if Henry Murphy was the man doing all the work in Albany, writing the legislation, "lobbying" with Tweed, and so forth, then just maybe Henry Murphy had exacted his own large fee for services rendered, and that that had accounted for a substantial part of the money Kingsley "invested" earlier on. If this were so, then perhaps some of the "facts and figures" put forth by Kingsley beside Murphy's fireplace that fateful winter night in 1866 were not quite what Alexander McCue implied in his account of the scene.

Kingsley had indeed purchased most of the stock and he had obviously handed it out where it would do the most good. He had every expectation of profiting from contracts with the Bridge Company, as the directors made no effort to deny, and it seemed pretty plain that he figured to recover what he had spent for political favors and for stock by the 15 per cent arrangement and quite possibly make a lot more on top of that. Had things continued as they were prior to the fall of 1871, there would have been no sudden unexplained changes in the 15 per cent arrangement and the bridge could have made him a very rich man. But the unexpected had happened. The Tweed Ring had crumbled like papier-mâché. Bargains with Tweed were no longer acceptable. Politically active building contractors were subjects of automatic suspicion. The whole bridge project was suddenly very suspect. Kingsley, just as suddenly, not to mention the whole Brooklyn Ring, was extremely vulnerable. His arrangement with the Bridge Company would not look good at all if it were to come to public attention. The management of bridge business might be put into other hands. The arrangement had to be changed, which it was, most hurriedly and clumsily.

So the rug had been pulled out from under him before he had a chance to do much. But as a result no one was able to pin anything on him either. And besides, since the bridge was being built by a private company, there was, as no one could dispute, little he or any major stockholder could not do, perfectly legally.

The importance of all this was not, however, the degree of Kingsley's greed or guilt. What mattered in the long run was that largely by a quirk of fate—Tweed's fall—the greatest municipal work of the age, the most inspired structure Americans had yet attempted, had been rescued from certain disgrace, and probable disaster, at the hands of the Tweed and Tammany rings. Moreover, the movement to make the bridge truly a public enterprise can be dated from this point. Two more years would go by before the old 1867 charter would be changed in Albany, but even so, the days when the Bridge Company could do entirely as it pleased were all over. No more could the Executive Committee and William Kingsley disregard public opinion or conduct their affairs with total immunity and no higher ethical standards to go by than those of a railroad.

At the same time, those people who were actually building the bridge now had to face up to the idea that the work was no longer viewed as an altogether noble and heroic endeavor. Too many seeds of doubt had been sown and the fact that Tweed was out or that a man like Hewitt was in and testifying to the conduct of bridge business certainly did not put an end to the rumors. Papers such as the *World* remained openly hostile to the bridge and by no means satisfied that its management was henceforth above suspicion.

One paper would list it as one of the "seven fraudulent wonders of the New World," along with Tweed's courthouse and the Northern Pacific Railroad. The bridge had been a subject of controversy since the beginning, of course, but always on technical grounds—i.e., was the engineering sound, would the finished structure stand or fall? Now there were other reasons to be skeptical and the bridge, as a result, became a subject of special fascination to a wholly different variety of skeptic, of whom there were a very large number.

The same rich opportunities for dishonesty were still there, it was agreed, the same people had the power, and the void created by Tweed's departure could well provide lesser scoundrels that much more room to maneuver in. A cloud of suspicion remained about the whole endeavor, in short, and come what may the bridge itself would be viewed by many as the very thing John Roebling had feared it might become when he contemplated its social and political setting ten years earlier: a grand and conspicuous aggravation to the general state of venality on both sides of the river. Despite anything said to the contrary, a good part of the public would remain convinced that every day the work continued some crooked somebody behind the scenes was getting rich on it.

13
The Mysterious Disorder

> Knowing from the reports of other similar works that compressed air was liable to affect some men unfavorably, every known precaution was taken to guard against this danger.
> —WILLIAM C. KINGSLEY

BY THE FIRST of June, 1872, when the Chief Engineer and the General Superintendent issued their annual reports, the Brooklyn tower stood one hundred feet above the East River at high tide, while on the opposite shore the lower edge of the New York caisson rested seventy-eight feet six inches below the same tidal mark. The General Superintendent in his report stated that 14,500 cubic yards of masonry had been laid on the Brooklyn tower in the year past and 13,075 cubic yards on the New York tower. The Chief Engineer, however, wrote as follows:

> To such of the general public as might imagine that no work had been done on the New York tower, because they see no evidence of it above water, I should simply remark that the amount of the masonry and concrete laid on that foundation during the past winter, under water, is equal in quantity to the entire masonry of the Brooklyn tower visible today above the water line.

It was an impressive way to picture what had been accomplished, if not quite accurate, according to Kingsley's figures. To be informed that something of comparable magnitude to the Brooklyn tower had been built unseen below the river was for most people to have all the abstract explanations of counteracting pressures and penciled diagrams of timber caissons replaced in an instant with a single vivid image that anyone could appreciate.

The massive, freestanding masonry tower rising at the edge of Brooklyn was still the only part of the bridge conspicuously on display. Through the whole of that spring, as charges of fraud and jobbery filled the papers and Brooklyn gossiped of bridge scandals, work on the tower had proceeded exactly according to schedule and the immense granite shaft was looked upon popularly as an irrefutable affirmation of all that had been promised and anticipated over the past several years. One look at something like this was enough to restore a person's faith in what man could do and to make crooked bookkeeping and the like seem both terribly petty and no more than a temporary nuisance.

In plan the tower was an irregular rectangle, its outside surfaces being broken up by heavy buttresses. It stood lengthwise against the shore, 140 feet long, 59 feet wide. So at a height of one hundred feet, it was still broader than it was high, still only a little more than a third as high as it would eventually go, and only nineteen feet short of the height of the roadway. But already it was considerably higher than anything else around it.

Moreover, the tower kept gaining all the time, as though it were coming up out of the river, growing organically, instead of being slowly, methodically added to stone by stone. The change was never enough to notice from one day to another. Like the movement of an hour hand its progress was best seen at intervals. A Brooklyn dock worker on Furman Street might one morning notice that the stonework had gotten up above every ship mast since the last time he looked that way or a homebound commuter at the rail of a ferry pulling away from New York might realize one evening for the first time that this great blunt shaft with its feet in the water now topped the rise of the Brooklyn skyline.

The intended purpose of the structure would have been rather hard to figure at this stage if one did not already know. In the very early morning, when the ferries still had their running lights on and before anyone was at work on the tower, it might have been taken for an ancient harbor defense, a gray solitary battlement standing guard over the swarm of ships to either side of it. And when the sun began coming up and lit the top of the tower, the derricks bristling there looked for all the world like medieval war machines, the trajectory of which, from such a height, would surely be enough to hit New York. But in the full light of day, with the sun glaring on its clean buff-colored granite, the tower looked very new indeed, and more like the beginnings of a gigantic astronomical observatory perhaps, or the pedestal for some breath-taking triumphal monument.

But everyone did know its real purpose, of course, and could do little but marvel at its growth and at the way it seemed to diminish the size of everything else nearby. The ferryhouse, the most imposing Fulton Street stores, the newest business blocks, did not look so grand any longer. At the end of day, when the sun was a red ball hanging low over New Jersey and the west face of the tower seemed to be glowing from within, the granite pink nearly, everything in back of the tower stood in shadow for a block and more.

When the *Eagle* claimed there was nothing on earth, save the Pyramids, to rival "this Brooklyn tower of ours," nobody thought that especially high-blown. And now Roebling had introduced a new vision to stir the public imagination. Now, one need only look at the Brooklyn tower and picture the same thing concealed below water directly across the river. As far as this tower reached above the river, the other one reached below, like a gigantic rock taproot. (Roebling undoubtedly wrote what he did a number of weeks before it was published, at the time when the Brooklyn tower was indeed at about seventy-eight feet.) For every stone the crowds on the ferry had seen hauled up the face of the Brooklyn tower, another had been added to the burden of the New York caisson. And those seemingly fearless figures working along the uppermost rim of the tower were no farther above the surface of the East River than the men in the caisson were below it.

Work inside the caisson was to be finished in another month. So it was the end of the first great stage in the building of the bridge, a clear dividing place. From here on the problems to be overcome, the work to be done, would be of an entirely different nature. Roebling made quite a point of this in his report, expressing congratulations to the Board of Directors "on the success which has attended the last of the two great tower foundations." At the start of the work, he said, the foundations had been the principal engineering problem. The work to come—the building of the towers, the cable spinning, building the superstructure—was all work that had been done before on other bridges, on a smaller scale, "but upon the tower foundations rests the stability of the entire work." Then he remarked almost as an afterthought, "Considerable risk and some degree of uncertainty was necessarily involved in their construction."

All the extensive preparations for receiving the caisson had been completed by the end of the first week in September 1871. The tower was to fill a space formerly occupied by two ferry slips, between Piers 29 and 30. The riverbed had been dredged out to a depth of thirty-seven feet, or a little more than twice as deep as at the Brooklyn site. A hundred feet of Pier 29 had been torn away, and a huge pile dock had been built, itself a bridge more than a hundred yards long between the new foundation and the shore. At the end of the dock a square enclosure for the caisson had been built of six-inch pine planks—this to break the force of the tidal current, which was decidedly stronger on the New York side.

Borings made from the end of Pier 29 indicated bedrock anywhere from seventy-seven to ninety-two feet down. How far the caisson would have to go or whether even it was essential that it go clear to bedrock were questions that had still to be decided. But in any case the strata appeared to be chiefly gravel and sand, with layers of quicksand from fifteen to twenty feet thick. It was very different terrain from that at Brooklyn.

The machinery needed was all standing by on the dock: three huge boom derricks similar to those used in Brooklyn, the same clamshell dredging equipment as before, hoists, steam engines, and pile-driving gear. Workshops and offices had been built, a blacksmiths' shed, sheds for cement, tools, general stores, a compressor house (the largest building) with its air-pumping machinery set up inside—thirteen Burleigh compressors ranged in a single row, each with its own steam boiler, as compared to the six compressors used for the Brooklyn caisson.

On September 11 the colossal wooden box was towed up from the Atlantic Basin, where in the four months since it was launched seven additional courses of timber, all laid with cement between, had been built on top. Once the pile enclosure was completed on the river side and the caisson confined to its permanent position, a final ten courses of timber were added, bringing the total height of the structure to just over thirty-one feet.

Particular care had been taken this time to guard against sea worms. The protection was needed only during the time the caisson was afloat and before it was entirely submerged below the riverbed, where the sea worm, the teredo, never penetrates. But this microscopic animal, less than a sixteenth of an inch in diameter, can bore into any crevice water can get through, so the precautions had to be quite substantial. Every outside seam was caulked. The entire outside surface was heavily coated with a composition of coal tar, rosin, and a hydraulic cement, which, all by itself, was supposed to have enough body and grit to dull the teredo's boring apparatus. Then this had been finished off with a sheet of tin covering all four sides and the top of the sixth timber course. All seams in the tin had been soldered airtight and layers of tar paper had been put in both above and below the entire sheet. Finally, the whole caisson had been sheathed in four-inch yellow pine saturated with creosote.

"The great timber foundation was now complete!" Roebling wrote. "It contains 22 feet of solid timber above the roof of the air chamber, seven courses more than the Brooklyn caisson, and since the strength of such structures varies as the square of the depth, we may consider it to be nearly twice as strong as its Brooklyn brother."

In their general features the two caissons were almost identical. The sides of the New York caisson were again of yellow pine and tapered from nine feet thick on top to an iron cutting edge eight inches wide. The timbers used in the roof were again a foot square. Headroom inside the work chamber was nine and a half feet as before. The base dimensions were 102 by 172 feet, making the new caisson just four feet longer than the one in Brooklyn. The heavier roof had been built to carry what Roebling figured would be a significantly greater load, since this caisson would have to go much deeper and therefore carry far more stone. But there were several other differences as well.

The light skin of iron boiler plate that lined the interior would not only provide fire protection, but make the caisson more airtight; and to improve visibility the whole inside had been given a heavy coat of whitewash. The water shafts this time, instead of being square, were round (they would be stronger this way Roebling had decided). In addition, some fifty iron pipes, four inches in diameter, had been installed throughout the work chamber as a way of removing sand.

In the Brooklyn caisson there had been no means of communication between the men below and those working up on the surface. But here Colonel Paine had devised a simple, ingenious mechanical signaling system. One of the sand pipes was capped below and an inch tube was passed through the cap with index pointers attached above and below. Underneath each pointer was placed a small plan of the caisson, showing the position of every pipe and shaft. By rotating this tube immediate attention could be called to any of the points. In addition, a small rod was passed down through the pipe and its weight offset by a weight above that was attached to a cord that passed over a pulley. Small indexes were then fixed above and below. These moved up or down on vertical boards on which were printed such messages as "stop," "start," "bucket is caught," and so forth.

The arrangement of air locks was also quite different. This time there were two double locks, each large enough to accommodate thirty men, which meant that a full shift of 120 could enter or leave at one locking. And instead of being mounted on top of the caisson, as had been done in Brooklyn, the locks were built into the roof of the work chamber, so they actually projected down into the chamber about four feet. Each set of locks was connected to the top of the caisson by a spiral stairway enclosed in an iron shaft.

The arrangement was essentially the same as Eads had used in St. Louis and it was over this particular feature that Eads and Roebling were to have their bitter falling out. The advantages to be gained were these, supposedly: the men could now step directly from the air lock into the work chamber, and at the end of the day they would not have to make the climb to the top while still under pressure.

On October 31 the last timber course was finished, the first stone of the new tower laid. By December 12 enough stone was in place to hold the caisson on the river bottom at high tide. The compressed air was turned on and Roebling, Paine, Collingwood, and a complement of some thirty men went down inside. (Since the water was thirty-seven feet deep and the caisson with all its timber courses stood thirty-one feet high, this left the top of the caisson just six feet below water at high tide and about two feet below at low tide.)

Tearing out the temporary floor took another two weeks. When the digging began, the work proved nowhere near so difficult as it had been in Brooklyn, but much more disagreeable, for the caisson was standing in the middle of what for years had been New York's principal dumping ground. Moreover, a street sewer was still emptying into the river close by.

The ground itself was a clay silt turned black by sewage and thick, with the putrid remains of animals, garbage, and what Farrington called "sewage abominations." All this had been odorless so long as it was embedded below salt water, but once the black muck was turned over inside the caisson, the smell, even in the compressed air, came forth in all its original strength, as Roebling wrote. The stench was such that several men were actually overcome and had to be helped back up to the surface. Only by keeping a skim of water over the entire caisson floor could the men keep on working.

But beneath this foul dock mud, which was only a few feet deep, they hit clean river sand and gravel, and things took an immediate turn for the better. The skim of water was expelled by compressed air, leaving a perfectly dry footing, and by now, too, the lights inside were fully operating and in combination with the white roof and walls they lit up the entire chamber as bright as day. From then on the great timber box descended into the earth extremely rapidly.

Above ground the scene was one of great energy and purpose. A reporter described it this way:

> At the foot of Roosevelt Street, where the New York tower is being erected, one of the busiest scenes in the city is met with and has been for months—dozens of workmen hurrying hither and thither with wheelbarrows and hods and spikes and shovels; engines puffing away, lifting huge blocks of stone with huge derricks from the barge at the side of the dock, drawing lumber from the foot of the pier, driving the piles of the cofferdam, and condensing and compressing the air to be used by the submarine workmen; men chopping and planing and sawing the immense timbers used in constructing the enormous derricks; others shoveling gravel and sifting sand for the cement; little knots of threes driving immense piles through the heavy timbers of the caisson with their sledges and kept steadily at work by an overseer who evidently enjoys his employment; some wheeling cement for others to lay between the large granite blocks, boring and hammering and cutting stone and carrying iron rails, everything indicating that the work is being pushed rapidly forward.

There were almost no boulders to contend with this time, indeed little else but sand. The average rate of descent would work out to about two feet a week, but at one stage, for several weeks, the caisson was sinking six to eleven inches a day. In the Brooklyn caisson, during the first discouraging month of excavation, the rate of descent had not averaged six inches a week. Now everything was working to perfection. The dredges had no trouble digging the pools below the water shafts and the sand pipes worked like a charm.

How the pipes were to be used exactly had been left undecided until it came time to give them a try. Either the sand could be forced out with water, using a new kind of sand pump devised by Eads, or it could be blown out by compressed air. The latter would be the simpler, less costly way, of course, if it worked. It would also greatly aid ventilation. An air chamber with an iron skin such as this one had inside would be practically airtight, but a certain quantity of new air had to be fed into the work area at all times to keep the atmosphere fresh enough to live in. In the Brooklyn caisson this had been no problem since air kept escaping under the shoe or into the timber roof. But compressed air lost that way did little work. Roebling's thought, therefore, was that with all the compressors he had at hand, why not allow air to escape through the pipes and take sand out with it?

So the air system had been tried and after that there was no more talk of sand pumps. For everyone who remembered how it had been in Brooklyn this was the smoothest sailing imaginable. Morale could not have been better.

The sand pipes extended down through the roof and into the chamber to within a foot or so of the work surface. Sand, loose earth, and fine gravel were shoveled around the pipe in a cone-shaped pile two to three feet high. When the pipe was opened, the pile vanished up the pipe. It was as neat and uncomplicated as could be, and the deeper the caisson went—the greater the pressure in the chamber became—the better the system worked. When the caisson was down about sixty feet, for example, the air was blasting out of the sand pipes with such force that fourteen men could stand in a circle around one pipe and shovel sand under it with all their strength and the sand would disappear as fast as they could shovel. At least three sand pipes were kept going all the time and some sixty men did nothing but feed these pipes, which was about the most tiring work imaginable. Time and time again the pipes had to be shut off to give the men a rest.

Up on top the sand blasted out with such velocity that it became a serious problem. At first, when there were only vertical discharge pipes, the sand was shooting four hundred feet in the air. To deflect this great geyser off at right angles, so it would feed into big scows tied up beside the caisson, iron elbows were fixed to the tops of the pipes. But the furious blast of sand would cut right through these, sometimes in a matter of minutes. Iron an inch thick would not last a regular workday. When elbows fixed with thick caps or patches of a special chilled iron were tried, they lasted only two days. Toward the end of the work the elbows would be dispensed with altogether and heavy granite blocks would be placed on supports directly above the mouths of the pipes. The sand would strike the blocks, then fall back. But the sharp, abrasive force of the sand was such that in three or four days' time it would cut a hole through the granite block.

Once, a man passing by in a rowboat, with one hand resting on the gunwale, had the end of a finger shot off by a pebble fired from a sand pipe. Another time a workman was drilled through the arm in the same way. And down inside the caisson Farrington at one point thoughtlessly placed his hand over the open end of a sand pipe and found he was unable to remove it. Only with several others pulling on his arm was he able to get his hand free and then found that his whole palm had been drawn up into the pipe like a stopper.

Boulders were encountered only on occasion and slowed things down but a little, except when one appeared directly below a water shaft. The shaft had to be capped then, the water removed, the boulder excavated out of the dry hole, the same as had been done so many times in Brooklyn. But relatively little time was lost that way. Paine's "mechanical telegraph" between the men on top and those below worked amazingly well. "The downward movement of the caisson has been under perfect control," Roebling wrote. Indeed, the work went smoother, faster than anyone could have hoped for. Everything functioned as it was supposed to in theory and as it seldom had in practice over in Brooklyn. Nothing very unexpected happened. The heavy, tiresome digging by the men inside and the noisy work of dredges and stone derricks up above continued day after day, six days a week, and on into winter. The caisson, itself as high as a four-story building, kept descending steadily, evenly, uneventfully, the lights inside burning twenty-four hours a day, and all the while an enormous load of limestone blocks piling up on its back just as steadily, evenly, and uneventfully.

As far as Roebling and his staff were concerned, there were only two problems to be considered—the effects of the compressed air on the men and the depth to which they might have to go before stopping. In Brooklyn every foot of ground gained had to be fought for and the physical discomfort of working under pressure had been but part of the problem. But here the mercury gauges on the big Burleigh compressors kept inching up just as steadily as the caisson was descending.

Every two feet gained meant a pound more of pressure. On December 18, when the caisson was grounded on the river bottom in thirty-seven feet of water, the pressure was at seventeen pounds. In Brooklyn the pressure had gotten up to only twenty-one pounds when that caisson was halted at a depth of forty-five feet. But here, where the water was so much deeper to begin with and the going so much easier, it took only about a month to reach forty-five feet. Even at that the bottom of the caisson was only eight feet into the river bed, which left twenty-three feet more, or the whole of the enormous timber roof, still surrounded by water. And bedrock was still a long way off. But at forty-five feet, just as at Brooklyn, some of the men began feeling a good deal of discomfort, and, in a few cases, severe pain.

The number of men employed at any one time in the caisson varied from fifty to 125 in the daytime and from fifteen to thirty during the night. At first the workday was divided into two shifts of four hours each, separated by an interval of two hours. But at forty-five feet Roebling ordered the workday for caisson men shortened slightly, to seven and a half hours, in two shifts. At fifty feet, with the pressure increased another two and a half pounds, the day was again cut by half an hour. The majority of men were having trouble by now. The climb up to the surface after each shift, for example, had become so terribly fatiguing that Roebling had one spiral stairway pulled out and a steam elevator installed, as Eads had also done in St. Louis.

It was not until late January, however, when the caisson reached a depth of fifty-one feet, that any serious effect among the men was observed. And it was at this point, when pressure in the chamber stood at twenty-four pounds, that Roebling decided there ought to be a doctor on hand.

His name was Andrew H. Smith. He was a New Yorker, a former Army doctor, a surgeon, and a throat specialist at the Manhattan Eye and Ear Hospital. He was a man about the age of Roebling. Nine years later he would achieve national prominence by performing the autopsy on President Garfield that revealed the much-debated location of the assassin's fatal bullet. Smith's pioneer work on the bends, however, would be of far greater importance. His title was Surgeon to the New York Bridge Company and, except for Dr. Jaminet, Eads's medical adviser, and some other St. Louis doctors, he was the only man in the country with any medical background to try to figure out what was causing the mysterious malady brought on by compressed air.

Smith took his assignment very seriously and went right to work. His first step was to prepare a set of rules, just as Jaminet had done two years earlier in St. Louis. These he had posted conspicuously about the dock and inside the caisson. They read as follows:

1. Never enter the caisson with an empty stomach.
2. Use as far as possible a meat diet, and take warm coffee freely.
3. Always put on extra clothing on coming out, and avoid exposure to cold.
4. Exercise as little as may be during the first hour coming out, and lie down if possible.
5. Use intoxicating liquors sparingly; better not at all.
6. Take at least eight hours' sleep every night.
7. See that the bowels are open every day.
8. Never enter the caisson if at all sick.
9. Report at once at the office all cases of illness, even if they occur after going home.

He next subjected every man to a physical examination, the idea being to exclude anyone suffering from heart or lung disease, anyone who struck him as too old for such work, and all obvious drinkers. Every new caisson man thereafter was required to have a work permit signed by him; and though only a few were actually rejected, the knowledge that an examination and a doctor's permit were required doubtless discouraged many who were unfit from applying.

In any case, Smith was convinced that the men he cleared were in the best possible physical condition. He also saw to it that each man got a strong cup of coffee every time he came up out of pressure. "It appeared to relieve, in a measure, the nervous prostration which marked the return to open air," Smith wrote. He did his best, too, to get the men to stay quiet a while after each shift, in a special resting room he had fitted out. But once out of the caisson, the next stop for most men was the handiest saloon. There the terrible, numb fatigue or the outright pain the work left them with could be cured considerably faster, they believed, than taking the doctor's coffee or spending time on a company bunk.

The young doctor had no misconceptions about the off-hours recreation of the men or the living conditions most of them put up with. Many, he knew, slept in "lodginghouses," as they were called, a damp cellar likely as not or one of the 14,872 tenements described in the 1870 census, where thirty people to a room was not uncommon, where the only light and ventilation came from a single passageway up to the street or to an ill-smelling common hall or kitchen. The ages of the men, as Smith noted, ranged from eighteen to fifty. They were of all nationalities, he found, but mostly Irish, immigrants who had known nothing else in New York but tenement life. The neighborhoods they went home to after a day in the caisson were famous as breeding grounds for measles, diphtheria, scarlet fever, the grippe. And in the teeming streets the one note of cheer was the saloon. "The habits of many of the men were doubtless not favorable to health," Smith wrote, "but everything which admonition could do, was done to restrain them from excesses."

Smith had been down in the Brooklyn caisson once or twice at Roebling's request. And now again, as at Brooklyn, he noted with much interest that when the men spoke to one another in the heavy air it was with strange shrill treble voices and that it was physically impossible to whistle. ("The utmost efforts of the expiratory muscles is not sufficient to increase materially the density of the air in the cavity of the mouth, hence on its escape there is not sufficient expansion to produce a musical note.") He noticed, too, that the men were breathing faster under pressure, and suspecting they were breathing harder as well, he wrapped a steel tape about his own chest, then compared the measurements he got when breathing inside the caisson and up on the surface. Under pressure, he found, his chest expansion was nearly twice what it was normally.

He studied the effect on circulation and discovered that while the normal pulse might rise sharply upon entering the caisson, after an hour or so it would drop back to normal or even below normal. The effect on the volume of the pulse was to diminish it. This, he thought, was caused by the pressure exerted on the artery. "Hence, the pulse is small, hard, and wiry."

He observed that the men coming out of the caisson all had a marked pallor that lasted twenty minutes or so and that their hands were slightly shrunken and the tips of the fingers shriveled, as if they had been in water for a long time. Inside the caisson he took the temperatures of several workers and found they were one, even two, degrees above normal. The whole force was running a fever, he concluded, and told them this was caused by the heavy, saturated air, which kept their bodies from cooling by evaporation as they would normally. It was the reason they were always wringing-wet with perspiration, he explained. It was not that they were perspiring so much more, but that the air was not drying them in the least.

Like others before him, Smith was also impressed by the way the work seemed to increase the appetite. This he believed was caused by a generally increased waste of tissue, which was the result of an increased absorption of oxygen. But he was not absolutely sure about that—or much of anything else. He had no reliable data to go by, as he said, nor were the men particularly cooperative. His means of testing his theories were quite crude at best. To find out what effect the heavy atmosphere had on the metamorphosis of tissue, for example, he took four healthy pigeons, cut a wound under the wing of each, took two down into the caisson and kept the other two in the temporary hospital he had established on the dock. But at the end of six days he could find no discernible difference in any of them. The wounds all had healed about the same; the birds appeared to be in comparable health.

A little later on he had a dog taken into the caisson and kept there for seven hours. Then he went down himself, killed the animal with prussic acid, opened its neck, took a blood sample, and carried that back to the surface to see what if anything had happened to its oxygen content. Small quantities of air injected into the blood stream of a dog would, he knew, normally escape through the lungs. But perhaps time under pressure produced air in the blood that could not be expelled in the normal fashion. His sample from the dead dog, however, indicated no such thing, so he abandoned that approach, little realizing how close he was to the truth.

But it was the suffering of the men that concerned him more than anything, and as the caisson continued downward, their suffering increased many times over. Smith was in daily attendance. He studied every symptom, kept careful notes, and though he was unable to put his finger on the exact chemical or physiological cause of the trouble, he began to have some ideas of his own about what the men were doing wrong and what might be done to help those suddenly "taken" by the effects of compressed air—"as if struck by a bullet" was the way they commonly described it.

One of the workers in the caisson about this time may have been an undersized Irish boy named Frank Harris, later to become a man of letters in England and author of the sensational autobiography *My Life and Loves*. Harris said he went to work in the caisson a few days after landing in America and was only sixteen at the time.

He never wrote anything about the experience until years afterward, which may explain the inaccuracies in his account. But he was also known his whole life for his inability to separate fact from fancy, so what he says may or may not be the way things happened.* Still, it is among the very few accounts written from the point of view of the men themselves and vividly conveys the terrible fear they had of contracting the bends:

> In the bare shed where we got ready, the men told me no one could do the work for long without getting the "bends"; the "bends" were a sort of convulsive fit that twisted one's body like a knot and often made you an invalid for life. They soon explained the whole procedure to me. . . . When we went into the "air-lock" and they turned on one air-cock after another of compressed air, the men put their hands to their ears and I soon imitated them, for the pain was acute. Indeed, the drums of the ears are often driven in and burst if the compressed air is brought in too quickly. I found that the best way of meeting the pressure was to keep swallowing . . .
>
> When the air was fully compressed, the door of the air-lock opened at a touch and we all went down to work with pick and shovel on the gravelly bottom. My headaches soon became acute. The six of us were working naked to the waist in a small iron chamber with a temperature of about 80 degrees Fahrenheit: in five minutes the sweat was pouring from us, and all the while we were standing in icy water that was only kept from rising by the terrific pressure. No wonder the headaches were blinding. The men didn't work for more than ten minutes at a time, but I plugged on steadily, resolved to prove myself and get constant employment; only one man, a Swede named Anderson, worked at all as hard. . . . Anderson was known to the contractor and received half a wage extra as head of our gang. He assured me I could stay as long as I liked, but he advised me to leave at the end of a month: it was too unhealthy: above all, I mustn't drink and should spend all my spare time in the open. After two hours' work down below we went up into the air-lock room to get gradually "decompressed," the pressure of the air in our veins having to be brought down gradually to the usual air pressure. The men began to put on their clothes and passed around a bottle of schnapps; but that I was soon as cold as a wet rat and felt depressed and weak to boot, I would not touch the liquor. In the shed above I took a cupful of hot cocoa with Anderson, which stopped the shivering, and I was soon able to face the afternoon's ordeal.

* Harris thought the caisson was made of iron. He describes the work chambers as small, when in fact they were quite large, and his figures for the wages paid, the hours kept, the time spent in the lock, etc., do not jibe with the records.

Still, he could make two weeks' wages in a day, he said. If he could last a month, he would have enough to live on for a year. But by the fifth or sixth day, he had terrible shooting pains in his ears and he was told he might be going deaf. An Irishwoman he was boarding with in a shanty beside Central Park fixed up a remedy—"a roasted onion cut in two and clapped tight on each ear with a flannel bandage." Harris said it worked like magic, relieving his pain in minutes. But not many days later he saw one of the men fall and writhe on the ground, blood spurting from his nose and mouth, and that was enough to decide young Harris, who quit soon after and who later took the literary license to say that the man's legs were "twisted like plaited hair."

Smith assumed medical charge of the caisson workers on January 25, 1872, and was on duty until May 31, when he resigned. During that time there were 110 cases of sickness that he could attribute directly to compressed air and that were severe enough to require treatment. Not by any means, however, did every man suffering pain or discomfort report to him—as he was well aware. The feeling was that a man might not get hired again at some future time if it was known he had had a dose of the Grecian Bends. (It was a feeling that would also persist among future and supposedly more enlightened generations of "sand hogs," a term not yet in use in the 1870's.) As Roebling would write in his own report, scarcely any man escaped without being affected by intense pain in one form or other. Martin and Collingwood both suffered attacks. Charles Young, the foreman who had collapsed in the Brooklyn caisson the same time Roebling did, had again become so much affected by the compressed air that on the advice of his own doctor he resigned, taking a job overseeing work on the dock instead.

But those cases Smith was able to treat and study, he described at some length in his notebooks:

> Case 11—E. Riley. Taken sick Feb. 16th, one hour after leaving the caisson. Pressure 26 lbs. Epigastric pain and pain in the legs. No loss of sensibility. Profuse cold perspiration. Pulse, when I saw him, two hours after the commencement of the attack, was 96. The pain, which at first was very severe, had by this time become much less. Gave him an ounce of brandy and a teaspoonful of fluid extract of ergot. In 10 minutes the pulse had fallen to 82. Was able to resume work the next day.

Case 12—Joseph Brown, foreman, American, aged about 28. Taken on the 28th of February, about an hour after coming up from a three hours' watch. Excessive pain in left shoulder and arm, coming on suddenly, "like the thrust of a knife." Pain continued until he went down again for the afternoon watch, when it ceased immediately. . . .

Case 13—Henry Stroud, a diver by occupation, began work on the morning of April 2d. Half an hour after coming up from the first watch, was taken with numbness and loss of power in the right side, also dizziness and vomiting. This was followed by severe pain over the whole body. Excessive perspiration. Was treated with stimulants and ergot, and in five hours was well enough to return home.

Case 14—John Barnabo, Italy, 42, reports on the 13th of March, while in a car returning home, he was taken with severe pain in both arms. This was followed by dimness of sight and partial unconsciousness. Extremities very cold. Remained in this condition for two hours. Was obliged to keep to his bed for three days. For a week afterward was unable to work, feeling very much oppressed about the chest. Had no medical attendance. Had a similar but less severe attack about a month previously.

Savage cramps in the legs were the most common first sign. Sometimes the pain lasted all night, in the knees mostly, and it felt as though the joint was being violently twisted apart and every muscle torn away from the bones—or worse. There was really no way to describe the pain, most men said. A modern medical textbook describes the pain as deep and relentless, and not throbbing. "When it is severe, local numbness, weakness, and faintness resemble the sickening pain of a blow on the testicle."

In one out of four cases the attack was accompanied by dizziness, double vision, and repeated vomiting. All of a sudden a man would begin to stagger, bend double, retch horribly, and fall. Sometimes there was no pain at all, just a massive numb feeling and an inability to walk or to stand upright.

The victim of an attack always looked the same, whether there was pain or not, the face a leaden color with cold beads of sweat standing out all over, which were probably signs of impending shock. Men complained, too, of excruciating pains in the chest and bowels. Some had their speech affected, as though they had had a stroke. In numerous cases the joints—knees, wrists, elbows—were swollen all out of shape, burning hot to the touch, badly discolored, and extremely tender.

By the first week of April the caisson was down past sixty feet, still descending steadily, and conditions had grown very serious indeed. The remedies Smith employed were all very simple. To alleviate pain he promptly administered ergot and often in quantity. Or he doled out whiskey and ginger. Or he gave injections of atropine, a poisonous alkaloid used as an antispasmodic. When nothing else worked, he used morphine. Since the average attack generally lasted only a few hours, his solution for severe cases was simply to drug the patient so heavily that he felt little or nothing.

He applied hot poultices to swollen joints. Paralyzed legs were soaked in hot baths, arms were packed in ice, spines were doused with ice water. Men with heaving stomachs were spoon-fed bits of ice or "a scruple of calomel," i.e., twenty grains of a white tasteless purgative. Sometimes these things worked, or seemed to. But the prevailing attitude among the workers and the engineers in charge was that it did not matter much what Smith did. As Collingwood noted at a gathering of the American Society of Civil Engineers later that spring, almost every man recovered eventually anyway, regardless, it seemed, of how much or how little was done for him.

In a few instances, when a man reported back to the job after recovering from an attack, Smith told him to find other work. Patrick Rogers, for example, a forty-year-old Irishman from Brooklyn, was on his way home on the ferry one night when all at once he had no feeling at all in his right side and very quickly after that was unable to stand up or move a muscle. When the boat docked, he was placed in a horse cab and taken home. As with most of the cases Smith recorded, the pain Rogers was in, terrible as it was, lasted less than twelve hours. But when he returned to the caisson, ready to go down again, he told Smith of a continued "trembling" in his chest and Smith advised him to go away and not come back.

A number of other Brooklyn men were sent by Smith to the Brooklyn City Hospital, where, interestingly, the cases became the special fascination of a young intern there, Dr. Walter Reed, later to be one of the best-known physicians in the world as a result of his research on yellow fever. Like Smith, Reed also kept extensive notes on each caisson victim to come under his care and these he subsequently turned over to Smith. As Roebling would write, it was hoped that Smith's efforts and conclusions would be made public eventually "for the benefit of future works."

Smith never used the term "bends." He called it the caisson disease, a name he was the first to employ and that is still used as the formal designation. He did a commendable amount of original research into the history of the subject and was thoroughly familiar with what Jaminet had written in a lengthy report published in St. Louis the previous year.* Smith described Jaminet's observations as "exceedingly valuable," but found it "especially to be regretted" that Jaminet's basic remedial routine had been merely to keep the patient lying on his back with his feet elevated slightly and to administer whiskey or beef broth.

From what he read and from his own observations Smith put together a number of theories, several of which were the same as conclusions reached in France some twenty-five years earlier. The disease, he decided, depended upon increased atmospheric pressure, but always developed *after* the pressure was removed. Attacks never occurred while the men were still under pressure, only afterward—as had become obvious to almost everyone. So there was a very good chance, he decided, that the principal cause of the trouble was "locking out" too rapidly. "Indeed," he wrote, "it is altogether probable that if *sufficient time* were allowed for passing through the lock, the disease would never occur." This, he knew, jibed with what the French mining engineer M. B. Pol had concluded in 1845 and later expanded on in a most interesting memoir published in 1854. "Experience teaches," Pol wrote, "that the ill effects are in proportion to the rapidity with which the transition is made from the compressed air to the normal atmosphere." In St. Louis Jaminet too had hit upon the same idea after being taken by a terrible seizure himself, but Jaminet thought it even more important to increase the pressure slowly when the men were going in. Smith it seems never suffered any discomfort from his time in the caisson.

To make things easier for the men after they emerged from the lock, Smith recommended that no climb up a long flight of stairs be necessary at that point. To have put the locks at the bottom of the shaft as Eads had was a serious mistake, he said. The arrangement in Brooklyn had been better. That way the climb was made "*in the*

* Eads had completed the east abutment of his bridge in early April of 1871, with his caisson an incredible 136 feet below the Mississippi. Eads too was having trouble with advancing expenses, with construction costing about double his original estimate, but in October 1871, before work had even begun inside the New York caisson, Eads had written that all the most formidable difficulties had now been surmounted.

compressed air, instead of immediately after leaving the lock, when the system is more or less prostrated." The elevator Roebling installed had been a wise measure, but it had not wholly alleviated the problem.

But Smith did not see how, in all practicality, the locking-out procedure could be changed much. What might be sufficient time in the lock for one man, he reasoned, would be too short for another, and far less work would be accomplished if the time in the lock were prolonged greatly. Delays would be very expensive for the Bridge Company. Besides, the men themselves would want no part of it. About all that could be done, he concluded, was to make the time required in the lock proportionate to the pressure. But even that time could only be "as great as is consistent with the circumstances." For the New York caisson Smith established a regulation that at least five minutes more in the lock would be allowed for each additional "atmosphere," or for every additional 14.7 pounds of pressure, which meant that for every three pounds of pressure added inside the caisson, Smith wanted the men to take a minute longer coming out. So at a depth of, say, sixty-five feet, with the pressure at thirty pounds, the men should spend five minutes in the lock on their way out; at seventy-five feet, with the pressure at thirty-three pounds, they would spend perhaps six minutes. Smith was not asking for very much, in other words. Even so, the regulation was followed only infrequently. "The natural impatience of the men to reach their homes," he wrote in a tone of despair, "makes the delay in the lock irksome, and great firmness is required on the part of the lock tender to prevent the escape cocks being opened more widely than is consistent with safety." One of the first steps in such work, he said, ought to be the employment of reliable lock tenders.

But despite his recognition of rapid decompression as the chief cause of the mysterious sickness, Smith remained convinced that susceptibility was still largely a matter of "special predisposition," as he called it. Some people, he said, were simply more susceptible than others. It was commonly known, he said, that certain people had a predisposition to pains in the joints just prior to a thunderstorm. These pains, as he said, were generally considered to be of rheumatic character and caused by dampness, but Smith now thought differently. The pains suffered by his caisson workers were precisely the same, he said, only, of course, immensely intensified. So very likely anyone who could feel weather in his bones was actually feeling shifts in atmospheric pressure, and just as some people could feel such things and others could not, so some people would fall victim to the caisson disease while others would not.

Fat people and heavy drinkers, he was convinced, were more susceptible than anyone else. Men new to the work also stood a greater chance of being hit by an attack than those who had been going into the caisson for a length of time and so had had the pressure build up on them slowly. New hands suffered worst during the first week he noted. The ideal caisson worker in his view was a spare man of medium height in his twenties or thirties, a description that would have applied to Washington Roebling, among others.

But of more importance was Smith's contention that the amount of pressure a man was exposed to and length of time spent in the caisson were as much a part of the problem as rapid decompression. So in this he agreed completely with Eads, Jaminet, and Roebling. "The testimony of all observers," he wrote, "is that the liability to attack is directly as the duration of the stay in the caisson." The common explanation given for Roebling's collapse in the Brooklyn caisson the night of the fire was not that he had been coming up too fast, but that he had been staying down too long. Smith concurred with the explanation. In fact, the explanation followed perfectly out of the conclusion he had come to concerning the real root of the problem.

14
The Heroic Mode

> As it is now demonstrated that the method of compressed air is applicable to a great range of engineering operations, and offers many peculiar advantages, it is extremely desirable that the principal objection to its employment, viz., the discomfort and danger to the workmen, should be reduced to a minimum. To this end I offer the following suggestions . . .
>
> —ANDREW H. SMITH, M.D.,
> *The Effects of High Atmospheric Pressure, Including the Caisson Disease*

SMITH said the caisson disease could be explained on mechanical principles. He said it was caused by the effect of abnormally high atmospheric pressure on the circulatory system. Under pressure, the blood was not distributed according to the normal physiological demands of the body, "but in obedience to overpowering physical force." As he saw it the envelope of heavy air in the caisson pressed against the surface of the body forcing the blood into the center of the body. The blood "retreats," he wrote, "from the surface to the center, and accumulates there until an equilibrium of pressure is produced."

Smith held, however, that a man's circulation could adjust somewhat to such unnatural conditions if the conditions were experienced by degrees. And the longer a man stayed down in the heavy air, and the heavier the air, the more the circulatory system would be affected. But when the pressure was removed suddenly—if a man were to waste no time getting through the air lock—then the blood vessels would fail to assume their natural condition in an instant, proper circulation would not be restored quickly enough, and "disturbance of function will result."

Smith was quite right that rapid decompression was the secret to the mystery, but his explanation of why was wrong. His call for slower decompression was a commendable step in the right direction certainly, and seemed radically cautious at the time. (In St. Louis, for example, Dr. Jaminet wanted his men to spend an extra minute in the lock for every six new pounds of pressure, indicating that perhaps Smith's convictions on the matter were twice as strong.) But even so, as a preventive measure, Smith's new locking-out procedure, even when it was followed, was so inadequate as to be of little real consequence.

The mystifying disease was, in fact, caused by the effect of too rapid decompression on circulation, but for reasons other than Smith had arrived at. The leg pains, the paralysis, the swollen joints, the agonizing stomach cramps, were caused by the liberation of nitrogen bubbles from solution in the blood stream and in the tissues of the body upon the sharp reduction of atmospheric pressure.

Under pressure the normal nitrogen gas in the blood dissolves to a high degree, then returns to a gaseous state—in the form of bubbles—when and if the pressure is suddenly relieved. Set free in the body fluids such bubbles can cause great damage. If liberated in the spinal cord, for example, they can cause total paralysis. But if pressure is relieved gradually, the gas comes out of solution slowly and is removed by the lungs.

The savage pains of the bends are caused by a stoppage of the oxygen supply in the blood stream. The nitrogen bubbles released by too rapid decompression create blocks in the blood stream—the same as mechanical blocks—that keep the oxygen in the red blood cells from reaching the tissue. The red cells fail to get past the nitrogen bubble, the tissue is denied the oxygen it must have, and the result is dreadful pain. This denial of oxygen-bearing blood—called ischemia—is much the same as what happens in a heart attack. So an attack of the bends might be likened to a heart attack in different parts of the body, most often the limbs and joints.

The level of pressure in the caisson and the time spent by an individual in the caisson do have a direct bearing on the problem and so Smith was right to keep the work shifts as short as possible, a policy that is still adhered to in caisson and tunnel work when pressures are extreme. Smith's nine rules were also sound policy, in that they contributed to the over-all fitness and health of the men and fit, healthy men are less likely to be victims of the bends than those who are not. Fat men are also more prone to attacks, as Smith surmised, since the nitrogen bubbles tend to collect and dissolve in fat tissue. And from all that is now known about the bends—from subsequent experience in construction work, from underwater research and the space program—it appears that some people are indeed more susceptible to the disease than others, just as Smith declared.

Smith was, in fact, a keen and intelligent observer and deserves great credit for his work. His thesis of slower decompression was the key to the puzzle. The only problem was he did not carry it anywhere near far enough.

A man's health did depend mainly on how sudden and great a change he was subjected to on coming out of the caisson back to normal pressure. If there was to be damage done it happened *then.* Time in the caisson, the amount of pressure the men were working under, their individual physical make-up and condition, even the temperature both in and outside the caisson, were all important contributing factors. But it was the speed of the exit that really mattered. And by modern standards the men in the New York caisson were making their exits disastrously fast—even when doing as Smith wanted and taking a few extra minutes in the lock. Today the accepted safe rate of decompression is no less than twenty minutes for each atmosphere, or more than a minute for every pound of pressure. So by that standard, at a depth of sixty-five feet in April of 1872, every man coming up from the New York caisson should have spent at least twenty minutes in the lock, instead of two or three as was the average, or five or six as Smith urged.

That bubbles of nitrogen were the true cause of the dreaded disorder had already been discovered in France by a professor named Paul Bert, and at about the same time Smith was conducting his research. But the discovery would not be published until August of that year, and although Smith would read what Bert had concluded before formally presenting his own conclusions, he would decide that Bert was mistaken.

Smith was quite right about one other very crucial matter: how to relieve the agony of the disease. Just when the answer dawned on him is not clear, but it appears quite noticeably twice in his case notes.

The first time was in February, in the case of the foreman Joseph Brown, already quoted. "Pain continued until he went down again for the afternoon watch, when it ceased immediately." Then in April, Smith described the case of another foreman named Card, who was hit by an attack of extreme trembling, followed by paralysis in the legs and bladder. Smith writes that the man remained in this state for two full days, but then adds, "After the paralysis had passed off in a measure, he went down again into the caisson and remained for a short time with decided benefit."

The quickest, surest way to relieve the pain was to send the patient right back into compression and when it came time for Smith to present his final report the next year he would write this:

> It frequently happened under my observation that pains not sufficiently severe to deter men from returning to work were promptly dissipated on entering the caisson, to return again on coming into the open air. Indeed, I do not remember a single exception to the rule, that any pain which may have been felt before, disappeared almost immediately on going down.

Smith was aware, too, that Pol had prescribed returning the patient at once to the compressed air and that Dr. Antoine Foley, also of France, had said the same thing in a paper published in 1863. Later that same spring of 1872, in his own annual report, Roebling would write that most men got over their troubles either by suffering for a time or "by applying the heroic mode of returning into the caisson at once as soon as pains manifested themselves."

But the puzzling thing is that Smith never seems to have actually prescribed this "heroic mode." Not once, according to the records, was a man suffering from a violent attack of the bends taken back down into the caisson. It was only the man who felt fit enough to go down on his own, to *work*, who ever benefited from this simplest and most effective of all remedies. Smith's explanation was that the means of access to the caisson were such that to take any but a comparatively healthy patient down inside would have been too difficult—"even if he could be comfortably cared for while there, or if his presence would not interfere with the work." The remedy was just too much bother, he seems to be saying. It would have deterred progress the same as would more time taken in locking out.

But equally important, it seems, was the fixed idea most of the men had that the pressure itself was the cause of the trouble. They could not get rid of that idea. The thought of going back for more pressure when they were in their agony—of getting back on the horse, so to speak—was more than any of them were up to and particularly if neither the doctor nor the engineers in charge ever insisted on it. The less risky course seemed simply to hold on and suffer it out.

At a depth of sixty-eight feet the caisson's steady plunge into the earth slowed abruptly. The men were into quicksand now and the going became extremely tedious. The big clamshell dredge buckets dropping down the water shafts were almost useless against the fine sand that, in combination with small stones and boulders, had compacted into a substance about as hard as rock. The teeth of the buckets made hardly any headway at all. The point of a crowbar could be hammered into the material, but just barely. And though the sand pipes still "answered admirably," as Roebling put it, even they were constantly clogging with coarse gravel and stones. The speed of descent was now perhaps a foot a week.

At a depth of seventy feet Roebling ordered that daily soundings be taken for bedrock. So a couple of men with sledge hammers and a ten-foot iron rod began probing the work surface and among the others there was talk of the caisson going twenty or thirty feet more before Roebling would call a halt. Then at a depth of seventy-one feet the first death occurred.

On the morning of April 22, a heavy-set German, a common laborer, went down through one of the air locks and into the caisson for the first time. Two days before, when he applied for work, the man had given the name of John Myers. Dr. Smith had judged him to be about forty and in good health. The pressure by this time was thirty-four pounds, and the workday, shortened once again, was five hours.

According to Smith's subsequent account of the case, Myers worked the morning shift, just two and a half hours, without any discomfort, and hung about the yard for nearly an hour after coming up, apparently heeding the doctor's rules about rest. But then he had complained of not feeling well and started for his boarding-house, which was quite close by. "As he passed through the lower story of the house," Smith wrote, "on his way to his room, which was on the second floor, he complained of pain in the abdomen. While ascending the stairs, and when nearly at the top, he sank down insensible, and was dead before he could be laid upon his bed." An autopsy at the city morgue showed that brain, heart, and kidneys were perfectly normal. The lungs, however, as Smith reported, were "congested to a very remarkable degree."

On April 30, just eight days later, with the pressure still at thirty-four pounds, Patrick McKay, of Ireland, age fifty, was listed as the second fatality attributable to the caisson disease. McKay had been working in the caisson for four months with no ill effect. On the afternoon of the 30th he had stayed down a half hour longer than usual, and on his way out through the lock, the others in the lock saw him suddenly slump to the floor, his back against the iron wall, quite insensible. He was at once carried into the open air and taken to Park Hospital, where Smith looked in on him some time later that evening. "He was there in an unconscious condition," Smith wrote in his notebook; "face pale and dusky; lips blue; pulse irregular and feeble. Under the administration of stimulants, he recovered some degree of consciousness, and begged incessantly for water." But only a little later the man went into convulsions and died. This time, however, the autopsy indicated Bright's disease in the kidneys and Smith would conclude that "the effect of the compressed air was merely to hasten an event which, at best, could not have been very long delayed."

Be that as it may, the word was out—not just among the work crew, but everywhere in the neighborhood of the bridge—that men were dropping dead of caisson sickness. A third man who had died some time earlier of spinal meningitis was also said to have been a victim, "if the truth were known." The stories became greatly exaggerated and spread like wildfire through the crowded tenements near the site of the New York tower.

One of the children to grow up on South Street in the 1870's was Al Smith, who would one day be almost as much a symbol of New York as the bridge itself. In later years he would describe his mother talking in tones of awe about the many workers who had died while struggling to sink the great caissons. "Perhaps if they had known," she had said, "they would never have built it."

On May 2 a man named Heffner began vomiting and despite everything done for him he was still vomiting twenty-four hours later. On May 8 the entire force of caisson men went out on strike.

They stood about in the street nearby, talking to newspaper reporters and anyone else who would listen. Conditions below had become so dangerous, so terrifying, they said, that they wanted three dollars for a four-hour day. By noon or thereabouts the Bridge Company had agreed to $2.75, but the men turned that down angrily and a man who tried to break through their lines was badly beaten. Negotiations dragged on for another three days. But then William Kingsley announced that if the men did not all go back to work immediately he would fire every last one of them and with that the strike ended.

There were more attacks of the bends during the next week and the caisson kept descending little by little. From the soundings Roebling had ordered, a picture of the underlying bedrock had begun to emerge.

"The surface was evidently very irregular," he wrote, "composed of alternate projections and depressions, the extreme difference in elevations encountered being 16 feet, and occurring chiefly along the water edge." Throughout the central section, however, and covering at least two-thirds of the entire area, the irregularities were much less, amounting to maybe no more than three or four feet. As near as he could tell, the caisson was about to settle on a broken ridge of rock running diagonally from one corner of the caisson to the other and having a dip of perhaps five feet in a hundred toward the land, but falling off rapidly toward the east.

Roebling now faced what would be the most difficult decision of his career. He himself was very near to a physical collapse. He had been spending as much time in the caisson as anyone, but going up and down through the locks, to check on this or that below, many times more often than the average laborer. He was on the job constantly, working twelve to fourteen hours a day, six days a week, frequently making three and four trips a day on the ferry, going back and forth from the Brooklyn offices. Most of his time was spent on the site itself. But it was the only way he could have worked. He was not an office engineer and had little regard for those who were.

At this point he could either keep the caisson descending until he had level rock on which to leave it or he could stop about where he was, before reaching rock. To continue deeper would mean enormous expense and time lost blasting the irregular rock ridge down to a comparatively even surface. It might also mean more lives lost. Already Smith had recorded more cases of the bends than Jaminet had in St. Louis. And whereas Eads had not suffered a single fatality until his first caisson was down ninety-four feet and the pressure was at forty-four pounds, Roebling, for some unknown reason, had already lost two men. So at this rate the New York caisson might take even more lives than the thirteen the St. Louis foundations had cost by the time they were in place.

Emily Roebling would remark later that her husband estimated it would take another year to go to bedrock and that it would cost another half a million dollars and possibly a hundred lives.

To leave the tower standing on anything other than bedrock, however, would seem to put the stability of the entire bridge in jeopardy. Yet Roebling, to the surprise of many, was now not so sure about that. The sand and gravel covering the rock was so compact, so very hard, he said, that it might provide as solid a footing as rock itself. Earlier, when they were down sixty feet or so, the men had uncovered the bones of a domestic sheep, and just below that fragments of brick and pottery, indicating that the strata at that level had changed within the time man had been around. But in the last ten feet and more, no such evidence had been unearthed; the strata showed no signs of having been disturbed since the time of deposit several millions of years in the geologic past and so in all likelihood it would remain perfectly stable. As Roebling noted, it was now nearly impossible to drive in an iron rod without battering it to pieces. The material, he would write in his forthcoming report, was "good enough to found upon, or at any rate nearly as good as any concrete that could be put in place of it."

If he was right about this, then the enormous stone tower could rest there as well as anywhere and his problems would be solved. But if he was wrong, then there was the chance that the tower might begin to lean or slip and the bridge would be a disastrous failure. Possibly, others noted, the tower might even slide into the river.

"The period of time at the end of the sinking of the New York caisson was," his wife would say, "one of intense anxiety for Colonel Roebling."

But the decision could wait a little.

At a depth of seventy-five feet the first spur of bedrock, the ordinary gneiss of Manhattan Island, was encountered under the shoe on the river side. "No part of its surface shows the rounding action of water or ice," Roebling reported. "On the contrary, the outcrop is in the form of sharp thin ridges, with steep vertical sides occurring in parallel ranges."

On May 17 one man became paralyzed in the legs and arms; another complained of savage pains in his legs; a third, an Englishman named Reardon, began retching violently after coming up from the afternoon shift. In minutes he was seized by excruciating leg cramps and pitched forward, unable to walk or stand. The vomiting continued all night and Dr. Smith had him taken to the Center Street Hospital, where he grew steadily worse. The following morning he died. Smith wrote in his notebook that Reardon had been "corpulent" and that the autopsy showed his spinal cord to be "intensely congested."

That same day, May 18, 1872, with the caisson at a depth of seventy-eight feet six inches, Roebling ordered that the digging stop. He had decided *not* to go to bedrock, staking his reputation and career on the decision. The New York tower would rest on sand.

The second and last great caisson was therefore in position, and as Collingwood noted, the differences of level at the extreme corners, as measured on the masonry above, was only three-fourths of an inch. It had been a spectacular feat of engineering.

The work of filling the air chamber began at once and Roebling finished his report to the directors. If anyone was upset about the incidence of caisson sickness, Roebling said only that the trouble had not been so serious as he had anticipated. He made no mention of the number of cases there had been and claimed that just two deaths could be charged directly to the effects of pressure. As for the unsung individual suffering there had been, he said only this: "The labor below is always attended with a certain amount of risk to life and health, and those who face it daily are therefore deserving of more than ordinary credit."

At the end of May, Dr. Smith resigned his position and went back to the Eye and Ear Hospital, satisfied his work was complete, his services no longer needed now that the caisson was at rest. But work inside the caisson continued right along, the concrete for filling it in being mixed above, then let down through the supply shafts. No brick piers were built this time; the caisson was quite strong enough on its own. (With 53,000 tons on its back, it showed not the slightest sign of deflection in the roof.) But about a third of the space was filled with stones, earth, and sand left inside during the sinking. With the concrete going in at the rate of one hundred cubic yards a day, Roebling figured to have the entire job done by early July. The saying was that the concrete would keep pouring into the caisson until there was room enough left for one last Irishman, who would make his final exit by one of the water shafts.

But some time before that happened, Roebling suffered another attack of the bends. There is nothing in the official record to indicate just when it happened, only that it was late spring, while the concrete work was going on. Apparently he collapsed again, as he had the night of the Brooklyn caisson fire, and he was immediately taken back to Brooklyn on the ferry.

Who was on hand to help him is not known. There would be nothing said of the incident in the papers, suggesting that perhaps he and the others wanted no more adverse publicity than they already had or that they thought the seizure would soon pass. He himself made only the briefest mention of what happened in a report published later that fall. The attack, he said, resulted from a stay of several hours in the caisson, suggesting that he still believed the time spent below was the determining factor and had never accepted Smith's theory on speed of decompression. "Relief from the excruciating pain," Roebling wrote, "was afforded in his [the writer's] case by a hypodermic injection of morphine in the arm, where the pain was most intense, and a further stupefaction by morphine, taken for twenty-four hours internally until the pains abated." According to Emily Roebling, however, in an account written a few years later, his condition was so serious the night of the attack that he was expected to die before morning.

There is no telling whether Smith was called back, whether the idea of returning to pressure ("the heroic mode") was even considered, or if so, why it was rejected in favor of drugging the patient into a stupor.

For several days more Roebling lay near death in the same Hicks Street house where his father had died. His assistants came and went. Somebody was with him at all times. Little hope was held out for him. In some of the things written about him a generation later, it would be said that Roebling remained painfully paralyzed, a total invalid from this point on. But the record shows this was not the case. In another few days, much to everyone's amazement, he went back to work.

Once when he was seventeen, his father had been faced with a cholera epidemic at Niagara Falls. More than sixty people had died in the first week and the doctors seemed incapable of doing anything to help. "The great secret," his father had written to Charles Swan, was to "keep off fear." His father, too, would have succumbed with the rest, according to one man who was there, had it not been for his uncommon powers of concentration. "He determined not to have it," the man wrote. John Roebling had spent one whole night walking up and down his room, fighting to rid his mind of the very thought of cholera. The incident made an enormous impression on the gentleman who witnessed it and on everyone back in Trenton when the story was told there. Now it seems Washington Roebling too had "determined not to have it." Other men might resign themselves to their fate, he could not.

Through the first weeks of summer the attacks kept recurring, however, and he suffered intensely. He made no public mention of this, nor did anybody else. It is only from comments made in private correspondence years afterward that anything is known of his extreme physical suffering. To judge by the Bridge Company's record books and occasional items in the papers, he was carrying on as though nothing were the matter. It was during this time, for instance, that his report exonerating the management of Bridge Company purchases was read before the board, and knowing this, one cannot but wonder if his physical and emotional torment, the anxiety Emily described, did not have something to do with the discrepancies between that report and some of the things he would say privately much later on.

On July 12 the filling in of the New York caisson was completed and apparently under the personal supervision of the Chief Engineer. The whole task of sinking the caisson had taken 221 days.

He took two weeks off and went with Emily to Saratoga. He was somewhat improved when they returned but that lasted only briefly. By September he was staying home two and three days a week. Still his condition remained a private matter. To judge by the official records and items in the papers the Chief Engineer was very much on the job.

On September 3 bills amounting to $50,000 were ordered paid, on being certified by the Chief Engineer. On September 17 the Chief Engineer was directed to solicit bids for the anchor bars for the New York anchorage and the Chief Engineer and the General Superintendent were authorized to award the contract to the lowest bidder. On October 8 the Executive Committee authorized the president of the Bridge Company to execute a contract with the Bodwell Granite Company of Maine, according to specifications prepared by the Chief Engineer. An agreement made by the Chief Engineer with Louis Osborne of East Boston, for building an expensive double hoisting engine, was also approved, and a number of substantial bills were ordered paid, after being certified by the Chief Engineer.

In late November it was Roebling who ordered that work on the Brooklyn tower be suspended for the winter. The tower by then had reached a height of about 145 feet, or well beyond the level where the bridge deck would be. It was no longer a solid flat-topped shaft. Now the beginning of the great archways could be seen thrusting upward like three immense teeth separated by the two gaping spaces left for the roadways.

It was in December, the same month the Committee of Investigation presented its findings, that work on the New York tower was halted, on account of the weather, at a height of nearly sixty feet. And it was in December that Washington Roebling found he was unable to go down to the bridge anymore. His condition was very serious now, extremely puzzling, and a closely guarded secret among the relative handful of men who were running things inside the bridge offices.

The sudden, violent cramps, the awful dizziness and vomiting had ended after the first horrible days in early summer, just as had been the experience of every other victim from the caisson, indeed as had been his own experience the time before. But the pains and the numbness had continued, coming and going, in his arms and legs primarily. He tired rapidly. He was sick at his stomach much of the time. He became extremely irritable and distraught over the slightest problems or inconveniences and slipped into moods of profound gloom that lasted for days. By December he was a very sick man. Still, he refused to give up. "He was never known to give in or own himself beaten," the men at the mill had said of his father.

Emily Roebling went to see Henry Murphy, to talk privately about the situation. Her husband was determined to continue as Chief Engineer, she said. Murphy told her that that would be agreeable with him, just so long as nothing went wrong at the bridge. She expected his troubles would last but a short time.

Total rest was the only cure prescribed for him through that winter, and Emily was apparently about the only person he wanted anywhere near him for any length of time. The doctors kept telling her that he had little chance of recovery, that she should be prepared for the worst, while he himself had become obsessed with the idea that he would not live to see the bridge finished. And knowing better than anyone how incomplete the plans and instructions for the remaining work still were, he spent that entire winter writing down, in his minute, meticulous hand, all that had to be done, filling page after page with the most exacting, painstaking directions for making the cables, for assembling the complicated components of the superstructure, and illustrating these with detailed freehand drawings and diagrams.

There was no work going on at the bridge all this time, other than paper work at the Fulton Street offices. Snow piled up in the yards. The two towers stood idle on either side of the ice-choked river. But Roebling in his bedroom on Hicks Street labored on, fighting with everything he had. In his condition writing for even half an hour was a terrible strain. He became extremely nervous and found he could no longer carry on an extended conversation with his assistants, who had been reporting regularly for instructions. His eyes began to fail. He thought he was going blind.

By early spring, when the weather was such that the men could return to the towers, it was common knowledge among the bridge workers that the Chief Engineer would not be resuming his command for some time, if ever. It had been decided that C. C. Martin would be authorized to certify bills and the masonry work would proceed as before. In April Roebling formally requested a leave of absence. His doctors had told him his only chance to live was to get away from his work. He and Emily had decided to go to Europe, to the health baths at Wiesbaden. The trip would be a frightful ordeal for him, but in this the darkest time he had ever known, he would turn for relief to Germany and the water cure. His feelings at the time can only be guessed at. But possibly in quiet desperation, everything else having failed, he had concluded that if the ways of his father had put him in this corner, then perhaps they might get him out as well.

Later that same spring of 1873, Dr. Andrew H. Smith presented the formal report on his experience as Surgeon to the New York Bridge Company, in which he included certain suggestions for future projects of a similar nature. The most important thing to do, he said, was to have some sort of facilities by which compressed air could be readily accessible *above* ground.

> My plan would be as follows: Let there be constructed of iron of sufficient thickness, a tube 9 feet long and 3½ feet in diameter, having one end permanently closed, and the other provided with a door opening inward, and closing airtight. This tube to be placed horizontally, and provided with ways upon which a bed could be slid into it. Very strong plates of glass set in the door and in the opposite end would admit the light of candles or gas jets placed immediately outside. This apparatus should be connected by means of a suitable tube with the pipe which conveys the air from the condensers to the caisson. An escape cock properly regulated would allow the constant escape of sufficient air to preserve the necessary purity of the atmosphere within.
>
> The bed containing the patient having been slid into the chamber, the door is to be closed, and the pressure admitted gradually until it nearly or quite equals that in the caisson. This should be continued until the patient indicates by signal, previously concerted, that the pain is relieved. The pressure should then be reduced by degrees, carefully adjusted to the effect produced, until at last the normal standard is reached. By occupying several hours, if necessary, in the reduction of pressure, it is probable that a return of the pain could be avoided.

The concept of the apparatus described by Smith in this proposal is precisely the same as the so-called "hospital lock" used for modern bridge and tunnel construction, whenever men are working in compressed air. Had Smith installed such a device at the New York caisson, and had it been used in the way he describes, there would have been little or no suffering from the bends by anyone, there would have been no deaths, and the subsequent life of Washington Roebling and the story of the bridge would have been very different.

PART THREE

15
At the Halfway Mark

Everything has been built to endure.
—FRANCIS COLLINGWOOD

IT TOOK three more years to complete the towers. The Brooklyn tower, started a year earlier than the other one, was finished a year before, in June 1875. And the last stone on the New York tower was set in July 1876, right when the country, with surprising exuberance and a seemingly insatiable delight in fireworks, was celebrating its one hundredth birthday.

Times had changed considerably. The terrible panic of 1873 had struck, worse than any ever before. The country had still not recovered from it and would not for some time. Incredibly, Jay Cooke & Company, the most famous banking firm of the day, had gone bankrupt. In no time one Wall Street firm after another had gone under. Thousands of businesses, mostly small ones, had been wiped out and thousands upon thousands of working people lost their jobs. The streets of New York had been filled with drifters and unemployed ever since.

But for all that the country kept growing, moving ahead. The poor and the hopeful kept streaming in from Europe, landing in New York, passing by the hundreds of thousands through the shabby, makeshift clearing point at Castle Garden.

Most Americans were anything but dissatisfied with the times.

Simple, ingenious devices were coming along one after another, changing the way people lived and the look of the land in the most astonishing fashion—barbed wire and ready-made windmills for settlers on the Great Plains, to name but two.

In Hartford, Connecticut, Mark Twain was busy working on *Huckleberry Finn.* Edison was working on electric light at Menlo Park, New Jersey. Carnegie had built the Edgar Thomson works, the biggest steel mill on earth, at Braddock, Pennsylvania, and was producing Bessemer steel in quantities unheard of at the start of the decade. Big corporations were growing bigger, and though some railroads were going bankrupt, other lines kept right on expanding. More and more railroad tunnels and railroad bridges were built, including the nation's most celebrated new bridge and its longest tunnel.

In St. Louis, Eads had finished his work. The bridge was hailed as an engineering marvel, which it was. A placard over one arch read: "The Mississippi discovered by Marquette, 1673; spanned by Captain Eads, 1874." "The love of praise is, I believe, common to all men," Eads had said in a speech that day, "and whether it be a frailty or a virtue, I plead no exception from its fascination."

Then in Massachusetts, later that same year, on Thanksgiving Day, the record-breaking Hoosac Tunnel had been completed, nearly twenty-six years after it was first begun, and at a cost of ten million dollars and an appalling 195 lives, most of them lost because of inexperience in using nitroglycerin.

In New York the nation's tallest office building had been completed, the ten-story Western Union Telegraph Building, designed by the architect George B. Post. With its tower rising 230 feet above Dey Street, it was still nearly 50 feet less than the top of the New York tower of the great bridge.

Mayor Havemeyer was dead and Tweed was out of jail and on the loose somewhere. One morning in December 1875, Tweed had left the Ludlow Street Jail, with his son and two guards, to take a carriage ride. Tweed had wanted some fresh air. On the way back, they had stopped off at his house and were sitting in the parlor when he asked if he might go upstairs to see his wife. The guards had agreed and that had been the last anyone had seen of Tweed.

Across the river, beside the Brooklyn tower, a new ferryhouse had been built, a costly expression of confidence in the future of the ferry system, bridge or no bridge. The building stood where the old one had, at the foot of Fulton Street, but it was a much more elaborate affair, with tall mansard roofs, a particularly elegant cupola, and in a niche over the main entranceway, a life-sized statue of Robert Fulton now gazed impassively over the throngs of commuters that swarmed in and out below. New business buildings had gone up in Brooklyn. New industries had been started, whole new residential sections had been built. The sound of hammer and nails was still one of the characteristics of the place.

But the event that had stirred people up more than any other, more really than anything that had happened in Brooklyn since the war, was the Beecher-Tilton Trial, which began in the City Courthouse in January 1875. Theodore Tilton had decided to bring suit against Beecher for alienation of his wife's affections. The show lasted six months and was the talk of the country. Beecher denied every charge against him. Not until late June, when the Brooklyn tower was being finished, did the jury retire. Then eight seemingly interminable days went by—the summer heat stupendous—before the jury returned to report no decision. Immediate tumult had broken out in the courtroom, with Beecher's parishioners rejoicing as though he had been completely vindicated.

Whether Beecher was guilty of adultery with Theodore Tilton's shy and decidedly neurotic little wife would never be proved. Beecher had turned out to be an inconsistent, fumbling witness. His testimony, like nearly everything said at the trial, had been treated by the papers as news of the highest importance, and a large part of the American public, not to mention the populace of Brooklyn and New York, had concluded he was just as guilty as could be.

Writers of later generations would decide that Beecher was absolutely guilty and, in general, a posturing fraud. He would be portrayed as the prime example of Victorian hypocrisy and his trial described as a watershed in the nation's social history. But it does not appear as though the people of the time saw it quite that way. For the millions of Americans who had read the word from Plymouth Church, week after week, for nearly thirty years, taking it as very nearly the word of the Almighty, Beecher's fall was assuredly a shattering blow. Still, nobody seems to have discarded his religion overnight, or his notions of right and wrong, because of what the Reverend Henry Ward Beecher allegedly did with Mrs. Tilton. Moreover, a great many people who thought Beecher might be guilty after all would continue to regard him as an extraordinary human being and felt he had suffered more than enough. And a very great many more people would believe in his innocence until their dying day. As a matter of fact, it is altogether possible that Beecher *was* innocent.

But apart from all that, something had very definitely happened in Brooklyn to the way people regarded the place. Brooklyn's cherished reputation for respectability had suffered irreparable damage. The name Beecher could be an emblem no more. Rightly or not, the man had been pulled down by a running tide of public sensation over what a pamphlet of the day called "Wickedness in High Places." No longer could the East River be viewed, as it had by many on the Heights, as the great dividing line between good and evil. No longer could Plymouth Church be regarded as the symbolic center of Protestant American virtue or as Brooklyn's answer to Tammany Hall. So by the time the talk had turned to other momentous events of the day, to the great Centennial Exhibition in Philadelphia, for example, Brooklyn had but one monument to take pride in. Brooklyn now had its bridge. The plain, barnlike church, never much of an architectural beauty, had been replaced by the immense Gothic bridge, which had been under way for seven years now, but still had a long way to go.

Building the bridge without the Chief Engineer on hand to direct things had posed no serious problems. The plans were very clear, his written instructions quite thorough, to say the least. Nor was there anything especially novel or complicated about the work. Most of it was straightforward masonry construction, the only significant difference being the immense size and height of the towers. (The towers were actually eight and a half feet higher even than John Roebling had initially said they would be. This was one of several changes that had been made as the work progressed. The height of the completed towers above the water was 276 feet 6 inches.)

But Roebling, it must also be remembered, was served by an exceedingly able staff of assistants, all of whom had been on the job since the beginning and each of whom had developed an uncommon loyalty both to the work and to the Chief Engineer.

". . . probably no great work was ever conducted by a man who had to work under so many disadvantages," Emily Roebling would write in time to come, when her own role in the engineering of the bridge had become the talk of Brooklyn. "It could never have been accomplished but for the unselfish devotion of his assistant engineers. Each man had a certain department in charge and they worked with all their energies to have the work properly done according to Colonel Roebling's plans and wishes and not to carry out any pet theories of their own or for their own self-glorification."

Martin, the senior man among them, efficient, pleasant, colorless, was still serving as Roebling's stand-in, supervising the work overall. Collingwood, the Elmira jeweler whose initial intention had been to stay with the job only a month, was now assigned to completing the Brooklyn tower, while the inventive and reliable Paine had charge of the New York tower.

Once the Brooklyn tower was finished, Collingwood went across the river to take charge of the New York anchorage, that four-square masonry pile, which, with its pair of deep arches, looked like the beginnings of a Roman bath. The anchorage was already as high as an eight-story building. It stood nine hundred feet inland from the New York tower and filled most of the block between Cherry and Water Streets. Four great cables descending from the tower would be secured to this mass on top, up near the end closest to the tower. The cables from the tower to the anchorage would carry what would be known as the land span of the bridge. They would be fixed to chains of huge iron bars that disappeared into the anchorage and were, in turn, tied to four great cast-iron anchor plates embedded deep within the granite, in the heel of the anchorage, as it were, down near the level of the street.

The dimensions of the New York anchorage and that of the one in Brooklyn were 119 by 129 feet at the base and 104 by 117 feet on top. For the time being the two structures stood just over eighty feet high, bristling on top with the same sort of timber derricks and other stone-hoisting apparatus used on the towers. Once the cables were finished, more stone would be added, taking the level up to that of the roadway, or to nearly ninety feet. The final total weight of each anchorage would be 120 million pounds, or sixty thousand tons.

Had either one of them been built someplace elsewhere, it would have been regarded as most imposing and awe-inspiring, in the way the Brooklyn tower had been when it loomed up all alone at a height of eighty feet. But standing in line with one of the towers, an anchorage did not look like much. The towers attracted all the attention, understandably. Still, in terms of the engineering involved, the anchorages were extremely interesting and their importance to the bridge was very great indeed.

Roebling wrote that there were but two factors to deal with in the anchorages—granite and gravity. The first he described as "a material whose very existence is a defiance to the 'gnawing tooth of time' "; the second he called the only immutable law in nature; "hence, when I place a certain amount of dead weight, in the shape of granite, on the anchor plates, I know it will remain there beyond all contingencies." *

The anchor plates—four to an anchorage, one for each cable—had been set in position during the early stages of the stonework. Their general shape, as Roebling said in his specifications, was that of an immense oval spider, 16 by 17.5 feet and 2.5 feet thick. They weighed 46,000 pounds, or twenty-three tons, apiece and just getting them into position properly had taken some doing.

The plates rested flat at the bottom of the stone mass, like mushroom-style ship anchors standing upright. Each of them had been cast with two parallel rows of nine oblong apertures into which eighteen of the iron eyebars were placed, set perfectly erect, making thereby two identical upright rows of nine bars each. The anchor bars, as they were called, stood twice as tall as a man and had an eye at each end. They had each been forged in a single piece, smooth, flat, and exactly like the others to be used in the same position—no easy trick at the time.

(Roebling had had to decide whether to use iron or steel for anchor bars. During his visit to the Krupp works at Essen, in 1867, the managers had forged an all-steel prototype for his inspection but could not guarantee the quality he wanted. So he settled on wrought iron. Several different mills produced the bars—the Keystone Bridge Company, the Edge Moor Iron Company, the Phoenix Iron Company—and as William Kingsley wrote, the finished products were considered "splendid specimens of what American manufacturers can do.")

Below the underside of the anchor plate, through the nine eyes of each row, all matched in position as one, big steel pins were inserted and drawn up against the plate, fitting into semicylindrical grooves, and thereby forming the first link, a double-tiered link, of a gigantic

* The anchorages were in fact built entirely of limestone, with the exception of the corners, front arches, and the cornice. There was also about 650 cubic yards of granite placed directly over the anchor plates.

double-tiered eyebar chain that extended up through the masonry in a gradual arc until it surfaced on top.

The anchor bars were of slightly different sizes, depending on their position in the chain, but they averaged twelve and a half feet in length, and in the first three links—those nearest the anchor plates, where the pull from the cables would be felt the least—they were seven inches wide and three inches thick, swelling enough toward the ends to compensate for eyeholes five to six inches in diameter. The bars of the fourth, fifth, and sixth "links," however, were increased in thickness, to eight by three inches, and from there to the top, as the bars became horizontal and so came directly in line with the pull of the cable, they measured nine by three, except for the last link, where the number of bars was doubled and their width was halved. The last link had in all thirty-eight bars, in four tiers, to catch hold of the cable wires. Washington Roebling had spent months working out the entire arrangement.

The anchor bar chains had grown apace with the masonry. Once the anchor plates were fixed in position, the stone was built over them and close about the first set of anchor bars in each chain. Then the second set of bars was put in place, the eyes of the new eighteen fitting into those of the first eighteen, and heavy pins were put through all of them, making joints like two parallel door hinges. Each new set of bars after the first two was then made to incline forward, toward the tower, a little more than its predecessor, forming a steady curve, or arc, so adjusted as to bring the end of the chain out on top of the anchorage at exactly that point where the end of the cable would be coming in—which was about twenty-five feet back from the edge facing the tower. By the summer of 1876 the huge iron bars, painted with red lead to guard against rust, were protruding out of the upper surface of each anchorage, ready to take hold of the immense load of the cables and looking, as someone said, like the clutching fingers of a giant imprisoned within the stone.

Wilhelm Hildenbrand was to design the approaches leading to the two anchorages. This assignment in itself involved a series of structures nearly half a mile long, all told, and a total of nine different stone or iron girder bridges to span the intervening streets. The amount of work required for these structures alone was enormous, as Roebling explained to the Board of Trustees.*

* In 1877 a group of architects would be called in as consultants on Hildenbrand's plans. The best known of them was George B. Post, who was then de-

The Brooklyn anchorage, begun in 1873 and finished two years later, was the responsibility of George McNulty, who by this time had also managed to grow an imposing handle-bar mustache. Though not yet thirty, McNulty was unquestionably one of the ablest men on the job. Roebling had assigned him to preparing hoists, drums, wheels, and other mechanical paraphernalia needed for cable making, all of which had to be made up exactly as Roebling wanted and mounted on top of the Brooklyn anchorage. As with nearly everything else about this bridge, the cables would originate in Brooklyn.

Since McNulty, along with each of the other assistant engineers, had had no previous experience building suspension bridges, every step after the stonework was a new one and there might have been costly delays or mistakes had it not been for Roebling's extraordinary written communications and for Master Mechanic E. F. Farrington, the one and only man among them who had ever worked with wire before.

The cable-making machinery was to be essentially the same as what had been used at Cincinnati, which Farrington had helped set up under Roebling's direction. Farrington was the one now who could train the men to do the all-important wire work out over the river, ship riggers many of them, but few of whom had ever before seen anything of this kind. Farrington had helped Hildenbrand do up a set of finished plans for the footbridge that was to swing from tower to tower and had built an amazing scale model according to Roebling's directions. The model was set up in a big room at the Bridge Company, where the men could gather about it. The towers were of wood and stood about five feet high and fifteen feet apart and from them were suspended small steel wires, miniature wheels, "cradles," the footbridge, everything exactly as it would be. Each part of the model was marked with a tag, explaining its function, and everything worked as it would once the cable spinning began.

Farrington was, in fact, the single person on the job who could speak from experience about the work that lay ahead and who had the clearest idea of the problems there could be.

In the time since Roebling's departure for Europe, the stonework had proceeded more or less as expected—slowly, gradually, stopping altogether during the winters. There had been a few brief,

signing a lofty new Queen Anne-style home for the Long Island Historical Society, at Pierrepont and Clinton Streets, and who would later do the New York Stock Exchange (1903).

unexpected delays—because quarries failed to deliver on schedule and once because funds ran low—and there had been more accidents. But generally speaking the towers had gone up about as smoothly and efficiently as could have been hoped.

The stone came from some twenty different quarries and of the thousands of shiploads sent only one was lost. The stone was unloaded at the yard at Red Hook and from there came up to the bridge on big scows that tied up at the tower docks. Then one by one, as needed, individual blocks were picked out of the scows by derricks and placed on small flatcars that ran on rails laid in various directions around the foot of the tower. A couple of men would roll the block around to the back of the tower and from there it would be hoisted up onto the tower, but no longer by the fifty-foot boom derricks mounted on top. The elevation was too great for that now. Instead the system worked this way:

On top of each tower, in the center and projecting over the edge, were two huge iron pulleys, in line with which, and running the full height of the masonry, were heavy timbers, laid up like tracks, to take the chafe of the block as it made its ascent. A steel wire rope, an inch and a half in diameter, attached to the drum of a powerful steam hoisting engine on the ground, passed first through one pulley, then down to the ground, then up again, through the other pulley, and down to the drum of a second engine. To both of the vertical sections of this continuous rope, running up and down the side of the tower, were attached big hooks that engaged the iron eyebolts inserted in the ends of each granite block. The hooks were fastened, in other words, so that one set was coming down while the other went up, or so that raising one block lowered the hooks needed to lift the next block in line.

Sometimes, when a particularly heavy block was being hoisted, the vibrations on the ropes, caused by the straining pulsations of the engine, became so violent that the block would have to be lowered again and extra rigging attached to it. Only once, however, did a block shake loose and fall, from two hundred feet up, demolishing the tracks below and burying itself halfway in the ground.

As each stone got to the top of the tower, it passed between the rails of a track laid lengthwise along the edge on a timber superstructure. When the stone cleared the track, another flatcar was shoved under it and the stone was unhooked and moved quickly to where a boom derrick could pick it up and swing it into place. On the arches of the Brooklyn tower, the keystones, huge blocks weighing eleven tons, were fitted in without any trimming, just as they came from the quarry.

The top of a tower was an extremely busy, crowded place, with perhaps eighty men at a time working up there. Every man had to know just what he was about. A stiff breeze blew nearly all the time, it seemed, and in late fall or early spring it blew bitterly cold. One November a magazine editor who ventured no farther up the Brooklyn tower than the base of the great arches wrote that from the finished span "a prospect will be afforded which, for grandeur, will have no rival in the world," but doubted there would ever be many sight-seers in wintertime.

Master Mechanic Farrington later wrote this memorable description of an early morning on top of the Brooklyn tower:

> There are times when standing alone on this spot, one feels as completely isolated as if in a dungeon. Some three years ago I had an experience of this kind by daylight. It was in the early morning, when a dense fog covered the whole region, that having occasion to examine some machinery, I went on the tower before the time for commencing work. I shall never forget that morning. I found the fog had risen to within twenty feet of the top of the tower, and there it hung, dense, opaque, tangible. It was what you might seem to cut with a knife. It seemed that I might jump down and walk upon it unharmed. It looked like a dull ocean of lead-colored little billows; vast, dead, immovable.
>
> The fog seemed to follow the conformation of the ground, rising to a certain height above it in all directions, and obscuring all below that line. The spires of Trinity in New York, and in Brooklyn, and the tops of the masts of a ship in one of the dry docks, with the roof of the bridge towers, were all that were visible of the world below. Here and there where the heat from the furnace chimneys rarefied the air, white cones would rise like boiling springs, and I could in one direction trace the cautious movement of a steamer by the same means. Rising through this misty veil was the confused crash and roar of busy life below.
>
> By and by the heads of the workmen began to appear, as they clambered up the stairway . . . The fog lost its density. A thin vapor seemed to rise from it—a fog upon a fog—like a mist from the ocean, and the whole began to settle and to melt away. Spires, masts and chimneys began to appear; boats were seen dodging about like porpoises, just below the surface of the mist. By 10 o'clock the fog had disappeared, and travel, which had been seriously interrupted, was resumed.

Perhaps the biggest problem on top was making signals to the engineers in the yard below to prevent overwinding of the hoisting engines. Shouting seldom worked because of the wind. At times fog made flag signals impossible, and a system of signal bells that had been rigged up was constantly breaking down. Men looking out over the edge had to be prepared always for sudden gusts of wind that could throw them off balance.

Thus far three men had been killed in falls from the towers. The engineers told reporters later that every precaution had been taken against such accidents. Anyone who experienced the slightest giddiness on top was immediately ordered down again and assigned to ground work. But a protective railing of the sort critics of the work had been calling for would have been more trouble than it was worth, the engineers said.

In 1875 a man named Reed and another named McCann fell from the Brooklyn tower and were killed instantly. Reed, it appears, had been subject to epileptic fits, a fact he had concealed when he applied for a job. A man who had been working nearby claimed he heard Reed groan just before he fell.

McCann had been standing at the edge, on the corner of one of the buttresses, as a box of mortar was about to be raised by a derrick. Instead of walking around the right angle of the corner, he jumped across it. But at the same instant, the mortar box swung into the air and hit him, knocking him beyond the edge of the tower. He fell the distance.

And it had been just that May of 1876 that John Elliot fell from the New York tower. His job had been to shove the flatcar under the blocks as they arrived on top. His foot slipped and he went through the opening in the track. He struck a projecting beam on the way down and landed inside one of the arches.

But these had not been the only horrible deaths. A man named Cope, who had the duty of guiding the rope onto the drum of a hoisting engine, tried to kick the rope into place when it was not winding to suit him. He had been shown how to do the job properly, where to stand and so forth, but ignored what he had been told. He kicked and he missed his mark. The rope caught his foot and wound his leg around the drum. His leg was crushed so badly that he died almost immediately.

Another man was crushed by a block of granite that struck him in the stomach. One of the carpenters was killed by a falling stone. And a man at work somewhere near the base of the Brooklyn tower was rolling a wheelbarrow loaded with dirt across a plank at a fairly considerable elevation, when, by accident, the barrow ran off the edge of the plank. Instead of letting go, he held tight to the barrow handles, falling to his death.

But there had been numerous narrow escapes as well, and one in particular would be talked about for years to come. Near the completion of the Brooklyn tower a man named Frank Harris (not the one who attained literary notoriety) fell head over heels down into one of the hollow spaces between the three main shafts of the tower —a drop of 186 feet, according to later accounts. His companions, assuming he was dead, started down for the body. But then they heard Harris calling for somebody to lower a rope. He had landed on an empty cement barrel floating in about three feet of rain water that had collected in the pitch-dark stone well and he had received only minor injuries. Eight days later he was back at work on top of the tower.

By the time the towers were finished, the bridge had already taken the lives of a dozen men, but in the early summer of 1876, the engineers were telling reporters that from then on the work would get even more dangerous. It was the sort of statement to double public interest overnight.

Since the summer the bridge began, there had never been a time when the public was not interested in it. But things were different now, with the two gigantic towers facing each other from opposite shores. There was no longer any problem picturing the immense scale of the highway to be slung between them. Nor did it seem very likely now that anything could bring the work to a halt. In the spring of 1876 there had been a much publicized hue and cry from some of the shipping interests. It was claimed the bridge would obstruct traffic on the river. Public hearings had been held; several warehouse owners had spoken out vehemently, calling the bridge a "nuisance" and claiming it would "cramp the commerce" of the river port. But a shipmaster named Leavy had argued for the bridge with great effect. It was too late in the day to start objecting to the bridge, he said. He said he never had had any trouble striking top spars on a ship; indeed, to those shipowners who were claiming it would cost five hundred dollars to strip a ship to pass under the bridge, he said it would be a pleasure to do the job for them at that price. Then he finished his speech by asking, if the bridge were not finished now, "What were they to do with the towers?" It was a question that appeared to dismiss the whole issue, which it did not, as things turned out, but for the moment the bridge had passed through one more trial, intact and in good part because the colossal twin towers seemed to provide a certain psychological momentum it had not had before.

To have the towers completed the summer of the Centennial also seemed especially appropriate, and particularly since the bridge now belonged to the people, quite literally.

On June 5, 1874, nearly two years after Demas Barnes first proposed changes in the original charter, the state legislature in Albany passed an amendment requiring that the cities of Brooklyn and New York be given increased representation among the directors of the Bridge Company. The mayor and the comptroller of each city were to pick eight directors and the mayor and the comptroller of each city were themselves to be directors. The bridge was formally declared a public highway and given a legal name at long last, The New York and Brooklyn Bridge. Then in May of the following year an act was put through dissolving the New York Bridge Company entirely and redefining the bridge as "a public work, to be constructed by the two cities for the accommodation, convenience and safe travel of the inhabitants . . ." Two-thirds of the cost was to be met by the city of Brooklyn, the remaining third by the city of New York. Private stockholders were reimbursed for their previous payments, with interest, and their title was extinguished.

But once again the old management survived—Murphy as president of the trustees (instead of the old Board of Directors), Kingsley now a member of the Executive Committee instead of superintendent. There was no change either in the engineering staff. So except for Roebling's physical absence from the scene, the cast of characters was no different from what it had been at the start.

Most of the business of the company was still being conducted behind closed doors. But like the immense, irrefutable presence of the granite towers, a law on the books making the bridge the possession of the people did something important to the way the people felt about the bridge.

Early in July of 1876 the New York papers carried accounts of a new apparatus constructed by Colonel Paine at the New York tower by which he could test the strength of steel wire. In Brooklyn the *Eagle* ran long, optimistic articles titled "The Present Condition of the Work." "Before winter shall drive the workmen from their positions," the editors wrote, "we shall see the first strands of the great cable stretching aloft, spanning the river." A contract for 120 tons of steel wire ("best quality") was being filled by the Chrome Steel Company of Brooklyn. The bridge, the *Eagle* reported, would be finished in three more years and added that the engineers in charge were the very best in the business. "One thing is certain, the Bridge Company have been exceedingly fortunate in securing the services of professional gentlemen who are without peer in their respective fields and whose talent and genius have enabled them to surmount every obstacle . . ." The leadership was all in the plural now. There was no mention of Washington Roebling.

Presently, about the middle of July, Paine reported that the gigantic saddle plates were in position, each weighing 26,000 pounds, or thirteen tons. There were eight of them, four mounted on top of each tower. They were the bases for the big iron saddles, so called, on which the cables would ride. The saddles went up soon after that. They stood about four feet high, rested on rollers, and were elliptical in shape, with a groove on top, about the size of a barrelhead, in which the cable would sit. Each saddle could work to and fro on the rollers, according to the pull of its cable, and thereby alleviate any lateral strain on the tower.

The movable saddles, like the big expansion joints that were to be built into the actual roadway, were essential to the stability of the bridge. Its capacity to move, the fact that it would *not* be perfectly rigid like a stone bridge, was the thing that would keep it alive, as the engineers said.

Then in early August, when everything seemed to be in order, it was announced that the first wire would be taken across. The bridge was half built.

16
Spirits of '76

DUTY—That which a person owes to another; that which a person is bound, by any natural, moral, or legal obligation, to pay, do, or perform.
—As defined in an 1856 edition of Webster's *American Dictionary of the English Language* belonging to Washington Roebling

FOR THE public the exact whereabouts of the Chief Engineer was a matter of considerable mystery. It was known that he was in a bad way, but nobody seemed sure just where he was or what was wrong with him or how much say he had in the bridge anymore. The papers did nothing to clear up the rumors.

Much nonsense would be written about Roebling in time to come. The impression given would be that he was still in Brooklyn all the while, living in a house overlooking the river, where, from an upstairs window, he kept watch over every move made at the bridge, sending his wife back and forth to tell the men what to do. But this was not the case, not during this particular stage in the story.

Roebling's original intention, it seems, had been to stay in Wiesbaden only a month or two. But he and Emily had stayed on in the old resort on the Rhine for nearly six months, hoping against hope that the warm alkaline springs would work a transformation for him. Not until late in 1873 did they return to Brooklyn and then they stayed only long enough to purchase a new house on Columbia

Heights, on the river side of the street, with rear windows overlooking the bridge, which was about half a mile away.

The journey to Wiesbaden had been to no avail and early in 1874, with the work at the bridge shut down for the winter, his doctors were urging still another change of scene. Roebling left Brooklyn for Trenton this time and there he stayed for nearly three more years.

So the entire time the towers were being finished, the anchorages built, the cable-making machinery assembled and set in position, the Chief Engineer was nowhere near the bridge and could see nothing of it. And in light of this fact his achievements seem all the more phenomenal, for a vigilance from Trenton was an even more extraordinary feat than it would have been from a bay window on Brooklyn Heights.

As it was, the day-by-day progress of the work, the changes in procedure and equipment, the advance preparations for the very different kind of work to come, all went on in his mind, supported only by letters from his assistants, or from Henry Murphy on occasion. His own orders and instructions had to be issued by return mail. The elaborate, formal specifications now required for all materials purchased he also drew up himself—an enormous task.

It was well after he left Brooklyn, for example, that he did the specifications for the granite for the New York tower, for the face stone, arch stone, and spandrel courses.

> . . . Above the arch is the spandrel-filling of varying thickness of courses, and covered by a broad band-course at the line of the keystone. The space between the keystone and the cornice is occupied by a recessed panel . . . The interior space above the spandrel-filling is not all solid, but consists of three parallel walls, separated by two hollow spaces. The middle wall is 4 feet 2 inches thick, the outer ones vary from 4 feet 2 inches to 5 feet 3 inches in thickness, and the width of the hollow spaces varies from 4 feet 3 inches to 4 feet 9 inches . . .

He described precisely how the stone should be cut and joined, how it should be unloaded at the dock, the requirements for delivery. This particular set of specifications was prepared in the fall and the winter of 1874, but at about the same time and shortly thereafter, he also drew up complete specifications for the granite and the limestone backing for the New York anchorage, for the anchor bars and anchor plates, the saddles and saddle plates, and for the several varieties of wire rope needed (steel footbridge rope, iron handrail rope, iron ropes for guy wires under the footbridge).

He had help with all this, of course, from whichever assistant he had assigned to that particular part of the work in question. Still, he had to provide general guidelines for them, evaluate everything they provided in return, make refinements, and make the final decisions on every item.

Inevitably certain details had to be discussed at length with the other engineers, or explained to various members of the Board of Trustees, and all this required voluminous, tiresome correspondence. But always, when it came to making his views known, his language was patient, plain, and to the point. There was never any doubt as to what he wanted done or why he wanted it done that way.

His knowledge of everything happening at the bridge, his total confidence about how each successive step ought to be taken, the infinite, painstaking care he took, seemed absolutely uncanny to the others back in Brooklyn. Had his communications on technical matters alone been written by a healthy man who was regularly on the scene, they would have been regarded as exceptional. But the idea that they were emanating from a sickroom sixty miles away seemed almost beyond belief.

He was attentive to more personal matters as well. He wrote to Collingwood to suggest remedies for a kidney ailment. He requested salary increases for Martin, from $5,000 to $6,000 a year; for Collingwood, from $3,000 to $3,600; and for Farrington, from day wages to "$3,000 per annum." (The raises went through.) He approved the hiring of an assistant for Hildenbrand, an RPI man named Theodore Cooper, who had worked for Eads in St. Louis. Cooper was to be an inspector of iron for the superstructure.* When the work stopped in winter and Murphy, to save money, began letting men go, Roebling urged that the best of them be kept on. How could they ever replace a man like Hildenbrand, he asked.

* Years later, at Quebec, a huge bridge partly designed by Cooper, by then an engineer of national prominence, would collapse during construction, killing seventy-five men. On hearing the news Roebling would write scathingly of engineers who design bridges but do not give the actual construction their personal attention. "It is one thing to sit in your office and split hairs," he would write, "but a different thing to get out and command men and meet the realities of great construction." Ironically, Roebling was unaware, it seems, that Cooper had not been at Quebec because of his health.

And in early 1875, when it had looked as though the work might have to stop altogether because money was running short, his seemed the one last voice of confidence. At the close of a long, persuasive letter to Murphy, he wrote:

> I would further add, *now* is the time to build the Bridge. At no period within fourteen years have the prices of labor and material been as low as at present. A rise of 10 per cent in these items during the year is within the experience of all, and is but little thought of; but a rise of ten per cent means a million in the cost of the Bridge. To build *now* is to save money!

His own condition was much more serious and complicated than generally realized at the time, or than would be said in print later. He was worse even than when he left for Europe. He was in pain much of the time, in his stomach, in his joints and limbs. He suffered from savage headaches. Some days he was so weak he could scarcely hold up his head. Still, miserable as he was physically, he was not so bad off as he would be portrayed. "There is a popular impression that Colonel Roebling has been for years a helpless paralytic," Emily Roebling would write in some private notes put down later. "This is a mistake as he has never been paralyzed for even one moment and there never has been a time when he has not had the full use of every member of his body."

The major problem was that his nervous system was shattered. The slightest noise upset him terribly. He was still hounded by visions of his own death before the work could be finished, of disastrous incompetence on the part of some subordinate, of precious days lost at the bridge over some technical problem he could solve in a minute were he there. He felt imprisoned within his own body. He grew extremely short-tempered. When visitors were with him he suffered the whole while. Talk of any kind tired him more than anything else. His eyesight had grown so dim he could neither read nor write nor sign his own name.

His troubles were not solely the bends anymore. That was clear to those who had any regular contact with him. The standard explanation then, and later, was that he was suffering still from the bends. While residual pains and discomforts of the bends can persist, occasionally over a lifetime, the bends were only part of his problem. It is extremely unlikely, for example, that the bends could, at this stage, have had anything to do with his failing eyesight or the terrible discomfort he suffered whenever people were around.

When describing his own condition in private correspondence, Roebling himself does not seem to have used the words "bends" or "caisson sickness." He spoke only of a nervous disorder and of his crippled physical condition. Farrington would later describe him as being a "confirmed invalid . . . owing to exposure, overwork and anxiety." There is, indeed, every indication that the strain he was under, the limits he had pushed himself to during the winter of 1872–73 to get everything down on paper, the anguish and massive frustration of knowing so much about what ought to be done but able to do so very little—all that on top of the physical torment of the caisson sickness, had brought on what in that day was called "nervous prostration."

He always told others his agony was his own doing. He had pushed himself too far he said. He longed for rest. It was the one and only cure he had faith in, but he simply had no time for that.

Collingwood, it seems, was also nearing a collapse of some sort and Roebling, gravely concerned about his old college friend, offered some revealing advice:

> Regarding your health my council would be sit down and keep quiet. . . . Above all don't let a fake ambition lead you on to undertaking tasks that will only break you down all the more. You are no doubt beginning to find out, as I have found out long ago, that nervous diseases are as intractable as they are incurable and only through mental rest of all the faculties and especially the emotions can they even be palliated in the slightest degree.

This letter, like all his correspondence with Brooklyn, the specifications and the rest, was dictated to Emily, who was in constant attendance as both nurse and private secretary. Gray-faced, he would lie propped up in bed or sit like an old man with a blanket over him in a chair by the window. She would sit close by taking down what he said in a letter book. When he had finished, she would read it back to him. He would make a few corrections, then she would do a final draft, in longhand, and read that back to him once more. As a result, week by week, month by month, she was learning quite a lot about the engineering of a wire suspension bridge.

The physical pain came and went. Frequently there were whole days when he felt well enough to be up and about the house. But everyone had to take extreme care not to upset him in any way.

Since childhood he had been interested in geology and in collecting minerals. Now they became a passion. He began sending to one place and another around the country for different specimens. How he was able to take any enjoyment from them, with his sight so impaired, is a puzzle. But he did. Once, with a check for some new specimens, he had Emily include a note of explanation. "I am an invalid confined to the house and minerals are the only things that do not tire or excite me."

A few incidents concerning the bridge upset him no end. Somebody in Brooklyn had suggested there was a secret connection between the Roebling wire works and Carnegie's Keystone Bridge Company, implying that had been the reason why Keystone, not the lowest bidder, got a contract for anchor bars. Livid, deeply insulted, Roebling had sent off an icy reply to Murphy, saying he had no interest in the Keystone company, financially, politically, socially, or any other way, and further stated that if the policy henceforth was to give "contracts for supplies to the lowest bidders, irrespective of all other considerations, I hereby absolve myself from all responsibility connected with the successful carrying on of this work." It sounded perilously like a letter of resignation.

The Eads lawsuit had also been a continuing aggravation for years now. In 1871 Eads had put in a claim for five thousand dollars, saying that Roebling, in the New York caisson, had infringed on the design he had used in St. Louis. Roebling called the charge absurd. But presently, after Roebling was stricken with the bends, Eads had angrily attacked him in the pages of *Engineering*, the esteemed English journal. The thing that had set Eads off was a harmless paper by Roebling published in *Engineering* in which Roebling had not, in Eads's opinion, credited Eads properly for placing his air locks at the bottom of the shafts, part way inside the air chamber, instead of on top. Eads claimed he had been the first to do it that way and said Roebling should have said as much. He accused Roebling of stealing the idea and in a rather snide, roundabout fashion dismissed the younger man for having no creative talents of his own.

Eads's letter was written in April 1873, but it did not appear in the magazine until later in the spring, when Roebling had just arrived at Wiesbaden, in no shape for any more emotional strain than he was already under. In a fury he wrote in answer to Eads's letter: "Its perusal has left only the one prominent impression on my mind, that his skill in blowing his own trumpet is only surpassed by his art in writing abusive and unjust articles about other people." Roebling said he had always had the greatest respect for Eads until then. He said he had designed his New York caisson before he ever saw anything of Eads's plans or went out to St. Louis. "My actual experience in the St. Louis caisson," he wrote, "consisted in nearly breaking my neck, and being half drowned in the bottom of a pitch-dark hole—certainly a forcible way of reminding one where the lock was located." He said, furthermore, that Eads had ridiculed his idea of using timber for the caissons, that Eads's prior interest in caissons had been scant and superficial, and that it was ridiculous to think that the position of an air lock was something that could be patented. "You might as well patent contrivances in a ship's rigging if she were loaded with grain or cotton, or entirely empty."

Then he wrote: "In conclusion I beg to assure Captain Eads that I feel perfectly competent to take care of the East River Bridge, and to overcome dangers and difficulties of which he has but little conception. . . . all of the St. Louis caissons together can find room in one of the East River caissons, with space enough left for several more like them. . . . And where would you go to find an easier material to sink through than at St. Louis, or a more difficult one than in the East River?"

It was an exchange of a sort not often witnessed in the profession, on the printed page at least, and might have been enjoyed as a memorable good fight had it not been known by many readers why Roebling was writing from Wiesbaden. Eads had never had any particular trouble with the bends himself, none personally that is, and this seemed hardly the time for him to be going after Roebling for what appeared to be a minor infraction, assuming even that Eads was within his rights. But Eads apparently was not to be put off by reports of Roebling's condition and refused to let Roebling have the last word, writing still one more letter to the editors of *Engineering*.

His time was too valuable, Eads said, to take part in petty arguments, but he spent several columns of small print picking apart the things Roebling had said, and to prove his point about the position of the air lock, he included drawings of both his and Roebling's designs, which, to be sure, looked remarkably alike.

Who was right in all this is difficult to say and not especially important. But the dispute deepened the division between the two men and caused them both considerable anxiety at a time when each had troubles enough to contend with. Each man believed his good name had been stained by the other. Neither was about to stand by and let that happen, or to have his bridge denigrated. The anger on both sides was all out of proportion to the issue. For Roebling, who let the matter drop, Eads's accusations were the cause of lasting mental torment, but for Eads, Roebling's amenity appears to have had some rather different consequences.

Eads needed every friend he could get just then, for it was in that summer of 1873, the summer of the exchange of letters in *Engineering*, that Grant's Secretary of War, William Belknap, convened a board of Army engineers to decide whether Eads's bridge was a hazard to navigation on the Mississippi and ought to be stopped. And it was in September, soon after Eads's last letter attacking Roebling, that the Army board issued its report calling the St. Louis Bridge "a very serious obstruction" to river traffic. By January the Army engineers were saying that probably the bridge ought to be torn down and what is so extremely interesting about all this is that the most outspoken member of the board was G. K. Warren, Washington Roebling's brother-in-law.

Very possibly Warren's opinion was a purely professional judgment. Had there been no bad blood between Roebling and Eads, Warren might have arrived at exactly the same conclusions. But the wording of his opinion suggests otherwise.

"I am convinced," he wrote in summary, "that a bridge suited to this great want [spanning the Mississippi], at an expense much less than has already been made, almost if not entirely unobstructing navigation, could years ago have been completed, upon designs well-known and tried in this country, had not the authors of the present monster stood in the way."

Since neither the construction cost nor the aesthetic merits of the St. Louis Bridge were at issue, Warren's comments on both were uncalled for, as well as quite debatable. The "monster," as he called it, happened, for example, to be regarded by many as one of the handsomest bridges in the world. Moreover, it is pretty obvious that the "well-known and tried" designs referred to were those by John A. Roebling.

But fortunately for Eads and his bridge, Secretary Belknap, General Warren, and the other Army engineers were overridden when Eads, as a last resort, went to the White House to see his old friend Grant. Congress had authorized the building of the bridge,

Grant told Belknap, so only Congress could decide to pull it down, which Congress did not do.

In another couple of years, with his bridge completed and being talked about everywhere, Eads had become a great popular hero. He was "the noble engineer." Early in 1876, a Presidential election year, it was discovered that Secretary Belknap had been getting kickbacks from the sale of Indian trading posts in the West. This on top of the sensational disclosure of a "Whiskey Ring" operating in St. Louis under the direction of Grant's supervisor of internal revenue had the whole country wondering where, from what walk of life, an honest leader might be found. The editors of *Scientific American* decided they had the answer. "In war and peace his commanding talents and remarkable sagacity have been devoted to patriotic labors . . . We nominate for the Presidency Captain James B. Eads of St. Louis. The man of genius, of industry, and of incorruptible honor."

Eads kept on pressing his claim against Roebling and Roebling got so he could not bear even the mention of the man's name. Finally, in May, Roebling gave in. "I am willing to accede to the proposition of Captain Eads in order to settle this matter," he wrote to Paine. "I give my consent more as a matter of expediency than from conviction. I am not in a frame of mind to stand any further worry about a lawsuit."

The issue was thereupon settled out of court and that at least was one less worry for Roebling to dwell on.

The previous winter had been a particularly bad time for Roebling. At one point, his nervous state had become so unsettled, his physical discomfort so acute, that in a moment of total despair he decided to give up on the bridge.

"My health has become of late so precarious a nature," he wrote to Henry Murphy, "that I find myself less and less able to do any work of any kind. I am therefore reluctantly compelled to offer my resignation as Engineer of the East River Bridge. The hopes that time and rest would effect a change have been in vain, rest being simply impossible."

The letter must have been ignored in Brooklyn, or perhaps it was never even sent. The one and only copy of it is in Emily Roebling's cardboard-backed letter book, written in pencil, in December 1875.

Maybe, just possibly, she wrote it herself, without his knowing it, at a moment when her own endurance failed, but then the moment passed. In any case, nothing came of the letter.

The extraordinary thing is that Roebling's mind through all this time seems not to have been affected in the slightest. If anything, his powers of concentration, his remarkable gift for recalling in every detail things he had seen, seemed greater. No longer able to work problems out on paper, as he always had, he did everything in his head, and when his younger brothers came to call on him, as they did from time to time, to inquire about his health or to discuss problems at the mill, it was he who seemed able to sort things out quickest and come up with solutions. In fact, his ability to direct the family business, *in absentia*, seemed no less than his ability to direct the bridge, and nearly as vital.

His brother Charles, who had come into the business in 1871, after finishing at Troy, had grown into a fastidious, intelligent-looking young man who wore a stovepipe hat and whose primary interest now was his work, which was the production side of the business. Ferdinand, who with his rimless spectacles and mustache appeared older than his age (thirty-four by 1876), was supposed to see to commercial matters.

Roebling thought highly of Charles. He admired his unfailing industry and his technical competence. "He was his father over again, to a far greater degree than any of the other children," Roebling wrote. "He inherited his temperament, his constitution, the concentrated energy which drives one to work and be doing something all the time." Charles, at twenty-seven, was still a bachelor.

When Charles had returned from college in 1871, everything had been prepared for him to take his place in the mill, and since Ferdinand knew comparatively little engineering, Charles had been more or less his own master from the start and old Charles Swan, his guardian, had turned over to him something like $300,000 "in good securities."

"Charles had one very strong point," his oldest brother would write admiringly, "he never copied; [he] tried to solve every problem according to the best of his ability. Every task was an education to him."

Roebling's relationship with his brother Ferdinand, the one John Roebling kept home during the war, had grown rather strained, however. Charles Swan, who had always looked after things at the mill whenever a bridge was being built, was not the man he had been, and in the time Roebling had been away from Trenton, Ferdinand, or "F.W." as he was called, had more or less taken charge. "He lost no opportunity to make me painfully aware of it," Roebling would write.

Now for the first time in his life, Roebling had become bitter over what he considered an unjust disparity between his own fate and that of his brothers. It was not at all like him. But his physical suffering, the endless confinement, the strain of everything he had on his mind, had begun to tell. And besides, his feelings were not without justification.

He had been the only one of them ready and able to carry on what his father had left undone at Brooklyn and he had been paying a terrible price for it. His brothers, both in perfect health, had had everything handed to them, as he saw it; they were getting rich speedily and effortlessly in a business that had been all set up for them in advance and that he felt he understood better than either one of them. He, on the other hand, and on top of all his other anxieties, was convinced he was nearly ruined financially. He never said a word about this publicly. The only record of his feelings is in private correspondence written years afterward. Nor is it possible to know whether his financial plight was quite so serious as he pictured it. His salary on the bridge remained at ten thousand dollars a year, but his expenses, he estimated later, were twice that. The medical expenses were the worst. The six months in Europe had been particularly costly. There had been Eddie's private schooling to pay for, as well as that of his own young son. The house in Brooklyn had cost him forty thousand dollars. Moreover, he saw his expenses growing greater in time to come. But realistic or not, his concern was still another severe strain, and his unspoken feeling of indignation was deep-rooted enough to last a lifetime.

Nobody outside the family would ever know anything of this. What is more, the respect he commanded inside the family appears not to have been diminished in the slightest, either by the rise of his brothers' fortunes or by the tragic turn his own life had taken. When it came time to incorporate John A. Roebling's Sons, in 1876—when Charles Swan was finally persuaded to retire—Ferdinand was made secretary and treasurer, Washington was made president.

The identification of the name Roebling with the bridge at Brooklyn was, of course, quite a good thing for the family business. Unquestionably the firm's reputation had already benefited. When a reporter came down from Brooklyn to tour the mill, Ferdinand showed him about.

"Their grounds cover fourteen acres," the visitor wrote afterward, "and within the walls are five wire rolling mills, and all the buildings needed for their three hundred and fifty workmen and office purposes . . . Their products amount to three-fourths of all the wire rope made in this country. It was a rare sight to watch these busy workmen taking blocks of red-hot steel in their tongs from white-heat furnaces, passing them through rolling mills which stretched them until they lay upon the iron floor like interlacing snakes in bizarre shapes, ready to be carried by other hands to annealing furnaces, and thence through other draw plates until the wire was prepared to bind together either the delicate handiwork of the jeweler or the two cities of New York and Brooklyn with their millions of inhabitants." The mill was then producing something like 450 miles of wire a day.

Once the cable spinning commenced over the East River the public would be treated to the most spectacular demonstration imaginable of the Roebling product—or so it was naturally assumed in Trenton—and during the last part of Washington Roebling's confinement there, he had the pleasant task of helping to plan a display for Machinery Hall at the Centennial Exhibition.

A variety of Roebling wire and cable, some of it on big spools, would be set out within a display area framed with iron rope draped as velvet rope would be customarily. But the centerpiece of the arrangement would be a sample section, or model, of the cable for the East River bridge, mounted on a little pedestal, like a piece of sculpture, and having as its backdrop an enormous drawing of the bridge by Hildenbrand that measured seven by twelve feet and presented "the noble proportions of the structure to great advantage."

The section was made up at the mill especially for the occasion and exactly as Roebling had decided the real thing would be put together. When finished, it looked like a metal drum about three feet long and fifteen and a half inches in diameter and bound with brass bands. The upper end was cut off and planed smooth so the position of the wires could be seen. In all it contained 5,282 steel wires, each a little over an eighth of an inch thick, and it weighed 1,200 pounds. The wires had been laid as they would be in the cables—parallel and in distinct stages: 278 wires bound together formed a strand, as it was called; 19 strands in one great bundle, all tightly wrapped in protective skin of soft iron wire, formed a cable.

Roebling had worked out the entire arrangement and in the early part of 1876 completed his final specifications. Each of the nineteen strands in a finished cable would be continuous wire some 185 miles in length, drawn from one anchorage to the other, up and over the towers, back and forth, back and forth, above the river. Each cable would contain just over 3,515 miles of wire and the wire in all four cables would come to more than 14,000 miles.

The whole process would begin with a single wire taken across by boat, then lifted up over the towers. After that a heavier steel rope would be pulled over, the "traveler" or "working rope" as it was known, which would do the job of hauling the cable wire itself back and forth. The trick would be getting the wires in each strand in exactly the right position.

Roebling's specifications called for 6.8 million pounds, or 3,400 tons of wire "of superior quality steel." The wire was to have a tested strength of not less than 160,000 pounds per square inch, which meant it would have nearly double the strength of the iron wire used at Niagara and Cincinnati. In addition, to guard against the corrosive salt air over the East River, the wire would be galvanized—coated with zinc—something that had not been done before and that a few later-day suspension-bridge builders would neglect to do to their regret.*

Sealed bids, the specifications stated, would be "received by the Trustees of the New York and Brooklyn Bridge, up to the 1st day of December, 1876." But it seemed a foregone conclusion that the Roebling company would get the contract, and when the Centennial Exhibition opened in May, the prototype slice of bridge cable set up in the Roebling display turned out to be one of the most popular items in Machinery Hall, along with Ben Franklin's old hand press, a first typewriter, and a telephone displayed by a courtly Scottish immigrant, Alexander Graham Bell. One day in Machinery Hall the fair's most popular visitor, Emperor Dom Pedro of Brazil, put his ear to Bell's device, then dropped the receiver, exclaiming, "My God, it talks!" The fair was a success from that moment on.

Machinery Hall was also the place to see the favorite attraction of the entire fair, the gigantic Corliss stationary steam engine. It

* Prior to this time, cables were made of "bright" wire, which was oiled, greased, or painted for protection against the elements.

stood just down the way from the Roebling display, taller than most houses, with two tremendous walking beams, a gigantic flywheel, several flights of stairs and little platforms for the mechanics and oilers. It had been erected in the central transept of the hall and provided the driving power for some thirteen acres of machinery displayed throughout the building.

On the opening day, the hall filled with spectators, every machine had stood motionless as President Grant, dressed all in black and looking pale and tired, stepped to the controls of the giant engine, along with Dom Pedro and George H. Corliss, its creator. Grant and the little Emperor each took hold of a lever. Then Corliss waved his hand, a signal to admit steam into the cylinders (the boilers were located outside of the building). "It was a scene to be remembered," wrote one reporter, almost overcome with excitement, ". . . perhaps for the first time in the history of mankind, two of the greatest rulers in the world obeyed the order of an inventor citizen." When the two men swung their levers, the engine hissed loudly, the enormous walking beams began moving, ever so slowly, the floor trembled. Then the walking beams were going up and down. The flywheel gathered momentum, belts moved, shafts and pulleys turned, and machines everywhere came to life—sewing cloth, printing newspapers (the New York *Herald*, the *Sun*, the *Times*), printing wallpaper, sawing logs, grinding out plug tobacco. The Pyramid Pin Company had a machine attended by a little girl that turned out 180,000 pins stuck in paper in a single day.

The giant Corliss itself required only one attendant, which greatly impressed most observers, including William Dean Howells, who wrote: "The engineer sits reading his newspaper, as in a peaceful bower. Now and then he lays down his paper and clambers up one of the stairways . . . and touches some irritated spot on the giant's body with a drop of oil, and goes down again and takes up his newspaper; he is like some potent enchanter there . . ." Americans liked their mechanical marvels done up on a grand scale, the bigger the better, and it was an age that adored pageantry. So a combination of the two was bound to please. But it was the contrast between man and machine that made the machine seem so monstrous big, the man so touched by some blessed new power, and the whole hall so enormously popular.

There were some, of course, who saw the Corliss engine as a menace, "ready at the touch of a man's fingers to show its awful power"; but most people went back to the cornfields of Indiana or the dry goods store in Fall River or wherever it was they came from filled with pride and admiration for all they had seen.

Two of the Roebling brothers went over to Philadelphia to attend the opening ceremonies. Charles probably considered the Corliss engine overly large for its purpose and inefficient, which it was, and Ferdinand must have been extremely pleased by the attention paid the section of bridge cable. The fair would be attended by eight million citizens by the time it ended in the fall, or about one American out of every five, a very large percentage of whom took some time to look over the Roebling display.

For Washington Roebling news of all this, like news of everything else happening beyond his walls, came to him second or third hand. The fair was an easy morning's train ride from Trenton, but for him it could as well have been on the other side of the world. The opening ceremonies in Machinery Hall and all the other attractions were described at great length in the papers. There was Old Abe, the famous eagle mascot of the Civil War, which, for fifty cents, could be seen dining on live chickens; or the gigantic hand and torch of the great statue *Liberty Lighting the World*, a one hundredth birthday gift from the people of France. These he could readily picture as Emily read aloud for him, just as later the following month he could see the gruesome scene on the high plains of Montana when she read about the slaughter of 264 federal cavalrymen and their commanding officer, George Armstrong Custer. Roebling and Custer were of about the same age. That the Little Big Horn and Machinery Hall were part of the same America said perhaps as much as anything about the sort of country it was after a hundred years if one stopped to think about it, which doubtless Roebling did.

And then, very gradually, he began to show signs of improvement. In July he was talking to Emily of returning to the bridge and he dictated a letter to Paine to tell him as much. The work he liked best, the work he knew best, was about to begin. When he had first arrived in Cincinnati after the war, the cable spinning had only just begun and he had been the one in charge from then on, not his father, as most people failed to appreciate. Now he grew keenly interested in everything to do with the footbridge. Farrington was the man to build it, he wrote Henry Murphy. Farrington had been through all this before at Cincinnati and knew just what do to. "He is a man of great resource when unforeseen troubles arise," Roebling told Murphy, who already knew all about Farrington and his abilities, "and he has the necessary coolness and perseverance and does not easily get frightened in time of danger . . ." It was what someone else might have said about Roebling.

Then on the afternoon of August 14, shortly past one o'clock, a telegram was sent up to the Roebling house from the Trenton depot. It was from Paine: THE FIRST WIRE ROPE REACHED ITS POSITION AT ELEVEN AND ONE HALF O'CLOCK. WAS RAISED IN SIX MINUTES.

Two other telegrams followed, one from John Prentice, treasurer of the Bridge Company, and one from Farrington late in the day. They reported what he hoped they would: after the first rope was in position, a second had gone across, the two to form an endless cable stretching from anchorage to anchorage. The whole operation had gone off without a hitch, exactly as planned. It was a moment Roebling had been anticipating for seven years and he had missed seeing it.

17
A Perfect Pandemonium

> With its princes of the lofty wire the Brooklyn Bridge is now the cheapest, the most entertaining, and the best-attended circus in the world.
>
> —New York *Tribune*

THEY all stood waiting for the river to clear—Martin and McNulty under the arches on the New York tower; Farrington and a carpenter named Brown out on a hoisting frame at the top of the tower, where they could signal to the engineman in the yard; foreman Dempsey and several workmen close by on the tower itself; and on the wharves below and across the river, in the rigging of ships tied up on both shores, perhaps six thousand spectators, many of whom had been waiting for several hours.

The idea at first had been to take the rope across on a Sunday or at night, when there would be little traffic on the water. On the average day as many as a hundred craft passed the line of the bridge in an hour's time. But Farrington had noted that frequently there would be clear water between the towers for stretches of four to eight minutes, even on the busiest days of the week, so the decision had been to go ahead just as soon as everything was ready.

A few days earlier the ends of two working ropes had been hauled up and over the top of the Brooklyn tower. Made of twisted chrome steel strands, these ropes were three-quarters of an inch in diameter, more than three thousand feet long, and were wound on a big wooden drum set at the base of the tower on the river side. To get the ropes over the top had been relatively simple. A heavy hemp rope had been put over first, then tied to the eyelet at the end of the wire rope. That done, the hoisting engine in the yard was started up. The wire rope was pulled to the top, where it passed through a set of pulleys, then down the other side.

From the yard the ropes were then hauled inland to the summit of the Brooklyn anchorage in much the same way, except that fenders and trestles had to be erected and men stationed on all intervening housetops to prevent any accidental damage. At the anchorage the two wire ropes were joined and passed around several oak wheels, the main one of which, the driving wheel, as it was called, was mounted horizontally in a massive timber framework and was a good twelve feet in diameter. Back at the base of the tower one of the reels was then put on board a stone scow and hung on a wooden axle, so when the scow started for New York the rope would be unwound by the strain from the Brooklyn shore, where the rope was temporarily lashed tight.

By nine that morning, Monday the 14th, everything was in order. Huge American flags had been raised on top of both towers and there was much excitement among the spectators. Slack water, the relatively calm interval between tides, would occur in the next hour. Martin, McNulty, and Farrington had gone on board the scow to supervise things, along with the white-bearded O. P. Quintard and two or three young ladies, the identities of whom were never given in later accounts.

Shortly past nine two steam tugs pulled alongside. One made fast to the starboard side of the scow; the other stood off slightly, ready to keep other craft at a distance during the trip across. At nine thirty the tugs sounded their bells, moorings were cast off, tugs and scow swung slowly out into the river. At the stern of the scow the wire rope trailed off into the water. The tide was still running out, and as the boats pulled away, the current carried them downstream some but not enough to matter. Slowly, steadily, they pushed for the opposite shore, the rope paying out and sinking to the bottom as fast as it unwound. Two-thirds of the way across the tugs had to stop to allow an English bark to pass upriver across their bows. "She came so close," wrote a reporter on the lead tug, "that a pebble could have been tossed upon her deck with the most perfect ease."

But that had been the single interruption. The whole trip took less than ten minutes and the arrival at the New York tower had been greeted by loud applause. The scow made fast in a very businesslike fashion; the balance of the rope was unwound and laid on the dock.

The next thing had been to get the rope over the New York tower. A hemp rope had been passed over the tower previously and was now attached to the end of the wire. But nothing more was done until Farrington had climbed to the top of the tower to make a few final checks. At ten twenty he signaled from above. The hoisting engine was thrown into gear and in a matter of minutes the wire was over the top and reeled part way onto the yard engine's big drum. The main body of the rope, however, still lay at the bottom of the river and there it remained as everyone stood watching for a moment with no boats in the way, or none about to be, when it could be pulled out of the water.

The waiting seemed interminable. Half an hour went by, three-quarters of an hour, and still there was no break in the traffic. Two barges and an excursion steamer moving out into the stream from Jewell's dock took forever getting under way. The excursion boat was bound for Oriental Grove, on the Sound, with a picnic party, and everyone on board appeared delighted by the grandstand view of the doings at the bridge.

At about half past ten, as a precautionary warning to passing ships, a little howitzer had been fired at the foot of the New York tower but that seemed to have no effect. It began to look, in fact, as though several hours would pass before the river would be clear enough to get on with the work. But as some of the subsequent newspaper accounts noted, the long wait did nothing to dampen anyone's spirits and the delay added considerably to the size of the crowd.

Then the break came. The river was perfectly empty from tower to tower. At twenty-five past eleven, from the archway on the New York tower, Martin shouted up to Farrington, "Go ahead!" Farrington had Brown signal to the hoisting engine. The cannon was fired a second time—to signal the men on the Brooklyn side to cut loose their lashings and as a warning to approaching ships.

"In a few seconds the rope began to move," Farrington wrote later; "there was a ripple around it in the water; it began to draw away from the dock toward Brooklyn, and soon we could see the other part coming from Brooklyn towards us. Faster and faster the space of clear water between the two parts narrowed, and in four minutes from the time of starting, it swung clear of the surface of the water, with a sparkling *swish*, amid the cheers of spectators, on the wharves and ferryboats, and the shouts of our own workmen."

This time the drum in the yard was wound by a thirty-horsepower engine that made 150 revolutions per minute (the engine used to pull the wire over the tower had been only half as powerful). As a result it took just two and a half minutes to pull the wire free from the water, and five minutes, all told, to get it into proper position for the time being, stretched from tower to tower at an elevation above the water of two hundred feet more or less. Almost immediately a boat passed by below, a lighter called *Comet* carrying a load of pig iron, and at least one reporter took the opportunity to go up on the Brooklyn tower to take a look at the view.

"When it is considered that one has to climb upward of thirty flights of winding stairway, the toil of the ascent on a close August day can be readily imagined," wrote the young man from the *Herald*, "but all this is instantly forgotten when the picture from the summit spreads out at one's feet." The buildings of both cities, he said, looked dwarfed beneath the overtopping height of the tower; the streets seemed narrowed down to lanes in Brooklyn and to mere pathways in New York. The view of the river and the bay, with their islands and with tiny ships moving restlessly this way and that, all looked extremely fine, he said. "What a splendid set of photographs could be obtained from this point! . . . Doubtless some enterprising photographer will seize the chance." *

With the first half of the working rope thus in place, the drum and hoisting engine in the New York yard had to be freed to haul over the second half. So a huge iron clamp was bolted to the end, near the enginehouse, about ten feet from the ground. A pulley block was made fast to the wharf close to the drum, another to the clamp, and a rope passed between them several times made a lashing strong enough to withstand the pull of the wire rope, the end of which was immediately cast loose from the drum.

The tugs and the scow, in the meantime, had returned to the Brooklyn tower and about noon they started back with the second rope. By half past three it too had been hoisted out of the river, everything going even more smoothly than the first time. The next step would be to take the ends of the two ropes back to the New

* At least one photographer had already been to the top of the Brooklyn tower, J. H. Beals, who earlier in the year had made the first great panoramic photograph of lower Manhattan, from the Battery to Rutgers Street, by taking five different views that he later spliced together into one panorama more than seven feet long.

York anchorage, splice them, and thereby form one immense loop, or endless "traveler," over the towers, reaching all the way from anchorage to anchorage. The entire length of the traveler when completed would be 6,800 feet, or considerably more than a mile, making it easily the longest belt connecting machinery anywhere on earth.

"WEDDED" was the one-word headline in the evening edition of the *Eagle*. "The thing is done," the article began. New York and Brooklyn had been joined at last. But no New York paper was willing to go quite that far. The *Herald*, for example, described the great endless rope draped over the river as only "the engagement ring in the marriage preparations of the two cities." All the same the event was an enormous popular success and talk of the bridge was everywhere as the papers reported that the next step would be to send a man across on the rope.

More than a hundred people appeared at the bridge offices to apply for the job, including a twelve-year-old boy who wanted to go hand over hand and a Long Island acrobat who considered it a once-in-a-lifetime opportunity. Nearly all of them volunteered to make the trip without pay and C. C. Martin told reporters there were at least a dozen of his own men who would give a month's wages to be the first one to cross the river.

To quiet things down some it was announced that the man picked to make the trip would be one of the most trusted employees and probably one of the engineers. The rope would first be run back and forth a number of times. Then the man would go over in a boatswain's chair, a seat and a sling made fast to the rope. He would start from the Brooklyn anchorage, the announcement said. (Henry Murphy wanted the historic journey to originate in Brooklyn.) He would ride up to the tower, climb out, cross over to the other edge, get back in his seat, and start across the river. "The object of this journey will be to see how the thing works."

All the machinery for running the rope was at the Brooklyn anchorage. At the foot of the great stone mass stood a thirty-horsepower steam hoisting engine that would drive the wheels. It was completely enclosed, as was its boiler nearby. Up above, across the face of the anchorage, secured just over the arches, was a line of shafting with several pulleys. A sixteen-inch-wide belt, ninety feet long, connected the pulleys on the shafting with the gears and cog-wheels that turned the enormous twelve-foot wheel that carried the working rope. The arrangement of cogwheels was such that the direction of the rope could be reversed without reversing the engine, an important feature since the rope was not to be revolved continuously, but worked back and forth.

On the New York anchorage the framework of the main pulley was adjustable, so it could be moved forward or back in order to give the rope the prescribed deflection, or sag. (At one deflection the rope would bear greater weight than at another, and thus adjusting the deflection just so would be a vital part of the work to come.) Had there not been some trouble with the delivery of one or two essential belts, the much heralded first crossing would have taken place almost immediately after the traveler was in place, but there were numerous other matters to attend to in any event, and the *Eagle*, ever the ardent champion of the bridge, wrote, "It is refreshing to see how the work is pushed forward, and yet the thoroughness with which everything is done, in these days of slighted work and ill-performed operations . . ."

To the surprise of almost no one who had had anything to do with building the bridge, the man chosen to make the first trip over the river was Master Mechanic E. F. Farrington.

Farrington, who would so soon become a subject of great public interest, was nearing sixty in 1876, but still agile, tough, and, of course, exceedingly knowledgeable about working with wire rope. Subsequent newspaper articles would reveal also that he came from Massachusetts originally, where he had been put to work in a woolen mill at age nine, that he had been a farmer, a carpenter, a machinist in England, a seaman, a gasworks superintendent, and was considered the best bridge mechanic in the country. On the morning of Friday, August 25, when he arrived at the Brooklyn anchorage ready to make his historic journey, he appeared "perfectly cool and collected"—a spare man of medium height, with gray beard and blue eyes, turned out quite formally for the occasion in a fresh suit of unbleached linen and a new straw hat.

An announcement that a man was to make the crossing that day had been published in the *Eagle* the previous afternoon. As a result the crowds had begun gathering since well before nine in the morning. Seen from Brooklyn, the piers adjacent to the New York tower looked black with people, and the gates to the Brooklyn anchorage and tower yard were jammed with spectators.

Up on the anchorage itself workmen were busy adjusting belts and pulleys, with Martin, McNulty, and Farrington supervising everything. By eleven all looked in order. The machinery was set in motion and the rope began moving across the river. To get every twist and kink out, it had to be worked back and forth several times. Otherwise anything attached to it, including a human passenger, would have been turned over and over. A stick tied to the rope as a marker and sent from the Brooklyn anchorage up to the Brooklyn tower twisted completely around several times while making its slow ascent. But after half an hour of working the rope to and fro, it moved along perfectly.

There was a break for the noon meal. The day was bright and very hot by then. Up on the Brooklyn tower, a small crowd of privileged spectators had gathered, including Senator Murphy and several ladies. The sun beat down on the exposed stonework and at one point some of the reporters in the group sent a note over the rope to their compatriots on the opposite tower asking for cold beer and sandwiches. An answer was returned by the same route, "Send the money and we will send the beer," but no money was sent.

Presently, about twenty past one, the huge American flag was again unfurled from the Brooklyn tower and minutes later another went up the flagstaff on the New York tower. Then two men with red signal flags were seen to wave to each other from the tops of either tower. Everything was set to go. Estimates were that more than ten thousand people were watching.

Farrington, all this time, had been supervising the preparation of his boatswain's chair, a simple board seat, two feet long and two inches thick, with rope holes drilled in each corner, like an ordinary swing, and with four ropes drawn through and tied to the wire rope just as they might be for a swing. The board itself had been placed so that only one end rested on the rim of the anchorage, while most of it hung out over the edge, eighty feet above the street. So when Farrington proceeded to take his seat, it was, in the words of one bystander, a somewhat delicate operation.

The men assisting him next passed a rope across his back, to form a rest of sorts, then brought it around, across his chest, and tied it securely to one of the corner ropes. All these precautions, however, appeared to make "the daring voyager" feel only more uncomfortable.

At thirty-two minutes past one o'clock, Farrington said he was ready. "Timothy McCarthy ran the engine," Farrington would write later, "and John D. Smallfield handled the starting lever most carefully, according to a system of signals previously agreed upon." Martin, who was standing close by, dipped a signal flag, John D. Smallfield in the yard below shifted his lever, and in an instant the master mechanic was on his way.

There was great shouting from down below, and up ahead, on top of the tower, people were waving hats and handkerchiefs. Then all at once, as he went swinging out over the housetops between the anchorage and the tower, Farrington freed himself from the rope about his chest and stood up on the seat. Holding on first with one hand, then the other, he lifted his hat in response to the continuing ovation. Then he sat down again. People were running through the streets beneath him now, shouting and cheering as they ran. He waved, he blew them kisses. Sailing steadily along all the while, his course was nearly horizontal at first, like that of a heavy bird taking flight, because of the sag in the rope. His light coat blew open and began fluttering in the wind. And then he was beyond the sag and climbing sharply, almost straight up, a coat-flapping, gently twirling form that looked very small, fragile, and very birdlike now against the granite face of the tower.

The rope had to be operated with the greatest of care at this stage, as Farrington neared the top of the tower, for if he were drawn suddenly against the coping, he might be knocked right out of his seat. A reporter described the moment this way:

> One of the most experienced engineers in the place held the lever [McNulty most likely], and as Mr. Farrington was seen to approach the top of the tower the engine was slowed. All eyes were now strained to discern the movements of the voyager. That he appreciated the danger was evident, as was also the reason for freeing himself from the restraints of the encircling rope, for he stood upright again with his feet upon the board and his hands ready to save himself by grasping the coping of the tower in case the wire was not stopped in time. The red flag was seen to drop, and simultaneously the wire was stopped. Two men stood by ready to help Mr. Farrington upon the tower, but he was still a little too low down to be reached. The red flag was held aloft, and the engineer, interpreting that signal to mean "go ahead," started the wire again very cautiously. It had moved but a few feet when the flag dropped again, and the engine was stopped instantaneously. Mr. Farrington was now nearly level with the top of the tower, and strong hands grasping

his, he was upon his feet and surrounded by an excited crowd of friends in a second.

A tremendous cheer went up from the streets and rooftops, followed quickly by a salute from the little cannon across the river. His time from anchorage to tower was three and three-quarter minutes. (Quite a number of those gentlemen with privileged vantage points on the towers and anchorages had their watches out through the whole of Farrington's aerial journey and the time he took from point to point would be a subject of the greatest interest among them and duly noted for the historical record.)

Farrington told those clustered about him on the Brooklyn tower that the trip thus far had been nothing at all. Murphy shook him heartily by the hand and asked how he felt. It was an exhilarating moment for the Senator. Farrington said he felt just fine.

The little sling seat was then carried across to the opposite rim and Farrington climbed down and seated himself once again for the long ride over the river. The rope he was traveling on did not look very big even up close. It was about as thick as a man's thumb. But to those who stood with him at the tower's edge, the rope appeared to trail off to no more than a thread, then to vanish altogether somewhere out beyond the middle of the river. It was all very well to know its tensile strength and the rest (it could carry the weight of ten men and more). Every instinct was still to pull back and shudder at the prospect of stepping off into such a void.

Again the signal flag waved and the rope started and the minute he swung away from the tower there was another outburst of cheering. This time all those crowded along the wharves were joined by thousands more on board the innumerable boats and ferries that had gathered for the occasion. All normal traffic on the river had stopped. From the towers it looked almost as though one could walk across just by stepping from boat to boat.

Farrington went sailing over the river, waving, lifting his hat, very obviously having a glorious time, but he stayed seated. Then a steam tug directly beneath him let loose with its shrill whistle. Instantly a dozen others joined in. In seconds every boat on the river was sounding its approval as the tiny figure of a man went soaring overhead, "to all appearances self-propelled," spinning around every now and then, the rope he dangled from all but invisible against the sky.

As he passed the center of the river and began his ascent to the New York tower, the reception from shore was louder even than his Brooklyn send-off had been. And a little less than seven minutes after leaving the Brooklyn tower, he made a flawless landing on top of the New York tower. Then with no delay whatever he was across the summit of the tower, back in his seat again, and on his way on the last leg of the trip, down to the New York anchorage.

Now the great mass of spectators along the river front surged inland toward the anchorage. Church bells were ringing, factory whistles screaming, along with all the boat horns, bells, and whistles that were still sounding forth from the water—"a perfect pandemonium" the *Times* called it. Indeed, Master Mechanic Farrington seemed the only one not carried away by the moment. It was as though he might be having second thoughts about the commotion he was causing, or that he was sorry the ride was over. "Despite the shouting and confusion that went on beneath him," wrote one onlooker, "he sat quiet with his hands folded, save when he waved them in response and showed every sign of perfect self-possession."

Then Farrington stepped lightly onto the New York anchorage, the first passenger to cross over from Brooklyn by way of the Great Bridge. The entire trip had taken twenty-two minutes.

After that, when Farrington climbed down from the anchorage, something close to a riot broke out. The crowd wanted to carry him through the streets in triumph. At first he had tried to make his way through, thinking naïvely that he could walk over to the ferry back to Brooklyn, but people were pressing about him so, reaching out to touch him with such fervor, that he was "obliged to seek refuge from their attentions" in an office in the bridge yard. The hope was that things might settle down if he kept out of sight. But an hour later the crowds had grown greater if anything. A rowboat was brought to the wharf under the tower. Farrington slipped out a back door and was rowed to the other side.

Farrington declared afterward, "The ride gave me a magnificent view, and such pleasing sensations as probably I shall never experience again . . ." But he thought much too much fuss had been made over the episode and told Roebling he was quite put out by the publicity he had received. He had had a natural desire to be the first man over, he said, but his real objective had been to demonstrate to his workmen, who would be doing the same thing under more hazardous conditions, his own complete confidence in the safety of the rope. He would ask no man to do anything he would not do himself.

Moreover, he allowed that he and the assistant engineers had been getting too much praise lately. Roebling was the hardest worker of them all, he told one reporter. "He does most of the brain work," Farrington said.

Be that as it may, Farrington had done something neither Roebling nor anyone else had. In the eyes of the public, for the very first time, he had transformed years of talk and expense and several million tons of granite into a bridge over the East River. He had shown the thing could work. And like it or not, he himself had been transformed by the act.

He said he had simply gone along for the ride. Anyone could have done it was what he told people; the only thing necessary was to sit there, all of which was perfectly true to a very large extent. But the more he went on that way, deprecating his own part in the spectacle, the more he seemed to be saying something else—that this bridge was a more miraculous affair than one might imagine. It had not only taken him over the river with perfect safety, it had transformed him into a hero. And, of course, the fact that he was a plain mechanic, but a man of natural good sense and courage, did nothing to diminish his popular appeal.

His crossing, very simply, had been a "public triumph," as *Harper's Monthly* said. Nobody who saw it would ever forget it. He could say whatever he liked.

The work to be done now, briefly stated, was this:

Two more three-quarter-inch wire ropes would have to be taken across and spliced to form a second endless traveler. Then a heavier rope, called the "carrier," would follow, this one to hold the weight of several still heavier ropes to be hauled over. These would be the two-and-a-quarter-inch ropes to hold the light frame platforms, or "cradles," upon which the men would stand when binding the wires for the great cables. Then supporting ropes for the footbridge would have to be laid up, the footbridge built, ropes for handrails strung, and two storm cables attached from tower to tower beneath the footbridge, in inverted arcs, to keep the footbridge from being carried off by the wind. All that accomplished, the real work of spinning the cables could begin and it would be then that the travelers would perform their vital role.

Work on the second traveler rope began the very next day, a Saturday. But this time the rope was hauled over by the first traveler, rather than going by water, and before the day was over, bystanders along the river front were treated to still one more memorable, but entirely unexpected, high-wire performance.

At eight that morning, first thing, a big reel of wire had been rolled into position on top of the Brooklyn anchorage. One end was lashed to the traveler. The traveler was started up. Slowly the reel unwound and the new rope started toward the Brooklyn tower, seeming to creep out over the other rope, but really moving with it.

When about fifty feet had run out, signal flags waved, the traveler was stopped momentarily, the two ropes were lashed together with heavy twine by men stationed next to the reel—to keep the new rope from sagging—and then the rope was started up again. After that similar lashings were made every fifty feet.

As the new rope crept out over the housetops, the news spread through the whole neighborhood and across the river by ferry, in advance of the rope. In no time the streets and wharves were once more jammed with spectators. Once the rope had crossed the river, passed over the New York tower and reached the anchorage beyond, it was secured at each end in a sort of monster vise. But then the lashings had to be cut loose from end to end and the one way to do that was by hand.

Accordingly, after the noontime break, two riggers began swinging themselves simultaneously from each tower, down the land spans, toward the two anchorages. From the New York tower came a former sailor with the appropriate name of Harry Supple, who had been working on the bridge for six years and had been among those injured when the derricks fell. He used a boatswain's chair, like the one Farrington had crossed on, which was hung to the traveler by a big iron shackle. Seating himself as Farrington had, only without any restraining ropes about his chest, Supple took two half hitches around the traveler with a short length of rope that he would use to check his speed on the way down. Then he pushed off into mid-air, kicked his feet to get the shackle started, and with sudden speed slid down to the first lashing, where he pulled hard on his rope and stopped.

A few fast slashes with a sheath knife and he had the knot severed. Instantly, bits of twine flew into the air, the wire ropes sprang apart with a terrific force, causing the new one to drop down in a big loop and the old one, which Supple was riding on, to vibrate violently along its whole length. Supple himself was seen to drop six feet in his frail-looking seat and bounce about wildly, but he appeared not in the least bothered by that and immediately cast off his gauntlet (as the stay rope was called) and continued on. He sliced open the next lashing, the next and the next, proceeding with incredible speed, a noisy crowd urging him on all the way to the bottom. To separate the two ropes from tower to anchorage, a distance of one thousand feet bound by twenty lashings, took him ten minutes. When his feet landed on the anchorage, the ovation was such that he ought to have taken a long bow.

In the meantime, however, spectators in Brooklyn had not fared so well. The other rigger, a German named William Kohrner, had been terribly nervous before stepping off from the tower and once under way he had been both awkward and maddeningly cautious. He held on to the wire rope with both hands, letting himself down ever so slowly and only short distances at a time. He took so long with each knot that there was some speculation on top of the tower as to whether he might finish the job that week. As it was, he took nearly an hour to do the same thing Supple had done in ten minutes. So when the time came to start on the main span, Patrick Timbs, the man picked to leave from the Brooklyn tower, was told by the others in no uncertain terms "to do better by them."

The plan was for Timbs and Thomas Carroll to slide down from the two towers and meet in the middle over the river, cutting the lashings in just the way the other two had. That done, they were to hitch themselves to the traveler, which would then be entirely free, and be hauled back up to the Brooklyn tower.

Timbs and Carroll were both Englishmen. Timbs was lithe and powerfully built. Carroll, a huge, portly man, would be testing the wire, it was said, with well over two hundred pounds. Timbs had come darting down his side at a great clip, recovering for Brooklyn whatever glory had been lost by the awkward Kohrner. But Carroll had run into trouble almost right away.

For some reason, probably to gain speed going down the rope, Carroll had hung his seat by a pulley, instead of the iron shackle used by the others. The pulley had worked fine at first. He shot away from the tower faster than any of them. But as he approached the second lashing, the pulley jammed between the two traveler ropes and try as he would he was unable to budge it loose or to reach far enough ahead to get at the next lashing.

It was at this point that young Supple, who had by now returned to the top of the New York tower, decided to go to the rescue. He swung himself out over the river, sailor-style, hand over hand, with his legs wrapped around the traveler rope. He reached Carroll quickly enough, passed him by, and cut the next lashing, which instantly freed the pulley. Then back he went, up to the tower, in the same way he had come down and carrying on an easy conversation with those on the tower all the while. The crowd below was ecstatic.

Carroll, meanwhile, slid on, only to get caught the same as before, again and again, and freeing himself only after the greatest effort. His progress was so slow, in fact, that he was no more than halfway down his side of the rope when Timbs, having passed the center of the sag, had started to haul himself by hand up the steep incline toward Carroll, cutting the lashings as he went.

Once they met and all the lashings were free, there was a new problem. The traveler would not move. Somehow the two ropes had gotten twisted around each other, with the result that it was impossible to haul the men in. So something had to be improvised.

A ring was put over the traveler and a heavy weight and one end of a hemp rope were attached to it. The weight, it was hoped, would be enough to carry the ring and the rope down from the New York tower to the stranded pair, who were perhaps four hundred feet distant. But the ring slid only a quarter of the distance, then stopped for good.

Once more Harry Supple went into action. Fixing a loop in the same hemp rope, he wrapped it about one leg and worked his way out toward Timbs and Carroll, both of whom, to the amazement of everyone, seemed quite nonchalant about the whole business. Timbs, swinging in his perch, his arms resting on the upper wire, looked as though he might be about to fall asleep.

Supple reached them with no trouble, tied the end of the rope he carried to Carroll's chair, climbed onto the chair with Carroll, and the two of them were pulled back to the tower, leaving Timbs hanging out there by himself.

The traveler was tried again then and this time it worked. Timbs began moving along back toward Brooklyn, whence he came, swinging his legs, as though on a joy ride, looking all about up and down the river. But before he was a third of the way to Brooklyn there was a sudden frightful jerk in the wire, as though something had snapped, and it was noted by those watching through glasses that Timbs suddenly changed his expression. A belt on the engine had broken and it took twenty minutes to fix it, which were twenty minutes during which Timbs had no way of knowing what the trouble was. Gradually regaining his composure, he just sat very still, watching the boats below and waving his hand in answer to cheers from passengers gazing up from passing ferries. Presently he was pulled to the tower and the rest of the day was devoted to getting the new rope into proper position.

The papers made much of all this. Even the *World*, which had seldom ever had a good word for the bridge, ran a long account, calling it, in a big headline, a "STUPENDOUS TIGHT-ROPE PERFORMANCE." And later, in a formal report to the Chief Engineer, Paine would write, "Mr. Harry Supple was all that could be desired as foreman of riggers . . ."

When another rope was taken over on Monday (the second half of the new traveler), the crowds were there again and the event was treated by the press as a major theatrical opening might be, or a new circus in town. People knew more what to expect this time. And this time the new men being given a chance at the work were out to break Harry Supple's record of one thousand feet in ten minutes, which one of them, William Miller, managed to do, going the same route Supple had from the New York tower to the New York anchorage in seven and a half minutes. "As he neared the anchorage," wrote a reporter, "the order was given, 'Stand by, men, to snatch him.' His face was firmly set, and his eyes had a queer light in them, his face shining with the galvanized iron dust that the iron shackle of his chair had ground from the wires, and his hands were in active use on the rope. When he came within reach the men caught him, and with a rousing cheer landed him on the stonework."

Two others, Frederick Arnold and James O'Neil, had also, by now, taken off from the towers and could be seen plummeting pell-mell down the extreme ends of the main span over the river. O'Neil, the man from the New York tower, appeared to be making the best time. But then he stopped abruptly, as though his chair had jammed the way Carroll's had. But when the engineers, Martin, McNulty, Paine, turned their glasses on him, they saw he was getting out of his chair and climbing up onto the wire above. Next thing, he slung himself by a strap—his belt apparently—to the wire he had not been riding on and it was then that everyone on the tower realized what had happened. Some way or other the two ropes had crossed and O'Neil had jumped off from the tower with his chair slung to the wrong one, to the new rope rather than to the traveler. O'Neil had discovered this in time, obviously, and with great nerve decided to make the change immediately.

Sitting there on the traveler, a good 185 feet over the river, he cut his chair clear from the shackle and tied the chair ropes to the proper wire. It took him about fifteen minutes to make the switch, "the hearts of many spectators beating fast and hard on witnessing his cool daring."

O'Neil had moved on again eventually and reached the middle of the river, where Arnold, meantime, had been sitting waiting patiently. Then, after dangling out there for a time, the two of them were towed back to the New York tower.

But the greatest excitement of the day had occurred a little earlier over on the Brooklyn side, where the descent from tower to anchorage was made by none other than E. F. Farrington, who apparently wanted to make up for Kohrner's poor showing on Saturday and to demonstrate to his men (and just possibly to the Brooklyn spectators as well) how the job ought to be handled.

On the Brooklyn side, also, quite a number of people had been admitted within the enclosure surrounding the tower yard and among them were several seemingly fearless young women who wanted the best view possible of the high-wire performers, and of the fatherly Farrington in particular. One pretty brunette in a pink summer dress, a girl of about eighteen, led the way. Followed by the others, she had climbed to the top of the tower and there, with a pair of opera glasses, waiting for the master mechanic to make his appearance, she was seen to study the vast panorama spread out below.

Farrington, it was understood, would commence his descent at two o'clock and at two o'clock he appeared, dressed like a doctor this time, in suit and tie and a white linen duster and carrying a large knife.

He took his seat, stepped off, and was on his way with no to-do. And he was making swift progress, with all the dispatch of a surgeon, when quite by accident he dropped his knife. It fell some two hundred feet, harming nobody, but it also left Farrington in a rather impossible predicament, or so it would seem from the ground. Farrington, however, proceeded right along as though nothing had happened. When he reached the next lashing, he simply went at it with his hands and to the delight of everyone had it untied in a matter of seconds. Though it took him quite a little while to get down, the crowd stayed with him the whole time and he still managed to do the job faster than had poor Kohrner.

And then the circus acts were over. Both travelers were in position, ready for business, mounted in such a way that the space between the sides of each enormous loop was about twenty-seven feet, or so wide apart that when they were seen from the waterfront or from out on the river, it looked as though four separate ropes had been strung over the towers, placed about where the four great cables were to hang.

"I have carried out your instructions to the letter," Farrington wrote to Roebling, ". . . and from my perfect familiarity with your plans, and my own experience, I shall expect the cables of this bridge to equal, if they do not excel, the best that ever were made."

18
Number 8, Birmingham Gauge

> Sealed proposals will be received by the Trustees of the New York and Brooklyn Bridge, up to the 1st day of December, 1876, for the manufacture and delivery in Brooklyn, N.Y., of 3,400 net tons or 6,800,000 lbs. of Steel Cable Wire . . .
>
> —From specifications issued over the signature of W. A. Roebling, Chief Engineer

EVEN among his political opponents Abram Hewitt was considered an honorable man. There was nothing very engaging about him. A nervous, brusque little person with an authoritative manner, he was anything but the ingratiating good fellow. But he was hard-working, not a politician by trade, and reputedly both intelligent and honest, all qualities that counted high with the electorate in the year 1876.

Hewitt had come quite a way since the late Mayor Havemeyer had asked him to take a look into the bridge management. He was a Congressman now, and more. His friend Tilden, who had become governor of New York chiefly because of his reputation as a standard bearer against Tweed, was the Democratic candidate for President. Hewitt was his campaign manager, and with the depression still gripping the nation, the Republicans divided, and ever more scandal in Washington, it looked as if Tilden might be in the White House come spring. Tilden's opponent was the mild, modest, and largely unknown Rutherford B. Hayes, governor of Ohio, who had been picked that June at a convention in Cincinnati, where the Roebling bridge had been a way for delegates "to take some air" as the balloting dragged on, and where the nominee himself had begun his law career about the time that bridge was being started.

Tilden, however, was not much as a candidate. Cold and secretive by nature, not in the least eloquent, he was also in poor health and rather reluctant to spend any of his sizable personal fortune on behalf of his own cause. As a result Hewitt had become more than just his manager. He was the driving force of the Tilden campaign and, with his brother-in-law, Edward Cooper, the biggest financial contributor. Hewitt also happened to be running for Congress again, but his reputation was so high and his Republican opponent so weak that his election seemed certain, despite the little time he could give to it.

"Hewitt was as true a patriot, as pure a man, as ever lived, in my opinion," wrote one admiring young campaign worker. Henry Adams, historian and man of letters, whose vision of his times would count so with future historians, wrote that Hewitt was among that New York school of politicians "who played the game for ambition or amusement, and played it, as a rule, much better than the professionals, but whose aims were considerably larger than those of the usual player, and who felt no great love for the cheap drudgery of the work." Everything considered, Adams judged Hewitt "the best-equipped, the most active-minded, the most industrious . . . the most useful public man in Washington."

Hewitt was known for his liberalism and sense of responsibility. He was the wealthy ironmonger who believed in labor unions. He was the Congressman who had the intelligence to appreciate scientific research, ardently supporting the geologic surveys in the West, for example. He was not simply a reform Democrat, he was what all respectable people longed for: a decisive gentleman in public life.

So it is not surprising that when Hewitt introduced a resolution at a meeting of the bridge trustees the first week in September, it was considered an exemplary piece of foresight and was adopted immediately and unanimously.

The principal piece of business to be attended to next was the awarding of the wire contract. It was plain a lot would be riding on the decision. Estimates were that the order for wire would come to somewhere near a million dollars.

Hewitt had arrived at the meeting a little late, just as the president of the board, Henry Murphy, was recounting the progress made since the previous meeting in July and predicting no more problems henceforth. The contractors for the stone had all been paid and the Chief Engineer's specifications for cabled wire had been approved by the Executive Committee. There was no reason, Murphy said, why the bridge could not be completed in a short time.

But then Hewitt, who was vice-president of the board, said that while he had read the wire specifications and found them to be "eminently wise," he had not found any provisions concerning those who should be allowed to bid for the contract. Hewitt expressed some surprise about this.

The Chief Engineer was also a manufacturer of cable wire, Hewitt reminded Murphy and the others. He himself was a wire dealer, but he did not consider it just that he should become a bidder, and would not, therefore. Bids from any firm, company, or individual interested in the bridge in any way should not be accepted, he said. The Chief Engineer *especially* should not be allowed to become a bidder. Hewitt said further that if Colonel Roebling was permitted to compete for the contract, he, Hewitt, would resign from the board and have nothing more to do with the bridge. It was quite a little speech. It made an issue of something that had never been considered an issue in all the years since John A. Roebling was first asked to build the bridge and apparently it made a great impression, for when Hewitt offered the following resolution, it was adopted without any further discussion:

> *Resolved*, That bids from any firm or company in which any officer or engineer of the Bridge has an interest will not be received or considered; nor will the successful bidder be allowed to sublet any part of the contract to any such person or company.

There was no specific mention of Roebling or the Roebling company, but just so nobody mistook his intentions, Hewitt later repeated what he had said for the benefit of the press. "I am very strongly opposed to the Roeblings having anything to do with the filling of contracts for the bridge," he was quoted as saying.

But before the meeting adjourned another man present stood up and asked to be heard. His name was John Riley and he too wanted to say something concerning the Chief Engineer. He said he had understood that Roebling was very ill and unable to attend to his duties, but now he had heard that Roebling's wife was doing his work for him. Clearly the time had come to give somebody else the job, Riley said. If there were mistakes in the construction of the bridge, the trustees would become liable, and he for one did not intend to be responsible for any such mistakes.

Murphy immediately answered that in the event of Roebling's death, Martin, Collingwood, or McNulty could take charge of the work (nobody had said anything about Roebling dying) and Kinsella commented that there were no more efficient engineers. That seemed to satisfy Riley and everyone else and the meeting broke up.

That was on September 7, the day after William Tweed, who had disappeared from his house on Madison Avenue nine months before, stepped off a ship at Vigo, Spain, and was immediately arrested. After hiding out in a farmhouse in New Jersey for three months, Tweed had moved to Staten Island and from there went by schooner to Florida, where he lived in the Everglades until he was able to sail for Cuba. In Havana he had booked passage for Spain on a bark called *Carmen*. On the trip across he had been so seasick that he arrived in Spain weighing a scant 160 pounds. Still, incredibly, the Spanish authorities, with only a Nast caricature to go by, had recognized him. The news of his arrest caused a sensation in New York, about the time Hewitt, the man who had taken Tweed's place in the running of the Bridge Company, was making news with his latest service in behalf of the great public work. His resolution was widely praised. And even though it was an intensely political season, and the pronouncements of candidates were pretty generally viewed with that in mind, Hewitt was taken at his word. Whether he or any of the other trustees anticipated the reaction in Trenton is impossible to say.

The New York and Brooklyn papers carried Hewitt's resolution and his remarks about the Roeblings on the morning of September 8. In Trenton, later that same day, Washington Roebling dictated a letter of resignation.

His health, he wrote to Henry Murphy, was such that he could no longer continue as Chief Engineer. His doctors had been urging him to give up for the past two years, but he had not, he said, because his personal direction seemed to be absolutely necessary to the success of the bridge. Now things were different. "All plans down to the smallest detail have been prepared by me for several years to come," he said; so the work would not suffer any if he were no longer in charge. He had been neglecting his private business. ("I have not been inside our mill for four years," he added, but then thought better of that and had Emily cross it out.) "My health has been undermined by my faithful attention to these duties [as Chief Engineer] and the extra expenses I have been subjected to during these years have far exceeded the recompense I have received and I therefore feel I have earned a rest."

Then he got to the heart of the matter. He had taken the full burden of responsibility for the engineering of the bridge, he had given the work his every energy, he had made financial sacrifices, he had endured years of physical suffering, but now his own honesty and integrity, and that of his family, had been questioned publicly and this he would not endure, and particularly at the hands of Abram Hewitt. For Roebling, as he would make abundantly clear in time, in his private correspondence, did not share the conventional view of Abram Hewitt.

> Although devotion to the success of the work has been my ambition throughout, it is only by strict adherence to this principle that I have been able to steer clear of the entanglement connected with the general management of the work and maintain the impartial position on which alone an engineer should stand. It is therefore with regret at the close of our pleasant relations I am obliged to resent the gratuitous insult offered to me by the Vice-President of the Board of Trustees . . . a man whose designs upon the cable wire and ironwork of the superstructure are only too transparent and whose nominal connection with the Board of Management has had from the first no object but his own personal advantage.

According to Emily Roebling's letter book the letter ended there. But a day or so later, still in a rage, Roebling wrote to the Brooklyn *Eagle*, explaining a little further what he meant by Hewitt's "nominal connection." Hewitt, Roebling charged, could resign his place as trustee anytime so as to evade his own resolution. The bridge itself meant nothing to Hewitt. But the letter never appeared in the *Eagle*. Either Roebling decided not to send it or Thomas Kinsella decided not to publish it.

Murphy did not keep Roebling waiting long for an answer. He said he could not accept Roebling's resignation, only the trustees could do that at their next meeting. In the meantime, he urged Roebling to reconsider, assuring him that his services were vital, and that Hewitt was motivated by only the noblest intentions.

Roebling was anything but pacified by this. Still hurt and angry, he was even more outspoken in his reply.

> I was publicly and specifically singled out by name by Mr. Hewitt, as if I had spent my whole life in concocting a specification which I alone could fill or as if I were a thief trying to rob the bridge in some underhanded manner and against whom every precaution should be taken. Coming from such a source this is an insult I cannot overlook and I am compelled to resent it by declining to remain in a position where I am at any moment liable to a repetition of such acts on his part.

In light of later events, however, the most interesting part of the letter, none of which was ever made public, was this single sentence:

> As you seem to be deeply impressed with Mr. Hewitt's action in declining to become a competitor for this wire, I desire to say that his magnanimity is all a show, as the firm of Cooper and Hewitt have no facilities whatever for making the steel wire, and if you receive a bid from a Mr. Haigh of South Brooklyn, it will be well for you to investigate a little.

What Henry Murphy thought of this is not known and there is nothing in the record to indicate that he followed up Roebling's suggestion. So presumably Murphy either knew more than he was ever willing to admit or he figured Roebling's accusations to be those of a man under a great deal of strain. Roebling and Hewitt had long been rivals in business after all and the Roebling brothers were all known to be staunch partisan Republicans just like their father.

How much Murphy knew about Haigh's business reputation one can only guess. It is possible that he and the other trustees had no suspicions about the man, but it is not very likely, for J. Lloyd Haigh had certainly not gone unnoticed during his time in Brooklyn.

Haigh had arrived in town some twenty years before. He took up residence on Columbia Heights, joined Plymouth Church, and commenced his social life as "a single gentleman." He was quite suave and handsome apparently, a fine vocalist and considered "a great catch." At Plymouth Church he "not only became an enthusiastic attendant, but was noted for his intense admiration for the pretty girls of the Bible class. His captivating manners and his personal attractions made him a welcome guest in many households, and his triumphs in winning hearts soon assured as great a success in that direction as his subsequent career in his business operations." In a short time he was courting the daughter of a prominent Willow Street family. An engagement was announced and the fashionable part of Brooklyn was "agog with the gossip of the approaching nuptials," until the father of the bride-to-be did a little checking into Haigh and found the man already had a wife and two children living "in rural retirement." The wife was brought to Brooklyn to confront Haigh and the whole affair was hushed up as quickly as possible by the Willow Street family.

Haigh, however, waited only long enough for the rumors to die down before setting off on another round of courtship, devoting his attentions this time to a young lady who lived only a few blocks from Willow Street, and again presenting himself as a bachelor. Again he was found out and again he became a suitor, on Henry Street this time, only now he was saying he had obtained a divorce. There was a wedding shortly, with Haigh's first wife again in Brooklyn claiming he was still her husband. Eventually there was a third wife and some question whether there had ever been a second divorce.

That there was a single Brooklyn man on the Board of Trustees who had not heard something of all this seems very doubtful. It is also doubtful that none of them knew, as Roebling did, that in the wire business Haigh was considered little better than a crook.

When the trustees met a month later, there was no talk of Roebling resigning. No business at all was taken up since the turnout was not enough for a quorum. The crisis had passed apparently, the rift had been patched up some way or other. Perhaps Roebling's assistants persuaded him to change his mind, or his brothers did, or Emily, or all of them together. Or perhaps he himself, with time to think things over, decided he had come too far, sacrificed too much, to quit over injured pride, that there was another honorable alternative, and that in truth the bridge really could not be built without him. Perhaps he was just incapable of giving up.

Whatever his reasons, he made two decisive moves in quick succession that October, both of which were taken as sure signs of his renewed determination to stay with the bridge.

About the middle of the month he left Trenton for New York. He was still in a very bad way, physically and emotionally. His condition, in fact, was so precarious that he was unable to make the trip by train, his nervous state being such that he could not endure that much speed or vibration or the crowds of people. So it was arranged for him to go the whole way by canalboat and tug, instead, and as he came up the bay and into the East River, he saw the bridge for the first time in three years.

It looked to him, he is reported to have said, exactly as he expected it would. The carrier rope was up by then, as well as the first of the cradle ropes. A newspaper item from about this time was clipped out and saved by Emily Roebling:

> There is something colossal in the look of the East River piers as they show in the morning sunlight; the ropes already connecting the two piers seem like slender threads, and as the vessels pass and repass under them some idea may be formed of what may be the effect when the graceful upper wire structure is completed, with the roadway crowded with passengers and vehicles of all descriptions and the high-masted clippers and coasting traders passing underneath.

He and Emily moved in with her brother, General Warren, and his wife, who were then living on West 50th Street. The intention, it seems, was to stay there just temporarily, before completing the return to the house on the Heights.

Then, only a short time later, Roebling notified Henry Murphy that he had sold his stock in John A. Roebling's Sons, three hundred shares, worth $300,000, thereby eliminating any possible conflict of interest. "Please acknowledge receipt of this letter," he wrote, "and oblige me by making the above fact known to the Board of Trustees of the New York and Brooklyn Bridge at their next meeting." This Murphy did, on November 6. According to the requirements of the Hewitt amendment, the Roebling company, the major wire manufacturer in the world, and the only one Roebling had total confidence in, could now stay in the bidding.

But the question of the Chief Engineer's health had not been resolved in the view of several trustees, including General Lloyd Aspinwall of New York, a most piously civic-minded figure, who urged that "some competent engineer be associated with him [Roebling] at once, in order to protect the future interests of the public."

Murphy replied that Roebling was greatly improved and able to see his assistants on a regular basis now that he was living in New

York. Stranahan, too, rose to Roebling's defense. But Aspinwall wanted a consultant just the same. There was a good deal more discussion, with much emphasis on the idea that nothing personal was intended toward Roebling, and in the end it was agreed that a committee be formed to select suitable candidates for the job.

Abram Hewitt did not attend this session, nor would he appear again for some time. The little Congressman had much else on his mind. The elections were over, and while he himself had won handily enough, there was some question about who had been elected President. Tilden, as the official returns later showed, had a plurality of more than a quarter of a million votes and was the rightful winner, but at that point the outcome in three southern states—South Carolina, Florida, and Louisiana—was still undetermined, and if they were to go for Hayes, then Hayes would have the electoral votes needed to win—which was what the Republicans were claiming. Hewitt, still the moving spirit of the Tilden camp, was doing all he could to rescue his man, writing speeches, sending prominent citizens off to the disputed states to see that a fair count was made (Grant, meanwhile, was sending his own set of "visiting statesmen"), and rallying his fellow Democrats to "boldly denounce all . . . fraudulent contrivances for the destruction of self-government." But in the year of the centennial of American democracy, the Presidency was about to be stolen by the Republicans, who were quicker and more efficient with their bribes than the other party. Hayes would win in the Electoral College by a majority of one. But it would be March before that happened, and until then Hewitt was spending most of his time in Washington. How he felt about Roebling's recent moves he did not say.

In the meantime there was work to be done and Roebling applied himself to it.

Still in confinement in New York, he was kept constantly informed by his assistants as the footbridge cables and the second cradle cable went up and he himself kept after Murphy not to let things slide while waiting for a decision on the wire contract. A whole force of men had to be trained for spinning the cables, he explained. It was work in which all would be novices, which would be immensely difficult at best and seem terribly dangerous to anyone not accustomed to it. Men would have to be taught the crucial techniques of regulating the deflection of individual strands. Others would have to be taught to oil and splice the wire. He would need good men to operate the various machinery to be used. "Our previous bridges," he said, "always came near enough together so that many of the old and experienced hands were to be found to initiate the new ones, but they are entirely wanting for this work."

Only a small amount of wire ordered now would be enough for the men to work with and would make a great difference he said. So Murphy ordered thirty tons of wire—ten tons each from John A. Roebling's Sons and two other firms, one of which was J. Lloyd Haigh of South Brooklyn.

Roebling sent off a steady stream of dispatches to his assistants—to specify how he wanted the oil kettles housed, to say that a sample ferrule joint sent over for his inspection looked a little short, to explain the differences in working with iron and steel wire (steel wire may crack, he warned). In a long letter to Murphy and Stranahan on the matter of a consultant, he expressed himself with customary bluntness. Any consultant would either be his superior, in which case he would resign, or his equal, in which case he would resign, or his inferior, in which case the man ought to take his place in the ranks. It was understandable that Aspinwall and some of the other New York trustees were concerned lest he die before the bridge was built. But there were things they ought to understand: "Man is after all a very finite being in his capacities and powers of doing actual work," Roebling wrote, "but when it comes to planning, one mind can in a few hours think out enough work to keep a thousand men employed for years. . . . Continuing to work has been with me a matter of pride and honor! You must however trust me in so far that the moment I am unable to do full justice to my duties as chief engineer, I shall give you ample warning . . ." He really did not want to be troubled by any more talk of consultants.

To better familiarize himself with what the Europeans were doing with steel, he had begun learning Danish and Swedish. He sent Ferdinand lengthy, highly technical instructions on steel-wire extrusion, and in one such letter, commenting on the deficiency of a Roebling product already in service, there appears what may possibly be a touch of his old humor: "Everybody is getting afraid of the carrier rope, so many wires are breaking in it and when they break they make such a noise you can hear it all over. It hangs right over the Trustees' office and if it breaks it might kill a dozen of them."

Sealed bids for the cable wire were to be received at the bridge offices in Brooklyn up until the first of December. In the meantime, on Pier 29, beside the New York tower, wire samples sent with each bid were being tested on various machines in the presence of the bidders or their agents, all of whom thus far had expressed total satisfaction with the procedure.

The specifications called for steel wire of what was known as Number 8, Birmingham Gauge (this was a diameter designation), with a breaking strength of not less than 3,400 pounds. The steel was to be of medium quality, neither too hard nor too soft, and the wire had to be "straight" wire, that is to say, when a ring of it was unrolled on the floor, the wire had to lie perfectly straight, without any tendency to spring back into coils.

The specifications called for 6.8 million pounds of wire. During the tests, the wire would be required to bear a certain amount of strain before it broke, and to stretch a certain number of feet, then recover a certain portion of the stretch when the strain was removed. "In the case of any dispute arising between the inspector and the manufacturer," the specifications stated, "the Engineer is to be the sole arbiter." But the way things were, with Roebling bedridden, the tests were actually being conducted, the records kept, by Paine, with Martin in over-all charge.

On Monday, December 4, the trustees gathered for the formal opening of the bids. There were nine bids in all, including three from European manufacturers. The highest bid, from a wiremaker in Worcester, Massachusetts, came to nearly fourteen cents a pound, which would bring the aggregate cost close to a million dollars. The lowest bid, from John A. Roebling's Sons, was for less than half that, at six and three-quarter cents per pound for Bessemer steel. The Roebling company had also submitted a bid for crucible steel, but it was higher than the one other bid submitted for crucible steel, that from J. Lloyd Haigh of South Brooklyn.*

None of the bids were released for publication when the trustees ended their meeting. All further consideration of the subject was to be deferred, reporters were told, until the tests were completed. Just

* Crucible steel, steel made in comparatively small quantities in crucibles, or casts, was considered the finest-grade steel and was used principally for tools. Bessemer steel, made in a "converter" according to a process developed by the Englishman Henry Bessemer twenty years earlier, was the least expensive steel on the market, the kind used in the greatest quantity in the 1870's and for rails chiefly. Between the two, crucible steel was thought to be markedly superior but the quality control of Bessemer steel had, in fact, been perfected to a remarkable degree by Carnegie and others. It could be produced in far greater quantity and was without question a perfectly respectable product.

the same, the rumor got about that the Roeblings were the low bidder and everyone assumed that was that.

But on December 13 the New York *Herald* published an interview with a man named Albert Hill, who was described as a consulting engineer with offices on Fulton Street in Brooklyn, and who had a number of very unflattering things to say about the wire specifications Roebling had drawn up and about the tests by which the wire was being judged. Nobody connected with the bridge would have believed that cable wire and its technical characteristics could ever become subjects of public interest, but that is just what was about to happen.

Hill considered the specifications very poorly written ("complex," "onerous," and "vague" were some of his adjectives). He thought there was too much emphasis on the manufacture of the wire and too little about the type of steel to be used. He objected strongly to the fact that Roebling had not specified what kind of steel he wanted. Hill's view was that crucible steel was the only acceptable thing for such a bridge and he said it was what Roebling himself had required in his earlier specifications for the different steel ropes already in use—which was quite true. What possible reason Roebling had for not demanding crucible steel this time was a great mystery to Hill.

Finally, Hill was not in the least happy that the Chief Engineer, the man with the final say on the tests, was a member of the famous Trenton wire family. Hill wanted the tests conducted by an impartial board of engineers, so as to place the awarding of the contract beyond all suspicion of favoritism.

The reporter who interviewed Hill went over to Pier 29 the next day to see how the tests were being conducted and to talk to C. C. Martin. But Martin sent him back to Brooklyn to see Murphy, who chose to make no comment.

Murphy wasted no time contacting Roebling. What was he supposed to say, Murphy wanted to know. Roebling answered that he attached nowhere near the importance to the tests that everyone else seemed to. There was nothing to guarantee that a bidder would supply wire of the kind submitted for the tests. "If one man's samples were too good he would be sure to reduce his standards, provided he got the contract, and another man, whose wire fell short of the standard, would have to make his wire come up to the mark before any could be accepted." The point of the tests, Roebling said, was to satisfy him that each bidder could produce the kind of wire called for and to satisfy the bidder that making such wire involved no impossible demands. When he drafted the specifications, Roebling said, he knew the contract was to go to the lowest bidder and he had considered it his duty to include the tests as a simple protective measure. "Of all known materials, wire possesses a shape most susceptible of being tested in every direction. If necessary, a whole mile of it could be tested for its elasticity, throughout every foot of its length, without injuring it in the slightest degree. It is not like a huge casting, which may be full of hidden flaws, or like a big gun which bursts at the first discharge."

Were he asked point-blank which were the finest samples tested to date, he told Murphy, it would be those from Richard Johnson & Nephew of Manchester, England. As far as the bids themselves were concerned, he remained "in total ignorance."

Hill, however, had still more to say. The mathematics in the specifications were not up to snuff, he next claimed, and the *Herald* presented his own computations, in several long, dense paragraphs, full of wire weights and measurements, diameters and principles of physics, very little of which anyone other than a professional engineer could or would wish to struggle through. But seeing it all set forth in print in one of New York's most powerful papers had a profound effect on the nervous system of those trustees who had had any prior misgivings about Roebling's ability to handle the job, most of whom knew next to nothing about engineering.

"Of course this is only a theoretical demonstration," Hill remarked in conclusion. Just how the *Herald* happened to come upon Hill or why the editors chose to give his opinions such a play was not made clear. He had never built a suspension bridge, as he admitted. None of the assistant engineers had ever heard of him, including Martin, who had been a Brooklyn man for more than twenty years. But the *Herald* called Hill's argument clear and lucid and claimed the errors in Roebling's calculations were so glaring that the specifications were worthless and no contracts should be made on them. *Herald* reporters looked up another engineer who agreed with Hill, a General Francis Vinton, professor of civil engineering at Columbia, who was interviewed at his bachelor quarters at the Racquet Club. *Herald* editorials demanded that the bridge trustees answer the charges and every one connected with the management of the bridge began getting extremely edgy.

The gist of all Hill's arithmetic was that the Number 8 wire being specified had a breaking strength of 3,600 pounds instead of 3,400 as Roebling had it.

In actual fact, there was nothing at all to Hill's charges, as anyone working on the wire tests could have shown, and as would be explained by Paine very shortly. But before that happened three badly informed trustees agreed to be interviewed on the subject. General Aspinwall made the silly comment that with 6,300 wires in each cable he did not see how a difference of two hundred pounds one way or the other mattered much; Thomas Kinsella said simply that he was going along with Hewitt on all this and that Hewitt, who was a wire manufacturer himself, had found nothing wrong with the specifications. And the third man, who refused to be named, said he could understand why Roebling might want his brother to get the contract. If he were Roebling, he said, he would want his brother to get the contract.

Hill fired back that a difference of 200 pounds per wire among 6,300 wires added up to a 1,260,000-pound difference and that he was not out to prove Roebling was no engineer. In conclusion he added a last gratuitous comment, which Roebling doubtless found about as revolting as anything said about him to date:

> I fully appreciate that Colonel Roebling would, to a certain extent, be liable to criticism for these errors, but, taking into consideration the facts that Colonel Roebling is only following out the work commenced by his father, and had also impaired his health . . . there are extenuating circumstances that the trustees should bear in mind. The errors that I have pointed out might have been made by some subordinate in whom Colonel Roebling had confidence, and were thus printed without his having really supervised the work. As for the gentleman saying that were he chief engineer he also would desire his brother to obtain the contract under him, that is a matter of taste.

To a great many people it probably seemed that an absurdly big fuss was being made over very little. But the effect on several trustees was quite serious just the same. Hastily it was decided that the specifications and tests were "worthy of investigation" and Murphy told Paine to come up with an answer to Hill's charges at once.

Ferdinand Roebling came on from Trenton to tell Henry Murphy that the Roebling family had had about enough of all this and to simplify things would just as soon withdraw their bid. But Murphy, who seems to have maintained his composure, talked him out of it.

Two days before Christmas Murphy called a private meeting of the Executive Committee to consider the bids and make a recommendation. The tests, he announced, were now completed and he had a report on the results from the Chief Engineer. He also had Paine's reply.

Hill's theoretical mathematics, Paine explained, were based on Hill's own figures for the specific gravity and diameter of Number 8, Birmingham gauge. These were different from the ones the bridge engineers were going by, which, he acknowledged, were round figures. "These specifications were intended for the guidance of practical wiremakers," Paine said, "and are written in plain language, easily understood by practical men, and are not incumbered by the formula employed, or the details of calculations necessarily used in their construction." He proceeded then to disprove each of Hill's charges, point by point, confirming the accuracy of the specifications to the satisfaction of everyone at the meeting. Paine did not, however, attempt any answer to the question Hill had raised about the quality of steel to be used.

Murphy next read Roebling's report on the tests.

The letter was dated December 18. Roebling still had not been told which firm was lowest bidder or what any of the bids were. He said that nearly every bidder had been able to meet the standards required. Except for a few cases, he had no information concerning the variety of steel used in the numerous samples submitted. Nor did he know whether the manufacturers had provided that information with their bids. Regardless, he said, "It would be very unwise to accept two special prepared rings, as an absolute guarantee of the perfection of 6,000,000 pounds." He then gave a brief account of each manufacturer's samples, describing the first on his list this way:

> Mr. J. Lloyd Haigh presented several samples of very good wire, apparently cast-steel, of three different stocks. The tensile strength exceeding the requirements, the elongation very good, the elastic limit up to the mark, the modulus of elasticity admissible. This wire is very straight, galvanizing smooth, the polish, though of no advantage, adds to the appearance of the wire. . . .

The best wire was from the English firm Richard Johnson & Nephew. The rings from the Cleveland Rolling Mill Company were quite good, but not well galvanized. A German wire was also rated as excellent, and one ring of Bessemer steel wire from John A. Roebling's Sons was designated very good, but two other Roebling rings, of cast steel, had not stood up to the bending tests satisfactorily.

Once Murphy had finished Roebling's report, the committee unanimously recommended that the contract be awarded to John A. Roebling's Sons and Murphy was requested to convene a special meeting of the board the following week.

But now things began changing swiftly behind the scenes. Several trustees, and most notably Thomas Kinsella, began playing for time. When the board met next, two days after Christmas, it was decided, on a motion from Kinsella, to postpone the final vote on the contract for two more weeks. The newspapers were informed that the results of the tests were still under consideration. Only the bids were released for publication, which made headlines but left the story still very much in the air.

These latest delays were the direct result of the Hill disclosures, the *Herald* quickly claimed, commending the "honorable members" of the board for their discretion. A little later, under an article headed "CHEAP STEEL INSURES A WEAK BRIDGE," the paper insisted that the whole issue at stake was the one Hill raised at the start: why Roebling had not specified crucible steel.

Then the night before the trustees were to meet to vote, the *Eagle*, after first demolishing Hill's attack (Hill was actually a Hungarian, the *Eagle* had earlier claimed, by way of a disclosure), suggested that perhaps Bessemer steel was *not* after all the best answer. "Unquestionably Bessemer steel wire is the cheapest," wrote Kinsella in a three-column editorial, "but whether the trustees should get the cheapest wire, or the best at the cheapest rate, is the question which they will be called upon tomorrow to consider."

Kinsella also pointedly raised the issue of Roebling's connection to his family's business, and even implied that perhaps Roebling's break with the business was not altogether certain or done for the most commendable reason. "He had recently, it appears, sold out his interest in the Trenton works, so as not to embarrass his brothers . . . There is no disguising the fact, however, that the whole subject is complicated by this consideration."

For anyone who had been following the story closely, it was clear the tables were turning. Never once before had the *Eagle* had a critical word for Roebling. Indeed it was Kinsella, more than anyone, who had made such a popular figure of the man. Moreover, Kinsella's call for the best steel at the cheapest price was clearly another way of saying that the contract ought to go to the lowest bidder for crucible steel, who, of course, was J. Lloyd Haigh of Brooklyn. But no one knew which of the other trustees Kinsella was speaking for or how many of them there were.

The meeting of the trustees on the afternoon of January 11, 1877, was held as usual in the board room of the bridge offices, where now the model of Roebling cable exhibited at the Centennial was prominently on display, along with Hildenbrand's mural-sized drawing in a mammoth frame. The meeting was the largest ever held. Nineteen were present, which was the entire board save one—Abram Hewitt, who was "unavoidably detained" in Washington, but whose presence would be very much felt all the same. Also in the room, sitting unobtrusively in the back and saying nothing, were a few privileged visitors, one of whom was J. Lloyd Haigh.

First on the agenda was the annual report from the Chief Engineer covering the year 1876, which was presented by Henry Murphy. The document included, among other things, Roebling's explanation of how and why the wire had been tested, and emphasized, as Roebling had to Murphy, that the tests should *not* be taken as a hard-and-fast guarantee. "The assurance of the correct performance of these tests must remain a matter of confidence and trust," said the Chief Engineer. "The building of the whole bridge is a matter of trust."

The board then proceeded to consider the resolution from the Executive Committee recommending that the contract be awarded to the Roebling company. General Slocum wanted Army engineers appointed to inspect the wire before it left the Roebling works. This he said would entirely remove all public suspicion about the Chief Engineer. Action on the resolution was deferred.

Then the chair was asked to read a letter from Abram Hewitt, dated Washington, January 8. The letter was addressed to Murphy and was quite long. Hewitt was still extremely concerned about who was to get the wire contract.

He began by saying that if the trustees were willing to rely on the specifications and on the kind of inspection called for in the specifications, then he did not see how the trustees could do anything but award the contract to the lowest bidder, the Roeblings.

> In this event, however, in view of the personal relation of the chief engineer to the stockholders of that company, and for the protection of the honorable reputation which he deservedly enjoys, it seems to me that it will be the imperative duty of the trustees to provide for the inspection of the wire entirely independent of the supervision and control of the chief engineer. In this particular I have no doubt I only anticipate a request delicacy and a sense of propriety would have led him to make to the trustees.

But, said Hewitt, there remained the very big question of whether the specifications guaranteed a suitable quality of wire, provided it were of Bessemer steel, and in his opinion they did not. He did not consider Bessemer steel of sufficient quality. He had had a great deal of experience in these matters, he said, and the kind of tests Roebling had designated were not enough to prove or disprove the quality of Bessemer wire.

> So far as I can see, therefore, a proper regard for the public safety requires that the trustees should either stipulate on the contract that Bessemer steel should not be employed for the manufacture of the wire, or if it be employed the wire should be subjected to different and more ample tests than are provided for in the existing specifications. Those tests should be made by engineers having no relations to the contractors . . .
> . . . I confess that I have such grave doubts that I would not venture to record my vote in favor of Bessemer steel upon the tests now provided for in the specifications, and I am convinced that the apparent economy involved in the use of wire made from this material should not weigh against the risk involved in its use, unless it can be more carefully guarded than it now appears to be . . .

The letter was a bombshell. This was the same Hewitt who, four months earlier, sitting in this same room, had called the specifications "eminently wise" and whose own much publicized resolution had supposedly resolved all ethical questions raised by Roebling's ties to the wire business. Moreover, Hewitt happened also to be the very one who had urged Roebling *not* to specify crucible steel this time, but to leave the bidding open for Bessemer steel as well.

Still, Hewitt was the expert, supposedly, and a looming figure these days, particularly among Democrats. And irrespective of politics or personalities, grave suspicions had been raised by the Hill attack and even the fairest, most impartial men in the room were quite honestly at a loss to know just whom to go along with: Hewitt or Roebling?

Furthermore, in the back of everyone's mind were two very recent sensational tragedies. On the night of December 5, the Brooklyn Theater, built by William Kingsley's construction company and owned by his partner, Abner Keeney, had caught fire and 295 people had lost their lives, many of them because the balcony had collapsed. It was the worst disaster in the city's history. Then on the night of December 29, one of the worst railroad disasters of the nineteenth century occurred when a bridge failed at Ashtabula, Ohio. The bridge was just eleven years old, a wrought-iron truss over a seventy-five-foot gorge. When a train pulled by two locomotives started across it in the middle of a snowstorm, the center span gave way. It was thought that the metal had failed somehow. Eighty lives were lost.

The newspapers were angrily crying for an explanation. *Harper's Weekly* in its latest issue asked:

> Was it improperly constructed? Was the iron of inferior quality? After eleven years of service, had it *suddenly* lost its strength? . . . Was the bridge, when made, the *best* of its kind, or the *cheapest* of its kind?

The chief engineer of the railroad, a man named Charles Collins, who had had nothing to do with the design of the bridge, but had examined it frequently and conscientiously, tendered his resignation, then committed suicide.

The Ashtabula bridge had not been cheaply built and the iron had not suddenly lost its strength in some mysterious fashion. As subsequent investigations would show, the bridge probably went down because the derailed wheels of several cars ripped the bridge floor, causing a violent pull of a kind the truss had not been built to withstand. But the idea of bad (cheap) metal failing had been planted in the public mind.*

After Hewitt's letter was read the bids were reviewed still one more time, at the request of "Honest John" Kelly, Comptroller of the City of New York, who had replaced Tweed as the head of

* The Ashtabula disaster was only the worst of hundreds of bridge failures of the time. Something like forty bridges a year fell in the 1870's—or about one out of every four built. In the 1880's some two hundred more fell. Highway bridge failures were the most common, but the railroad bridge failures received the greatest publicity and cost the most lives.

Tammany Hall. General Aspinwall said the history of crucible steel was too well known to need further consideration. The whole matter resolved itself, he said, into the question of whether they would put into the cables of the bridge a wire made from steel, the strength of which might be in doubt, as was the case with Bessemer steel, or use crucible steel, about which there could be no doubts whatever. Emphatically he was in favor of using crucible steel and nothing else.

Mayor Ely of New York said this was the most important question put before the trustees in the entire history of the bridge and he personally wanted more time to familiarize himself with the subject. He therefore moved for adjournment. But Stranahan said now was the time to discuss the issue, while there were so many of them present, and the meeting continued.

At about that point a trustee named William Marshall, who was a wealthy cordage manufacturer and one of Brooklyn's most prominent citizens, recalled a conversation he had once had with John A. Roebling, during which Roebling talked about a testing strain for the wire that was half what his son had specified. So it did not seem to Marshall that anyone ought to get very worried about the standards called for in the specifications. The important thing, he said, was to buy wire that came up to standards. Thomas Kinsella, who had kept very quiet so far, said he thought no undue weight should be attached to the informal remarks of the elder Roebling. Kinsella did not think the lowest-price steel would be the cheapest. "It was the duty of the trustees to do for the bridge, as they would do for themselves," he said. He was not interested in any special kind of steel, he wished them to understand. However, he did have an interest and pride in his own city and said he had a natural wish that the contract might come there. (There were two Brooklyn firms in the bidding, J. Lloyd Haigh and the Chrome Steel Company, but the Chrome Steel bid worked out to more than $200,000 higher than the Haigh bid and so was, for all intents and purposes, quite out of the running.) He would vote, Kinsella said, for using crucible steel.

Henry Murphy read some extracts from engineering papers, extolling the superiority of steel made by the Bessemer process. Then there was a long discussion about what crucible steel was or was not, how Roebling's earlier specifications called for crucible steel and why that was. William Marshall reminded everyone that the change had been made at Hewitt's urging. "Mr. Hewitt was something of an expert and ought to know something about steel," Marshall said. The problem seemed to be that Hewitt could be quoted to substantiate either side of the argument.

Comptroller Kelly said he wanted the bids for Bessemer steel referred back to the Executive Committee, Kelly wanted the other manufacturers to have the chance to bid on the lower quality of steel (as though they had not in the first place) and he moved the Executive Committee open up the bids again. Aspinwall seconded the idea. The motion carried and that might have ended things for the time being had Kinsella not said that they ought to test the prevailing mood of the meeting on the question of which kind of steel to use. He would offer a motion, he said, to make the contract with the lowest bidder for crucible steel.

Kelly said he hoped the resolution would not pass. Aspinwall said he did not want to be trapped into committing himself. Kinsella answered that he had no desire to trap anybody. The only object was to call a test vote. Stranahan said the motion, if carried, would pledge them to use crucible steel.

The vote was taken and the motion lost, 8 to 7, with four abstaining. After a few further comments, the meeting broke up. By that time it was nearing five in the afternoon. But then the Executive Committee met, privately, and instead of reopening the bids as directed by the board, the contract was immediately awarded to J. Lloyd Haigh.

There is no way of knowing what happened, since everything said in the meeting was kept secret. All Murphy said later in a letter to the board was that the committee's decision had been the direct result of Kinsella's test vote. "They [the committee] regard that vote, although wanting one of a majority, still as decisive against the use of Bessemer steel; for in so important a matter as the main cables, it would, in their opinion, be unwise to adopt a material which is distrusted by any considerable portion of the trustees. The question of cost is an important one, but it is subordinate to that of safety, and the difference of expense between the two is comparatively too small to permit such difference to prevent unanimity and entire confidence." (The difference between the Haigh bid and that of the Roeblings for Bessemer steel came to $132,600.) The official record of the committee meeting states there were seven men present—Murphy, Stranahan, Slocum, Van Schaick, Motley, Marshall, and Kingsley.

How close was the vote? Who voted which way? The record provides no answers.

Since its meeting of December 23, the committee had done a complete about-face. But because everything was done in private, the public, to whom the bridge supposedly belonged, would never know anything about that. Four days later, on Monday, January 15, another special meeting of the board was called. Murphy announced that J. Lloyd Haigh would post $50,000, or about 10 per cent of the contract, as surety, and he read a letter in which Haigh promised to supply crucible steel of the same quality as his samples. Then a resolution giving Haigh the contract was adopted by a vote of 16 to 1, the one dissenting vote being cast by William Marshall.

So the wire in the bridge would not be Roebling wire. It would be made in Brooklyn by the one man Roebling had specifically warned Murphy not to trust.

The news was warmly received in Brooklyn. Thomas Kinsella called the decision "most satisfactory" and said it was a "matter for congratulation" that a Brooklyn manufacturer had won out over the leading wiremakers of America and Europe (he did not specify which he meant). The resolution of this whole wire controversy was a great triumph the *Eagle* contended. "It is, we suppose, admitted on all hands that the cables which are to sustain the bridge structure are the most important features of this great undertaking. These failing, all fails."

The *Union* wrote that the bridge trustees had honored themselves and said, "We shall try to forget as soon as possible that they were ever brought to discuss so absurd a proposition as the use of Bessemer steel." The impression left was that a catastrophic blunder had been narrowly averted. Someone had not known what he was doing and that someone had to be Roebling. The *Union* wanted prompt action taken.

> . . . They [the trustees] can help us and the public to forget this by taking the next most necessary step in their great undertaking, the selection at once of a suitable and eminent consulting engineer. We know the exceeding delicacy of this point. No one, and not we, certainly, desires to be unconcerned or lacking in sympathy with the physical troubles and disabilities of the present Chief Engineer . . . But we must deal with things as they are; the subject is too important for sentiment, and the bridge needs the live attention of a man in his

best powers. It is almost such a case as that where General Winfield Scott used to sit in lethargy over the early business of the war, when the great rebellion at its outbreak found him with his great powers masked and half useless by the infirmities of age. It seemed to be unkind and treasonable to say of this old hero, and in his presence, that the duties of the Commander-in-Chief must be done by someone who could take the field, endure the hardship, and live in the saddle. . . . So now the great bridge enterprise needs an active consulting engineer, bringing to his duties the best qualities of natural fitness and training, with physical powers equal to every emergency. It is loading a great and difficult undertaking to an unnecessary strain, this carrying with it its disabled chief engineer, and keeping down its discussions to the atmosphere and the hush of his sick room . . .

There had been no comment from Roebling since the wire decision was announced, nor any from either of his brothers in Trenton. But a few days later, the following letter appeared in the *Eagle*. It was signed "Tripod." Quite possibly it was written by Washington Roebling.

My attention has been called to an article in the *Union*, relating to the appointment of a consulting engineer for the New York and Brooklyn Bridge. I know not what power behind the throne dictates the spirit of this, and similar articles, though I am forced to the conclusion that there is one as there was in the case of Mr. Hill, who professed to expose inaccuracies in the specifications for the wire, a matter by the way, in which he has finally failed . . .

The *Union* calls for a consulting engineer who will "endure great hardship" and practically "live in the saddle." If the writer understood whereof he wrote, he never would have used those expressions in that connection.

Consulting engineers seldom seat themselves in any other saddle than a cushioned office chair, or expose themselves to any greater hardships than a few hours' quiet office chat, per day, and the labor of signing a monthly receipt for their salary.

The hardships of a campaign usually fall on the subordinate officers, as they have in the construction of the Bridge, since the illness of Colonel Roebling commenced. If Colonel Roebling had thrown more of the details of the work on his subordinates, in its earlier stages, he would not now be taunted by the *Union*, with breathing the air of a sick room, nor insulted by comparison with a superannuated general of armies.

Neither would the present Assistant Engineers, who under the immediate direction of the "invalid," have successfully brought this great work thus far on its way toward completion, with unsurpassed skill, fidelity and endurance, be told that they were of no account, and that they must give way to a consulting engineer whose "natural fitness, training and endurance" qualify him to lead "in the saddle."

Will the *Union* kindly tell me where such a one is to be found? Can it point to any living engineer outside of the "sick room" who has had sufficient training in this specialty of suspension bridge work to guarantee to the trustees and the taxpayers that he could do the work as well as the "invalid" assisted by those who may be said to have grown up with the work, under Colonel Roebling's own eye, who are familiar with his plans, and devoted to their success?

. . . The fact is, there is no better talent in the country in this specialty than is now engaged in the construction of the New York and Brooklyn Bridge. It is ample for all its needs, present and to come. And if chronic grumblers and those who have private "axes to grind" would let the work alone, they might wake up some morning and find it completed, and be ready to take part in the opening ceremonies.

If the letter was indeed from Roebling, then it is the one and only time his feelings ever appeared in print. But by then there were people in Brooklyn talking about more than just a consulting engineer. The move had begun to get rid of Roebling entirely. On January 18 another editorial appeared in the *Union*.

THE CHIEF ENGINEER

It has become the deepest of mysteries in the Board of Bridge Trustees, too solemn for the keenest reporter to penetrate, and far too solemn for gossip, where the chief engineer is, and what is his condition. For aught any public act or appearance of his may indicate, he may have been dead or buried for six months. He is surrounded by clouds impenetrable. . . . We declare the great East River Bridge in peril, because it has no head, because its wires of control run into somebody's closely guarded sickroom, because it is certain that a sick depressed tone runs through all its engineering discussions, from this cause. . . . The sooner we have a live, active chief engineer in full powers on the bridge work, the better the public of two cities will be pleased with the prospect.

When Henry Murphy read this, he must have figured, knowing Roebling's sensitivity to such charges, that another letter of resignation would be in his hands the next day. But no such letter arrived and there would be no more talk of resignation from Roebling. He had decided he would stay with the job, and fight for it, if need be.

Some time in 1877, when things had quieted down a bit, Washington Roebling made some extraordinary private notes in his letter book.

The whole maneuver to take the wire contract away from the Roeblings and give it to J. Lloyd Haigh had been the work of Abram Hewitt, he said, just as he had warned Murphy. Haigh, a known scoundrel, was in fact Hewitt's man. Hewitt, Roebling noted, held a mortgage on Haigh's wireworks and he had made a deal with Haigh not to foreclose so long as Haigh turned over 10 per cent of what he made from the bridge contract. When his first attempt at exempting the Roeblings from the bidding had failed (because Roebling sold his stock), Hewitt had then manufactured the crucible steel issue. Roebling never said Albert Hill was working for Hewitt or for Haigh, but that would seem to be the case and what is implied by "the power behind the throne" reference in the letter signed "Tripod." Hill did not interest Roebling much, but Hewitt did: "In laying this plan, he [Hewitt] well took the calibre of the men in the board, for when a demagogue wants to effect an object he always raises the cup of public virtue—and under cover of the smoke he raises, slips in himself. It is on such low and crafty tricks that the honor of a Hewitt rests," wrote the engineer.

Roebling never bothered to speculate in his notes on why Kinsella turned on him and worked so hard in Haigh's behalf. Maybe the editor was sincerely convinced crucible steel was the superior product. He also very much favored the idea of the contract going to a Brooklyn firm, as he said. But there is a further point to consider. No paper in the East had so strongly supported Samuel Tilden for President that fall as had the Brooklyn *Eagle*. Kinsella's efforts in behalf of Abram Hewitt's candidate had been extremely helpful and much valued by Abram Hewitt. And that January of 1877, with Tilden very likely about to become President, the times were ripe with possibilities for a brilliant, politically ambitious and cooperative editor.

19
The Gigantic Spinning Machine

I never saw better days for bridge work.
—C. C. MARTIN

THERE was now one continuous path from Brooklyn to New York. The temporary footbridge, finished in early February 1877, was a sort of hanging catwalk strung from city to city, draped above the river at an elevation sixty feet higher than the actual roadway would be. Farrington had been in charge of the work and it was carried out with the greatest dispatch, even during days of extremely cold weather. No sooner was the footbridge in operation than the newspapers sent reporters to make the crossing, which a few of them managed to do, with Farrington going along each time as an escort. His own men were never bothered by great heights, he was quoted as saying. "No sir, no man can be a bridge-builder who must educate his nerves. It must be a constitutional gift. He cannot when 200 feet in the air, use his brain to keep his hand steady. He needs it all to make his delicate and difficult work secure. They must plant their feet by instinct . . . and be able to look sheer down hundreds of feet without a muscle trembling. It is a rare thing for a man to lose his life in our business for loss of nerve."

But few of Farrington's first visitors were so constituted. One reporter described proceeding along, step by step, nearly frozen with terror, as though his feet were fixed to the slat floor by Peter Cooper's glue, as he put it. Another wrote, "The undulating of the bridge caused by the wind, which was blowing a gale, the gradually increasing distance between the apparently frail support and the ground, the houses beneath bristling all over with chimneys, looking small enough to impale a falling man, the necessity of holding securely to the handrail, to prevent being blown off, produced sensations in the reporter's head—and stomach—never experienced before. In vain he glanced furtively into his companion's face to detect any signs of flinching on his part. Stolidly the master mechanic kept on, and the reporter fancied once that he caught a backward glance of enjoyment at his discomposure."

The customary visitor's entrance to the footbridge was from the top of the Brooklyn anchorage. Beside the short flight of steps leading up to the footbridge, a big sign had been posted.

SAFE ONLY 25 MEN AT ONE TIME.
DO NOT WALK CLOSE TOGETHER. NOR RUN, JUMP,
OR TROT. BREAK STEP!
W. A. Roebling, Engr. in Chief

From there the footbridge swept upward to the tower, at an angle of about thirty-five degrees. The width of the bridge was just four feet. There were wire rope handrails on either side, at hip level, but there was nothing to prevent a person from tripping and falling under the handrail and there were spaces between the slats, big enough to look through, put there intentionally to give the wind less hold. Actually with its guy wires and storm cables, the bridge was amazingly solid. Though men walking on it experienced a slight rocking motion, in ordinary weather there was very little horizontal swing. Still Thomas Kinsella was telling people that old John Roebling had said the thing would probably be blown down a dozen times and to judge by the looks of it nobody found that hard to believe.

Halfway up the walk, between the anchorage and the tower, was the first cradle, a narrow platform, a hundred feet long, with wooden handrails, that was hung on cables, like a slender scaffold, at right angles to the cables. Five such cradles had been put up, one between each anchorage and its companion tower and three over the river, at equal distances. By all reports they were a good deal more stable underfoot than the "sensitive" footpath, the main purpose of which was, in fact, to provide access to the cradles, where men would be stationed to see that the wires were hung precisely right and to bind them into strands.

Once when Farrington and a reporter from the *Tribune* reached the top of the Brooklyn tower, the reporter sat down to rest and to take in the view. But it was then his troubles began.

> Trinity Church steeple was fencing with Grace Church, the City Hall was bumping into the [Central] Park lake, Governor's Island, guns and all, was playing shuttlecock and battledore with Harlem, Beecher's Church shook its windows on the top of St. Paul's, the top of the *Tribune* tower had fastened itself somewhere and was swinging the building pendulum fashion, and the reporter leaned against the solid tower in dread lest his weight would push it over.*

On Washington's Birthday, about nine in the morning, passengers leaving from Brooklyn on the Fulton Ferry suddenly spotted two young ladies out on the footbridge. "There was no hesitation or misgiving in the demeanor of the ladies," according to one account. "On the contrary, they stepped out boldly . . . without the use of the handrail." Everyone on the boat began waving and calling, as the girls, accompanied by a man and two boys, headed for New York. As was learned later, the girls were the daughters of C. C. Martin, who was the man seen accompanying them (the boys were his sons). They were, as the papers all noted, the first women to make the crossing, but the fact that they had been allowed to do so struck many people as utter lunacy.

"While Revs. Drs. Storrs and Buddington and several excellent ladies are moving in the matter of providing a new insane asylum for this city," wrote the *Eagle*, "a considerable number of our people are providing the necessity for such an institution and their own fitness to be life occupants of it . . . by crossing the footbridge . . . without call, without necessity, out of no business or artistic impulse, and from sheer foolhardy and peripatetic 'cussedness.' "

Something like a hundred people crossed the footbridge that same day. They were able to go right up onto the anchorage and out onto the footbridge. There were no gates to stop them, no guards on duty. But Martin's daughters were the only ones to cause

* One of these firsthand accounts of crossing the footbridge was an entire fabrication, Farrington said later, but he never indicated which one it was. The day the reporter appeared at the bridge, Farrington had told him it was too windy for an inexperienced man to go out. Immensely relieved the reporter had returned to his paper, only to be told by the editor that he was to get the story wind or no wind. So he had retired to a quiet place, sat down, and drawn on his imagination.

any kind of popular stir, "BEAUTY ON THE BRIDGE" ran one headline the next morning and the *New York Illustrated Times* published a panoramic engraving of the two, their silk scarves and heavy skirts whipping in the wind, stepping nimbly out from the tower, as high as the clouds, a gentleman in a derby showing them the way.

Enough of a fuss was made over the incident that Henry Murphy decided all visitors would henceforth be required to apply for a pass. This was supposed to put a stop to the traffic, and it did, temporarily.

For now there was too much going on in preparation for the cable spinning for there to be room for anyone on the catwalk who did not belong there. Outside the Brooklyn anchorage yard, in the vicinity of James and Front Streets, workmen were tearing down old houses to make room for an expanded storage yard for the wire. The air was filled with dust and noise. Rubble was piled in immense heaps, enough brick it seemed to build twice the number of houses being torn down. Old women in shawls and street urchins came daily to gather whatever firewood they were able to carry off.

Inside the anchorage yard, both back and front, every foot of space was taken up with heavy timber frames, about six feet high, where the wire coils were hung out to dry after being coated with oil. The wire came from the factory galvanized but not oiled. This was done inside a low shed on the Front Street side of the yard. The coils were simply dipped into a trough of linseed oil—a two-man job.

On top of the anchorage, inside an enormous covered shed, was a wilderness of big wooden drums mounted upright in vertical timber frames, like a convention of water wheels, as someone remarked. Each drum was about two feet in width and eight feet in diameter, but mounted as they were, clear of the floor, they stood nearly twice as high as a man and they had handles all around their outer rims, exactly like a ship's wheel. Also, standing to the rear of the drums, on the floor, in a horizontal position, were a number of smaller reels, built along the same lines, but only half the diameter.

Once a coil of wire had been dried out sufficiently in the yard, it would be hoisted to the top of the anchorage, where it would be wound first onto one of the small horizontal reels, then onto one of the big upright drums, the wire going on as smoothly as thread around a spool. It was from the big drums that the wire would play out over the bridge, in much the way a fishing line goes from the reel at the handle out along the rod.

Since a coil from the factory constituted only a few hundred feet of wire, innumerable splices had to be made before the wire was wound onto the drums. It was essential, of course, that every splice be as strong and weather-tight as the wire itself. It had taken two years of experimenting and testing to develop the system settled on. A galvanized steel ferrule two inches long and about as thick as a lead pencil was double-threaded inside, at both ends, one thread to the right, the other to the left, and corresponding threads were cut on the ends of the wires to be joined so that the same turn of the ferrule would screw both ends at once. The ends of the wires were also mitered, so that once the wires had been screwed tightly they could not twist. With the help of a small viselike apparatus, the wires were held together and the ferrule was put on, great care taken to screw it straight. The sharp edges of the ferrule were then beveled, the joint was cleaned of dirt and oil and dunked into a small ladle full of melted zinc to give it an all-over galvanizing. That done, the joint was coated with red paint.

In this way coil after coil was spliced and run onto the big drums as a single continuous wire. On each drum there were fifty-two coils, or nearly ten miles of wire. Once things really got going, it was expected that the cable-making machinery would consume some forty miles of wire a day, or about four drums a day. So for months the work crews were kept constantly busy "drumming up" wire.

On Tuesday, May 29, things were far enough along to send a first experimental wire across the river. (Just to see that everything was in proper order, and that the wire was strung at exactly the right deflection, Farrington, for one, crossed over the footbridge a total of fourteen times in that one day.) On June 11, 1877, the spinning of the great cables was begun. The way the system worked, two cables, those on the downstream side, were built simultaneously.

The impression among most people was that the wires were to be twisted, like the fibers in an ordinary rope or like the wires in the different steel ropes already in use on the bridge. But this, of course, was not the case.

In the first place it would have been impossible to twist such a mass of steel over such a distance, and even had it been possible, twisted strands would have less strength than those laid up parallel, all in line, as these were, like a bundle of rods, and compacted into what would, in the end, be essentially a great curved bar of solid steel.

The traveler rope was now working back and forth across the river day in, day out, the big horizontal wheel upon which it revolved turning overhead on the Brooklyn anchorage, first this way, then that, and all the other smaller pulleys and belts and innumerable cogs keeping up a low, steady rumble.

The wires were taken across the river by what was known as a "carrier," a big iron wheel that looked like an oversize bicycle wheel with six spokes. Its axle was fastened to the working rope by an iron arm, or gooseneck, and was weighted to make it stand out perpendicular from the rope so as to clear the cradles and supports on the towers. At the Brooklyn anchorage, the end of a wire would be drawn off one of the big drums and a loop of it slipped over the carrier wheel; the end of the wire would be drawn back taut and secured around a hefty iron brace, or "shoe," that was shaped roughly like a horseshoe magnet, about two feet long and little more than a foot across with a groove around the periphery for the wire to ride in—as a skein of yarn is held on one's thumbs. The shoe was secured flat on the back end of the anchor bars, or at the end opposite from where the strands would be finally attached. The engineer would then start up the working rope and away would go the carrier, trundling off toward the Brooklyn tower, then over the tower and out across the water, towing the loop of wire behind, which meant that two wires were being strung at once.

In the meantime another carrier wheel would be coming back from New York, riding on the other half of the endless working rope. So by the time the first carrier was approaching New York with its load of wire, the empty carrier would be arriving in Brooklyn to pick up another loop in exactly the way the first one had. When the outgoing carrier reached the New York anchorage, the engine would be stopped at a signal from flagmen and the loop would be slipped off and drawn taut around a shoe there. Then the engine would be reversed, the empty carrier would start its return trip, while the other one would be starting out from Brooklyn with two more wires. And so it went, always with one carrier going out as the other came back, the two of them in turn constantly towing over big loops of the same unbroken wire that kept playing off an enormous upright drum, until a whole strand was built up—

hundreds of wires in unbroken continuity, with uniform tension and with exact parallelism between all of them.

With everything working right, it took the carrier about ten minutes to make the full trip from anchorage to anchorage. Along the way men were stationed on the towers and cradles to watch the progress of the traveling wire and to see that each wire was positioned with the proper sag and tension. As the running wire went over the tower, pulled by the carrier, a man would lightly guide it with his hand to keep it from chafing against timbers or masonry. Another man would catch it with a great pair of clamps that were attached to a block and tackle and with this he would draw up the slack until the wires from the tower back to the anchorage hung with exactly the same sag, or deflection, as the others. From the cradle halfway between the anchorage and the tower, men called "regulators" would signal just when to stop, then fit the new wires up against the others, signal again, and one of the towermen would immediately mark the wire with red paint where it passed a similar mark on the other wires, exactly at the point of crossing the axis of support. Similar marks would also be made at the cradle.

Then the towermen would turn their attention to the river side, where the same system would be repeated, with the regulators on the three river cradles going through the same motions as their turns came up. So by the time a loop of wire reached the New York anchorage, it would be thoroughly "regulated"—its sag properly adjusted all along the line—and the paint marks provided a ready index of any slip or strain that might need correcting.

Once a strand had been completed, pairs of workmen would go riding down from the towers in "buggies," compressing the wires into a cylindrical form with big clamp tongs and applying temporary "seizings," bindings of soft wire, every fifteen inches or so, to hold the strand together until all nineteen strands of the cable had been strung and could be clamped into one compact unit. The buggy was nothing more than a pine wood trough, about 10 by 6 feet, with a side rail, and was suspended from overhead trolley wheels that rolled nicely along the bundled wires on which the work was being done. The men would merely let themselves down from the towers by letting out a long rope.

During the time a strand was being made, it hung higher than the ultimate position of the cable it was to be part of. At mid-span over the river the difference in elevation was sixty feet. This not only kept the wires well above the topmasts of passing ships, but nearly doubled the tension the wires would have at the lower level —the deeper sag—and that helped straighten any crooks, or kinks, there might be and further tested the strength of the wire. Once the seizings were completed, the strand would be unhitched from the temporary fastening at the anchorage by a powerful block and tackle, let forward carefully into permanent fastenings at the end of the anchor chains and also lowered into the saddles on the tops of the towers.

It was basically the same system used at both the Niagara and Cincinnati bridges, only here, as with everything else, the work was on a far bigger scale. Judging by previous experience, Roebling estimated that the time needed to make the four cables would be about two and a half years, taking into account that much would depend on the weather.

Sometimes wires would break when part way over the river. The loose end would have to be hauled in and a splice made. Sometimes the delay would be only a matter of minutes. Other times, when the break occurred on the New York side, more than an hour might be lost. "These delays often occurred in the midst of a promising day's work," Farrington wrote, "and were very vexatious."

High winds and fog could make the delicate business of aligning the wires virtually impossible. Extreme temperature changes would cause significant expansion or contraction in the wires that would have a pronounced effect on their deflection and in the early stages this could complicate things enormously.

Before the first wires went across, the engineers had four guide wires strung for the men in the cradles to go by when adjusting the deflection of the first wires. To everybody's surprise the two land spans had not hung the same. The difference in the deflections could be readily seen just by looking at them. But only after considerable trouble was the cause found. There was a slight difference in the diameter of various lengths of wire and to solve the problem hundreds of coils had to be stretched out, measured, and enough wire selected of uniform size and weight to make up the required lengths. After that the weather had to be watched for periods of perfect calm, during which time the necessary adjustments could be made, to a hairsbreadth. As a result of all this, about six weeks were used up.

But the wire stringing, once it got going, went faster than had been anticipated. The weather was just about ideal. With a little practice the men were laying up fifty wires a day, which was not bad for a start and would have been better had the wire manufacturer been delivering on schedule. The *Eagle* was now calling the bridge "The Gigantic Spinning Machine."

By July 2 the first of two strands was completed for the two cables on the downstream side of the bridge. The work of lowering the strands into position then began. At the anchorages the strands were drawn back by a hoisting engine until the shoe was released from its fastening. Then shoe and strand were lowered slowly, carefully forward, twelve feet, the hoisting engine and a block and tackle holding the immense pull of the strand. Because there was a twist in the tackle, the shoe turned up on edge as it came forward and slipped neatly in between the eyelets of the anchor bars. The forward motion was stopped then and a seven-inch steel pin was passed through the eyelets and the shoe.

On the towers, too, the strands had to be lowered into the groove of the saddle, a distance of about three feet. This was done by eight or ten men working a capstan on a platform built over the saddle. The capstan turned a nut on a screw that lowered the strand. Once the strand was properly attached at the anchorages, and at rest in the tower saddles, then it was also at the desired altitude over the water. The whole operation was "difficult and delicate," as the newspapers reported, requiring "nice calculations." The great danger, of course, being that the strand might get away. The strain exerted by each strand at the anchorages was about seventy tons.

In the meantime, the first two strands for the two upstream cables were begun. So by the end of the first week in July all four cables were being strung simultaneously; all four carriers were shuttling back and forth high over the river, as regular as clockwork. Paine and Farrington had been assigned by Roebling to be certain everything was done just so. Collingwood and McNulty had been put to work on the approaches. And Roebling, too, was now watching the work himself once again, for at the start of the month he and Emily had returned to Brooklyn, to the brick house on Columbia Heights. With a pair of field glasses, from a bay window overlooking the river, he could at last follow the day-by-day progress being made.

That was the summer of the Great Railroad Strike and for much of the country it was a dark, discouraging time. Half a dozen cities were hit by walkouts and violence. In Baltimore twelve people were shot down by militia. Pittsburgh was in the grip of a mob for two straight days. Millions of dollars' worth of railroad equipment was destroyed in Pittsburgh alone. The Union Depot was burned, stores were looted, and a pitched battle between rioters and soldiers took the lives of fifty-seven. It was the bloodiest labor uprising the country had ever known and it left much of the populace wondering what in the world was happening to life in America.

But at the bridge things had never gone better. Not in eight long years had the work advanced so smoothly. Even the newspapers seemed satisfied with the way things were being handled and could find fault with no one. "The network of wires across the East River is rapidly beginning to look something like a bridge," commented the *Herald* in mid-August. By then four strands had been completed and a new feature added, "regulation cradles," as they were known, long, narrow, flimsy-looking scaffolds suspended fifty feet below the regular cradles, which put them in line with the lowered strands and made possible a closer surveillance of the strands as they lined up alongside one another.

Somber-looking trustees in stovepipe hats climbed the stairs on the James Street side of the Brooklyn anchorage to pose for group portraits at the start of the footbridge, or they went off to Saratoga or the White Mountains with their families, confident the bridge was in good hands. And for thousands of New Yorkers and visitors, the footbridge had become one of the city's greatest summertime attractions. Virtually anyone could go up and sample the view and test his nerve if he cared to. In fact, so many people were now applying for passes to take the walk that Henry Murphy was spending an hour every morning just listening to what the applicants had to say.

"People from every corner of the globe have crossed the bridge," he said, "Australians, New Zealanders, a man from the Cape of Good Hope, and persons from every country in Europe and Asia, from every state in the Union, from Canada and South America. The Governor of Bermuda went across the other day, and the officers of the Russian man-of-war gave us a visit. Captains of steamships and merchantmen are frequent applicants, and we like to pass them, because they will have to sail under the bridge, and we desire their friendship." As a general rule, five classes of applicants were granted every courtesy—foreign visitors ("They may never come again, and it is natural they should desire to cross the bridge while they have the opportunity," Murphy explained), newspapermen, engineers, and all politicians and preachers.

Daily at the Bridge Company's Brooklyn offices the crowds jammed the hallways and lobby waiting to see Murphy. Everyone had his particular reason for wanting permission to make the walk.

"I am a stranger here," explained one applicant.

"Where are you from?" asked Murphy.

"From New York," the man replied gravely, and the story was soon all over Brooklyn.

A Connecticut couple, both in their seventies, had walked over. Murphy even allowed a doctor and his wife to carry their newborn baby across. But when a Miss Mazeppa Buckingham requested permission to ride over on horseback, he said no. (Her agent proposed to hoist the horse up onto the Brooklyn anchorage with a sling.) "It would have made a great sensation," Murphy told reporters, "but you see that's just what we want to avoid. We don't want to turn the bridge into a show."

By the middle of August two or three thousand people had made the crossing, and most all of them went home to tell how he or she had been "one of the very first" to cross the Brooklyn Bridge and thus the claim would be passed along proudly to many thousands of grandchildren and to their progeny.

Amazingly, there were no accidents. Several men became so dizzy that they got down on their hands and knees and crawled back, hugging the slat floor for dear life. At least one woman fainted and had to be carried off; many started out, then turned back. Several people had gotten about halfway out over the river with no trouble but then suddenly froze with fear, unable to move one way or the other. One of these was a Brooklyn hatter who figured such a conspicuous display of daring would be good for his business.

Among the children to cross was Al Smith, whose father had been employed as a sort of guard to keep unauthorized people off the bridge and who "gave himself permission to take Alfred across," as Smith's sister told the story years later. Smith himself would often describe the hazardous journey over the footbridge as the most thrilling experience of his boyhood, while his sister would remark, "I remember Mother sitting at home, saying ten rosaries all the time they were gone! But my father was determined to take the boy across the bridge so he could say he crossed it before it was built."

Murphy saw no reason why anyone should not be allowed to travel the footbridge, providing he had a comparatively valid reason and looked as though he would not do anything foolish out there. He was annoyed by the way people were cutting the wires and taking off pieces for souvenirs, but then it was the people's bridge. None of the workmen seemed bothered by the sight-seers traipsing along. The thing they found most interesting was the number of women who passed by and how fearless they appeared. Murphy admitted to one reporter that he himself had not been across as yet. "I started to go once," he said, "and while I looked upward or ahead I was all right; but I chanced to look down, and . . . and I determined that I couldn't afford to lose the President of the company just then, and so I went back."

And then it was September and a broken and aged-looking Tweed was standing before the New York Board of Aldermen telling the whole truth and nothing but the truth, on the understanding that if he did, "Honest John" Kelly would see that he was set loose from prison and be henceforth immune from further prosecution. It was at this time that Tweed described in his own words the part he had played in getting the bridge started and Henry Murphy and William Kingsley spent the better part of several days denying everything Tweed had to say about as fast as he said it.

And before the month was out an English seaman walking the footbridge was seized by an epileptic fit and it was all several workmen could do to hold on to him as he writhed in convulsions. In desperation they finally tied him to the narrow floor of the bridge, with his arms and legs hanging over the side. The man recovered shortly and was helped back to the ground, but the story was made so much of by the papers that Murphy promptly stopped issuing any more passes. The fun was over.

In early October workmen digging foundations for the Brooklyn approach turned up some old Spanish coins worth about sixteen cents at most and the story spread through town that Captain Kidd's treasure had been discovered. One paper commented that if they kept digging they might find enough to finish the bridge. Another expressed great pleasure that the New York approach required the demolition of one of the city's worst neighborhoods, and the *Eagle*, with nothing better to say apparently, ran a long macabre essay on the bridge as the coming place for the truly artistic suicide. "It is hardly necessary to point out to thoughtful men the splendor of a suicide committed from this virgin height." Hanging, poison, blowing one's "vulgar brains out" with a pistol, were all condemned for their "despicable lack of originality." The river below is swift and treacherous, the editors wrote, and there would always be a good-sized audience on board the ferries. If jumping did not appeal, there were other choices. "Let us imagine a man addicted to hanging and think of the unique picture which early passengers would behold should they turn up their eyes in the ghostly dawn and see a man hanging by his neck fifty feet from the water's edge! A little ingenuity would enable him to so affix one end of his rope that he could not be cut down for hours and could oscillate before the eyes of an admiring though horror-stricken crowd of thousands." Even poison, shooting, and stabbing would have some style, the editors concluded, if done from the Great Bridge.

In October a well-to-do New Yorker named Henry Beers, a spokesman for the so-called Council of Political Reform, announced that he was joining the cause to have the bridge stopped. It was nothing but a flagrant waste of public money, he contended, and a serious hazard to navigation, which was the same claim being made by a warehouse owner named Abraham Miller, who had decided to sue the Bridge Company. English law, said Beers, who was assisting Miller, always held rivers and oceans sacred to seamen.

But the old buildings kept coming down inland from the New York anchorage as the path for the approach pressed toward City Hall Park. And out over the river the carrier wheels, glistening in the sunshine, kept spinning along, faster than ever. Never had he seen such weather for bridgebuilding, C. C. Martin said. By November, twenty-two strands were in place. On the cradles, little sentry boxes had been built where the men could get out of the cold while waiting for the carriers to come by. In the sheds over the saddles, wood stoves had been installed and the men were wearing heavy mittens and shoes of buffalo hide.

On Thanksgiving Day one of the wires snapped. It was considered a small matter. Nobody was injured and the newspapers heard nothing of it, but Paine decided to send a section of the wire over to Roebling for him to take a look. "It is as brittle as glass," Roebling wrote back to Paine. ". . . The first question arises is how much of this same brittle wire has been going into the cable without our knowledge and secondly what steps must be taken to prevent its reoccurrence. Is it due to a wrong system of inspection or what is the reason in your opinion?"

Roebling also broke off several pieces of the wire and put them in with a letter to Henry Murphy.

"This is what Mr. Kinsella is pleased to call the best," he wrote angrily. "In reality it is worthless . . . and the most dangerous material that could be employed. How much of this poor wire has been going into the cables I do not know. Can I be held responsible for that? It is scarcely right that the engineers should have to be acting as detectives. I see but one way of preventing such wire being run out and that is to double the number of inspectors at the contractor's works."

If Murphy read Roebling's letter to the members of the Executive Committee at their next meeting, or to the trustees, there is no mention of it in the official records.

Kinsella and Roebling were running head on once more.

Bessemer steel had become a bone of contention again, exasperating the engineers no end. No issue it seemed ever stayed resolved for very long. This time the argument was over the suspenders, the wire ropes that would hang down from the cables to the roadway. Roebling had decided that Bessemer wire would do perfectly well and that was what he specified, but at the last minute, at Murphy's urging, he had had the word "Bessemer" scratched out of the printed specifications. Again "steel wire" was all that was called for. Kinsella had been furious, exclaiming in the editorial columns of the *Eagle* that cost was no factor here, that no chain was stronger than its weakest link, etc., etc. The engineers had no business deciding such matters alone, he said. And when John A. Roebling's Sons, the lowest bidder for the suspenders, was awarded the contract, Kinsella wrote that it was solely because Washington Roebling wanted it that way and that the contract should have gone to a Brooklyn firm (unnamed).

Henry Murphy was quoted as saying that Roebling had nearly died when the earlier contract was awarded to Haigh and that he, Murphy, had no wish to see that happen again. "All of which is bosh," responded the Brooklyn *Union and Argus*. "We have as much sympathy for Mr. Roebling as other people . . . But, we submit, that this work is entirely superior to any man or all of the men concerned in its construction, and it cannot, nor any part of it, be subordinated to the whims, fancies, or caprices of a sick man."

The paper refused to let the matter drop, writing scornfully of Roebling's power and of the stupidity of the "stupendous enterprise being wholly committed to a single brain, which is extremely liable at any moment to be stilled forever."

The sick man, meanwhile, had had a powerful telescope mounted at his window and trained on the bridge. As for the things being said about him in the papers, he had no comment. He would not see reporters.

Late one Saturday afternoon, shortly before Christmas, there was a bad accident behind the Brooklyn anchorage. Masons were finishing up a series of arches, set on big, square brick piers, that would support the roadway of the approach inland from the anchorage. The foreman noticed a great crack in one arch, about twenty-five feet above the street, and immediately ordered the men down off the work. But one man standing below never heard the warning and when the arch gave way he was buried.

The men started digging frantically through the rubble and in about ten minutes they found him, so badly crushed that he would have been difficult to identify had they not known who it was. The body was covered with a sheet of canvas and carried to a tool house. A big crowd had gathered around by then. The area was one of seeming chaos even under normal conditions, with heaps of brick and stone all about, swinging derricks, and their countless ropes, cement machines, scaffolds, great half-dug pits, and sixty or seventy men busy at one task or another. But now things were out of hand. Somebody began saying the other arches were coming down. There was a panic and the crowd went surging in every different direction and nobody seemed to know what was happening. Then somebody was saying something about one more man trapped under the debris. So half a dozen volunteers started digging again and the crowd rushed back to watch, fully expecting to see the rescue workers buried next.

By this time, too, the news had spread over to Fulton Street that a lot of men had been killed and a crowd coming from that direction was so big that the police had trouble holding them back.

No more arches fell and no other bodies were found, but an investigation was immediately called for and there was great public sympathy for the victim and his family. He was Neil Mullen, a Brooklyn man and a widower with six children.

A coroner's inquest established that the centering, the temporary wooden supports used under the arches, had been removed before the mortar had set properly. The Brooklyn approach was McNulty's domain and McNulty, who testified at the inquest, looked to be pretty much at fault. Roebling was infuriated by the whole affair. It was exactly the sort of thing he might have prevented had he been on the job. "The brick arch fell because it had a right to fall," he wrote bitterly to Henry Murphy, who felt, understandably enough, that he ought to have an explanation on hand from the Chief Engineer. "Every arch, be it round or flat, must fall if its thrust is not met by an adequate lateral support," Roebling lectured. ". . . The real accident was not so much that this arch fell, as that the other one stood."

> As to the matter of responsibility I am primarily responsible because it is my business to see that everything goes along right. Mr. McNulty is secondarily responsible because he was the engineer directly in charge of the construction and because he did not sufficiently heed the special warning I gave him about this very thing some weeks before its occurrence.

McNulty had told Roebling he did not know why he had removed the centering. "Ambitious natures are apt to be overconfident and to shrink from asking counsel of more experienced persons for fear their infallibility might be impugned," Roebling wrote Murphy. "Time and age cures all this." But then he added that the real explanation might be simply that McNulty was overworked.

Roebling could appreciate the problem. He himself was doing more now than he had since the long, difficult winter before the Centennial. For a great many people it might have appeared that his real work was nearly done. The engineering involved, the planning, and the decision making ought to be all but over, it would seem, now that the towers were up and the wire was going across. But it was not that way. Nor did Roebling by any means have everything all figured out.

In the public mind he had become a thorough mystery, the tragic victim of his own wondrous creation, cursed perhaps, like his father before him, remote, hidden, maybe a little mad, seeing everything and yet never seen. It was said he was so crippled that his wife had to feed him, which was true partly. It was said the disease had affected his mind, which was not true. And still, from a chair behind a distant window he could raise towers of granite and spin steel through the sky.

But for the man himself every detail was a personal concern and no answers came easily, despite the things said about his genius. Nothing could be taken for granted, especially now after the accident. Nobody could be trusted, completely. Anybody might let him down, including his father.

At the moment he was wrestling with the design of the enormous truss that would stiffen the roadway and wondering whether to make it of steel, instead of iron as his father had specified. He was not sure either if his father had made the truss big enough. He delegated Paine to find out all he could on the latest advances in steelmaking. He wrote to Hildenbrand, day after day, pouring out his own thoughts, his doubts and questions, for pages.

> There are so many points to be considered, so many conflicting interests to be reconciled on the parts of the truss that it is perfectly bewildering to pick out the best thing. For example, I want to reduce the aggregate weight so as to keep down the pressure on the masonry. I want to simplify the superstructure so as to make work in the shop easy and erection easy and safe and I also want to keep down the wind surface as much as possible. On the other hand I want the truss sufficiently strong to resist a reasonable amount of bending, and this goes against the other points. But the only possible way in which I can reduce pressure on masonry and wind surface is by reducing the height and weight of the trusses and increasing the strain per square inch on the iron. I do not see that any reduction of weight is possible in any other parts of the structure. By making the truss rods as far as possible of steel we make some reduction in weight but it is only in the low truss that the rod section is great enough to enable us to attain any appreciable advantage by the substitution of steel for iron. In the high truss with rods through two panels the section is hardly sufficient to make it worth while to change. This therefore would be one argument in favor of again reducing the weight of the intermediate truss and leaving the rods in all the trusses within one panel. This includes the two central trusses even if they are arranged with a square bar in the middle of two flat ones outside.

He was working toward another momentous decision. And he was feeling his way. But days like this were what he enjoyed most.

His concern for incidentals was perhaps the most extraordinary thing of all. For him there were no incidentals. Everything counted. Nothing could be left to chance or for someone else to decide. Hildenbrand, Martin, Paine, Farrington, all heard from him daily now. It seemed he wanted them to know his every thought.

The following is only an excerpt of just one of the letters Roebling wrote to his assistants during this time. It was to Farrington.

> I want you to help me get out a specification for all the timber planking for our bridge floor, and it must be done by the first of April or sooner. There is a tremendous quantity of it to be got out, and most of it has to be planed, all of which takes time. It should also season for a while. You know we cannot hang up any of the ironwork unless we have planking to follow right along, so there is no time to lose.
>
> The bulk of it is yellow pine. First: There is the planking for the promenade; next, the planking on the wagon tracks; then the longitudinal stringpieces under the tramway, and also under the regular rails on the Rail Road track. Then a lot of short pieces of bridgings of yellow pine between the floor beams and lastly short oak planking laid crossways where the horses walk, of different thickness, and also some spreaders between the safety rails on the railroad track. Our flooring here differs from the Cincinnati flooring in having only one thickness of plank. I don't propose to treat or preserve or tar this lumber in any way, because I am pretty well convinced, judging from past attempts on previous bridges have cost more than they are worth and have often done positive harm. And as our planking is of but one thickness it can season from above and below. By the introduction of the light intermediate floor beam I have been able to reduce the thickness of the floor beam such that is from 6 to 5 inches.
>
> First of all I want you to go down to the back office and consult with Hildenbrand and Paine about the best means of fastening down this planking and securing the ends to the floor beams. As there is but one thickness generally it is more difficult to do. The track stringers should all be spliced by halving them at the ends. They can be [illegible] down to the top chord of the floor beam by either one or two light bolts and a cross plate underneath. I want your opinion as to whether it will pay to splice the 5″ planks or only butt them. The promenade plank will be too thin for splicing. I think the Cincinnati plan of fastening will answer best. That is a little round-headed bolt having on one side a square washer underneath which catches under the flange of the 6″ channel. The head of the bolt can be sunk in pretty well and the hole filled with hard cement. This will

answer very well on the wagon tracks where much is covered. But it will make a nasty-looking promenade. Yet I hardly see how I can help it. You know we have on the promenade alternately a double channel and an I beam for floor beams. Now it occurs to me that we could fill in between the double channels with a pine filling piece and fasten the planking into that with wooden nails. To the I beam we fasten with little bolts.

I believe yellow pine won't warp as badly as oak. The promenade planking must be very long and very narrow—nothing over 4″ wide. The other planking can be 5 x 5 or 5 x 6 as the space demands. (Would it pay to caulk it? Hardly I guess?) The ends must butt over the center of the floor beam. Shall we therefore order them exact lengths or make allowances and saw the butts here? How much allowance for waste? Must everything be planed? These stringers could be let in ⅜″ on the floor beams. The bridging can be ordered in long lengths and then cut to suit. The promenade suspenders run through the floor. Here we must have two 5″ streaks of plank because 4″ would be too narrow.

In regard to length of planking, stringers and so on, it must run from 3 panels to 5 panels in length . . . The timber must be good sound clear stock free from sap, cracks, splits, shakes, wind shakes, slivers and wavy edges, knots, black-knots, work holes. No bush timber or dry-rotted timber or dead timber, etc., etc. The timber must be planed up true, full and square with sharp edges . . .

And the letter continues on in the same fashion for pages. Twice Emily, who was taking it down, had to sharpen or change her pencil. The letter must have taken a good hour to dictate, perhaps longer considering his condition. Only three words in the whole thing were crossed out. The rest was put down with total certainty and no second thoughts.

On January 8 the Executive Committee held its first meeting of the new year, during which a request from J. Lloyd Haigh was considered. According to the record of the meeting, "Mr. Haigh, the contractor for furnishing the steel wire for the cables, applied to be allowed to substitute the personal obligation of Messrs. Cooper and Hewitt in place of the percentage retained under his contract, amounting now to $29,277, in order to save himself interest upon it." Mr. Haigh's proposition was declined. This bit of information appeared in several newspapers the next day, along with a report on various other items taken up at the meeting.

If any of the bridge officials or trustees had been ignorant of Abram Hewitt's interests in the fortunes of J. Lloyd Haigh and his wireworks, they were no longer. But yet there is nothing in the record to indicate that any of them thought this the least bit out of the ordinary, nor did the papers in either city make any editorial comment on it. Nothing was said either of Haigh's generally unsavory personal reputation. Abram Hewitt made no comment.

The *Union and Argus* did, however, pick up another item concerning certain legal fees authorized by the committee. "Of course more or less legal information is required by the bridge trustees," wrote the paper, "but it does seem as if there might be something more than coincidence in the twin facts that law costs the bridge $7,500 a month, and that the cheapest establishments at which the article is purveyed are those of E. M. Cullen and H. C. and G. I. Murphy. Inasmuch as H. C. Murphy is the President of the Board, the effect of the figures is an impression that the gentleman in question is overduly given to taking counsel of himself and pays a little highly for his soliloquies."

H. C. and G. I. Murphy were Henry Murphy's sons. They were doing as competent a job as could be done and were charging no more for their services than would any other reputable firm, or so said Henry Murphy by way of explanation.

On August 25, 1876, Master Mechanic E. F. Farrington made the first crossing of the East River by way of the Great Bridge, riding in a boatswain's chair that was tied to an endless wire rope called the "traveler." An estimated ten thousand people were watching from both shores.

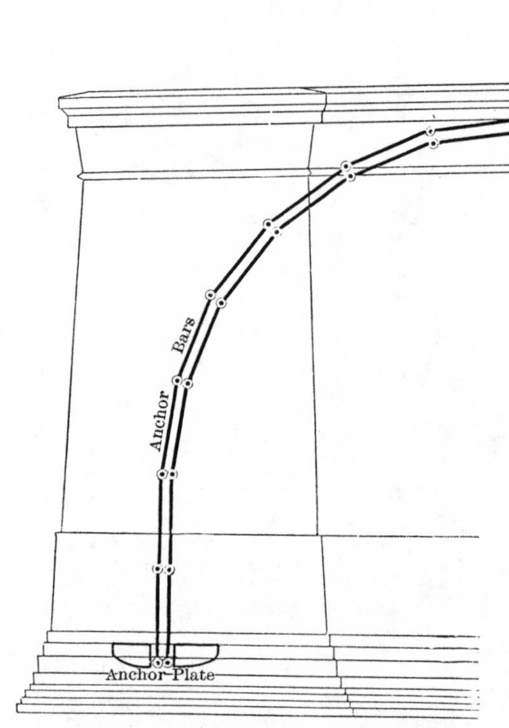

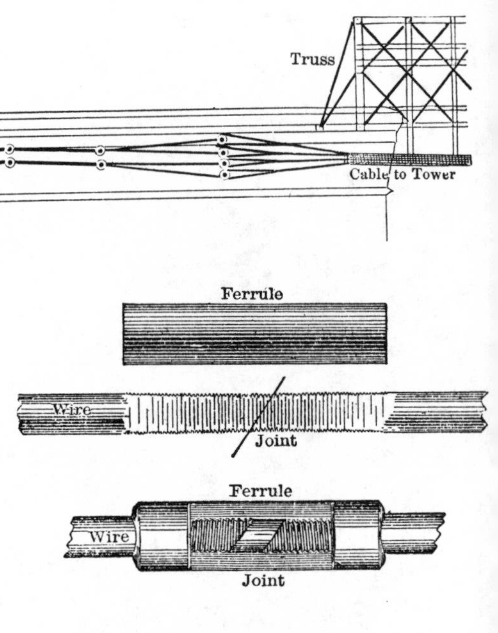

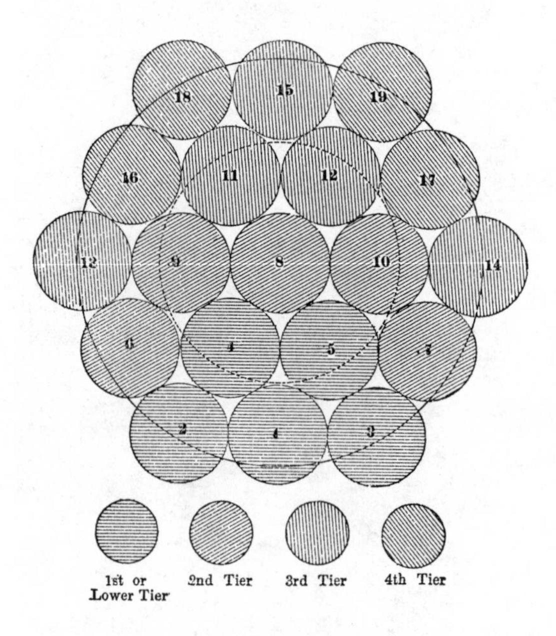

The large diagram above shows how the chains of iron anchor bars, built into the masonry of each anchorage, were fixed at street level to enormous iron anchor plates and attached to the cables at the top of the anchorages on the sides facing the towers. Individual cable wires were spliced by means of a custom-designed ferrule joint (*above*) and "spun" in place into strands. Nineteen strands made a cable 15¾ inches in diameter. The diagram at left shows the ingenious arrangement of the strands worked out by Washington Roebling.

To make way for the long approaches to the bridge, blocks of old houses, shabby stores, saloons, and the like ha to be demolished. The view at right, drawn just prior to the start of the cab spinning, shows the Brooklyn anchora and, in the distance, leading up to the summit of the tower, the newly install footbridge.

FRANKLIN COTTAGE
AGER BEER

Once the cable making got under way it became a subject of great fascination to the public and the bridge trustees alike. Several of the latter are shown above posed at the Brooklyn entrance to the footbridge. How many of them ventured farther is not known. (The large eyebars in the foreground are the anchor bars.) The one tragic mishap during the cable making occurred when a cable strand broke loose from the New York anchorage (*at right, above*) killing two men. The group portrait on the right was taken on top of the Brooklyn anchorage at the point where the cables tie into the anchor bars. Henry Cruse Murphy, at the center with his arms folded, is seen chatting with various members of the engineering staff. William Paine is seated at center looking directly at the camera. Wilhelm Hildenbrand is the clean-shaven man standing fifth from the left, on the right. To his left is E. F. Farrington and the man beside him, with the watch chain, is C. C. Martin. The dapper figure seated on the right and wearing the high silk hat is probably George McNulty.

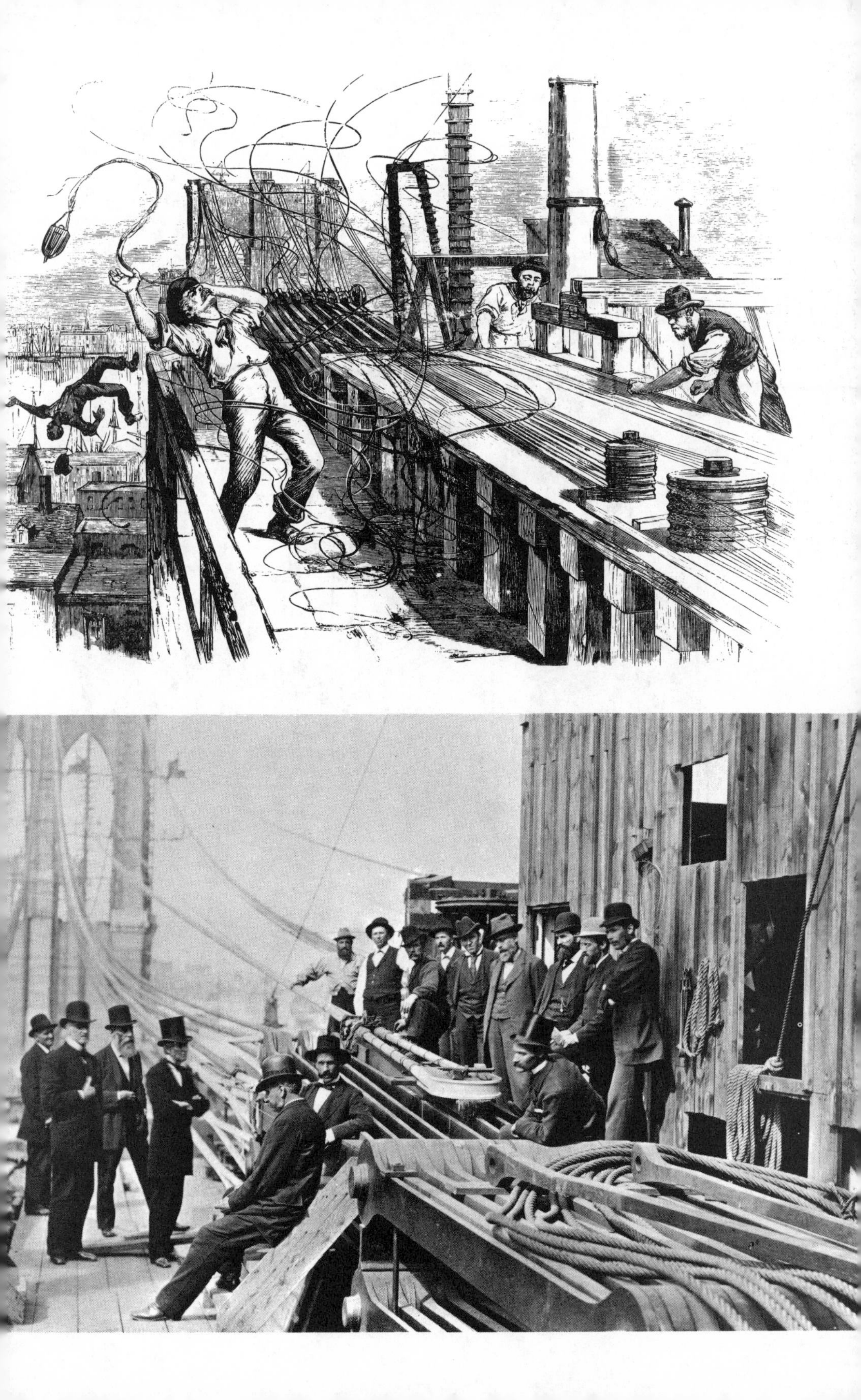

Once the cables were completed the suspenders were attached, wire ropes that were strung like harp strings from the cables down to the bridge floor. The men, many of whom were former sailors, performed their tasks with no apparent fear and to the fascination of the crowds on board the Brooklyn ferries. After the suspenders and deck beams were in place, the inclined, or diagonal, stays were installed.

FRANK LESLIE'S ILLUSTRATED NEWSPAPER

No. 1,444.—Vol. LVI.] NEW YORK—FOR THE WEEK ENDING MAY 26, 1883. [Price, 10 Cents.

Confined to his house on Brooklyn Heights during the final years of the work, the Chief Engineer suffered intensely from the aftereffects of the bends and from a nervous condition brought on by overwork and anxiety. Never seen, he became a figure of great mystery. It was reported that he was totally paralyzed, that his mind was gone, that his wife was the one who was really in charge.

For several years the temporary, slat-floored footbridge (*above*) was one of New York's most talked-about attractions. For a time anyone could go up on it—to sample the view or to test his nerve. This photograph, taken from the Brooklyn tower in 1881, also shows how the steel deck beams, strung to the suspenders, progressed out from the towers to join over the river, at the center of the swooping catenary curve of the cables. One of the original Tiffany invitations to the opening day is shown at right, along with a *Harper's Weekly* view of President Chester A. Arthur making his ceremonial march from New York to Brooklyn.

The Trustees of the
New York and Brooklyn Bridge
request the honor of the presence of
Col. William G. Wilson.
at the
Opening Ceremonies
to take place on Thursday, May twenty fourth at two o'clock P. M.

Committees,

On behalf of the Board of Trustees. | William C. Kingsley, Pres.t Henry W. Slocum. Jenkins Van Schaick. James S. T. Stranahan. John T. Agnew. Otto Witte.

On behalf of the Cities. | Seth Low, Mayor of Brooklyn. Franklin Edson, Mayor of New York.

On behalf of the Engineers. Washington A. Roebling.

The opening of the "Eighth Wonder of the World," on May 24, 1883, was the biggest celebration New York had seen since the opening of the Erie Canal nearly sixty years before. Some of the Irish were unhappy because the day chosen for the ceremony happened also to be Queen Victoria's birthday, but almost everybody else had a splendid time. Both cities went on a holiday and the fireworks on the bridge that night lasted a solid hour. In all, some fourteen tons of rockets and flares were set off from the center of the river span and from the tops of the towers. Bands played on board excursion steamers on the river and the celebrating lasted until dawn.

The first day that the bridge was open to traffic 150,000 people went across. But two days later, a beautiful May Sunday, the count was more than 160,000. The illustrations include various views of the promenade, the bridge trains (which did not begin running until September of 1883), and the two terminals, with the New York terminal at bottom left and the Brooklyn terminal on its right.

The first to jump from the bridge was Robert E. Odlum (*above*), a swimming instructor from Washington, D.C., who fell to his death on May 19, 1885. Others would jump and live, but the only one to become famous, Steve Brodie (*left*), probably never did any such thing. He said he did and had his own witnesses. Later he became the star of a hit play in which his leap from the Brooklyn Bridge was the big scene.

The worst tragedy to take place on the bridge occurred on May 31, Memorial Day, 1883, only a week after the bridge had opened. Some twenty thousand people were on the bridge when a panic began. Reports differ as to how it started, but at a narrow stairway twelve people were trampled to death. Others had their clothes torn off. In places, it was reported,

people were jammed so tight that blood oozed from their noses and ears. The newspapers blamed the Bridge Company for employing too few police. The Bridge Company blamed the newspapers for creating public doubts about the stability of the structure. These illustrations, though perhaps somewhat exaggerated, do justice to the horror of the moment.

"The beautiful and stately structure fulfills the fondest hope," wrote Mayor Seth Low of Brooklyn. ". . . The impression upon the visitor is one of astonishment that grows with every visit. No one who has been upon it can ever forget it. . . . Not one shall see it and not feel prouder to be a man."

THE TRIBUNE
THE TRIBUNE
TRAVELERS INSURANCE COMPANY

20
Wire Fraud

> Yet the existence of evil in human life is a fact too patent to be ignored or to be denied. There is evil and plenty of it, the world over . . .
>
> —JOHN A. ROEBLING

FROM his window the Chief Engineer watched the wind gather force through the early morning, driving snow almost horizontally and whipping up whitecaps on the river. New York was barely visible. By ten a regular gale was blowing and the effect on the bridge was tremendous. The wind, as he noted in a subsequent report, was up to sixty-five miles per hour. He could see the half-finished cables tossing about wildly, like a child's skipping rope.

Roebling could pick out tiny dark figures moving up the footbridge from the Brooklyn anchorage and he knew what they were setting out to do, but from where he was he could not hear the sharp clashing of the strands striking against one another or the eerie moaning and whistling of the wires. Down by the bridge the noise was loud enough to be heard for blocks, and the cradle inland from the tower was slamming about so violently that people in houses below were terrified it might snap its lashing and come plunging down on them.

Farrington had detailed a force of men to go out and secure the wires as best as possible. "It was not a pleasant thing for them to contemplate," according to one account, "and yet there was not a murmur of dissatisfaction." Carrying the little boatswain's chairs, they started up the footbridge, moving very slowly, almost bent double against the wind and snow. The bridge was swinging like a pendulum and the slats were sheathed in ice. But by hanging on to the handrail, they were able to keep their feet and eventually reached the tower.

For the next two hours they worked their way up and down the cable strands, lashing them together every fifty feet or so. The wind never let up during that time, and when two or three of them reached the middle of the river span, and caught the full brunt of the wind, it looked as though they might be carried off at any instant, their frail swings tossing about even more than the cable strands.

But they all came back and none of them complained. "Our men deserve credit for the way they do their duty on such occasions as this," Farrington told a reporter. The bridge itself, he also pointed out, had held up just fine. The footbridge had not lost a single slat.

The storm struck on the last day of January and for the rest of the winter and on into spring the work proceeded without a hitch. The wire spinning was going faster than it had at Cincinnati, as Henry Murphy announced with pleasure, predicting the entire bridge would be finished by 1880.

In February Roebling reported to the trustees on plans for the bridge trains. Everything would be as his father had described it he said. The trains would be hauled by an endless cable, powered by a gigantic steam engine located on the Brooklyn side. "These two tracks, therefore, will be treated exactly like an inclined plane, an operation perfectly simple and perfectly well understood," his father had written. "There is no novel feature and no experiment involved in its arrangement." The elder Roebling had proposed an effective running speed of twenty miles per hour, but said that could be stepped up to thirty in the center of the bridge, or even forty, with absolute safety. Each train could have as many as ten cars, with each car fifty feet long and having seats enough for a hundred people. There were to be suspended sliding doors on opposite sides of the cars, one for coming in, the other for going out. These would be worked by conductors. As one train went over to New York, the other would be coming back, just as the carrier wheels worked.

It was possible too that the rope might revolve constantly, and to start or stop, the cars would simply catch on or let go of the rope. "An ingenious arrangement for attaching cars to a moving rope, devised by Col. W. H. Paine, has been successfully at work for more than a year on the Sutter Street Railways in San Francisco," Roebling informed the trustees. The great virtue of Paine's grip was that it took hold of the cable in such a way that the car did not start off with a violent jerk. The San Francisco cable car operated with perfect ease, and certainly, as Roebling said, the grades were considerably steeper than those of the East River bridge.

His father's plans still appeared to be the best possible solution, Roebling said; "and I am now making every arrangement to carry them out substantially as indicated." This he knew was in direct opposition to what Stranahan and some of the others had been talking about in recent months, and what Kinsella had begun hinting at in the *Eagle*. The scheme was for regular passenger trains on the bridge, linking up with Vanderbilt's New York Central—so it would be possible to go to sleep in Brooklyn and wake up in Buffalo, as they put it. That Henry Murphy's Coney Island line might also benefit from such an arrangement had also become a topic of conversation among Brooklyn businessmen.

Roebling explained that the grade of the bridge would be too great for any but heavy locomotives. The bridge had not been designed for such loads, he said. Possibly a narrow-gauge locomotive could be used, drawing a few light cars, but that would cut passenger loads to a sixth of what the cable system could handle. Moreover, according to his calculations, in a storm such as the one of January 31, narrow-gauge cars would blow right over. "Neither, must we overlook the effect of a puffing, snorting locomotive on horses already sufficiently startled by the novelty of a very elevated position," wrote the engineer.

Kinsella seemed to take all this very graciously. If they had reached the point where all there was to argue about was the size of the doors on the passenger cars, then they had come a long way indeed. That was not the issue, of course, and the issue was still very much alive behind the scenes; but for now an atmosphere of peace settled over the bridge offices and morale among the men actually building the bridge was very high.

Early in March the full-rigged ship U.S.S. *Minnesota*, passing under the center of the bridge, clipped one of the cables with the tip of her topmast. The mast went down with a crash, taking flag and halyards along with it, but the bridge suffered no damage at all. Workmen on the footbridge cheered and waved their hats.

Ever since he returned from Spain Tweed had been telling people he wanted to die. On the morning of April 12, 1878, his wish began to come true.

A few weeks before, on his way from court back to the Ludlow Street Jail, he had caught a cold, which developed into pneumonia, complicated by heart disease. "They will be preaching sermons about me," he had said. Gray, sunken-cheeked, actually gaunt now, he grew steadily weaker, virtually all alone. According to those few who were at his bedside, he died just as the Essex Market clock struck noon.

There would be some dispute later over just what his last words were. A lawyer who was there claimed Tweed said faintly, "I hope Tilden and Fairchild are satisfied now." (Charles Fairchild was then the Attorney General of New York.) Some of the newspapers said Tweed had died talking of angels. But a man named S. Foster Dewey, who was Tweed's secretary and at his bedside, denied this vehemently. "He never thought of angels in his life." Dewey asserted that Tweed's final words were these: "I have tried to right some great wrongs. I have been forbearing with those who did not deserve it. I forgive all those who have ever done evil to me, and I want all those whom I have harmed to forgive me."

It was decided by the family that Tweed would be buried in Brooklyn, at Greenwood Cemetery. "If he had died in 1870," said one old crony, "Broadway would have been festooned with black, and every military and civil organization in the City would have followed him to Greenwood." As it was, the funeral was extremely modest indeed, attended by the family, a few friends, and maybe twenty politicians, among whom there was no one of consequence except "Honest John" Kelly. A procession of just eight carriages followed the hearse down Fifth Avenue and then Broadway, to the tip of Manhattan, where they took the Hamilton Avenue Ferry to Brooklyn.

At Greenwood, Tweed was laid to rest by twenty Freemasons, as he had requested, and wearing a white apron of lambskin—"the emblem of innocence."

"Alas! Alas! young men," cried the Reverend T. De Witt Talmage from his Brooklyn pulpit the following Sunday, "look at the contrast—in an elegant compartment of a Wagner palace car, surrounded by wine, cards and obsequious attendants, going to his Senatorial place at Albany; then look again at the plain box . . . behold the low-studded room, looking out upon a mean little dingy court where, a prisoner, exhausted, forsaken, miserable, betrayed, sick, William M. Tweed lies a-dying. From how high up to how low down! Never was such an illustration of the truth that dishonesty will not pay!"

But Godkin in *The Nation* commented, "A villain of more brains would have had a modest dwelling and would have guzzled in secret." And young Chris Magee, Republican Boss of Pittsburgh, made a special trip to New York to spend several months studying the reasons for Tweed's downfall and returned home to tell his associates that a ring could be made as safe as a bank, and he would do just that.

Asked by a New York reporter what he thought of Tweed and the part he had played launching Brooklyn's bridge, the distinguished editor of the *Eagle* said, "Well, the Brooklyn people have no right to find fault with the Tammany Ring, so far as we are concerned . . . they favored the bridge project, and always acted fairly and liberally with us."

On June 14, at about five minutes past twelve noon, people in Tweed's old neighborhood surrounding the New York anchorage were suddenly startled by what many thought to be the report of a cannon, followed by a loud scraping, hissing noise that sounded, one man said, more like a skyrocket taking off than anything else he had ever heard. A candy vendor on South Street was nearly struck by stones falling about him. A telegraph pole was snapped in two and a chimney was clipped off a nearby house as something went caroming overhead and crashed out of sight over near the bridge tower. People rushed into the streets, including, it was noted in one account, "several harlots" from the Water Street dance halls who supposedly got down on their knees and commenced praying. A bridge cable had snapped, it was said, something had happened to the tower, the whole bridge was coming down, nobody knew what to believe.

Below the north side of the anchorage lay the body of a man, his chest torn open, his back, arms, and legs broken. He was unconscious but still alive. On top of the anchorage a dead man lay sprawled on the stone and two others were lying nearby, groaning pitifully. The only man on the anchorage who had not been hurt, except for a small scratch on one hand, was Master Mechanic E. F. Farrington.

Farrington had been supervising the "easing off" of the sixtieth strand, which had been finished a day or two before. Some thirty men had been working on the anchorage during the morning, but when the noon break came, he had kept only a few of them on to help lower the strand into position. Thomas Blake had been standing where the strand tied onto the shoe, near the pulleys, so he could see that everything went right at that end. Harry Supple and Farrington had been about four paces forward, on either side of the front ends of the anchor bars, at the point where the finished strands for the upstream cable were attached and where the new strand was to take its place. Two other men named McGrath and Arberg were just opposite Blake.

When everything was ready Farrington told Blake to remove the fastenings and the order was passed back to the hoisting engine to begin lowering away. The steel "fall rope" that held the strand began moving through the pulleys and the strand started forward. It had moved about four feet when one of the men cried out that a segment of the fall rope had parted. But the words were no sooner out of his mouth when the whole thing let go. The fall rope had snapped with a deafening report.

It was all over in an instant. Farrington, who had been knocked down by something, but not hurt, looked about to find that only the jagged ends of the fall rope remained. Blake was dead. McGrath and Arberg were bleeding badly and clearly in terrible pain. The remainder of the rope, the pulleys, and the strand had disappeared. And so had Harry Supple.

Blake, it seems, had been killed instantly, struck by the flying shoe more than likely. Supple had been hit by the rope and knocked off the anchorage, falling eighty feet into the yard. The rope had knocked Arberg down and it had caught McGrath by the feet, ripping open the soles of both his shoes, and throwing him as it had Supple, but in the other direction, twenty feet across the top of the stonework.

The strand, and everything it was dragging behind it, had shot away into the air. With one enormous leap it had landed in the bridge yard behind the tower, a good five hundred feet away. Except for the coping on one house, the telephone pole, and the chimney, it had struck nothing in its violent flight and harmed no one. At the bridge yard it had come down on top of a stone pile, shattering some rowboats lying there and barely missing a group of men who were sitting out in the sunshine enjoying their noontime meal. Instantly the great weight of the strand midstream had sent the free end shooting up over the top of the tower. And when the whole strand had gone plummeting into the river, the splash had shot fifty feet in the air and stretched from shore to shore, like a wall suddenly raised up. Passengers on the Fulton Ferry had been drenched, the strand had hit so close by, but nobody was hurt and no boats had been hit.

By the time all the excitement cooled off and it was clear what had happened, everyone realized what a miraculously close call it had been.

Harry Supple, the one who had performed such heroic high-wire feats two summers before, never regained consciousness and died in less than twenty-four hours. Several papers immediately charged that the rope that failed was made of Bessemer steel and that it had been manufactured by the Roeblings. Both claims were true.

There was no explaining what happened, Martin and Paine told reporters as they walked about the anchorage yard where Supple fell. Henry Murphy had rushed over from Brooklyn at first word of the accident and told the reporters they could go up on top of the anchorage to look about for themselves if they wished. The Bridge Company had nothing to conceal. But he refused to offer any possible explanations. The rope had been used maybe fifteen times before this for the exact same purpose and had been tested for a strength six times the load it had been carrying.

In another couple of days the engineers had completed their investigation and solved the mystery, to their satisfaction at least. The rope, they said, had somehow slipped out of place as it was running through one of the pulleys and the sharp edge of the pulley had cut into it, damaging it badly enough to cause the break. Blake, the dead man, should have seen this, but obviously he had not. The steel rope was not to blame, perhaps even Blake was not wholly to blame. The consensus was that it was one of those chance things that happen.

Still the episode had put a scare into people that they would not soon forget and made a number of those who had some say in bridge matters even more skeptical than they had been before. The work went right on. Farrington was back seeing to other things that same afternoon. But the critics would grow increasingly louder now, and more numerous, and Roebling's burden of worry, which was supposed to be lessening as the final phases of the work grew nearer, became greater than ever.

Only a week or so before, the New York *World* had questioned in big headlines whether the bridge was a failure. For some nine millions of dollars, the paper claimed, the people of New York and Brooklyn had acquired nothing but a lot of disgusting scandal and two stone towers with a few wires dangling between. The *Times* had joined in saying that for all the money poured into the bridge, the ferries could have been offered free to the people for a lifetime.

But of far greater seriousness was the hostility growing in that old seedbed of bridge enthusiasm, Tammany Hall. Anxious to disown any previous connection with Tweed, Connolly, and Sweeny, and reportedly exercised over how much the bridge was costing, the new boss of the Tammany, "Honest John" Kelly, was letting it be known that the city of New York just might refuse to spend anymore money on the bridge. The move was seen by many as nothing more than a political maneuver to replace some of the bridge trustees with Tammany men and to subjugate Boss McLaughlin. No one had taken Kelly very seriously at first. But an installment from New York of half a million dollars was already three months overdue. (Brooklyn had met its obligation of one million dollars right on schedule.) Kelly, regarded as "a warm advocate" of the bridge only a few years before, was now making public remarks about withholding New York's payment until he was sure the bridge was being managed competently, and sensational accidents killing innocent laborers only aggravated his "grave concern."

Kingsley, Stranahan, and Henry Murphy claimed no knowledge of Kelly's motives, nor did they care even to speculate on the subject. But if he persisted, Henry Murphy said, they would take him to court, since the law required that New York meet its financial obligations. Part of the problem, Murphy said, was that too many people still failed to comprehend the sort of bridge this was going to be and were listening to a lot of baseless nonsense from second-rate engineers whose only motivation was publicity. "It will not sway from side to side nor rock up and down," he said. It was to be "a great street," he said, solid and stationary. Kelly claimed to have heard from reliable sources that another immense pier would soon be needed to prop up the center span of the bridge. Besides, Kelly said, he was listening with increased interest to the arguments of Abraham Miller and others who were predicting that the bridge would destroy commerce on the East River.

Perhaps in his quiet room overlooking the river, Washington Roebling recalled something his father had written when the same issue was raised at Cincinnati. "I have no fears of those who *honestly* believe the bridge to be injurious to the navigation," John Roebling had said, "the opposition of cavilers I most dread."

By the end of June, New York had still not met its payment. July came and went and still there was no money from New York. By the first week in August, when the trustees convened for their monthly meeting, it was obvious that something would have to be done soon. There was virtually no cash on hand.

At the close of the workday, Saturday, August 10, Murphy took what he viewed as his only course of action. He shut the work down except for the strand making. Approximately a hundred men were kept on. Some six hundred were laid off.

Times were still hard, jobs scarce, and the idea of six hundred men suddenly idle and a great and costly public work standing unfinished was not going to be very well received by the public, as Murphy fully appreciated. The decision, he said, was entirely his own. He did not want to bankrupt the Bridge Company by carrying on, and he did not wish to see Brooklyn spend any more of its money until this trouble with New York was straightened out once and for all.

Kelly said he was now convinced the bridge would do New York little good anyway and that it had been a great mistake for the city to get so financially involved in the first place. And since the costs were running ahead of what had been projected in the original agreements, then New York was no longer legally bound to its side of the bargain. As far as he was concerned, he would be very happy to settle the issue in court. So he held out, as tempers in Brooklyn kept mounting.

With another accident to explain away and "Honest John" Kelly testing his newly gained power, Henry Murphy appeared to be coping with about all the trouble one man could handle during that summer of 1878. And yet this was but part of the story. For on July 22 he had been presented, privately, with what must have appeared to be the most devastating piece of news in all his nine years of administering the business affairs of the bridge.

He was informed by the Chief Engineer that J. Lloyd Haigh, contractor for the cable wire, had been perpetrating a colossal fraud.

The deception had been suspected by the engineers as early as mid-June, but they had had no real proof until July 5. Four days later, when the whole pattern was clear, Roebling had written a long letter to Murphy disclosing what had been going on, but then put off sending it for two weeks, to be absolutely certain his case was solid. So the letter Murphy received on the 22nd was dated July 9 and it told the following story.

"From the known reputation of this man [Haigh], I deemed it necessary from the first to test *every* ring of wire made by him . . ." (instead of every tenth ring or so, as had been planned). Roebling had also warned Paine and the others that Haigh would probably try to bribe the inspectors, which was exactly what Haigh had done, without success.

But as Roebling expected he might, Haigh then tried another maneuver. Inspection of the wire was carried on by Paine and his assistants at Haigh's big brick mill at Red Hook, near the Atlantic Docks. Once the wire was passed, it was loaded onto wagons and hauled up to the bridge. But in June it was found that rejected wire was also getting to the bridge. Wire that had been accepted, but held in the mill overnight, would be replaced before morning by rejected wire, which then went off to the bridge. The trick was discovered by secretly marking the good wire. Haigh was informed of the discovery and given a strong warning. Paine was assured there would be no more of that, but he remained suspicious. The rule from then on was that no more wire was to be inspected than could be delivered on that same day.

But a little later it was noticed that the great pile of rejected wire, instead of increasing as it should have, with rings failing to pass inspection every day, was growing steadily smaller. The bad wire was going somewhere obviously and the assumption was that it was going to the bridge. But how, since all departing wagonloads were being carefully watched to see that they carried good wire only? The solution, the engineers decided, had to be that wagonloads were being switched on route, somewhere between Red Hook and the bridge.

"A watch was therefore set on the morning of the 5th of July," Roebling wrote, "and the trick discovered. The wagonload of wire as it left the inspector's room, with his certificate, in place of being driven off to the bridge, was driven to another building, where it was rapidly unloaded and replaced with a load of rejected wire, which then went to the bridge with the same certificate of inspection."

When these rings reached the bridge—eighty in all—they were immediately tested. Only five out of the total were up to standard.

Two days later the Haigh people were caught trying the same thing again. This time Paine and three others, concealed behind a fence, had watched a wagonload of good wire being unloaded, then replaced with rejected wire, which was all very carefully weighed to be sure the weight of the shipment tallied exactly with what it had been when it left the mill. The good wire was then returned to the mill, where it was submitted to an inspector once more, who, supposing it to be new wire, tested it all over again, gave it another certificate of approval, and sent it on its way, only to go through the same routine.

"The distressing point of this affair," Roebling told Murphy, "is that all the rejected wire which has come to the Bridge has been worked into the cables, and cannot be removed." How long Haigh had been practicing his little scheme, Roebling could not say. "We know that it has been going on for two months, and the probability is that it extends as far back as last January." According to the inspectors' books, nearly five hundred tons of wire had been rejected to date. Most of this, Roebling suspected, had gone into the cables. To determine the precise quantity would be extremely difficult he said. "An engineer who has not been educated as a spy or detective is no match for a rascal."

For the time being he had ordered that a man on horseback accompany each load of wire from Haigh's mill to the bridge and he had instructed Paine to withhold his signature to Haigh's monthly estimates and thereby prevent Haigh from receiving any more money until the extent of the fraud had been more thoroughly investigated.

This, Roebling said, was the first instance of deliberate, incontrovertible fraud that had come to his notice in the nine years he had been Chief Engineer and he urged the trustees to make it known publicly without delay. But in the brief covering note to Murphy that he included with the letter, Roebling made a rather different point: ". . . in case a want of strength shall in the future be found in the cables I wish the responsibility to rest where it belongs, with the Board of Trustees." And this appears to have troubled Murphy about as deeply as the fraud itself.

Murphy took three days to answer Roebling. His concern was very great, clearly enough, but so too was his instinctive caution. No action ought to be taken he said until they had a better notion of the real damage done and until he was clear on the technical remedies Roebling might have in mind. It was a lawyer's letter. "I have waited with much anxiety for the report of Col. Paine," he wrote, ". . . in regard to the wire on hand and not used, which he has been engaged in retesting, since the suspicions arose in regard to the action of Mr. Haigh, and to the possible extent to which any rejected wire has been foisted upon us in the cables. . . . It is manifestly proper, before any definite course be taken by us, that we should know the nature and extent of the injury, and that so far as the work itself is concerned, we should have your distinct recommendation in the premises." In the meantime, he wanted to know what possible responsibility Roebling could conceive of resting with the trustees.

"The responsibility of any weakness that may be found in the cables," Roebling answered, "rests with the old board of trustees, because they awarded so important a contract as the cable wire to a man who had no standing, commercially or otherwise, and the same responsibility must be assumed by the present board if they fail to at once put an end to Mr. Haigh's contract . . ." Out of tact, perhaps, or out of sympathy for the position Murphy was in, Roebling did not remind Murphy that he had warned him about Haigh, quite explicitly, well before the wire contract was awarded.

Given a week's notice, the Cleveland Rolling Mill or Washburn's of Worcester could supply all the wire needed to finish the cables, Roebling assured Murphy. The price would be about the same and the quality could be relied upon. As for Haigh, a committee of trustees and an engineer should be appointed to assess the damage to the cables and fix the value of the condemned wire. That sum should then be deducted from the money currently being withheld from Haigh as security. "This," said the engineer, "is a straightforward way of dealing with a dishonest contractor."

But the trustees chose not to do that. The trustees, in fact, decided not to do anything at all about Mr. Haigh. The whole unfortunate affair would be very neatly and quietly swept beneath the carpet.

. . . They gathered for their regular monthly meeting on the afternoon of August 5, and the record states that the president read letters from Engineer Roebling "relating to alleged frauds practiced by the contractor for cable wire"—no more. And nothing on the matter was released to the press. Several years later, however, William Marshall, the one trustee who had voted against granting Haigh the contract in the first place, said Paine appeared before the board that afternoon and told the entire story of what Haigh had been up to. "I was in favor at the time," Marshall said, "and so said in the Board, of giving the whole history of the matter to the public, but was overruled."

According to an item in the *Eagle*, there was a meeting of the Executive Committee immediately following the board meeting. "The executive session lasted until 6 o'clock, but the subject matter under discussion was not divulged by the members," the paper reported. But Marshall said Haigh himself appeared at this session and that Haigh denied any intention of trying to deceive the bridge people, professing to be wholly uninterested in any money that might be in question. "All I am anxious about," he said, "is lest the trustees may entertain a poor opinion of me." That they certainly did, responded several men in the room. "I am sorry for that," declared Haigh. "Do you know that is what I was afraid of? Indeed, it was the only thing I was afraid of." Haigh spoke the whole time, Marshall said, "with imperturbable coolness."

Interestingly, the record kept by the Bridge Company carries no mention of an executive session being held that afternoon.

The following morning Roebling wrote once again to Murphy to answer several questions Murphy had sent along. Most of the letter was taken up with technical explanations of current cable strength, assuming, as Roebling now did, that some 221 tons of rejected wire had actually been laid up. But in closing, Roebling reminded Murphy that the cables had been designed to have a margin of safety of six, that is, they were six times as strong as they had to be. And he recalled for Murphy that his report of January 1877 had stated specifically that such allowances would have to be made "for any possible imperfection in the manufacture of the cables." So even with Haigh's bad wire hanging up there, the cables had a safety margin of at least five, Roebling concluded, and that he regarded as perfectly safe, provided no more bad wire was used.

Roebling's say on the matter was quite comforting for Murphy and for the other trustees apparently. The whole unpleasant business could now be very conveniently forgotten. Wasting no time, they reconvened the next day, August 7. When the meeting adjourned, the president had been directed "to continue the contract with Mr. Haigh for the wire required to complete the large cables, on such conditions and terms as he deems proper under the circumstances."

It was just as though nothing had happened. The papers carried no mention of any of this. The public remained ignorant of the entire affair.

The Chief Engineer, however, after a great deal more thought on the problem, ruled that the contractor would have to supply additional good wire for the cables, at his own expense, to make up for the calculated deficiency of the bad wire already in place. As a result each of the cables would contain some 150 more wires than originally planned.

From Washington Roebling's private day journals, kept by his wife, a few further pieces of information emerge to complete the picture. Haigh's original samples of crucible steel wire were made by somebody else, while a good percentage of the wire he delivered was of Bessemer steel after all, but sold to the Bridge Company at the crucible price. Roebling estimated that Haigh netted $60,000 on this bit of deception alone and that he had also cheated his supplier out of several hundred thousand dollars. In all Haigh's illegitimate profits came to $300,000.

Some years later, after the bridge was finished and the story of Haigh's swindle had leaked out, the radical economic theorist Henry George, who had set out to resolve the paradox of progress and poverty, wrote of the bridge as a prime example of the good and evil of the age.

> We have brought machinery to a pitch of perfection that fifty years ago could not have been imagined; but in the presence of political corruption, we seem as helpless as idiots. The East River Bridge is a crowning triumph of mechanical skill; but to get it built a leading citizen of Brooklyn had to carry to New York sixty thousand

> dollars in a carpet bag to bribe a New York alderman. The human soul that thought out the great bridge is prisoned in a crazed and broken body that lies bed-fast, and could only watch it grow by peering through a telescope. Nevertheless, the weight of the immense mass is estimated and adjusted for every inch. But the skill of the engineer could not prevent condemned wire from being smuggled into the cable.

Come what may the Brooklyn *Eagle* would not be diverted from its main theme—accomplishment.

> The thousands who daily cross the ferries and look up to the lofty towers that rise on either hand above the water, and note the strands that stretch across the intervening space, hardly realize that the cable making of the great structure is nearing its completion. But such is the fact, and with a fair degree of success, by the time the cold weather sets in we shall see the four great cables completed and ready for the superstructure or roadway of the bridge. It has been steady and patient work—wire upon wire and strand upon strand—through heat and cold and storm and calm, and now this branch of the great enterprise nears the end, and another department of the work of construction appears in the near future.

Nothing belied talk of political scheming, bankruptcy, labor unrest, vicious rumor, or plain despair quite so much as the great work itself.

Progress on the cables was in truth very far along. Seventy strands had been completed, which meant there were only six more to go.

An explanation offered at the time to show the interested layman how the strands were arranged to form a cable was to take seven nickels, place one at the center with six around it, all touching, and then twelve more around the outside of the six. This illustrated the pattern quite rightly, but it was somewhat misleading in that it implied that the first strand put into position was the center one—the middle nickel—then six more were compacted about it, and twelve more around that. The system did not work that way, however. The strands were being laid up in four different tiers and these were arranged in a most ingenious pattern, so that they stacked one on top of the other, like building blocks, rather than being built outward from a center strand, and still they wound up forming the cylindrical shape wanted for the cable. The first tier, consisting of five strands, had three strands forming a bottom row (the middle strand of these three was the first strand put in place) and two put on top, forming half of the next row. The next tier, of five more strands, placed one at each end of the second row (making four strands to that row) and three more on top. Then tier three, also five strands, added one more strand to each end of the third row, two more on top in the middle, and one on top of those. The fourth and final tier consisted of the last four strands stacked two to each side of the three upper strands of tier three.

The arrangement was quite ingenious and it was entirely Roebling's doing. "It has pleased the average penny-a-liner," his wife would write, "to remark that there is nothing new in the East River Bridge and that Colonel Roebling only copied his father's plans. The fact is there is scarcely a feature in the whole work that did not present new and untried problems." His arrangement for the strands was a perfect example, she said, comparable to the water-shaft system he had worked out for the caisson, or his use of double tiers of anchor bars, which had been necessary to handle the number of strands required for such large cables (the earlier bridges had had only seven strands to a cable, not nineteen).

Regulating the strands was found to be the most tedious and time-consuming task of all. The strength of the finished cable would depend on getting each strand into its exact, particular position, and since those positions were at different heights within the cable (there was a difference of about fifteen inches between the first strands, say, and those in the top tier), the length of the strands therefore had to vary. "Each must hang in its own peculiar length and curve to a mathematical nicety," as one magazine article explained; "for if left but half an inch too long or too short for its true position, it will be too slack or too taut for its fellows, and it will be impossible to bind them solidly in one mass, and make them pull equally together."

In the abstract this was simply a matter of mathematics. But in practice there were a number of variables to contend with, just as there had been when stringing the individual wires. Temperature was again a prime factor. Even ordinary temperature changes during a day were such that the length of a strand was seldom the same from one hour to another. And to further complicate the problem, one span could be affected more than another, depending on how the sun was striking it. One strand might be in shadow, while another was taking the full glare; one might be exposed vertically to the sun, while the other was at a more oblique angle. So periods of strong sunshine, like days when the wind was up, were not the easiest times to regulate strands. The best progress was made when the weather was calm and a little overcast or between the first light of day and sunrise.

Studies made by the engineers showed that the deflection of the cable strands from the towers at a temperature of 50 degrees was 127.64 feet, while at 90 degrees it was 128.64 feet—which was a variation of nearly a third of an inch for every degree of temperature. So it was not uncommon to find the cable strands varying as much as half a foot in height in the course of a single day.

The way things were going, the two downstream cables would be laid up several weeks before the other two in order to give the men some practice with the wrapping machinery. To bind the strands of each cable into one compact unit required that the cable be tightly wrapped from end to end with iron wire. The work would begin at the towers, with wrapping machines proceeding down the cables toward the center of the river span and toward each anchorage. So ultimately there would be sixteen machines in operation.

First a powerful iron clamp would be used to bring the strands into an exactly cylindrical shape. This was composed of two semicircles that placed together formed a ring the prescribed diameter of the cable. The clamp would be screwed up tightly to compress the strands and directly behind it would come the wrapping machine, an iron cylinder—about sixteen inches long and cast in halves that were bolted together about the cable—encircled by a reel of wire that wound off the drum through a hole in the rear end of the cylinder where it passed with one turn around a small roller attached to a disk and then to the cable. The reel had handles around it, like a ship's wheel. Men riding on a "buggy," a small platform hung to the cable by big trolley wheels, would turn these handles, thereby revolving the reel and winding the wire onto the cable as tight and close as thread on a spool. Once the wrapping machine reached the clamp, the clamp was moved forward and the machine then advanced again, and the process would be repeated until the entire cable was clamped and wrapped. After that the cables would be oiled and a coat of white paint would be applied.

The system worked well and without mishap except for one close call when the captain of an outward-bound full-rigged ship neglected to trim his top masts. The men in one of the buggies, working over the center of the river, did not see the ship until she was nearly upon them. Then they scrambled out onto the great cable above them and the ship clipped the buggy an instant later, sending it spinning and knocking a shower of tools into the air.

Now there was a great push on to get the cables finished and wrapped before winter. It was expected that the job would take three months. The next step would be to hang the suspender cables from which the roadway was to be hung.

In September, as directed by Henry Murphy, the contract for the wrapping wire, awarded to J. Lloyd Haigh at the start of the summer, was quietly changed and awarded to John A. Roebling's Sons. No explanation was given for the change. No voices were raised about the Chief Engineer having a conflict of interest. The *Eagle* remained silent. Abram Hewitt remained silent.

On October 5, 1878, at 4:45 P.M. by the clock on City Hall, the last wire went over, one year and about four months after the cable spinning had begun, or eight months sooner than Roebling had expected. "This desirable event," wrote E. F. Farrington, "was marked by no demonstrations, save the sounding of a steam whistle, and the raising of a United States flag on the Brooklyn tower." The greatest length of wire laid up in one day had been eighty-eight miles. The white carrier wheel, which had crossed the river some twenty-three thousand times, would be crossing no more. "The end, then, is near at hand," announced the *Eagle*. But a month later, the cable wrapping not half finished, Henry Murphy declared that the work would have to be shut down entirely. "Honest John" Kelly was still holding out on New York's quota and the money was all gone.

21
Emily

> At first I thought I would succumb, but I had a strong tower to lean upon, my wife, a woman of infinite tact and wisest counsel.
>
> —Washington Roebling

SHE HAD been born and raised in a house much like this one and her whole life, until she married, had been spent in the upper Hudson Valley, where the river was not only a major event in the landscape, but a central part of everyone's way of life. Talk of tides, of winter freeze-overs and the spring breakup, had been part of ordinary conversation for as long as she could remember.

The town dock at Cold Spring stood at the foot of Main Street. In summer when the "up" boats from New York stopped—the *Mary Powell*, the *Emeline*—it was always a grand occasion. And at night, from her bedroom as a child, she could hear the steam whistles of the great side-wheelers trailing off through the Highlands.

She had grown up on that part of the Hudson where, for some fifteen miles, it cuts a deep narrow channel through thickly wooded mountains, the most picturesque and fabled part of the whole valley. She could still see part of the river now, just the very broad lead-gray final end of it, emptying into the Upper Bay, beyond the tip of Manhattan. And now, as then, she could stand at her window and watch the afternoon sun go down and the lights come on across the water. The sun set earlier in Cold Spring than it did here, and with the mountains crowding all around, the evening skies had never been so spectacular as these. Then, the lights had been few, from West Point only. Still and all there was enough that was the same to make her feel very much at home here.

The house stood at a prime spot on the Heights. It was tall, stately, spacious, built before the war in the Greek Revival style, and it was located at the northern end of Columbia Heights, the street running parallel to the river, about half a mile from the bridge. The address was 110 Columbia Heights, in the block between Pineapple and Orange Streets, on the west side of the street, the side with the view. Like nearly every house on that side, it had a deep garden in back, extending out over the top of a carriage house and stable built below the brink of the bluff, fronting on Furman Street beside the wharves. Moses Beach, the publisher and a pillar of Plymouth Church, lived next door. Henry Bowen, who had done much to stir up the Beecher scandal and whose deceased wife was said to have been another paramour of the famous divine, lived just up the street in a colossal white mansion with a two-story Corinthian portico, and Beecher himself lived in the next block.

From her front windows, overlooking the street, she could see the old Turkish baths that John Roebling had patronized and directly across the street stood a row of three-storied brick houses with beautifully arched doorways and long, polished plate-glass windows, much like her own. The houses fronted directly on to the brick sidewalk, as hers did, giving the street a nicely balanced, orderly look. With the sun casting tree shadows on the pink brick walls, everything looked secure and private, as in a courtyard. But from the back of the house, from the big bay windows on every floor, the whole of the harbor, the river, the bridge, and the city beyond were spread before her.

For six years in all, this would be the center of her universe. She was anything but a recluse by temperament, and unlike her husband she could come and go at will, but when she did it would be for his sake nearly always and for his sake she would do everything in her power to keep this place of theirs both private and utterly tranquil, like the eye of a storm.

She was thirty-five years old now. She had been married for fourteen years. For nearly ten of those years her husband had been working on the bridge and for more than half that time he had been an invalid, for a long while very close to death and always greatly dependent on her.

Their son now was nearly an adolescent and apparently she had known for some time that there would be no more children. She had had a bad fall in Germany shortly before he was born, and afterward had bled for nearly a month in the little inn in Mühlhausen, her husband calling in one German doctor after another. The trip home across the Atlantic had been an agony. But she seems to have kept in almost perfect health thereafter, despite everything she had on her shoulders. And she seems to have made an enormous impression on everyone she met.

One newspaper article said, "Mrs. Roebling is a tall and handsome woman, strikingly English in style and shows not only in her face, but in her graceful carriage, an aristocratic ancestry." She was considered an exceptional horsewoman and known for both her "scientific bent of mind" and decided opinions on many subjects.

Among the best physical descriptions of her is one Washington Roebling wrote during the war, in a letter to his sister Elvira:

> I would send you a little tintype [of Emily] if it didn't happen to be a horrid picture, not doing a particle of justness to the subject. Some people's beauty lies not in the features but in the varied expression that the countenance will assume under various emotions, etc., etc. . . . She is dark-brown eyed, slightly pug-nosed, lovely mouth and teeth, no dimples in her cheeks, like Laura the corners of the mouth supply that, and a most entertaining talker, which is a mighty good thing you know, I myself being so stupid. She is a little above medium size and has a most lovely complexion . . .

He would never be satisfied with any photograph of her no matter how many times she tried. They gave no idea of her "peculiar grace of carriage," he said.

Six weeks after he had met her he had bought a diamond ring and gone off on a flying visit to Baltimore, where she was staying with her sister-in-law, the general's wife. He had never had any second thoughts about her and apparently her feelings were the same. By April he was addressing his letters, "My good Mrs. Wash," and telling her, "You know, darling, that your presence always made me feel so good, a kind of contented feeling pervaded me if you were only near. It was not necessary to say anything, perfect silence was as much companionship as the liveliest chatter."

She had written him steadily through the rest of the war, long, affectionate letters full of the everyday details of her life. But he had destroyed them all, almost as soon as he read them, telling her they made the separation that much more difficult for him. She, however, had saved everything he wrote, more than a hundred letters from the front in less than a year's time.

"This full moon evening would be delightful if I only had someone to enjoy it with," she read in a letter from Virginia, shortly before the Battle of the Wilderness. "In fact I would not care how the evening was if I only had you with me. I do wonder which of us two can be called the most lovesick; I am disposed to yield the palm to you because you used to consider such a thing so utterly impossible in your case. How long will it be before we shall get tired of each other, in other words what is the length of the honeymoon among people raised around Cold Spring, just ask your friends about it and tell me dearest."

He told her about the things he loved, dogs, astronomy, Thackeray. He told her about a Trenton girl named Gussie Laveille, who, he warned, was coming down to visit his camp if she did not. He told her about the boredom and futility of war, and it seemed he had an infinite number of names for her. "Dearest Emmie," he called her, or "Sweet Em" or "My good Emily" or "Dearest Girl," "My dear old woman," "My charming Miss Warren," "My loved one," "My Darling," "My darling Emmie," "My Lazy Darling," "My own particular Darling," "My own Emily," "Sweetest Love."

"After all, dear Emmie, pray tell me what is love," he asked. "Is it kissing each other, is it tickling, hugging, etc. one another? Is it writing billy duxes, kicking each other's shins under the table? That must be it I think—the shins!"

"Look for a big thief next winter," he wrote in July, "he proposes to steal the only valuable thing in Cold Spring and intends to escape detection by changing the name of the stolen article which will render identification impossible."

"Does the *Mary Powell* run when the river is frozen?" he asked later. "When I visit you at Cold Spring I am supposed to fly on wings of love so anything short of the railroad will be too slow. Isn't it curious that although I was nearly four years at Troy and traveled ever so many times up and down by rail and boat I should recollect every place except Cold Spring. I dare say I must have seen you often when the train passed, rolling a hoop along the street in short frocks."

She had gone down to Staten Island soon after that to meet some of his family for the first time—his sister Laura, her Mr. Methfessel from Mühlhausen, and their children. Apparently the experience was something of a cold bath for her, as he learned soon enough, to his great pleasure.

> Your letter describing the visit to all the Dutch uncles and cousins, etc., was very amusing to me; your heart must have sunk within you as you seem to take it for granted that your life henceforth was to be spent in a Dutch atmosphere. The tone of your letter is one of sad resignation and even your Wash seems of scarcely sufficient weight to counter-balance the scale. And well might it be so if your life were doomed to be spent among that Dutch crowd on Staten Island. . . . However you must take heart my dear; all of our family is as much American as you could wish with the exception of Mother and she never had the opportunity. And again my dear you must remember that in course of time you will be the one to take the lead and be at the head of the home circle.

It was after that that she made her first trip to Trenton to see the "home circle" where she was expected to "take the lead." Then his two brothers had gone to Cold Spring before he did, at her invitation. They were all looking each other over. Ferdinand especially had taken a great liking to her, a little too great, Roebling wrote to her, and only partly in jest one suspects.

"When the two hopefuls of the house of Roebling come I hope you will take good care of them and keep them out of temptation and danger," Washington kidded her. "Their youthful minds are just at that stage now that their visit to you at Cold Spring will never be effaced from their minds as long as they live." How many days would he be expected to stay at her house before the wedding took place he wanted to know. He hoped one would do.

"I still entertain a lively remembrance of the promise you exacted of me to stay in my own room the first night, but I forget whether I assented or not—how was that?"

He had come to Cold Spring on leave, in the fall, when the weather had turned suddenly sharp and raw. She had met him at the depot, just back from the boat wharf, and they had driven up Main Street, a straight steep climb back from the river. He was in uniform and if there had ever been a handsomer couple seen in Cold Spring nobody could remember when.

Everybody in the little town knew her. The Warrens were one of the prominent families in the county. Her grandfather was John

Warren, who, according to one of the old Putnam County histories, "aspired to no higher distinction than that of a plain, practical farmer, which he was. The purity of his motives, and the honesty of his heart, were never questioned; and in all the relations of life he never gave just cause of offense to his neighbor. . . . His children, so far as we know them, inherit his virtues." Her father, Sylvanus Warren, had been the youngest of old John's seven children, a distinguished, learned man and a close personal friend of Washington Irving's. Her mother had been Phebe Lickley before she married.

Old John had kept a well-known tavern that was still standing on the Albany Post Road and he had prospered until steamboats began plying the river and the Post Road was no longer the fastest route north and south. Her own father had also provided well for his big family, having invested in the famous West Point Foundry, an ordnance works and Cold Spring's sole industry, which stood by the river's edge. With the foundry testing its Parrott guns, and officers coming and going from West Point, the war had never seemed quite so far removed to Cold Spring people as it had to most Northerners. Often as she sat at her desk writing to Washington, she could hear the big guns pounding away. In a yard beside the foundry they were loaded to full capacity, then fired at the rocky face of Storm King Mountain on the far side of the river, upstream. The shells, when they hit, threw up enormous masses of earth and stone and the impressions made in the side of the mountain would be plainly visible for years to come.

The most notable member of the family, however, was her brother, the general, G.K. as he was called, who had been named after Gouverneur Kemble, an erudite, convivial Cold Spring man who had started the foundry. There had been twelve Warren children, but only six had survived beyond childhood, of whom G.K. was the oldest. He was nearly fourteen years older than Emily, who was next to the youngest, and for her there was no more dashing heroic figure. Except for Hamilton Fish, Grant's aristocratic Secretry of State, who kept a country estate in nearby Garrison, General Gouverneur Kemble Warren was Putnam County's most famous citizen.

His graduation from West Point had been a momentous event in the Warren family, but Emily had been too young to remember much of it. When he had entered the Academy at sixteen, he had gone with a lofty admonition from Gouverneur Kemble: "We expect you to rank, at graduation, not lower than second." And he had done just that, finishing number two in a class of forty-four. For the next ten years or so he had returned home only rarely. He had been assigned to the Mississippi Delta first, to work on flood control projects, then to the West. He fought the Sioux under General Harney and mapped Nebraska and the Dakotas. When he came home again, in 1859, to become an assistant professor of mathematics at West Point, he was known as the first explorer of the Black Hills, a slim, black-haired, deeply tanned young man, who, except for his mustache, looked remarkably like an Indian himself, and who to his younger brothers and sisters seemed to have stepped from some exotic other world.

It would be said of him later, by numerous people who knew him, that he was notably gentle and kindhearted for a soldier. The distressing thing about Indian fighting, he had said, was that quite often one shot women and children and when it came time to tend to their wounds one found them to be not at all unlike other women and children. In the West he had been known as "the good Lieutenant."

When his father died, the year he began teaching at the Academy, G.K. assumed most of the responsibility for his younger brothers and sisters, looking after their interests and health with uncommon care and faithfulness. For Emily, then just sixteen, he was much more than a brother only, and it seems his influence had much to do with her orderly ways and subsequent interest in science, and in botany in particular. Like a number of other celebrated soldiers before and after, he was passionately fond of flowers.

He was also an engineer with a particular interest in bridges, he seemed to have no fear of physical danger, and he had an obvious contempt for pretense of any kind, all qualities he valued in his young aide, Roebling, and which she too must have recognized soon enough after their first meeting at the Second Corps Officers Ball.

The Warrens were not wealthy people by Hudson River standards, but were considered gentry. Cold Spring had become a gathering place for a small but distinguished group of artists and literary people and the young Warrens were part of that society. Once, during the war, the artist Thomas R. Rossiter painted a hypothetical *Picnic on the Hudson*, which would one day be considered among the finest works of the Hudson River school. It is supposedly a representative portrait of Cold Spring elite, twenty ladies and gentlemen gathered with picnic hampers on Constitution Island, just down from Cold Spring. Dressed in elegant summer attire, bathed in sunlight, they pose formally beside the great river, sailboats and Storm King in the distance. Among the group are Gouverneur Kemble, looking very robust for his years; Julia Fish, daughter of Hamilton Fish; Robert Parrott, inventor of the gun; Robert Weir, the painter; and old white-bearded George Pope Morris, editor of the New York *Mirror* and author of *Woodman, Spare That Tree.* In the center foreground, looking rather stiff and self-conscious in a half-reclining pose, is G. K. Warren, the trousers of his uniform providing the only splash of bright red in the composition. But in the background, on the left, there is a young woman in a big flowered straw hat who has never been identified for certain, but who is probably Emily Warren.

Roebling's first encounter with Cold Spring and her family was brief and apparently went very well. "I think we will be a pair of lovers all our lifetime," he wrote to her soon after returning to Virginia.

They were married on January 18, 1865, in a little brick church on Main Street. It was a double ceremony, with her brother Edgar Washburn Warren, a major in the cavalry, marrying Cornelia Barrows of Cold Spring. There was a good-sized crowd gathered in the cold outside her house when she and Washington came out the door at the end of the reception, and years later old women in Cold Spring would tell how as children they had seen Emily Warren come down the steps on her wedding day, as though they had witnessed an occasion of state.

But even in January of 1879, for Emily all that seemed a long time back. Her stricken husband was in constant torment, his work a nightmare instead of the inspiration and source of pride it had once been for them both. Any chance for a normal life together was now beyond recall if she was to believe what the doctors were saying.

She had seen her husband all but destroyed before her eyes, his spirit as well as his body. And by uncanny coincidence, she had seen much the same thing happen to her beloved brother.

Only a few months after she was married, Gouverneur K. Warren's brilliant career had run into a puzzling, tragic snag at Five Forks, the last decisive battle of the Civil War. The strain of war had begun to tell on the young general. Always fussy about details, he had grown increasingly engrossed in things he should have left to subordinates. He was taking a little longer with everything, and quite a little longer than Grant, for one, thought acceptable at this stage. Grant was in a great hurry. Warren was then commanding the V Corps, one of the most famous infantry units of the Federal Army, and Grant had put the V Corps and Warren entirely under the flamboyant Phil Sheridan, a cavalryman. Grant also told Sheridan to remove Warren if he saw fit. Grant never quite explained his reasons for this and historians still differ on who was right or wrong or to blame for the outcome.

After receiving conflicting orders on which route to take, Warren had marched his men all night through pitch-black, rain-flooded country to give Sheridan the support he needed, but he had arrived a little late. An attack planned for that morning had to be called off until afternoon. When the fighting started, Warren was again not where Sheridan wanted him. Warren had done his best, but that had not been good enough for Sheridan, who in a violent rage suddenly ordered that Warren be relieved of his command.

Warren's subordinate officers were incredulous, furious, and would defend his reputation for the rest of their days. Roebling, who was out of the Army by then, would always feel things would have gone differently for Warren had he been there. He and Emily were both people of "decided temper," as they said, but on this particular subject they were quite decided indeed, never seeing but one side of the argument.

"Just imagine Sheridan sitting on a fence, sending a staff officer every five minutes to Warren to hurry up and save him and his cavalry from being captured by Lee's troops," Roebling would write indignantly. "And when Warren does come (after wading through an icy creek up to their middle), saves Sheridan and wins the battle, then Sheridan turns on him and cashiers him."

After Five Forks, Warren was put in command of defenses at Petersburg. Later he went to Memphis to command the Department of Mississippi. When the war ended, he decided to stay in the Army, serving as an engineer on the upper Mississippi and as a member of the commission assigned to examine the Union Pacific Railroad. He was also in charge of the survey of the Gettysburg battlefield, where he and his young brother-in-law had had their day of glory.

But in 1869, when Roebling was getting started on the Brooklyn caisson, Warren had been put in charge of building a bridge over the Mississippi at Rock Island, Illinois, where before the war an earlier railroad bridge had been the issue in a historic lawsuit involving Abraham Lincoln. In 1856 a new steamboat called the *Effie Afton* rammed into a pier of the Rock Island Bridge, caught fire, went down, and left part of the bridge burning. The bridge, a big timber truss belonging to the Rock Island Railroad, was repaired quickly enough, but the steamboat people decided they had a case and went to court. Lincoln represented the railroad, arguing that the east-west "current of travel has its rights as well as that of the north and south." The fact that the jury failed to make up its mind on the matter was taken as a signal victory for railroads, for bridges, and for the notion that the manifest pattern of American commerce and growth was to be east-west.

But the new Rock Island Bridge was meant to satisfy the river interests as well, and Warren had labored so hard over it, between 1869 and 1870, that his health broke and in all the time since he had never quite recovered. Moreover, he kept struggling to clear his name of the Five Forks incident, repeatedly and futilely requesting a board of inquiry to examine the case. His duties never lessened; he continued with river and harbor work, on the upper Mississippi, along the Atlantic coast, on the Great Lakes. He served on Humphreys' review board during the Eads bridge controversy, and just that October of 1878, he had been put on the advisory council of the Harbor Commission of Rhode Island. But overwork and exposure at Rock Island and the refusal of official Washington to grant him a hearing had all but broken him physically and drained his spirits. Pale, hollow-eyed, he looked more like sixty than forty-eight.

His problem was that in Washington the men implicated in his version of Five Forks were the ones in power now, and apparently he wanted more than just his name cleared.

"I have heard men like Humphreys and others say that Grant was inclined to give Warren an investigation," Roebling wrote, "but that Warren demanded that Sheridan should be publicly reprimanded for having done a cowardly and unsoldierlike act—and in choosing between the two he finally shielded Sheridan. Grant was then at the height of his popularity and could do what he liked."

The restoration of his honor had become an obsession with Warren. He refused steadfastly to admit defeat, and the effort was costing him dearly, financially as well as in other ways. Never a man of wealth, he had come out of the war all but penniless. Quietly, on occasion, his sister and brother-in-law were providing financial help as well as moral support.

For Emily it was a heartbreaking thing to see the men who mattered most in her life victims of such dreadful misfortune. It must have seemed as though the two of them, with their pride and decency, their old-fashioned sense of duty, were somehow out of step with the times and paying an awful price for it. Everywhere about her, lesser men, witless, vulgar, corrupt, men of narrow ambition and the cheapest of values, were prospering as never before, grabbing up power, money, or just about anything else they hungered for. This Gilded Age, as Mark Twain had named it, seemed to be tailor-made for that sort. It was the grand and glorious heyday of the political bribe, the crooked contract, the double standard at every level. It seemed the old verities simply were not negotiable any longer. Good and brave men who had a legitimate claim to honor, respect, position—at least according to every standard she had been raised by—were somehow in the way now and so got swept aside.

But if ever she let such thoughts plague her for long, or get her down, there is no suggestion of it in the record. And like the men she so loved and admired, she quite bluntly refused to give in. More, she seemed to gather strength as time passed and gradually she began exerting a profound and interesting influence in bridge matters.

There would be all kinds of stories told about her later and the part she played, and quite a number of them were perfectly true. She did not, however, secretly take over as engineer of the bridge, as some accounts suggest and as was the gossip at the time.

But it is not at all surprising that the stories spread. As was apparent to everyone who met her, Emily Warren Roebling was a remarkable person. And since every piece of written communication from the house on Columbia Heights to the bridge offices was in her hand, there was, understandably, a strong suspicion that she was doing more than merely taking down what her husband dictated. At first she was credited only with brushing up his English, which may have been the case. But by and by it was common gossip that hers was the real mind behind the great work and that this the most monumental engineering triumph of the age was actually the doing of a *woman*, which as a general proposition was taken in some quarters to be both preposterous and calamitous. In truth she had by then a thorough grasp of the engineering involved. She had a quick and retentive mind, a natural gift for mathematics, and she had been a diligent student during the long years he had been incapacitated.

Trustees grumbled over her reputed influence. Newspapers made oblique references to it. And the fact that she had assumed such importance was often used as a basic premise for the argument that Roebling was not right in the head.

Even Farrington was said to be partly her creation. Farrington had been giving a number of highly popular lantern-slide lectures on the bridge at the Brooklyn Music Hall and at Cooper Union (several thousand people had turned out to see and hear the illustrious master mechanic) and the New York *Star* remarked, "It is whispered among the knowing ones over the river that Mr. F's manuscript is in the handwriting of a clever lady, whose style and calligraphy are already familiar in the office of the Brooklyn Bridge." Maybe this was so. In any case a very great many people took it to be the truth and that was the important thing.

She had also become so adept at shielding her husband from visitors that many of them went away convinced she knew as much about the technical side of the bridge as any of the assistant engineers. When bridge officials or representatives for various contractors were told it would be acceptable for them to call at the Roebling house in Brooklyn, it was seldom if ever the Chief Engineer who received them. She would carry on the interview in his behalf, asking questions and answering theirs with perfect confidence and command of the facts. Most of them left quite satisfied that her husband would be correctly apprised of everything said. But so impressed were some that they went out the door convinced they had met with the Chief Engineer after all and their future correspondence would be addressed directly to her.

At one point in 1879, for example, a controversy developed over the honesty of an important contractor, the Edge Moor Iron Company. Ugly insinuations were traded back and forth in the papers and it began to look as though there might be still another drawn-out investigation. To assure the engineering department of their honesty and good intentions, the firm addressed a formal written statement to that effect, not to the Chief Engineer, but to Mrs. Washington A. Roebling. And there was no mention in the letter of conveying any of its contents to her husband, or to ask for his health or to solicit his response or opinions.

Her services as his "amanuensis," as he called her, were enormously important, as he said later. She kept all his records, answered much of his mail, delivered various messages or requests to the bridge offices, went to the bridge itself to check on things for him, and was his representative at occasional social functions. She was quite literally his eyes, his legs, his good right arm. And the more she did, the more the gossips talked.

Half a dozen New York and Brooklyn papers were delivered to the big brick house regularly each day. For Roebling they were still the only access to the world beyond the bridge and his own four walls. They still had to be read aloud to him. His eyes were greatly improved, but he had trouble reading for more than a few minutes at a time. So the two of them would sit together in the room on the second floor that was his office, sickroom, command post, where the days dragged by, one by one, ever so slowly for him, and where week by week, month by month, year after year, as he talked, she saw the bridge take form and grow on paper just as clearly as its progress could be seen from the window.

A day rarely passed during that winter of 1878–79 when the newspapers did not carry something about the bridge, and after she had finished reading them to him, she would clip out whatever there was on the bridge and paste the articles in a big scrapbook, just as neatly and methodically as her brother had done with items on his campaigns all through the war.

But the clippings must have seemed the top of the iceberg only, knowing what they did, feeling as they did about certain people. Never during this time was anything written about the anguish of their years in that room. No journalist or magazine writer was permitted to interview them there. Nor did either of them write anything about the experience, beyond the briefest, most factual statements. There would be no soul-baring memoirs. Their privacy was total and strictly enforced.

Only in the letter books that have survived are there any chinks in the wall of privacy they built about themselves—brief, sudden bursts of emotion sandwiched in with page upon page of technical detail—and even these are frequently illegible, her penciled lines having become badly smeared after so many years. The frustration, the sharp, bitter indignation, the rage expressed are always his, however. What she felt, what she said or did to keep him in balance, to be ballast for them both, can only be guessed at.

His worst time since leaving Trenton had come in the spring of 1879, in early May. For nearly a month it had looked as though everything was back on course again. Comptroller John Kelly, for all his Tammany bluster, had been put in his place by the courts. Murphy had hired William M. Evarts as counsel for the Bridge Company. He was the celebrated and expensive New York attorney who defended Andrew Johnson in the impeachment trial and Beecher in the adultery case and who had just been made Secretary of State. The central issue, Evarts argued, was whether a great public work was to be pulled down because it did not quite conform to some early bookkeeping or in order to save a few ships from making minor adjustments to pass beneath it. Henry Murphy had been a persuasive witness, to no one's surprise, and the judges, first in the Supreme Court of New York, then in the Court of Appeals, decided in favor of the bridge. The city of New York which meant Comptroller Kelly—was ordered to continue its payments without further delay. "There is, of course, great rejoicing in Brooklyn," the New York *Herald* said. "The success of the bridge is assured, and the work upon it which has been interrupted for more than six months, will be resumed within a few days." And that was what had happened. Six hundred men went back to building the approaches and the cable wrapping was resumed at once.

The so-called Miller suit, to remove the bridge altogether, had also been settled at long last and again largely as a result of the tireless, determined efforts of Henry Murphy. The State Committee on Commerce and Navigation had held hearings in the Metropolitan Hotel in New York and one by one Murphy, Stranahan, old Julius Adams, C. C. Martin, Paine, McNulty and Collingwood had all gone over to appear as witnesses for the bridge. The opposition had rounded up a number of harbor pilots, shipmasters, shipbuilders, warehouse owners, and a few engineers to testify against the bridge. Abraham Miller, the warehouse owner who was the plaintiff, said he had not brought suit until as late as he did because he never expected the bridge would be finished. He was convinced the cities would fail to get up the money it would take. A representative from Standard Oil warned that the bridge would divert trade to Philadelphia; some harbor pilots complained that the cables were already hazard enough, as did several ship captains. The total testimony taken, the exhibits presented, the charts, tables, statistics, and the like, all printed up and bound together eventually made one great doorstop of a volume weighing a full five pounds. But the opponents of the bridge achieved nothing. As with Lincoln's *Effie Afton* case or Eads's victory over the Corps of Engineers review board, the pattern of east-west travel prevailed, the bridge was the victor.

But in May came what for Roebling was the lowest blow to date. He had decided on another major change in the bridge. After receiving Paine's report on the ability of various manufacturers to produce steel in certain desired shapes and quality, he had decided to substitute steel for iron in the trusswork. This meant that it was to be virtually an all-steel bridge now, and with the approval of the trustees he had called for new bids. General Slocum announced at a meeting of the trustees that the assistant engineers (and Paine in particular, it was understood) had been taking bribes from steel manufacturers, which at different times, Slocum said, amounted to sums of as much as ten thousand dollars. Except for Roebling's assistants, just about everyone who had had any real say in bridge matters had been accused of something or other by this time. Now it was their turn. Slocum's charges were omitted from the official record, but two days later the papers had the story, with the result that the trustees met again in secret session on May 5 to discuss the matter and this time their comments were released to the press.

Slocum said he had been told of the bribes by a man named Marshall P. Davidson of the Chrome Steel Company in Brooklyn. Slocum said he wanted it understood that these were distinct charges, not more idle rumors. He also said there was some question about certain transactions of the Roebling company. "And I want to say right here that I think it is indelicate that the brothers of the Chief Engineer should be engaged in furnishing us materials."

It was William Kingsley, interestingly, who stood up at this point and, looking Slocum in the eye, said he regretted to hear such statements made about "gentlemen who were not present to defend themselves." Furthermore, Kingsley said, no firm in the country had a reputation for honor and business integrity exceeding that of John A. Roebling's Sons.

All the same a special committee was formed to investigate the charges and this committee met the following morning to hear Davidson speak for himself and to listen to Ferdinand Roebling, who had come on from Trenton. Davidson said Slocum had misquoted him. He had made no such remarks about the engineers. What he had said, he believed, was that there were rumors of bribes but that he himself did not believe them.

Ferdinand Roebling, for his part, said he thought the time had come for him to put the matter in the hands of a lawyer and begin suit against "somebody" for libel. He said the end had been reached so far as the abuse the Roebling company was willing to endure. His family's connection with the bridge had been anything but advantageous, he reminded the trustees. His father had lost his life, his brother had sacrificed his health, the family reputation had been assailed. And so far as making money from their contracts was concerned, his company would be perfectly satisfied to produce the rest of the wire at cost.

Ferdinand almost certainly spent some time at his brother's house while he was in Brooklyn and it seems Henry Murphy was going and coming from the Roebling front door rather frequently. The level of emotions Emily Roebling had to contend with can be gauged from this letter from her husband to General Slocum, dictated the same day Ferdinand appeared before the committee:

> I hope I have heard for the last time your oft repeated remark that you think it indelicate in me that I should allow my brothers to do any work for the bridge while I am the Chief Engineer. Did it ever occur to you that my brothers act independently of me without consulting me and that I have no control over them even if I wished to prevent them bidding on any contract for the bridge? Or did you ever consider that the John A. Roebling's Sons Company hold the first rank in this country as manufacturers of wire rope—and the word "*fraud*" has never been coupled with their name save in your board? Would it not be at least probable that my reputation as an engineer is as dear to them as it is to me and that I should feel better satisfied to have work that I know requires care and skill entrusted to them rather than to some rascally contractor without capital or reputation who after he has been again and again detected in fraud is allowed to go on with his contract.
>
> You should have been very sure of Mr. Davidson's meaning before you brought the subject up in the way you did, and you should not, if you really had any desire to know the truth, have been contented with his simple assertion that you misunderstood him.
>
> The course of a true gentleman would have been to come to me first with a lie that had been whispered behind my back and at least heard what I had to say, whether you believed me or not. . . . I

have the right to think Mr. Davidson never said anything to you, but you merely gave the board the benefit of your own opinions. . . .

The investigation committee presented its findings at the end of the month and the engineers were completely exonerated, as was the Roebling company. It had all been an unfortunate misunderstanding, it was explained. On Roebling's orders, Colonel Paine had spent some time in Pittsburgh with Andrew Kloman, Carnegie's former metallurgist, who was now in business for himself, and had rolled the steel for General William Sooy Smith's new railroad bridge at Glasgow, Missouri, the first bridge in America built exclusively of steel. Kloman had a new way of making steel eyebars and Paine had gone to Pittsburgh to study the process. This, apparently, was what gave rise to the bribery stories, since it appeared that Paine was giving Kloman preferential treatment.

The newspapers assured the public that all was fair and square inside the engineering department. The work went right ahead. But Slocum never apologized to Roebling or to Paine or to any of Roebling's staff or to his brothers. Roebling never would forgive him for what he had done and the deep-seated animosity between the two former war heroes would prove to be no minor issue.

Then, almost immediately, there was a change-over in the bridge trustees, the first real realignment since the bridge began. Thomas Kinsella declined to serve again because of "pressing business engagements" and several new faces were to be seen now in the board room, most of them quite young faces by Bridge Company standards. It was also quickly noted that there was a decidedly dubious look in their eyes whenever some of the more notable older members commenced to talk.

Among the new men were William G. Steinmetz, who automatically became a member of the board when he became Comptroller of the City of Brooklyn. He was an engineer by profession and a German by birth, with a thick head of wild black hair and one wooden leg. Alfred C. Barnes of Brooklyn, the oldest son and business partner of A. S. Barnes, the book publisher, was "one of the most cultured and affable gentlemen in the city," according to one account. Edward Cooper, son of Peter Cooper and brother-in-law of Abram Hewitt, was mayor of New York. And Robert B. Roosevelt, wealthy New York lawyer, was an energetic politicial crusader and noted sportsman, whose favorite nephew Theodore was then in his last year at Harvard.

All four had come in as a result of the elections of 1878. The two Brooklyn men, Steinmetz and Barnes, were Republicans, while Cooper and Roosevelt were Democrats. But they were all reputed champions of reform, and with the exception of Steinmetz, they were the gentlemen sons of wealthy, prominent fathers—city-born, expensively educated, urbane, public-spirited, and politically ambitious. Despite the party labels they had much more in common with one another than they did with a Kingsley or a Stranahan, the self-made men of another generation, who, with their back-country origins had grown up with the city, as it were, and had acquired, somewhere along the way, what the younger men found to be a reprehensible degree of patience with what the older men would call human failings.

The new men were determined to set things in order. But from the start it was Steinmetz who made the biggest fuss. Right off he wanted Kingsley removed, for one thing, and he made no bones about it. Kingsley was the keystone of the old regime, as the Brooklyn Comptroller saw it, and the reason the bridge had taken so uncommonly long to build was because the old regime either wanted it so or because they did not know how to run things. Either way Kingsley could no longer remain a trustee.

But Steinmetz grossly underestimated the power the contractor had. The mayor of Brooklyn, a man named Howell, who was a Democrat and doubtless beholden to Kingsley in innumerable ways, said Kingsley would stay. So Henry Murphy was removed instead—temporarily. No sooner was Murphy out, taking the blow very graciously, chatting affably with reporters as he packed his things, than another trustee resigned so Murphy could be reappointed in his place. That done, the others promptly voted Murphy president again and made Kingsley his vice-president. All of which left Comptroller Steinmetz, a testy, excitable man at best, so furious he was barely able to speak when the reporters came around to get his views.

But Steinmetz kept pressing the attack through that summer and into fall, opposing the use of Bessemer steel for the superstructure, opposing the awarding of the contract to the Edge Moor Iron Company, the lowest bidder, trotting out every old argument for crucible steel, and being so silly and tiresome about it much of the time that the other young men who had come in with him were left with no choice but to side with the opposition. They were just as eager as ever to clean house but they were looking for something more important to battle over.

But in December, just as had happened three years before, a sensational bridge disaster seemed to add credence to every rumor of shoddy steel and poor engineering. The new Tay Bridge over the Firth of Tay, in Scotland, one of the biggest, most famous bridges in the world, gave way in a gale and collapsed into the sea, taking with it a train carrying seventy-five people, all of whom were killed. The bridge was the work of Britain's leading engineer and a disciple of the great Stephenson, Sir Thomas Bouch, who, along with Henry Bessemer, had been knighted by Queen Victoria that June. His bridge, a series of trusses, had been built mostly of wrought iron, however, not steel, and subsequent investigations of the disaster indicated that he had not calculated his wind loads accurately. The conclusion was that the engineer was mainly to blame. (His health and mind broken by the ordeal, Bouch died in less than a year.)

As might be expected, the news of the disaster caused a great stir in New York and Brooklyn. McNulty and Paine, interviewed at length in the papers, did their best to assure the reading public that the East River bridge was an entirely different kind of structure. But who was to say? Had not the word of the ill-fated Bouch been as respected as any in the profession?

By curious coincidence, the same papers that carried the Tay Bridge story also reported that J. Lloyd Haigh, "the well-known wire manufacturer" who had supplied the wire for the great cables, had just gone bankrupt. And to add one further note of doom, still another "noted engineer" was claiming the East River bridge would not hold a fifth of the weight that was liable to be put upon it. "WILL THE TAY DISASTER BE REPEATED BETWEEN NEW YORK AND BROOKLYN?" asked one big headline as the new year 1880, and the new decade, began.

There must have been moments in those early weeks of the new year when Emily Roebling stood alone at a window thinking of the Tay tragedy as she watched the tiny doll-like figures working up among the cables and suspenders. She knew enough now to appreciate the countless number of things that had to be taken into consideration and the immense weight of responsibility every calculation entailed. She knew enough to know how very many things could go wrong. The East River was not the stormy Firth of Tay, standing wide open to the sea, as some were saying, still it was salt water, and for all the shelter Long Island provided, winds on the river could be savage. When the Tay Bridge went, the papers said, the train had dropped nearly ninety feet.

But there must also have been moments during those same weeks when she went about the house or drove along the snow-covered streets of the Heights with her heart lifted as it had not in years. In December, G.K.'s request for a board of inquiry had at last been granted, fourteen years and eight months after Five Forks.

In late February she pasted into her scrapbook a large illustration of Ferdinand de Lesseps, in top hat and overcoat, standing with a group on the summit of the New York anchorage, "inspecting" the bridge, according to the caption. "The Great Engineer" (who was no engineer at all, but a diplomat) had arrived from France to promote what he intended to be the triumph of his career, a sea-level canal across the Isthmus of Panama. In another week or so he would deliver an impassioned speech before the American Society of Civil Engineers (Collingwood and McNulty would be in the audience) and be lionized at a sumptuous banquet at Delmonico's at which Richard Storrs, the Brooklyn pastor, would deliver the welcoming address and she herself would be among the ladies accompanying De Lesseps when he made his grand entrance into the dining room. The grandfatherly Frenchman was greatly impressed by the bridge, he told reporters, but in the illustration Emily saved, he appears more interested in an unidentified young lady in the foreground.

A full page from *Frank Leslie's Illustrated Newspaper* went into the scrapbook, a close-up view of workmen out on one of the cables attaching iron suspender bands and to these the wire rope suspenders that would hold the steel floor beams of the roadway. The cables by this time had been thoroughly wrapped from end to end. Once the suspenders were in place, the men could start laying up the crossbeams of the deck, beginning with those closest to the anchorages and towers and working out. The nearest suspenders had only to be pulled back, the beam attached and swung out into place. Planks would be put down over the beam then for the men to stand on as they launched the next beam, and so on out over the water.

Once the steel deck began to take form, she would be able to walk out on it herself to look things over as her husband directed.

In April she cut out two articles about St. Ann's Church, which eleven years before Roebling and Paine had used to sight the center line. The historic old building was about to be demolished—to be "swept away by the march of modern improvement," said the *Eagle*.

Her scrapbook tells the full story of the bridge that year as the public saw it. There was the usual wrangling over finances, a minor accident or two, a great deal of complaining and explaining about the steel contract (the Edge Moor Iron Company was maddeningly slow on delivery), periodic reports on the progress of the work (about the building of a big skew arch over William and North William Streets in New York, for example, and the steady advance of the bridge deck). One tiny item, a clipping not much bigger than a postage stamp, reports that J. Lloyd Haigh of Brooklyn was breaking rocks at Sing Sing. He had been convicted of passing bad checks.

In June Henry Murphy said it would be all right for two Plymouth Church musicians to take their coronets out to the center cradle on the bridge, and there "to the delight of hundreds of upturned faces from the ferryboats and the Fall River boat *Newport*," they played "Jesus, Lover of My Soul," "Rock of Ages," and "Old Hundred."

Summer and fall were uneventful on the whole except for one incredible scene that took place at a trustees' meeting in October, and which had all Brooklyn talking.

The meeting, involving only routine matters, had been about to adjourn when Comptroller Steinmetz announced that he had a communication for the president, whereupon he handed Henry Murphy an unusually lengthy printed document and a messenger burst into the room and distributed additional copies to everyone present. William Kingsley slowly got to his feet, unfolded his copy, and moved to have it tabled.

"It is an insult to the dignity of the board that this man should present any communication in this way," he said.

"Mr. Chairman, Mr. Chairman," exclaimed Steinmetz excitedly in broken English. "I demand my communication shall be read." Kingsley said he had the floor and Steinmetz was out of order.

"Ever since this individual has been a member of this board," Kingsley said, "he has never suggested a single practical or intelligent idea, but has continually been bringing before the board such claptrap stuff as has just been presented."

"But the communication has not yet been read," protested Steinmetz.

Kingsley, "glaring at the Comptroller," exclaimed, "You must realize, sir, that you are not in the slums of politics."

The motion to table the document was then quickly carried, while Steinmetz kept protesting. "I represent the citizens of Brooklyn," he shouted. "I would not be here if I was a private citizen. . . . Mr. Kingsley's suggestions have always been heard and accepted, while mine have not."

"This is all buncombe," Kingsley snapped back, shaking Steinmetz' printed letter. "It's ward politics brought into the board."

Murphy promptly shut off any further discussion of the subject and took up another matter. But when the meeting adjourned, Kingsley marched over to Steinmetz, who was standing among some of the others in an adjoining office.

"You are acting the part of a demagogue," Kingsley said to Steinmetz, who was nearly a head shorter. "What or whom do you represent?"

"Mr. Kingsley's manner was determined," according to one man in the group, "and as he stood facing Mr. Steinmetz, he looked steadily at him. The latter turned pale, and the other members of the board crowded into the room."

"I represent the people of Brooklyn," Steinmetz answered.

"You represent nobody," said Kingsley.

"You are no gentleman," said Steinmetz.

"You are a blackguard," responded Kingsley.

"I am Comptroller and I represent the citizens of Brooklyn."

"I say you represent nobody—nobody at all."

"Well, Mr. Kingsley," said Steinmetz backing off, "I can afford to take that from *you*."

"Of course you will," said Kingsley.

Steinmetz then turned his back on Kingsley and left the room, but Kingsley followed after, "continuing to express his indignation."

The story was all over town by nightfall. Steinmetz said later that Kingsley had threatened to strike him, but nobody else who had been in the room agreed with that. But neither could any of the trustees justify Kingsley's behavior, and the papers made a point of the fact that Steinmetz was a cripple. Even Kingsley's own

Eagle allowed it would have been better had the Steinmetz letter been read. As it was, the whole thing had taken on much more importance than it deserved. Every afternoon paper carried the Steinmetz letter in full and there was hardly enough in it to have attracted any but passing interest under normal circumstances. Steinmetz made a number of wild accusations—about the steel contract chiefly—none of which could be supported, and although a few editorial writers took them at face value and made some foolish charges as a result, the whole issue blew over in a matter of days.

Nonetheless, the confrontation in the board room was indicative of the strong feelings developing between various personalities on the board, and half a lifetime later, when Roebling would be asked what part his wife had played in building the bridge, it would be "her remarkable talent as a peacemaker" among these gentlemen and during these particular years that he would praise highest, telling people with customary deadpan understatement how she had a way of "obviating personal friction with her tact" and could smooth over difficulties that were "naturally inherent in a work somewhat political in its conduct."

Apparently just about everyone involved with the work liked her enormously and held her in great respect, regardless of his politics, profession, age, or particular feelings about her husband. That she was welcome among them, her opinions regarded seriously, was considerable testimony in itself, in a day and age when a woman's presence in or about a construction job except as a spectator on special occasions was absolutely unheard of.

How many ruffled feathers she smoothed, how many times she sat patiently listening first to one side of an argument, then another, how many tactful words of caution she offered a Henry Murphy or a Ferdinand Roebling or a C. C. Martin before they entered her husband's sickroom, how frequently she herself dealt directly with a Steinmetz or a Kingsley, is not indicated in the record. But the impression is that she was very busy indeed at just such tasks. Roebling would describe her role as "invaluable."

In February Henry Murphy, James Stranahan, and the Reverend Dr. Storrs put on formal attire and went over to New York with Martin, Collingwood, and three younger Rennselaer graduates recently hired as assistants to attend a gathering of the alumni of the famous Polytechnic Institute. Collingwood was to be the main speaker of the evening and his subject was the bridge. Murphy and Stranahan had been invited as representatives of the Bridge Company, and Storrs, it seems, was becoming something of a fixture at such occasions, a sort of unofficial chaplain to the somber-looking technical men who talked so matter-of-factly of improving on God's handiwork and who, since the war, had already changed the look of the country more than any army ever had.

A great deal was said during the course of the evening about what had been accomplished by the 739 engineers the Institute had sent into the world in its fifty-five years, about the countless dams, canals, bridges, railroads, and water works they had built. The strongest testimony of all, however, according to a Professor Greene from Troy, was to say simply that the East River bridge was the work of RPI men—which was not altogether the case but which pleased the professor's audience no end.

Collingwood spoke a little too long about the staggering quantities of brick, stone, steel, and iron that had gone into the bridge and then announced to great applause that the work was nearly done. The real hero, he said, was Roebling, who had never lost his hold on the bridge. "The men who have come from the Institute to the bridge had come to stay," he said, "they seem to have a wonderful sticking power."

There was much conversation about the Chief Engineer during the dinner that followed, almost as much apparently as there was about his wife, who by this time had become an idolized figure among the assistant engineers.*

In the spring the steel floor beams started going up and the great structure began to look like a bridge. By summer, even with the contractor behind on deliveries, the superstructure had advanced well over the river, coming from both directions. After studying the work from a boat out on the river, one admiring engineer told re-

* There was so much talk about her, in fact, that at the next alumni gathering, a year later, a Brooklyn engineer named Rossiter W. Raymond, who was not an RPI graduate, but was widely known as an afterdinner speaker, was asked to come and give a special toast. (Raymond had such a grandiloquent platform manner that he would one day be invited to succeed Beecher at Plymouth Church, an invitation he declined.) "Gentlemen, I know that the name of a woman should not be lightly spoken in a public place," he said to his hushed audience, ". . . but I believe you will acquit me any lack of delicacy or of reverence when I utter what lies at this moment half articulate upon all your lips, the name of Mrs. Washington Roebling."

porters that the way the bridge was being built it would be as immovable as an enormous crowbar and would last a thousand years. That was the summer Garfield was shot and Chester A. Arthur, who had been collector of customs for the Port of New York back in the early years of the bridge, became President of the United States.

The dome on the Custom House was one of those New York landmarks Washington Roebling could pick out quite easily from his second-floor window. His vista on the world was the same as it had been since the summer of 1877, when he returned to the house on the Heights. It was the only way he had seen the city or the bridge in four years.

From the Battery to the New York tower, all the best-known landmarks stood out like points along a ruler. The steeple on old Trinity Church was the highest point downtown. Then over to the right, farther uptown, was the tower on the Western Union Building, which was almost on a line with the New York landing of the Fulton Ferry and with his own house. Then came the Post Office, with its flags and heavy mansard roofs, the shot tower built thirty years before by James Bogardus, pioneer in the use of cast iron. Just to the right of that, before the bridge tower, was the Tribune Building, with its long spike of a clock tower. With his field glasses or telescope, it was possible to read the time on the Tribune clock.

There were any number of things to read through the telescope, written large in different colors on the sides of New York buildings —THE EVENING POST . . . ABENDROTH BROTHERS, PLUMBERS IRON ENAMELED WARE . . . HOME FOR NEWSBOYS . . . Roebling had seen the skyline steadily changing year by year, reaching a little higher always. Still the granite towers remained in full command. Trinity Church and the Tribune tower were actually taller, in measured feet and inches, but they did not look it, being farther in the distance, and by comparison they were mere needles against the sky.

From Roebling's window both of the bridge towers stood out plainly. Indeed, very nearly the entire structure could be seen, silhouetted against that segment of sky that seemed to dome Harlem and Blackwell's Island. It would be said that his telescope was so powerful he could examine the very rivets being put into the superstructure, which was nonsense. But the bridge was certainly close enough for him to see a great deal of what was going on. Interestingly, his wife would remark later that he actually spent little time peering through his telescope. One glance of his "practiced eye," she said, was enough to tell him if things were being handled properly. His troubles with his vision, it seems, were confined to things close up.

In the fall C. C. Martin was saying the bridge would be finished by the next summer. By the time the first snow was flying, steel beams and suspenders had been strung all the way from tower to tower, and in early December Emily had her own first moment of triumph in a very long time. Presumably, from his window, the Chief Engineer watched the entire ceremony.

She had left the house in the early afternoon and was driven directly to the foot of the Brooklyn tower, where, accompanied by Farrington, she climbed the spiral staircase as far as the roadway. A gentle wind was blowing as she waited. The air was as mild as April nearly, with a smell of rain.

Presently a delegation of trustees arrived in the yard below, climbed the stairs, and were next tipping their silk hats, shaking her hand, commenting on the abnormal weather, and saying nothing of the meeting they had come from. It had been, she knew, one of the most crucial sessions since the bridge began. But neither she nor they made any mention of it.

Then they had set off, strolling along on the plank walk that had been put down on the steel superstructure, a plank walk no more than five feet wide that now stretched the whole way over the river. The sensation was apparently only a little less nerve-racking than on the footbridge.

She walked out in front, leading the way, escorted by Mayor Howell of Brooklyn and the new mayor of New York, William R. Grace, the wealthy steamship owner. The others, about a dozen in all counting the assistant engineers and two or three reporters, followed behind. The views were exclaimed over. A strong wind was blowing and one reporter noted how it played with the long white locks on the head of the venerable James S. T. Stranahan. Sea gulls were everywhere, gliding past cables and stays in big arcs, swooping beneath the bridge, tiny flashes of white far down below against the deep gray-green of the water. Boats passed beneath, white sails close-hauled. The water swirled and turned with great turbulence, from the wind, from the boats plowing through in both directions, from the churning tide. There was much animated conversation within the little procession. Everyone was having a fine time on this first crossing of the bridge by the actual roadway.

On their arrival at the New York tower several bottles of champagne were uncorked and the trustees drank to the health of Mrs. Roebling and "to the success of things in general."

22
The Man in the Window

> The best way to secure rapid and effective work is to get a new Chief Engineer.
>
> —New York *Star*

NOBODY would say who made the decision and that greatly aggravated the gentlemen from New York, and the mayor in particular. There was no possible way, they said, by which their city could benefit from Pullman cars and freight trains crossing the bridge and they wanted to know by whose authority the engineer's plans had been suddenly and secretly changed to accommodate such traffic. But none of the other trustees would give a direct answer. There was mounting tension in the board room, whereupon James Stranahan, addressing himself to William R. Grace, said, "There has been authority for everything that has been done . . . No personal considerations and no personal feelings have entered into the matter in any way, and I say distinctly there has been nothing done on which a charge can be made that it was invested with secrecy or in any way outside the legitimate duty of the members of the board."

Stranahan, obviously angry, was resorting to a defense he seldom used. His customary method was to assume his opponent was perhaps half right, well intentioned beyond question, but still, on the whole, altogether mistaken in the premises from which he started. He was always willing to rely upon "the more mature judgment" of his opponent, and to give him opportunity for maturing his judgment, he was generally willing to accept a delay. But there was no easy way to delay the direct question and so the usually amiable Stranahan, who reminded people of an English statesman and who was looking more dignified than ever now that he was in his seventies, had been filled with great righteous indignation, basing his case, as it were, on the excellence of his personal character and past services.

The tactic worked. Mayor Grace said he meant no harm and asked no further questions. Stranahan said there was really no reason for the gentlemen from New York to be apprehensive. He told them how ten years earlier he had talked with Cornelius Vanderbilt about linking up with the New York Central and how Colonel Roebling had met with Vanderbilt's engineer to figure a way to handle the problem. It was thought that a sunken line could be run from Grand Central Depot south to the bridge, then the trains could be raised by hydraulic lifts to cross over the bridge. After all, Stranahan asked, was it so unnatural for Brooklyn to want a depot of her own? Consider what is happening all around, he said. "Already there are three trunk lines to the West terminating in Jersey, and more lines will be completed in twelve months. In New York the only trunk line going west is the New York Central. New Jersey sends Annex boats for passengers, and floats carry the freight past your shores, and of what benefit is that to you gentlemen of New York? Sooner or later, gentlemen, you will consider this matter, and when that time comes I trust you will see that we have not labored in vain, either for our own benefit or that of the City of New York."

And then it had been remembered that Mrs. Roebling was up on the Brooklyn tower by this time, waiting to make the walk over the bridge, and the meeting had adjourned, the question still hanging.

As things turned out, there never would be an official answer. Those responsible for the decision preferred to remain anonymous, and so the one person left to account for the change was Roebling.

Had it not been so near the end of the work, had there been less talk in the papers about the ever-mounting cost of the bridge, the change would probably not have mattered a great deal. But in October of 1881 when Roebling submitted his request for an additional one thousand tons of steel to put into the trusses to make the floor rigid enough to carry heavy locomotives and cars, several individuals saw it as the moment they had been waiting for.

Why the sudden concern about the strength of the bridge, they wanted to know. Who decided on railroad trains? How was such an immensely important decision arrived at and why was there no explanation of this in the record? And always, between the lines, was the larger question: Did the Chief Engineer know what he was doing?

The trouble was, mainly, that everyone was getting extremely anxious to see the work completed, and to know exactly, once and for all, just how much the whole thing was going to cost.

As of the start of the new year, 1882, total expenditures on the bridge would come to $13,377,055.67. But as several editorial writers noted habitually, the original estimate had been for about half that, and although Henry Murphy said he thought another $600,000 should be enough to finish with, even he was unwilling to be held to that figure. The old memory of Tweed's courthouse was returning to haunt honest men on the board, the cost of the bridge now being up to what had been spent on the courthouse.

In addition, only a small minority of the trustees understood very much about the engineering of the bridge, or its history, and fewer still had ever had any personal dealings with the Chief Engineer. In fact, more than half the men now charged with managing the great enterprise had never even seen Washington Roebling. And there was no way to communicate with him except in writing, which few wanted to do, which was a nuisance for those who did, and which occasionally led to misunderstandings.

Robert Roosevelt, the most outspoken of them, asserted foolishly at the trustees' meeting of October 13 that the added weight of the extra steel would weaken the bridge. Exactly how much more weight could the cables carry, he demanded to know. He wanted Roebling to present a complete report promptly. His motion had carried and Roebling had done as directed, explaining in a letter dated January 9, 1882, why the added weight meant added stiffness in the deck and so even greater strength, not less. Roebling did not say who had directed him to revise his plans, he said only that it had become "incumbent" upon him to do so. He had been against the idea of conventional trains on the bridge, as they knew, but he had gone along with it. "When I consented to make this change," he said, "it was not so much owing to personal solicitation as to the reflection 'of what benefit had it been to erect this bridge at a vast expense unless we use it for every possible purpose to which the structure will lend itself.' " They were now to have a bridge so rigid that ordinary traffic would have no visible effect on it. And to settle the nerves of anyone concerned about the cables, he concluded by stating rather dramatically that the cables were strong enough to uproot the anchorages.

The original plan for a cable-drawn bridge train had not been abandoned. It was just that the bridge would now have enough added strength to accommodate the future, which, it was quite naturally presumed, would involve some sort of railroad. The presumption was mistaken; but for different reasons nobody could have guessed at then—the advent of the internal combustion engine and the automobile, mainly—the decision would turn out to be the right one. As a result, the bridge would remain in service for another fifty years before any major alterations in the superstructure became necessary.

It was an extremely important and fortunate decision, in other words, and from some things said later Stranahan appears to be the one who made it, with Kingsley, Murphy, and, ultimately, Roebling backing him up. Perhaps Stranahan had reached an "understanding" with old Commodore Vanderbilt, who was dead by this time, and the New York Central was still serious about the idea. Perhaps Stranahan and the others simply figured the decision was in the best long-range interest of Brooklyn, that a bridge could not be too strong after all, and rather than risk a noisy fight over the thing—a fight they might very well lose—they just went ahead and did it.

The decision was made in Brooklyn, that much is certain, and whatever the explanation, they were pressing their luck. Delivery of steel already on order was way behind schedule; any additional steel would mean further delays, as well as greater cost. So the immediate result of the announced change in plans was a clamor to find out who was mismanaging things in this the final stage of the work.

Ironically, the bridge was all but built. Certainly the difficult and demanding work was done with. It was largely a matter of finishing things up now—the final masonry on the approaches, the last of the trusswork, the plank flooring for the roadways and promenade, and a lot of painting. Already the men had begun taking down the footbridge.

For the first time since his father's death in the summer of 1869,

Roebling could relax a little. For the first time in thirteen years his services were no longer vital. For the first time he was not really needed any longer.

Like Roebling, those trustees who had been in on the work since the beginning were all thirteen years older. A few, like Henry Murphy, were showing their age. But now, with the completion of the work at last almost within reach, it was the newcomers on the board who were the most impatient and the most frustrated by delays. As before, Robert Roosevelt made the greatest noise, claiming with some justification that nobody ever gave him a straight answer to his questions. But he also had an important new ally, the very young new Republican mayor of Brooklyn, Seth Low.

Seth Low said hardly anything at the first trustees' meeting he attended in January 1882. He simply sat and listened, an alert, serious expression on his boyish face. Only toward the end did he inquire of Henry Murphy when the bridge would be finished.

Murphy answered the question as directly as it had been asked. The bridge would be finished in the fall, he said, providing there were no further complications about money.

At age thirty-two, Seth Low was young enough to be Henry Murphy's grandson, but still not quite so young as Murphy had been when he was elected mayor of Brooklyn. Low was the son of A. A. Low, the wealthy silk merchant, whose clipper ships had once lined the Brooklyn wharves, and he had grown up with "all the advantages," in the Low mansion down the street from the Roebling house on Columbia Heights. Perhaps Henry Murphy saw something of himself in the young man. Without question he understood the background Low came from.

Like Murphy, Low had done extremely well at Columbia College. President Barnard had called him "the first scholar in college and the most manly young fellow we have had here in many a year." Since then he had taken his place in the silk business, married the daughter of a United States Supreme Court Justice, and organized a Young Republican Club in Brooklyn to help elect Garfield and Arthur. As political clubs went, it was quite a departure, there being no smoking or billiard rooms, no bar, no social occasions, nothing but hard work.

When Seth Low ran for mayor it was as "the people's candidate" and the overriding issue had been home rule for Brooklyn. His well-scrubbed, well-brushed good looks and "tender" age had been in his favor. Politicians did not much care for him, but the people responded in no uncertain terms. He roundly trounced Boss McLaughlin's candidate, winning with a majority of four thousand votes, and without using his own money. Out-of-state papers were already calling him a figure of national importance and there was talk he would be governor before long.

But by spring the completion of the bridge looked no nearer, Murphy's assurances were no longer good enough, and Low had joined forces with A. C. Barnes and the New York contingent—Mayor Grace, Roosevelt, and the Comptroller of New York, a man named Campbell—"to clean the stables," as the papers put it. The young men were all in a very great hurry now and the papers and the public seemed very pleased with that.

Low made his first move on June 12. He asked that the Chief Engineer be required to submit a regular monthly report on the work accomplished and that he include an opinion on when the bridge would be completed. Then a motion was made that the Chief Engineer be requested to appear in person before the trustees at a special meeting two weeks hence, "to consult on matters appertaining to the bridge."

But later that very day Robert Roosevelt had thrown up his hands and in a great pique announced his resignation. If his purpose, in part, was to draw attention to the mounting animosity within the board, he succeeded. The problem with the bridge was lack of leadership, Roosevelt declared in a long open letter to William Grace. A conscientious trustee could never get straight answers on anything. The management of the bridge, he said, was in the grip of a solid phalanx that "even death itself could not break in upon." The delays in steel delivery were scandalous and left little doubt that things never would have come to such a pass had there been a full-time Chief Engineer on duty, not just to supervise the work but to meet with the trustees whenever they needed explicit information.

The papers made much of what Roosevelt had to say. Henry Murphy immediately called in the reporters to present the other side of the story. If Roosevelt felt in the dark on bridge matters, the elderly lawyer said, it might be because he had attended less than half the meetings there had been in the time since he became a trustee. Nor had he ever once taken his technical questions to the assistant engineers, who were always available for just that purpose. As for Roebling, his place could not be filled by any twenty engineers.

Another trustee who was with Murphy at the time, John T. Agnew, interrupted to make the point that Roebling's mind was as clear as ever. From his window Roebling could see everything going on at the bridge, Agnew said. Why, with a glass he was even able to distinguish the faces of the different men at work. "His plans and diagrams are all about him, and nothing is done in the work until it has first been submitted to him. He cannot walk about the streets as you and I can, but he moves about his room, and on some days can go out. He has promised to be present at a future meeting of the Board."

Roebling had made no such promise. It was true he had improved somewhat. His eyesight, in particular, was much improved. He could write again, in a dreadful, childish scrawl that was just barely legible, but he could do it. And by this time he had gone out of doors once or twice. But irrespective of its accuracy, Agnew's statement put a very different light on things.

Seth Low had also by this time accepted the chairmanship of a committee to investigate the failure of the Edge Moor Iron Company to deliver according to contract. When the head of the company appeared before Low's committee to explain himself, he said the fault lay not with his firm but with the Cambria Iron Company at Johnstown, Pennsylvania, suppliers of the steel blooms Edge Moor rolled into eyebars and other pieces for the bridge. To the delight of the newspaper editors the man's name was Sellers, the same as the central character in Mark Twain's *The Gilded Age*, which had become a hit play in New York. The editors had a fine time with this right away, sarcastically emphasizing that the Sellers in question was a Pennsylvania gentleman and not, of course, Twain's archetype Southern promoter whose stock expression was "there's millions in it."

Sellers' contract was the issue to be taken up in the presence of the Chief Engineer at the special meeting scheduled for June 26. But when the appointed hour arrived and everyone was gathered, the Chief Engineer did not appear. Not only that, as the reporters and most of the trustees suddenly learned for the first time, to their extreme surprise, the Chief Engineer was not even in Brooklyn any longer. He was in Newport, Rhode Island.

"Cannot meet the trustees today" was all Roebling said in the telegram that arrived that morning, which Henry Murphy read as soon as everyone was seated.

The meeting continued as planned, but not before a good deal had been said about Roebling's lack of manners. Slocum remarked acidly that Roebling was an employee of the board, that he should therefore appear before it when so directed, and that he could have at least sent a letter of explanation. Mayor Low commented ominously how much better it would have been had Roebling simply attended.

More than a few people were exceedingly upset now.

"The curt indifference displayed in sending such a message is not excusable by reason of any illness, . . ." wrote the New York *Sun*. "It is natural to sympathize with a distinguished engineer who has sustained severe physical injuries while engaged upon a great public work; but he should not rely upon this kindly sentiment to excuse conduct on his part that is unbecoming and hardly civil." The civic-minded Seth Low, too, was leaving no doubts as to his feelings on the subject. He criticized Roebling whenever he got the chance and called the bridge nothing more than the "unsubstantial fabric of a dream."

All kinds of intriguing stories were going about. It was said Roebling had been shipped off to Newport against his will, that he himself had nothing to hide but that Stranahan, Murphy, and Kingsley dared not let him be questioned by the others. Roebling knew too much about the original contracts, it was said. He was the one man who could tell the story from beginning to end.

There was another rumor that he had become hopelessly paralyzed by this time and that the trustees certainly did not want this known. It was said that he lost all control over his mind, that he was raving mad, that he was "really as one dead," that his wife, without anybody knowing it, had been deciding everything, directing the entire work for months. Soon most of the papers were saying as much.

Roebling was again requested to appear before the board. But again when the day came around, he failed to appear. However, this time he did send a letter of explanation. He was too ill to come he said. He could only talk for a few minutes at most and could not listen to conversation if it continued very long. He had gone to Newport on the advice of his doctors, who hoped, he said, that being out of doors some and away from the noise of the city might lessen "the irritation of the nerves of my face and head." He was now able to be out of his room occasionally.

He had not explained his absence from the previous meeting because he had assumed everyone knew he was sick and he figured the trustees must be getting as tired as he was of seeing his health discussed in the papers.

Not a day went by that he did not do some work on the bridge. His assistants could refer to him for advice at any time. The work to be done that summer was "very plain routine." If the contractor, Sellers, would supply the steel as fast as it was needed, the work would proceed with no delays whatever.

But then he wrote to Henry Murphy to say he was powerless to speed Sellers up. Making the various shapes required was a difficult job, Roebling stressed, but even more to the point, since Sellers was making no profit out of the contract, he was in no particular rush about it. If all the steel needed were at hand, the superstructure could go up in three months. As it was, there was no chance of the bridge being finished that year, as Murphy had been saying it would. A more realistic date would be late in 1883. For Murphy and those others on the board still loyal to the Chief Engineer, this was extremely discouraging news.

"Newport has never looked more attractive than it does at present," reported the Brooklyn *Eagle* in early July. "A large number of summer residents have already arrived. The business people seem to be satisfied with the outlook, and the hotel proprietors anticipate a bigger season than they have had for years." The National Lawn Tennis Association was to hold its tournament there by invitation of the governors of the Cassino, and the meets of the Queens County Hunt, yachting, shooting and horse races were also to be included in the program of outdoor sports.

Newport had its reputation, of course, and when it was reported that Colonel Washington A. Roebling had taken "the Meyer Cottage" for the summer, people quite naturally had a definite picture in mind of the life he was leading.

The picture was decidedly mistaken, however. The house Emily had rented was not the sort of "cottage" Newport was famous for. It was large and comfortable, but located in what was known as "the other Newport," the older, less fashionable section of the old seaport, out near the end of Washington Street, which runs parallel with the shore of the bay and which was described in the *Newport Guide* of the time as "shady, quiet, and a favorite resort of persons of literary character." The house suited their needs perfectly. The air came right off the bay, there was little noise or distraction and a good deal of privacy. The broad front porch and the front bedrooms upstairs offered a spacious view of the water and the Newport Harbor Light. Most important, the house was an easy, level, ten-minute ride from the New York steamboat landing, so bringing him up from the boat had been about as uncomplicated and painless as possible.

The yachting, tennis-playing, lawn-party side of Newport was not only out of sight several miles away, but was an entirely different and separate world from the one they experienced that summer, quite as distant in spirit as it had been when they were still in Brooklyn.

They had picked Newport because G. K. Warren was now stationed there. He had been put in charge of all Corps of Engineers activities in New England, the principal work at the moment being the construction of the breakwater at Block Island. Whether Emily was aware of how much her brother's health had failed by this time is not clear, but more than likely that too was on her mind when she arranged for the house.

It was the first time she and her husband had been away from Brooklyn in five years. And it might have been a first real vacation since the trip to Europe in '67 had it not been for the clamor for him to return to face the trustees. As he saw it, there was little cause to have remained in Brooklyn. His instructions had all been prepared long since, the work was quite routine, as he said; there was really nothing more to be decided of any major consequence. Indeed it must have begun as about the most hopeful summer they had seen in a very long time. By now it was clear to both of them that he was going to pull through. The bridge was all but built, and except for an unfortunate falling out between Martin and Farrington, the work was going perfectly smoothly. (What the fight was about is not known. But Farrington was so angry he quit, much to Roebling's regret, and there is no record of what became of him afterward.)

A sudden return to Brooklyn would have been a tremendous strain for Roebling, physically and emotionally, but he would have done so immediately, without reservation, had anything serious gone wrong at the bridge. But at this late date he did not propose "to dance attendance on the Trustees," as he said in the private notes he dictated to his wife. "I never did it when I was well and I can only do my work by maintaining my independence." If the trustees were angry and irritated with him, the feeling was mutual. He saw their request for him to appear before them as no more than a political ploy at his expense. Important elections were coming up in the fall and there were ambitious men of both parties on the board. The bridge might well decide who would be the next governor. How seriously Seth Low was taking the talk about his becoming governor was anyone's guess. But there was no doubt at all about Slocum. He was a prime contender for the Democratic nomination. It was a long-awaited opportunity for Slocum, and for William Kingsley, his great benefactor, the time was ripe to become something more than the man to see in Kings County.

Roebling refused to be "dragged into the board and put on exhibition," as he said, simply to serve the purposes of political ambition. He had had his fill of politicians. "I am not a politician and I have never tried to conceal the contempt I have always felt for men who devoted their lives to politics," he wrote privately. At another point he said there was not a self-respecting engineer in the country who would have put up with what he had over the years. He was seething with indignation, and when the *Sun* editorial appeared, charging him with irresponsibility, his patience ran out. He drafted a long letter in response, a letter he never sent. The one copy is in pencil, in Emily Roebling's handwriting, in one of her letter books, and it is the single piece of evidence that perhaps there was, after all, a grain of truth to the whispered story that he was staying away because he knew too much.

Roebling said in the letter that over the years he had had to deal with "no less than one hundred and twenty politicians" on the board. But now he found it particularly infuriating that the "virtuous Slocum" had been among those demanding that he appear in person before the board and that the "virtuous Slocum" had been seconded by "the still more virtuous Kingsley."

"This is the same General Slocum," he wrote, "who joined with the request that I absent myself from any meeting of the board because my presence may embarrass Mr. Kingsley's proposed operations of putting a couple of millions in his pocket, millions which have not yet reached their destination." His "patience at an end," Roebling said flatly that General Superintendent William Kingsley had been paid $175,000 for work he had never done—for work that he, Roebling, had in fact done—and that it had been Henry Slocum who stood up in the conference room years back and exclaimed that no man could name a sum that would compensate so eminent a man as Mr. Kingsley for the services he had rendered.

In another note he commented that Kingsley had also been in line to get a granite contract, but the fall of the Tweed Ring had put an end to that. Which of the granite quarries Kingsley had an interest in, the note does not say.

So the Chief Engineer had been aware of what had been going on. Had he not been there, he seems to say in his notes for another letter, this one to Comptroller Campbell, things might have been worse. ("I have always had bitter enemies in the Board for no reason except that I was in the way of any schemes for robbery.") But so far Roebling had held his tongue.

Farther along in his notes for the Campbell letter, Roebling wrote, "I have over and over again been interviewed by trustees who when they found themselves face to face with me and found me a live man and not the driveling idiot they had expected, had very few questions to ask and scarcely anything to say about the bridge in any way."

Toward the end of July, Seth Low decided that if Roebling would not come to Brooklyn, then he, the mayor of Brooklyn, would go to Newport and see for himself. Low was one of those trustees who had never laid eyes on the fabled Chief Engineer.

It is only from comments Low and Roebling made later that anything can be deduced about Low's flying visit and the confrontation. The papers said merely that the mayor of Brooklyn would be out of town briefly.

Low arrived by boat, took a carriage down Washington Street, and apparently was ushered directly into Roebling's room. Few words were wasted. Low told Roebling that it was time for him to step aside, which would be a perfectly honorable move for him to make. History would still remember him as the builder of the bridge, Low said. He could remain on as a consulting engineer and his salary would stay the same. C. C. Martin would be made Chief Engineer.

Roebling flatly refused to do any such thing. If Low and the others wanted him out, he said, they would have to fire him outright. He would not step aside of his own accord and his decision on that was final. Low answered that if Roebling insisted on being stubborn then fire him they would. Why, Roebling wanted to know. Low tried to explain, but according to Roebling's notes on the interview, the young mayor's reasons "were so weak and childish he finally abandoned all attempt at argument and said, 'Mr. Roebling, I am going to remove you because it pleases me.' " Whereupon Low walked out of the house and was back in Brooklyn in less than twenty-four hours.

A few days later Emily and Washington Roebling were struck still another crushing blow. On August 8, after a sudden, severe illness, G. K. Warren died at his Newport home at the age of fifty-two. The military court appointed to examine his ignominious relief at Five Forks had by this time reached its decision, not only exonerating Warren fully and applauding him, but casting serious doubts as to the manner in which he had been treated. But tragically, for him and his family, the findings of the court would not be published for another three months. He had not lived to see his name cleared.

On the afternoon of August 17 Mayor Low sent each of the Brooklyn trustees a note saying there would be a meeting of the board on Tuesday the 22nd. "Please make it convenient to be present, as business of importance is to be considered," the mayor wrote. Right away reporters at the Brooklyn City Hall wanted to know what this meant, but Low replied that there was nothing more to be said in advance. In another couple of days, however, before the meeting, it was reported on good authority that Low planned to move for the dismissal of the Chief Engineer.

The trustees met as planned promptly at three in the afternoon. It was noted by one reporter that Henry Murphy "did not look nearly so eager to proceed as did the youthful, bright-faced Mayor Low of Brooklyn . . . Nor did the venerable member of the Board, Mr. James S. T. Stranahan . . . appear as thirsty for information as did the affable business-intending Mayor Grace of New York."

Low made a little speech. "I am convinced," he said, "that at every possible point there is a weakness in the management of the Brooklyn Bridge. The engineering part of the structure—the most important—is in the hands of a sick man." He went on to say what a serious handicap this was and told how he had gone to Newport to reason with Roebling but had failed. He had made an eminently fair proposition to Roebling "in order to facilitate the completion of the work," but the man "would not accede." Therefore action had to be taken. So Low said he had held a private conference on the issue with two other ex officio members of the board (the mayor and comptroller of New York) and they were "of the opinion that the change ought to be made, notwithstanding Mr. Roebling's unwillingness." He had, therefore, Low said, prepared certain resolutions, which he immediately presented:

> WHEREAS, The Chief Engineer of this Bridge, Mr. W. A. Roebling, has been for many years, and still is, an invalid; and,
>
> WHEREAS, In the judgment of this Board the absence of the Chief Engineer from the post of active supervision is necessarily, in many ways, a source of delay; therefore,
>
> *Resolved*, That this Board does hereby appoint Mr. Roebling Consulting Engineer, and Mr. C. C. Martin, the present First Assistant Engineer, to be the Chief Engineer of the New York and Brooklyn Bridge; and,
>
> *Resolved*, That in doing so, this Board desires to bear most cordial testimony to the services hitherto rendered by Mr. Roebling, and to express its regret at the necessity of making such a change at this time.

Mayor William Grace rose to his feet and said he "heartily seconded" the resolutions of Mayor Low. "It is our duty to set aside all other considerations in seeing that nothing shall interfere with the progress of this work." The elected heads of the two cities building the bridge and paying for it had decided the time had come to dispense with Washington Roebling.

But at that point the Comptroller of Brooklyn, Ludwig Semler, a new man on the board, asked for the floor. "I also think the work should be pushed," he said, "but I do not think it would be using Mr. Roebling justly to oust him from his position now that he is about to reap the full benefit of his labors. If he had been in any way guilty of delaying the bridge, I should be in favor of retiring him, but there is not a shadow of a charge upon which to base such action. In fact, Mr. Roebling has done much toward pushing the bridge along. Let us not act summarily toward him after his thirteen years of service. If someone will tell me how an engineer is going to build a bridge without material I shall be pleased. Mr.

Roebling did not have the material and so he is made a scapegoat for others' sins."

Semler, who had had almost nothing to say at the few previous meetings he attended, spoke with great feeling and was strongly commended for it by Stranahan, who had gotten up from his customary seat on a sofa that stood against the wall and began addressing the group, as one man said later, "with the manner and voice of a speaker at a funeral."

"I cannot but remember at such time as this," Stranahan said, "the eminent abilities of the elder and younger Roebling. When I think that that ability has given us the finest bridge in the world I cannot help but feel that there is something yet due to the Chief Engineer. True, his health is not as sound as we could wish, but it is as good as it has been for many years past—years wherein we could not dispense with his services because of the accurate knowledge of the enterprise which he alone possessed. The older trustees will comprehend the truth of what I say. If this matter is to be referred to another meeting I should be much obliged, so far as I am personally concerned, to have it come up at the next regular meeting. I have traveled one hundred and eighty miles to attend this one, and I must travel one hundred and eighty miles tomorrow to join my family."

Stranahan, who had returned to Brooklyn from Saratoga, where he customarily spent part of the summer, was, in his usual fashion, stalling for time, banking on a "maturing of judgment" among those trustees who had not as yet made up their minds about what to do.

It was then moved that the subject be "laid over" until September 11 and the motion carried.

All Brooklyn was talking about it by the time the sun went down and in the next several days virtually every newspaper on both sides of the river began taking sides. The New York *Tribune* said the bridge had been "tainted at the start" and linked Roebling with Murphy, Kingsley, and Stranahan as the proper parties to blame. Things would have been far better, the paper said, had they all been pensioned off years before. "This is the day for sharp decision and vigorous action."

The *Star* said it was obviously time to get a new Chief Engineer. "Had Mr. Roebling done his duty instead of becoming the cat's-paw of the Bridge Ring, he might have saved millions of dollars to the two cities, and his zeal would not have gone unrewarded."

The New York *Evening Post* said Mayor Low's resolutions had been made not a moment too soon. The *Daily Graphic* reasoned that since Roebling's special knowledge of suspension bridge construction was no longer essential to the work, and the bridge was too far along for anyone to mistakenly alter the plan, there was no real reason to keep him on. "That a man has been supervising a structure like this from its beginning gives him no right to delay the completion, as Mr. Roebling is now delaying this bridge," concluded the editorial writer of the *Daily Graphic*.

The Iron Age said the only thing wrong with Low's action was the idea of replacing Roebling with C. C. Martin. "If new life is to be infused in that comatose engineers corps, it must come from the outside—for inside it there is no leaven of redemption left. No wonderful or even exceptional engineering talent is required to bring the work to completion; but what is required is some man with good executive ability, of great force of character, and complete independence of all possible future political preferment."

Even the Newport *Daily News* felt obliged to report on the situation. "Mayors Grace, of New York, and Low, of Brooklyn, are getting tired of waiting for the completion of the celebrated East River Bridge," the little paper announced the morning of August 24. "The monster connecting link between the two cities has already cost them too much and it is about time that something was done to prevent anymore such delays as have hindered its completion."

Only the Trenton *Daily State Gazette* and the Brooklyn *Eagle* rose immediately and angrily to Roebling's defense. "His spotless integrity and high sense of honor are unquestionable," wrote the *Daily State Gazette*, "his great skill as an engineer is established, and his devotion to this work has been attended by the sacrifice of his health."

The *Eagle*, not surprisingly, had more to say. How, asked Thomas Kinsella in a two-column editorial, could Low present a proposal to appoint Roebling the consulting engineer, when Roebling, as Low himself reported to the trustees, had said explicitly he would not accept the job? And what earthly good would such a change accomplish? "Is it possible that the existence of Mr. Martin himself is now in the nature of a discovery, while the fact is that he has had practical executive control of the work for many years past? If he is not a 'natural channel' of information, who is?"

As for Roebling's competence, that was not even an issue, Kinsella asserted. His past contributions to the work were enormous.

There had not been a single failure on any part of the work that could be chargeable to the engineers. Furthermore it would be a rotten thing to degrade the man and deny him his rightful honors on the very eve of his triumph.

And that pretty much summed up the case for Roebling as it was being presented in Brooklyn—except for the strongly and privately expressed opinion of several on the board who claimed to know Roebling personally that he had been kept alive all these years only by his intense interest in the work and the desire to see it finished properly. The implication was that a vote to discharge him would be as good as a death sentence.

A very important question, of course, was whether the voice of Kinsella was the voice of William Kingsley. A few years before, the immediate assumption would have been that it was, but not so now, for the influential editor had grown increasingly independent. Not long before, when he felt there was too much interference in editorial matters on the part of the owners, Kinsella had threatened to resign and start a rival paper and ever since he had been left to decide things pretty much on his own.

The one other argument in Roebling's behalf was, of course, the actual progress still being made on the bridge itself, progress that Henry Murphy kept presenting in formal reports to the two mayors. A hundred and fourteen intermediate chords had been put up in a week, 72 diagonal stays, 60 posts, 21 intermediate floor beams, 21 bridging trusses, 16 intermediate promenade floor beams, 12 lower chord sections, 2 upper floor stay bars. At the end of the Brooklyn approach, work had begun on the foundations for the iron viaduct and terminal station building.

But the question of Roebling's mental and physical health had not been put to rest. If anything, rumors and innuendo on the subject were more plentiful than ever before. The three most prominent trustees of long standing—Murphy, Stranahan, and Kingsley—said they could vouch for Roebling's state of mind, but their word was not only not enough any longer, it was outright suspect among the younger trustees. The only one of the younger men who had seen Roebling and talked to him in person was Seth Low and he would make no comment on Roebling's condition one way or the other.

Toward the end of the month of August Comptroller Ludwig Semler received the following letter from Newport, from Mrs. Washington Roebling:

> I take the liberty of writing to express to you my heartfelt gratitude for your generous defense of Mr. Roebling at the last meeting of the Board of Trustees. Your words were a most agreeable surprise to us as we had understood you were working in full sympathy with the Mayors of the two cities and the Comptroller of New York. Mr. Roebling is very anxious for me to go to Brooklyn to convey to you . . . a few messages from him. Can you see me at your office some morning . . . ? I will go to Brooklyn any day you can give me a little of your time and see you at your own house or your office just as you may prefer. . . .
>
> As you are a stranger to Mr. Roebling all that you said was doubly appreciated. There are some few old friends in the Board of Trustees who know him well and who have always stood by him in the many attacks that have been made on him in the past ten years, but we never expect such consideration and kindness from those who have never seen him.

On Tuesday, September 5, a week before the trustees were to meet to vote, Comptroller Semler suddenly announced he was leaving for Newport that evening to see Roebling and judge for himself.

"Nobody should be convicted before he is tried," Semler said. "As I have undertaken to defend Mr. Roebling to a certain extent against the attempt to remove him, I want to make his personal acquaintance and see what impression he makes on me. It seems from certain statements made in connection with this matter that an impression has gone abroad that he was not only suffering physically but that his mental faculties were also impaired. Of course, if this were so, his plans should not be relied upon, and the work should be suspended until an investigation could be had; but physicians tell me his intellect is all right. I am today more convinced than ever of the great injustice of displacing a man of his merit and standing without giving him an opportunity to defend himself. The idea that he could not do his duty without being at the bridge office is preposterous."

De Lesseps had been in Paris while the Suez Canal was being built, Semler said, and so why was it so unreasonable for Roebling to remain in a house a few blocks from the bridge. "There is nothing sentimental in my feelings in this matter. The question is simply one of justice." Semler told reporters he would be back in his office Thursday morning.

All at once Semler had become a most important figure. And a great deal seemed to hang on what kind of report he would come home with. As things looked now, Roebling had at least four sure votes against him and four for him. Low, Grace, Campbell (the New York Comptroller), and A. C. Barnes were clearly committed to ousting him, while Murphy, Stranahan, Marshall, and John T. Agnew could be counted on to vote the other way. But the rest were undecided, or appeared to be, and Semler's evaluation might therefore be the deciding factor, at least among the younger men.

But there was also by now very particular interest in which way William Kingsley and General Slocum might go, for there was no longer any doubt that Slocum was the front-running contender for the Democratic nomination for governor and the convention was only two weeks off. On the surface it would appear both men would naturally go along with Murphy—to stand solidly behind the Chief Engineer. But Slocum had attacked Roebling on too many occasions and with no little public fanfare about it and to side with the old regime and vote against such known champions of reform as Low and Grace, to vote for what might appear to be further delays on the bridge and greater expense (greater graft and corruption was the implied idea), could be extremely foolish politics for a serious candidate at this particular moment and very hard to explain to the electorate later on. Still Slocum was Kingsley's man, it was pretty generally believed. Kingsley had the power, Kingsley would be the one to decide. Kingsley was the man to watch.

In a note to Paine written about this time, Roebling remarked that if his position as Chief Engineer depended on Kingsley's vote, then he would just as soon "be out of the bridge."

On Thursday morning, as good as his word, Ludwig Semler was back at his desk in City Hall, just a few doors down from Mayor Low. The reporters were called in and the interview commenced.

Semler said he had been very kindly received by Roebling, whose acquaintance he had not made previously, and that a full, frank talk had ensued between them. He said he found the engineer suffering from a severe nervous affection, but his intellect was perfectly clear and strong. "If his intellect has been impaired," he said, "I should consider myself a happy man if I had what he lost. He spoke to me with clearness, and exhibited a memory which was something astonishing."

Semler was asked what Roebling had said about the proposition to supersede him.

"He said that under no circumstances should he take any other position than Chief Engineer," Semler replied, and quoted Roebling as saying, " 'If they want to remove me, let them do it absolutely. They know I will not take any other position. Why don't they say they do not want me anymore. That would be the straightforward way to do.' "

"Did you ask him his opinion as to the motives which prompted the opposition?"

"I asked him if he had any cause to believe that there was unfriendly feeling toward him. He said he did not like to say anything upon that point."

"Now, to do away with the driver who has brought us very nearly to shore, I think would be shameful," Semler continued. "Suppose that the resolution offered by Mayor Low should be adopted and Mr. Martin should not accept the position of Chief Engineer?" (It was commonly being said by this time that none of Roebling's staff would assume his title, if offered, but this was the first time anybody in a position of authority had said so publicly.) "Suppose it should be offered to Mr. Paine? He will not take it. We should then have to have another man. That will cause further delays in the work. . . . He might commence to meddle with the work and defer the completion of the bridge ten years."

A reporter for the *Eagle* then asked whether Mr. Roebling had said anything about what passed between him and Mayor Low during Low's flying visit to Newport? Semler answered that Roebling had indeed quoted from his conversation with Low, but Semler said he did not think he ought to repeat Roebling's remarks for publication.

As it was, very little more would be said for publication by anyone until the trustees gathered to cast their votes the following Monday, September 11.

But between Semler's return and the crucial meeting of the board, Roebling had still one more visitor at Newport—a reporter for the New York *World* who somehow talked his way into the house and managed to get an interview with Roebling, something no other newspaperman had been able to do in ten years. The agreement was that no direct quotes would be used. Roebling apparently trusted the man and in the course of the conversation made some bitter remarks about the Board of Trustees being full of candidates for governor. On his way out the front door the reporter had again promised Emily Roebling that he would not print a line of what had been said. But the man had not kept his word and when the article appeared it did little to further Roebling's cause in Brooklyn. His loyal backers on the board felt he had dealt himself the worst blow possible.

Emily was shattered by what had happened, felt she was to blame, and wrote a long letter of apology to William Marshall, the one man who had voted against the J. Lloyd Haigh contract and one of those few long-time trustees who was not a politician. Now she too was full of despair. There was no doubt, she said, of her husband's "perfect sanity and ability as an engineer, but he certainly is unfit to be on the work where so many political interests are involved." He had, she said, no capacity for doing anything for the sake of politics: "I thank you very much for all your efforts and do not think I shall be greatly disappointed when the bridge controversies are ended, even against us. It has been a long hard fight since Mr. Roebling first took sick and if this chance reporter's visit changes everything I shall see in it the hand of God, that all my care could not direct or change."

Seventeen were present, the entire board but three—John G. Davis, Henry Clausen, and H. K. Thurber—all of whom were in Europe. Henry Murphy sat in the president's chair, as usual, and ranged in a semicircle before him were Mayors Grace and Low, Comptrollers Campbell and Semler, General Slocum, A. C. Barnes, Kingsley, Stranahan, Agnew, J. Adriance Bush, Thomas C. Clarke, Jr., William Marshall, Charles McDonald, Jenkins Van Schaick, Alden S. Swan, and Otto Witte, the secretary. A half-dozen representatives of the press had also been admitted.

The first ten minutes were spent on routine matters. Minutes from the preceding meeting were read, C. C. Martin made some comments on the delays on the two terminal stations being built at either end of the bridge, and it was announced that several hundred tons of steel still remained to be delivered by the Edge Moor Iron

Company. Mayor Low requested that the president inform the company that the trustees were "in a hurry," to which there was great laughter.

Then C. C. Martin left the room, the doors were closed, and Seth Low, having risen from his chair, began speaking in a very deliberate manner.

"If there is no other business before the meeting I will call up the resolution which I presented last month, and in doing so there is very little which I wish to add to what I said then. As I said at that time the resolutions bring the Board of Trustees face to face with the question as to whether the existing engineering arrangements of the bridge are the best that are within reach for the work that lies before us. If the majority of the board will take the responsibility of saying they are I shall feel that I have done my duty in bringing the matter to this issue, and no one will be more glad than myself—if the majority does decide that way—to find the facts justifying the judgment.

"On the other hand I have presented the resolutions which suggest making Colonel Roebling consulting engineer because I believe sincerely that that would be the best pledge we can give the public that nothing whatever shall be allowed to interfere with the speedy completion of this work. I think the effect would be instantaneous not only upon the employees of the trustees, but upon all the people with whom they are dealing. It would convince them that from this time, whatever may have been the case in the past, these trustees must be dealt with upon the theory that they mean business. I do not mean by my wording to reflect upon the trustees in their intentions in the past. . . . For myself, I repudiate entirely the idea that there is anything in this proposition that reflects upon Mr. Roebling either directly or indirectly. I—"

Henry Murphy interrupted. "Will you allow me to say that I have here a communication from Colonel Roebling, which perhaps ought to be read before you proceed with your remarks. In all events, it is on this question and—"

"I think it would be better if I finished now," Low snapped back, "if you will allow me.

"I wish to say," he went on, "that I repudiate the idea that there is anything in the proposition that reflects on the engineer, either directly or indirectly. I offered it believing that it was a proposition which he could honorably accept. I think it is one which he ought to accept. More than that I think it is one which, in its essence, is kind, because in my judgment it would take him out of a false position and place him in a true one with reference to this work, and by so doing we will relieve him of the criticism to which he has been subject.

"One other thing I will say is that I offered the resolutions which I presented last week without consulting Mr. Martin. I felt that if that question could not be settled by me without consulting a subordinate I had better not offer the resolutions. He was entirely unaware, so far as I know, of any such resolution being thought of, and in all of his utterances he has acted the part of the loyal friend of the Chief Engineer.

"But the real question is not that which concerns the engineer as much as it does the trustees and the people. By our action today we say: 'We have nine months before us. We have charge of the expenditure of a million and a quarter of dollars and an interest and expense account of about three thousand dollars a day. The question is, shall we have a sick man or a live man—a man who is responsible and with whom we can come into contact day after day?' I say this without any reference to what has been done in the past—which has no further claim to my attention than that of historical interest."

Here, according to one of the reporters present, Mayor Grace broke into a broad grin, as if Low had just said something extremely funny.

"I have only to say," Low continued, "that if the majority will take the responsibility of leaving things as they are it will give me the greatest pleasure to work with them in justifying their judgment and in bringing about the result we all want—the finishing of the bridge."

With that Seth Low sat down and Henry Murphy handed Roebling's letter to the secretary, Otto Witte, who read it aloud.

The letter was brief and said pretty much what was already known—that he would not accept a position as consulting engineer, that his absence from active supervision had in no way hindered progress on the bridge, and that it was his personal wish that the vote be taken simply as to whether or not he was to remain in command.

Mayor Grace was immediately on his feet. "At the last meeting I seconded the resolution offered by Mayor Low, and I have since found no reason to change my mind and I again second it now."

Then William Marshall spoke. He had never had a great deal to say at board meetings before, the way most of the others had, and he was not known as a speaker. So it is doubtful anyone in the room was quite prepared for the speech he gave.

"Mr. President," he began, "I am sorry this resolution has been introduced here. I see no reason why Mr. Roebling should be removed. I know that this bridge has been kept back time and time again by many, but I never knew that Mr. Roebling had kept it back one day or one hour. The very gentleman that you propose to put in his place is honorable and honest enough to say he doesn't believe the bridge has been kept back by the engineer who is his chief. Furthermore, he says that to his knowledge the engineer and his assistants have not made a mistake upon the bridge."

Then turning to Seth Low, he said, "You want to remove him; to drive him out. For what? Why? As a bridgebuilder he has not had his equal on the face of the earth. I defy contradiction!

"There are two bridges across Niagara. He built the largest of them and it stands there today—a perfect success. When I say 'he' I mean his father and himself—the father who sacrificed on this bridge. There are two bridges across the Ohio, one built by Mr. Roebling and one by a man who is ashamed of his name. The one at Wheeling fell into the river; the other, at Cincinnati, is an honor to the man who built it. I never heard he made a mistake and kept the bridge back, and he must be sacrificed. For one I would take the arm off my shoulder before I would permit myself to vote against a man standing here without a blemish upon his character or ability. If you search back to the time of the sinking of the caisson to the present moment you will find that he has not kept the bridge back a moment."

Again he turned to the Mayor of Brooklyn and this time angrily. "But our friend Mr. Low goes down to Newport and demands his resignation! By what authority?" There was absolute silence in the room. "Have you any law for it? If you have I should like to see it. I should like to know by what parliamentary usages three or four trustees, meeting in the Comptroller's office in New York, claim to represent the wishes of this board?" The two mayors and Comptroller Campbell kept looking right at Marshall, smiling all the while.

"I consider you are bringing an innocent man and holding him responsible for the delay and losses we have gone through here. . . . If I know this board, I know there is too much honor on this floor to enable you to remove the Chief Engineer. If there is any fault to be found we should begin where the fault belongs. If there has been any fault in the board for the last ten years, for one I am willing to assume the responsibility for it, but I don't want to sneak out and place it on the shoulders of the Chief Engineer. It would be mean and contemptible for me to do that, and I don't propose to do it."

There was a long silence when he sat down again. Then Secretary Witte read a short prepared statement supporting the Chief Engineer and J. Adriance Bush said he also would vote to keep Roebling. Martin was not the only member of the engineering staff to speak up in Roebling's defense, Bush said, they *all* had. The core of the issue was the change in the plans and that had not been Roebling's doing and they all knew it. "I think," he said, "the question is one that we ought to approach . . . with just the same solemnity as we would approach the trial of a man accused of high crimes and misdemeanors. We ought to look at it in the same light that you and I would regard the impeachment of a judge or anyone in authority. We ought to look at it free from public clamor on the one hand and free from any personal feeling on the other."

A. C. Barnes said he had a very sincere admiration for Colonel Roebling, but that he did not think Roebling ought to feel wounded if he were to be retired with every honor and his present pay. "Why, sir, if Mr. Roebling were a regular Army officer he would have been retired long ago, with half pay and nothing to do, and nothing would be thought of it." Barnes said he was simply unable to understand why Roebling did not accept Mayor Low's resolutions "in the same kindly and considerate spirit which I am sure animates every one of us who would vote for it."

Ludwig Semler, who apparently had decided that the meeting was running against Roebling, moved the vote be postponed and the whole issue referred to special committee. But William Kingsley, who like Murphy, Stranahan, and Slocum had said nothing thus far, had done some figuring it seems and decided that this was exactly the time to vote.

"I feel the same way," Mayor Low said instantly.

A few others asked to be heard and were, briefly, including Stranahan, who said simply and rather weakly that Roebling was needed still. Then a voice vote was taken on the Low resolutions, with the following result:

Yeas—Mayors Grace and Low, Comptroller Campbell, and Messrs. Van Schaick, Clarke, McDonald, and Barnes.

Nays—Comptroller Semler and Messrs. Murphy, Bush, Witte, Marshall, Stranahan, Agnew, Swan, Kingsley, and Slocum.

The count was 10 to 7. The resolutions had lost; Roebling had won by a majority of three, including William Kingsley and Henry Slocum. But had the two of them, or any other two, voted differently, Roebling would have been out.

"ENGINEER ROEBLING RETAINED," "A MAJORITY FOR ROEBLING," "ROEBLING NOT TO RETIRE," were some of the headlines the following day.

Mayor Low told reporters that he was quite content with the decision of the board and he told several others on the board who had backed Roebling that he had secretly been hoping Roebling would win all along. He would rather see the bridge finished under Roebling, Seth Low told Ludwig Semler, than any other engineer. Semler repeated this in an interview, thinking apparently that it would put the mayor in a more flattering light, and he offered his own personal diagnosis of Roebling's mysterious malady, something he had not done before the voting began. "I actually believe that all that ails him is a nervous affection which prevents him from mingling with numbers of people."

23
And Yet the Bridge Is Beautiful

And yet the bridge is beautiful in itself.
—*Scientific American*

IN THE early spring of 1883, about the time the weather had turned warm enough for the Chief Engineer to spend some time outdoors in the garden, the bridge was finished. There was no one moment, no particular day, when he could have said as much, nor would there be. Bridges did not end that way. There was always something more to finish up, some last detail to attend to. The final touches at Cincinnati, for example, had dragged on for nearly six months after the opening ceremonies and it looked as though the same might happen here. But the bridge he saw standing now against the sky half a mile in the distance was the finished bridge for all intents and purposes. In another few weeks it would be open and in use.

It had taken fourteen years. In another few weeks he would be forty-six years old. He had spent nearly a third of his life on this one bridge, nearly as much time as his father had given to all his major bridges combined.

His health was much improved. The nearer the end came the better he felt. He could get about the house much more easily than before or go out into the garden. His eyesight had returned. It was a little as though he himself had returned from a long absence. He could read the papers again, for one thing—about General George Crook chasing Apaches across the border into Mexico that spring or about the housewarming party given by Mrs. William Vanderbilt at her limestone palace on Fifth Avenue, the most lavish and costly fancy-dress ball ever put on in the United States (an estimated $250,000) and reputedly the greatest social event of the age. According to the *Times* the costume problem alone "disturbed the sleep and occupied the waking hours of social butterflies, both male and female, for over six weeks." Abram Hewitt came as King Lear, "while yet in his right mind," the paper noted, a remark that doubtless cheered Washington Roebling.

In the time he had spent on the bridge, the telephone and the electric light had been introduced. (What a difference they would have made during the work inside the caissons.) Now at night he could see hundreds of electric lights burning over in New York, directly across the river, in the blocks Edison had first lit the summer before, when Roebling was at Newport.

Instead of one transcontinental railroad, there were now four and a fifth was under construction. There were ten million more people in the country than there had been in 1869. (Brooklyn had grown by 180,000; New York by more than 200,000.) The buffalo had been all but exterminated on the Great Plains and Chester A. Arthur had installed modern plumbing in the White House. Robert E. Lee was dead. Horace Greeley, Jesse James, Brigham Young, Emerson, Crazy Horse, Peter Cooper, they were all dead now. A whole era had passed. His own son would be entering RPI in the fall.

The bridge had taken nearly three times as long as his father had said it would and it had cost $15 million, which was more than twice what his father had estimated. It had taken the lives of twenty men, not including his own father.* The price he himself had paid was long since past reckoning.

Henry Slocum had failed to become governor, because of the

* There is no official figure for the number of men killed building the bridge. The Bridge Company compiled no list, kept no precise records on the subject, which is characteristic of the age. In a booklet made up from his Cooper Union talks and published after the bridge was built, Farrington says between thirty and forty men died in the work, which is especially interesting if it is remembered that Emily Roebling may have done Farrington's writing for him. The Chief Engineer and William Kingsley, however, both said twenty had died and from the deaths reported in the papers and mentioned here and there in the minutes of Bridge Company meetings that seems to be a realistic figure.

bridge mainly. Now Henry Murphy, too, was under the ground at Greenwood.

Would they all have gone ahead with it anyway back in 1869 had they known what was involved? It was a question neither Washington Roebling nor anyone else would ever be able to answer.

When the Democratic state convention opened in Syracuse late the previous September, it had been obvious that the nomination for governor was going to be worth a very great deal. A few days before, in Saratoga, the Republicans had picked a lackluster candidate named Charles Folger, who was generally taken to be what he was—a stooge for President Arthur and the infamous Jay Gould. Not even the Republican faithful had been able to get very enthusiastic about Folger. So in Syracuse, as the Democrats gathered inside the Grand Opera House, spirits were running high and especially among the Kings County delegation, for it had looked even to impartial observers as though General Henry Slocum would be the party's choice. But instead the convention had picked an unknown upstate lawyer, the reform mayor of Buffalo, Grover Cleveland, who had been considered strictly a local candidate before the balloting began.

Slocum's only serious rival had been Roswell P. Flower, a debonair Congressman from Watertown who had made a fortune on Wall Street and who, like Folger, was handicapped by his friendship with Jay Gould, as well as by a bad lisp that was sometimes linked unkindly with his name. But just as the convention was about to open, the *World*, then owned by Gould, began a series of sensational articles on the "Bridge Frauds." The Tweed disclosures were published in full still one more time, as though they were all new history. Every prior example of corruption within the Bridge Company, documented or alleged, was assembled into a massive attack on the Brooklyn men who had been behind the project. It was charged that three million dollars had been stolen outright during the early years of the work, that Kingsley and Stranahan were little better than common crooks, but, unlike Tweed, so skillful at covering their tracks that proof of their guilt would be next to impossible to come by. Once again Henry Murphy was asked for a statement and once again he denied the charges, as did Stranahan. Kingsley refused to see reporters. Only Thomas Kinsella was willing to talk and he said the fanfare was nothing more than politics.

"I consider it a very bold movement on the part of Jay Gould to get control of the Democratic party, just as he has already got control of the Republican party," Kinsella said. "In my opinion it is all politics and very bad politics, too. It is as transparent as glass, and anybody can easily see through it. Kingsley, one of the men attacked, is quite likely to appear as a delegate to the Syracuse convention, while General Slocum, one of the trustees, is the man who will undoubtedly be placed in antagonism with Congressman Flower for the nomination for governor. The object, it appears to me, is to cast odium on these men, and break them down in advance of the convention." Kingsley was not only a delegate, he led the Kings County delegation, solidly committed to Slocum, and Kinsella would be the one to deliver the speech putting Slocum's name before the convention.

The great issue at first had been the seating of the Tammany delegates, supposedly representing 45,000 voters and headed by that old enemy of the bridge, "Honest John" Kelly. Until late the night before the balloting was to begin, it had looked as though the Tammany delegates would not be let in. But then some of Kelly's people came calling on the Kings County Democrats to beg for their support. The meeting lasted far into the night, when Kingsley at last agreed to go to work for Kelly. As a result the Tammany delegates were seated and Slocum's backers went into the convention confident Slocum would get the nomination on the second ballot.

But a man named Ira Shafer, a Flower supporter, gave a speech referring at some length to Slocum's connection with the bridge and its "gross frauds," and this, on top of the *World* "disclosures," was, as the *Times* reported, "the means of frightening many of the country delegates who were friendly to General Slocum, but feared the effect his association with the original promoters of the bridge enterprise might have . . ."

Slocum was also strongly opposed by an influential reform faction from New York City known as the County Democracy, the head of which was none other than Abram S. Hewitt.

As a result Slocum and Flower were deadlocked on the first ballot. Kelly had not delivered for Slocum, but split his votes, biding his time. The Brooklyn men felt they had been double-crossed. There was a fierce scramble to pick up votes but on the second ballot Slocum and Flower were deadlocked again. But this time Hewitt and Kelly had both gone for Cleveland and that decided it. The Flower delegates immediately abandoned their man and Cleveland won on the third ballot. As a consolation prize, Slocum was later nominated for Congressman-at-Large.

William Kingsley and his delegation returned to Brooklyn bitterly disappointed and talking freely with reporters of their disgust with Kelly, who, as they said, would never have been seated in the first place had it not been for their help. Kelly answered later that he himself had never said a word about supporting Slocum, which was quite true.

In November Cleveland defeated the Republican Folger by the largest majority ever given a candidate for governor. Slocum was elected to Congress, but among Brooklyn Democrats there were wistful reflections on what might have been. In an inverse way the bridge had made the Buffalo man governor.

Once the elections were over, the papers had little more to say about bridge frauds. The clamor over corruption ceased instantly. But on November 20, determined to settle the issue once and for all, Mayors Low and Grace, with the approval of the trustees, appointed two accountants to examine fully all receipts, vouchers, papers, payrolls, and all other documents connected with the bridge, "and especially to investigate the books and papers of the Bridge as they bear upon the charges made by the New York *World* and others against the Trustees." It would be a year before their findings would be made public, and they would cause no stir whatever. There were vouchers on file, the accountants reported, for total expenditures of $15,211,982.92. Due to certain clerical errors overpayments to contractors came to $9,578.67 and those were the only discrepancies found in the company's books, all of which had been "honestly and neatly kept."

As it happened, and as no one realized at the time, November 20 was also Henry Murphy's last day at the Bridge Company. That night he came down with a bad cold. On Friday, December 1, 1882, he was dead.

As the papers reported, Murphy, at age seventy-two, had been in good health up until the night he took sick. As regular as clockwork he had left his house each morning at nine and walked to his law offices on Court Street, where he and his two old partners, Lott and Vanderbilt, both dead now, had long been such fixtures. Later in the morning he would go over to Montague Street to the Coney

Island Railroad offices, stay perhaps an hour, then walk down to the Bridge Company, where he generally spent the remainder of the day. On the 20th he had gone home about five.

His cold had turned to pneumonia a day or so later, but it was his heart that killed him, according to the papers, and from what his son said in an interview, it seems he died in terrible agony. The time of death was six in the morning.

"At the clubs and other places where men gathered, the deceased was the general topic of conversation," wrote the *Eagle*, the paper he had founded. The feeling was that the day marked the end of an era. Murphy had been a historic figure, everyone felt. He was the closest thing to a Founding Father Brooklyn ever had, both in his personal grace and in the things he had accomplished. To many he had seemed a last holdover from a vanished golden age. His passing was like the tolling of a bell, as almost everyone who wrote about it tried to express in one way or other.

The bridge trustees called a special meeting and with Kingsley presiding, sitting in Murphy's old chair, a long formal statement of grief was drawn up, saying, among other things, that the bridge would remain a memorial to Henry Cruse Murphy.

The major things still to be seen to by mid-May were these: the electric lights were not yet fully installed; the big iron terminal buildings at either end of the bridge were nowhere near ready; and it would be September at least before the bridge trains would begin running between the terminals. But there were no more specifications to get up, no more contracts to sign, and everything was being handled with the greatest dispatch by Roebling's immensely capable assistants. Amazingly, they were all still on the job, after fourteen years, even Collingwood, who had signed up originally for a month only. Except for Farrington, not one, in all that time, had quit out of discouragement or frustration or to take a better-paying position, several of which had been offered. Not one had been relieved of his job. Roebling's own sense of duty and determination had been matched in kind. Every man he had hired had proved up to the work. For some, such as McNulty, it was the only work they had ever known.

The bridge itself looked now about as it did in the drawing Hildenbrand had done for the Centennial Exhibition, except that there were no crowning capstones on the towers, as John Roebling had wanted, and there never would be because of cost. The towers, of course, had been standing there for nearly seven years now and were an accepted part of the landscape. But with the last of the timber falsework removed from inside the archways, they looked now as they were supposed to, like colossal Gothic gateways to the two cities.

But it was the finished span between them that made the towers seem so much more important and purposeful than ever before. It was the finished roadway, arching slowly, gracefully upward over the river to meet at the center with the great downward swoop of the cables, that made it a suspension bridge at last—and the greatest on earth. And finally, now, the diagonal stays were in place, hundreds of them, radiating down from the tower tops, angling across the vertical harp-string pattern of the suspenders, and forming what, at close range, looked like a powerful steel net, or, from a distance, as Roebling saw it, like a vast, finespun web. The bridge now, as never before, was a thrilling thing to see.

More even than the other modern structures people flocked to gape at in New York, the bridge could look extremely different at different times of day, and depending on one's vantage point. From the narrow, low-lying streets on the New York side, for example, one got relatively little sense of its long reach over the river, or what it might be reaching out to. The impression, instead, was one of fantastic upward magnitude, of breath-taking elevation, the tower in the foreground and the roadway it carried within its arches upstaging whatever else the bridge might be achieving beyond the tower.

The very shabbiness and stunted scale of the old neighborhood beneath the tower worked to the advantage of the bridge, which by contrast seemed an embodiment of the noblest aspirations, majestic, heaven-directed, lifting into the light above the racket, the shabbiness, and the confusion of the waterfront, the way a great cathedral rises over the hovels of the faithful. And the twin archways in the tower, seen from street level, looked like vast vacant windows to the sky. For a child seeing it at night, the tower could have been the dark and mighty work of medieval giants. Where on earth could one see so many stars framed in granite?

The roadway to the tower was finished now but was still closed off at Chatham Street by a high board fence plastered thick with theater posters and handbills. And along the roadway unsightly heaps of rubbish stood waiting to be carted off. Even so, there was nothing in the average person's experience to compare to this spacious, beckoning, empty thoroughfare. It climbed up and out of the city like something seen in dreams. It was a highway people just naturally wanted to travel, even if they had no interest in the smaller, more sedate city they knew to be at the other end. To the New Yorker who lived within its shadow, it was not just a bridge to Brooklyn—few New Yorkers had any special desire to go to Brooklyn—it was a highway into the open air. When the day came when everyone could go out on it, when people by the tens of thousands could go up that road and through those colossal arches, they would go, they knew, not to Brooklyn, but to a place where sailing ships would glide like toys beneath their feet, where they could look down on the tallest buildings and their own mean, narrow streets and the people in them, where they could gaze out over land and water and everything man-made. "What a relief it will be from the ill-smelling streets and stuffy shops," one man wrote. "What a happy escape from those dreadful cabins on the ferryboats! What a grand place to stretch your legs of a bright winter's day after toiling through the streets! To go from shore to shore in one straight and jolly tramp, with the sky for a roof and the breeze for good company."

Even before the bridge was opened it had become a symbol of something impossible to define that made New York different from every other city on earth. The bridge dominated the imagination the way it dominated the skyline, as Al Smith would say when reminiscing about his boyhood.

The view from the water, from the deck of a passing ferry or excursion steamer, a view being enjoyed daily by many thousands, was, of course, very different still. From there the elaborate and extremely interesting steel understructure was plainly visible overhead. From there, for example, one could see the wind braces Roebling had put in, something that would be hidden from the view of travelers on the bridge itself. These were cable stays designed to prevent horizontal vibrations. They were anchored to the corners of the towers, beneath the deck of the bridge, and extended diagonally under the deck to attach to the opposite side. The longest of them reached a third of the way across the central span and there were similar braces on the land spans.

But it was the over-all arc of the entire bridge that impressed people most when they saw it from the river, and again it seemed somehow above and beyond ordinary experience. Even the most sophisticated and analytical observers felt this. An editor from *Scientific American* wrote, ". . . the bridge is a marvel of beauty viewed from the level of the river. In looking at its vast stretch, not only over the river between the towers, but over the inhabited, busy city shore, it appears to have a character of its own far above the drudgeries and exactions of the lower business levels."

Still, the finest view of all, perhaps, was from Brooklyn Heights. Visually the bridge belonged to the Heights as it did to no other point on land.

The bridge was just far enough distant and the elevation of the Heights such that the scale of everything seemed more manageable. The towers did not loom up all out of proportion as they did from the streets of New York. There was little foreshortening of the great span. Moreover, the other essential components—the cables, suspenders, trusswork along the deck, the anchorages—could be plainly seen and in proper perspective, their function and relationship to the rest of the bridge being neither concealed nor distorted. Were one to draw a picture to explain how the bridge worked, about the easiest, clearest way would be to show it as it looked from the rear window of a house on Columbia Heights in the spring of 1883, when there was nothing on the Brooklyn side blocking the view and the skyline of Manhattan was, by later standards, quite restrained.

From the Heights it was perfectly clear why the bridge had been built. Its practicality, no less than its grandeur, was unmistakable. There below was the sparkling river and there beyond was New York, stretched out before the eye like an enormous scale model. The bridge was the way to get there. It was the great highway to New York, just as had been intended from the start. And while Brooklyn in the mind of the average New Yorker might remain an indeterminable, even dubious, destination, for everyone living in Brooklyn, New York was a known quantity and the reasons for wishing to get there—and to get back again—were quite clear.

The river now looked very different than it had in the early days of the work. Traffic was heavier, it moved faster, and there was a good deal more coal smoke trailing in the wind. The river was change itself—ships coming and going, sails turning in the sun, cloud shadows crossing over from New Jersey, gulls circling and diving, the water changing color with the sky, the other shore now very near, now distant, depending on the light or atmospheric conditions, the tides running. But over it all, triumphant and immovable, stood the bridge, seeming to hold the land in place against all change. On the one hand it was a vaulting avenue over the river, defying space and gravity like some weightless natural phenomenon (". . . high over all the Great Bridge swept across the sky," a novelist would write), but it was also fixed, deep-rooted. It was as though the two cities might drift apart were the bridge not there. The bridge kept things in place. It belonged.

There were some, of course, who for the rest of their lives would see the bridge in other ways. For them it would be an emblem of colossal greed and deception, of hideous physical torture and unbearable grief. There were people on both sides of the river who would look at its commanding silhouette and see the faces of Tweed, "Brains" Sweeny, and the rest, leering, with black eyes full of sly deception, as Nast had drawn them. There were those who would follow with their eye the path of the great cables in the sunshine (". . . the arching path/ Upward, veering with light, the flight of strings of wires, . . . telepathy," the poet Hart Crane would write) and think only of the bad steel woven into them forever. And there were those, in tenements back from the river, for whom the lofty towers would remain a day in summer when the broken corpse of a husband or father was brought to the door in a spring wagon.

The only major parts of the bridge that Roebling could not see from his house were the two terminal buildings and once, in late April, he had been taken by carriage to have a first look at the one in Brooklyn that McNulty had designed. The big, curved two-story building, which was to be twice the size of the station at the other end, was little more than half built then. Still it had looked most impressive, even on dingy Sands Street, where the main entrance was to be and where Roebling's carriage stopped. A reporter who visited the New York terminal about the same time wrote of "an almost deafening din of a hundred workmen hammering away for dear life." In Brooklyn the noise must have been twice as bad.

From here, on the upper level, the bridge trains would leave for New York. The commuter would pay his fare at one of several ornamental iron toll booths ("pretty enough for opera boxes"), then climb a broad iron stairway to the waiting platforms. The cavernous building itself, as was already apparent, was to be an extremely ornate affair, like the elevated railroad stations in New

York, with all kinds of fancy ironwork, panels, pillars, molding, and row on row of plate-glass windows. Once finished, the whole building was to be painted a dark red.

The bridge trains would be much like the newest cars on New York's elevated trains. They would have large windows, double sliding doors, open platforms in front and back, and they would appear to be self-propelled, unless one were to look between the tracks and see the steel traction cable they hooked on to. (At the time Roebling paid his visit to Sands Street, workmen were installing the two 300-horsepower horizontal steam engines that would supply the power for the cable. The engines were located beneath the Brooklyn approach, but their boilerhouse, at Washington and Prospect, with its very conspicuous smokestack, was one of the other changes brought to the neighborhood by the bridge.)

The trip across on the bridge train was to take five minutes, as John Roebling had said it would, and the fare would be five cents. A horse and rider using the roadway would also pay five cents, a horse and vehicle ten cents. The charge for cattle would be five cents each, sheep and hogs two cents. Anyone wishing to walk over by the elevated promenade would have to pay a penny, although there was a movement in Albany, started by William Kingsley, to make the bridge free to pedestrians.

Roebling did not get out of his carriage the day he came to inspect the Brooklyn terminal. After he had seen enough he was driven home again. So not once in all fourteen years did he ever set foot on the bridge.

The terminal buildings were among the several things he had had to occupy his mind during these final months, and like the electric lights or the iron railings for the promenade or the plans for the opening celebration—all things that would have seemed very much after the fact, trivial even, in times past—he gave them his full attention, concerned over every last detail, as always, the totally disciplined professional to the very end.

A contract for lighting the bridge with seventy electric arc lamps had been awarded to the United States Illuminating Company, as he had recommended. The cost was to be eighteen thousand dollars, which was several thousand less than what the Edison Company had bid for doing the job with incandescent lamps. But cost had not been the deciding factor. Roebling had concluded that the sputtering blue-white arc lamps would be superior to the Edison type for lighting large areas.*

The dynamos to furnish power for the lamps were set up in the engine room of the Brooklyn terminal. A reporter who visited this generating station as it was about to be put into service gave the following description:

> The scene suggested the subterranean laboratory of a magician. Blue lights burned, invisible engines shook the ground with ponderous stroke, and a dozen grim and anxious men toiled in the ghastly glare. Around these perspiring men stood two or three directors, giving orders and hastening the work. Great belts, a yard wide, ran over dynamo pulleys at a frightful speed, and eight or ten other pulleys were awaiting new belts which were hanging slack over their shafting.

It was as though the monstrous bridge was about to be jolted to life by a sudden massive charge from this eerie laboratory, like the creature in Mrs. Shelley's story of Dr. Frankenstein.

The lamps themselves were being mounted on posts set on top of the steel trusswork, beside the promenade, at intervals of about one hundred feet. When the night came to turn them on, it would mark the first use of electric light over a river.

As before, Emily was serving as her husband's principal contact with the work. She was still going to the bridge regularly, and some days two and three times. There were the usual messages to deliver, answers to bring back, and things he had asked her to keep an eye out for. Once when a manufacturer had been puzzled as to how a particular part of the superstructure should be formed and had come to the Roebling house to get some questions answered, she had made a drawing to show how it could be done, carefully explaining each step. Now she could see to its proper installation as well, and any doubts there may have been among the men about her ability to pass judgment on such matters had long since vanished.

Then in early May, when the last of the superstructure was in place, the roadway at last completed, and the time had come to send

* Apart from any interest he had in the lighting contract, Thomas Edison was enormously fascinated by the bridge and spent hours watching its progress. He also took some extraordinary movies, among the earliest he made, of the final weeks of construction.

a carriage across—to test the effect of a trotting horse—Roebling had asked that she be the first person to ride over. The others on the staff and in the bridge offices agreed wholeheartedly. So one fine morning she and a coachman had crossed over from Brooklyn in a new victoria, its varnish gleaming in the sunshine. She had taken a live rooster along with her, as a symbol of victory, and from one end of the bridge to the other, the men had stopped their work to cheer and lift their hats as she came riding by.

Now, the week before the bridge was to be opened, Roebling had agreed to another interview. Apparently he and Emily had decided there was little damage that could be done at this late date, provided they kept the conversation brief and pleasant, and that is the way it went.

The man was from the *Union*, Brooklyn's Republican paper, which had recently made the bridge the dominant pictorial element in its logotype. His name is not known, but he wrote later that this was the first time he had seen Roebling in eleven years.

Emily received him in the library. Colonel Roebling was resting in his room upstairs, she said. He had spent the morning sitting for a sculptor who was doing a bust for the opening ceremonies and he was feeling a little tired just now.

When the reporter inquired for the Colonel's health, she told him not to be surprised if he found her husband looking a good deal healthier than he might expect. "He is not so sick as people imagine," she said. "The difficulty with him is that it wearies him to talk for any extended time. Any unusual exertion is sure to be followed by prostration, and the effort of talking or listening for any extended time has a very debilitating effect."

The reporter wanted to know if Colonel Roebling would be taking part in the grand opening. No, he would not, she said. The excitement would be too much for him. "After the ceremonial at Sands Street and the procession are over, we will receive our friends here," she continued. "Colonel Roebling will take part in the reception as long as he can stand the strain . . ."

She handed him an engraved invitation, a large white card from Tiffany & Co. In the upper left-hand corner was a small portrait of Roebling, resting on a laurel branch. His name and professional title were on a scroll underneath. To the right of the portrait, extending across the top of the card, as though seen in the distance, was the bridge, "in perfect detail," as the reporter noted. The invitation itself read as follows:

THE EAST RIVER BRIDGE
will be opened to the public
Thursday, May twenty-fourth, at 2 o'clock.
Col. & Mrs. Washington A. Roebling
request the honor of your company
after the opening ceremony until seven o'clock.

110 Columbia Heights
Brooklyn

R.s.v.p.

The reporter asked if Colonel Roebling was likely to undertake any other great work, now that the bridge was finished. According to the article he wrote later, "Mrs. Roebling elevated her brows and said decisively, 'Oh, no. This is his last as well as his greatest work. He will need a long rest after this is over. He needs it and he has certainly earned it.' "

Then she excused herself to go upstairs to see if her husband was ready to receive him. When she returned, she asked if he would please follow her.

"The writer found the Chief Engineer of the greatest suspension bridge in the world walking about his room and wearing a light spring overcoat and a soft felt hat," he wrote. On a side table he noted "an imposing array of medicine phials." But Roebling did indeed look better than he had expected—much better. He had put on weight and was noticeably fuller in the face and his hair and beard were streaked with gray. He was much paler, too, than he had been, and when he came forward to shake hands, his step appeared "short and a little uncertain." But to judge by appearances, time had not been altogether unkind to Roebling, the reporter decided. "Seen at a standstill or sitting in an easy chair, with one leg thrown over the back of another, no one would suppose that this robust-looking gentleman, with massive forehead, without a wrinkle, and keen gray eye that lights up wonderfully in conversation, was a victim to one of the most terrible diseases known to medical science." It was only in the lines around the eyes, the reporter said, that Roebling's face revealed any traces of past suffering.

The three of them sat down and the only thing serious touched on in the conversation that followed was the question of locomotives on the bridge, a subject about which there was still some curiosity in Brooklyn but little reliable information to go by.

"There will be no difficulty about running such locomotives as they use on the elevated railroad in New York across the bridge," said Roebling, who seemed to be having no difficulty speaking. "It was built to sustain them, and there would not be a particle of risk in it." If anyone wanted to transfer two or three passenger cars from a railroad in Brooklyn to one in New York, using a small locomotive, that too would be possible. The bridge had been built to sustain such weight. But he was still "unalterably opposed" to full-sized locomotives. The reporter wanted to know about Pullman cars.

"Oh, don't say that you would not consent to Pullman cars," Emily Roebling said. "You know you promised Mr. Stranahan that Pullman cars could go across."

At which, according to the reporter, Roebling laughed and replied, "Do you know what Mr. Stranahan wants? He wants a Pullman car to go right up to his back yard. He wants to be able to step into it at his house, ride across the bridge and up to Saratoga without changing his seat. That's what he wants."

Then, according to the reporter's account, Roebling suddenly began to look very tired. Perhaps he had said too much. "I congratulate you on the successful termination of your great work," the reporter said, standing up to leave. "I suppose this will be the last of the kind you will undertake."

"I don't know," Roebling answered. "If I get well there is lots of big work in the world to do yet." And with that the interview ended.

Plans for the great occasion were now complete. According to all accounts it was to be the biggest celebration in New York since the opening of the Erie Canal. For Brooklyn, said the *Times*, it was certain to be "the greatest gala day in the history of that moral suburb." It was to be known as "The People's Day."

Mayor Seth Low was the one chiefly behind the idea. He had proclaimed it an official holiday in Brooklyn. He would decorate and illuminate his own home, he said, and urged all his neighbors to do the same. He called on Brooklyn business establishments to close for the day and already most of them had sent out neatly printed cards saying they would. Schools would be out, most stores would be closed.

About thirteen thousand tickets from Tiffany had been issued by the trustees. Seven thousand of them, a pale-blue color and the size of a theater thicket, were good for admittance to the bridge on the opening day. The rest, large, stiff white cards with an engraved view of the bridge, were for the ceremonies to be held inside the Brooklyn terminal.

President Arthur, Governor Cleveland, and their parties were to walk over the bridge from New York, escorted by Mayor Franklin Edson, the Seventh Regiment, and a seventy-five piece band. They would be met on the bridge by an official delegation from Brooklyn and proceed to the Sands Street terminal for the formal ceremonies. James Stranahan would preside. William Kingsley, as acting president of the bridge trustees, would formally present the bridge to Mayors Edson and Low, each of whom was expected to make a few brief remarks. Then the "principal orations" of the day were to be delivered by two gentlemen selected as fitting representatives of their home cities, Abram Hewitt for New York and for Brooklyn the Reverend Dr. Storrs. After a private reception at the home of the Chief Engineer, there was to be a dinner for the President and the Governor down the street at the home of Mayor Low. A fireworks display at the bridge would begin at eight and would be followed by a public reception for the President at the Brooklyn Academy of Music.

The North Atlantic Squadron—the *Tennessee*, *Kearsarge*, *Saratoga*, *Yantic*, and *Vandalia*—had been ordered to Brooklyn to take part in the celebration. The *Minnesota* had already arrived and was anchored off the Battery. Guns would be fired from the Navy Yard, from Governors Island, from the warships. A theatrical promoter, Commodore Joe Tooker, had chartered the mammoth excursion steamer *Grand Republic* and planned to steam up and down beneath the bridge and fire guns from his deck as well. "Bell-ringing, steam-whistling, and band-playing are among the incidental attractions offered the patrons of this boat," the papers reported.

Brooklyn wharf owners were inviting select friends to spend the evening with them on their piers. Innumerable New York business firms with offices overlooking the river were inviting favorite customers to watch from their windows. Janitors in the tallest buildings on Printing House Square were overrun by applicants for admission to their roofs. All tenants in the Morse Building, at Nassau and Beekman Streets, had been told they could watch from the roof. The tops of the Temple Court Building on the opposite corner and the Mills Building nearby were also to be open. Richard K. Fox, the flamboyant proprietor of the *Police Gazette*, had sent out ten thousand invitations to watch the show from his new building on Franklin Square, thinking possibly several hundred recipients might appear.

Popular interest in all this was considerable, to say the least, and the press made much of it, including the New York *World*, which had changed hands just the month before. Jay Gould had sold the paper to young Joseph Pulitzer of St. Louis, and the new owner not only considered the Brooklyn Bridge a historic event, in the way the Eads bridge had been, but thought the *World*'s previous hostility to the mammoth new structure was just plain bad publishing. The *World* now loved the bridge.

Indeed, the only people who seemed displeased with the arrangements being made were some of the more militant Irish, who in mid-April had suddenly realized that the 24th happened also to be Queen Victoria's birthday and so began angrily protesting the date selected. The Central Labor Union issued a statement calling on "all good men and women in both cities to remember this latest insult of the would-be aristocratic element in our midst." The *Tribune* answered that "it would be difficult, perhaps impossible, to fix upon a day that did not commemorate something or other unpleasant for Ireland," and as the appointed day drew nearer, there was talk of Irish fanatics, "Dynamite Patriots," attempting to blow up the bridge.

The idea of a grand celebration did not much appeal to Washington Roebling either. Kingsley, too, Roebling understood, was of a like mind and had suggested to the trustees that once everything was in order they simply put up a sign saying "The Bridge Is Open." But the other trustees had been against that, and Seth Low especially. As early as March a committee had been formed to make the arrangements.

When he heard later what was being planned, Roebling had grown extremely uneasy. If there were to be fireworks, he wrote to the trustees, then the bridge must be cleared of all spectators. If there were to be soldiers participating in parades on the main span, then they must not march in step. He was also concerned about how many people might be permitted onto the bridge once the ceremonies were over, and what the consequences might be if a mob ever got out of control.

People had been getting onto the bridge for several months now, despite the precautions taken to stop them. One evening in April a mob of boys from New York had broken through the barriers at Chatham Street, crossed over the bridge, and started hurling rocks down on the houses near the Brooklyn tower. Police converged on them from both ends of the bridge and the boys had shinnied up the suspenders and climbed down under the flooring. "The officers used their clubs in an effective manner," according to a newspaper account the following morning, "and soon cleared the structure of the roughs." But the whole incident was a very dangerous sign Roebling said.

There were other things he found annoying. He had been receiving inquiries, for example, from Abram Hewitt, who wanted help with his speech. In a letter written in early May (a letter in which Hewitt, or his secretary, misspelled Roebling's name), Hewitt said he intended to take up "the social and political considerations involved in the creation of new avenues of transportation." He wanted the engineer to send him "comparative examples of great engineering works, which would show that by scientific appliances the cost of the bridge is very much below what would be possible in any preceding age." He wanted a table of wages from Roebling and other technical information. Hewitt planned to present the bridge as a symbol of progress.

But the builder of the bridge did not see it that way. Or perhaps he was in no mood to be of any assistance to Abram Hewitt. In any event, Roebling's answer went as follows:

> To build his pyramid Cheops packed some pounds of rice into the stomachs of innumerable Egyptians and Israelites. We today would pack some pounds of coal inside steam boilers to do the same thing, and this might be cited as an instance of the superiority of modern civilization over ancient brute force. But when referred to the sun, our true standard of reference, the comparison is naught, because to produce these few pounds of coal required a thousand times more solar energy than to produce the few pounds of rice. We are simply taking advantage of an accidental circumstance.
>
> It took Cheops twenty years to build his pyramid, but if he had had a lot of Trustees, contractors, and newspaper reporters to worry him, he might not have finished it by that time. The advantages of modern engineering are in many ways over balanced by the disadvantages of modern civilization.

It was the sort of thing he had doubtless wished to say to Hewitt for quite some time and that he had somehow refrained from saying to the *Union* reporter. His concept of energy consumption was also well in advance of his time.

A week or so later Hewitt wrote again to ask for the names of all those men he ought to "particularize" in his oration and for a brief explanation of what their individual contributions had been. This request Roebling willingly answered.

For Emily it was as busy as any time since the bridge began. The reception would be all her doing. She had drawn up the guest list, decided on the design and wording of the invitation, commissioned the bust of her husband as well as an oil portrait, ordered flowers and bunting to decorate the house, planned her own entourage to attend the ceremonies, and did the best she could to protect her husband from any more last-minute nervous strain than was absolutely necessary. She was also making arrangements to vacate the house almost immediately after the reception. She had rented the house, starting in June. She, her son, and husband would return to Newport for the summer.

The reception had been her idea. If her husband could not participate in the day Seth Low and the others were planning, then she would bring them to him—the trustees, the mayors, the Governor, the President of the United States.

Her own party would ride in twenty-five carriages. She would be in the first of them, in the same victoria in which she had crossed the bridge. She would ride with her son, John A. Roebling, II. Following would be Ferdinand Roebling and his wife, Charles Roebling and his wife, Professor and Mrs. Methfessel, Emily's brother William Warren from Washington, her sister-in-law Elvira Stewart, her sister, Mrs. Hook, and Eddie Roebling, now a twenty-eight-year-old bachelor living in New York. The rest were mostly personal friends.

"I wish you would make one of my party of ladies to attend the public ceremony of opening the big bridge," she wrote to a Mrs. William G. Wilson of New York. "I want the ladies to meet at my house at one o'clock on Thursday and go in a procession down to the bridge—sort of opposition to the Presidential procession on the New York side you know!

"Wear short dresses and bonnet—as I shall even at the reception. I want you to help me receive after the public performance is over."

The mail arriving at the house was full of notes of congratulations and grateful acceptances for the reception. On May 18 came still one more letter to Roebling from Abram Hewitt. To the Chief Engineer it must have seemed one last absurd insult to end on. Hewitt had hoped to be able to pay his respects following the ceremony, he said. "But as I am to dine with the Mayor it is barely possible that the interval will not be sufficient, in which case I pray you and Mrs. Roebling to accept the will for the deed." Then he said, "Will you kindly give me the full name of Mrs. Roebling . . ."

On Saturday night, May 19, to test the lights before the opening day, the hidden dynamos were turned on, and people returning to Brooklyn by ferry between eleven and midnight were suddenly astonished to behold overhead a great display of light across the bridge from city to city. Whether they knew it or not, they were looking up at the future—steel and electricity.

24
The People's Day

A festival so unique New York has seldom seen . . .

—*Harper's Weekly*

ESTIMATES were that fully fifty thousand people from out of town came into the city that morning by train alone. But probably that many again were arriving by boat. The *Mary Powell* had all she could carry. A Fall River steamer that docked at eleven had six hundred passengers on board and the boat from Stonington had that number or more. One iron steamer from New Haven carried a thousand people. How many private boats and "special excursions" came into the harbor was anyone's guess. By midday all the major hotels were sold out.

The weather was perfect. "A fairer day for the ceremony could not have been chosen. The sky was cloudless, and the heat from the brightly shining sun was tempered by a cool breeze." Countless flags snapped overhead all up and down Fifth Avenue and along Broadway, where the President was to pass. Buildings were draped in red, white, and blue, with banners and bunting floating from rooftops and window ledges, and most stores had some sort of display in their windows. In Madison Square, across from the Fifth Avenue Hotel, where the President and several of his Cabinet had spent the night, thousands of people were waiting, milling about under the trees or walking round and round the enormous torch and hand of the Statue of Liberty, which had been brought up from Philadelphia after the close of the Centennial.

At nine the fence across the Chatham Street entrance to the bridge had been torn down by workmen and replaced by a solid line of police. In another hour it was almost impossible to get within two blocks of the bridge. The streets leading to the river were packed solid with people. City Hall Park and Printing House Square were overrun. Every available rooftop and window was filled and along the river front there was scarcely a place left to stand.

The huge wagons that hauled milk and produce into the city had arrived as usual during the night, but loaded instead with country people, as many as twenty to a wagon, and now with their sunburned faces and bewildered looks they stood out plainly in the sea of people. "One moment they were clambering clumsily up the sides of stoops or balancing themselves insecurely on fences, and next they were pushing their way, with half awe-struck faces, through the crowds in the gutter out into the street itself," wrote a reporter. "The crowd impressed them with awe, the buildings and flags with admiration, but the consuming desire of their heart was to see the President, the Governor, and other political magnates."

There were as well, it seems, an abnormally large number of "symmetrical, shapely, graceful, elegant, neat, bright-eyed and comely women in brilliant costumes and resplendent colors," and these remarkable creatures were "pushed, jostled, and inextricably mixed with ungainly, uncouth, and ill-favored women." And in turn, they were all swayed back and forth and "jumbled up" with ragged men, with "rural swains" in frock coats and green ties, and with the unperturbed, self-contained New Yorker in dark suit and derby. "Embroidery, lace, fringe, trimmings and skirts were rent and torn by the friction of the crowd," and the large corsages, which many of the women wore when they started out that morning, were, after a half hour's experience in the crowd, crushed and torn to pieces. And everywhere, with or without adults in attendance, were "myriads of all sorts of children," none of whom, for some miraculous reason or other, was trampled to death.

Vendors hawked gumdrops, bananas, flags, pictures of John A. and Washington Roebling, bridge buttons and commemorative medals that sold for fifteen cents. It was a great day too for circulating all kinds of advertisements. Thousands upon thousands of pamphlets, fans, and handbills having the bridge as a decorative element or part of the text were handed out and tucked away in dresser drawers later on as mementos of the historic occasion. By noon, down by the bridge itself, the blue tickets issued by the Bridge Company were selling for five dollars apiece. Liquor stores and saloons were doing three times their normal business. And at the *Police Gazette* building a riot was under way. Better than a thousand "sporting men" having responded to publisher Richard Fox's invitation, the place was packed to the rooftop. Already the "guests" had consumed several hundred bottles of champagne and whiskey, devoured a barbecued ox, and were busily smashing up the furniture for fun when the police arrived to clear the building.

Schools were officially open in New York that day, but it would have been difficult to find a classroom that was not empty. And although the Stock Exchange too was open, the half-dozen brokers still on duty there had little to do but watch the visitors in the galleries. Elsewhere any business not closed by noon had been left in the charge of a few lonely clerks.

At the Custom House, Chester A. Arthur's old domain, things were extremely quiet. The Post Office was open, but it too was as still as Sunday. Federal courts were closed and the only people inside the County Courthouse on Chambers Street were twelve jurors who had been locked up all night trying to agree on a verdict. At noon, gongs clanged on the floors of the Produce, the Cotton, the Maritime, the Mercantile, and the Coffee exchanges and all business promptly ceased. In less than an hour these buildings were empty, their doors locked. About the only place in town where business continued as usual was Castle Garden. It was remarked that only a storm on the North Atlantic ever stemmed the tide of immigration.

Why there was quite so much excitement in New York, some observers were at a loss to explain. For Brooklyn people the bridge had a great deal of importance obviously enough, but for these throngs, the *Times* noted, "there could have been no special cause of congratulation, since not one in one thousand of them will be likely to have occasion to use the new structure except for curiosity."

No one will ever know when or how the story started. But possibly it was that morning, while the city waited for Chester A. Arthur to emerge from the Fifth Avenue Hotel, that one or more of those numberless countrymen in the crowd "purchased" the Brooklyn

Bridge from some new-found city friend and thereby made an everlasting contribution to American folklore. Or perhaps it was one of the dark-eyed, mustachioed men being processed at Castle Garden, some brand-new aspiring American with his belongings tied up in string, who at the end of his very first walk in the New World that bright morning arrived somewhere in the neighborhood of the bridge in time to be taken. Or maybe the story simply started in the imagination of some contemplative onlooker who, after studying the people pressing by, concluded the large part of them would believe anything, buy anything, even the Brooklyn Bridge itself.

The sun was barely up in Brooklyn before the streets were swarming with people. Virtually every single house and building downtown had a flag flying from its rooftop or hung from a window. Or if not a flag then a string of Chinese lanterns. Along Fulton Street and on the Heights most buildings were covered with streamers and bunting. There were flags in all 120 windows of City Hall. The dome of the courthouse was "gorgeous in its dress of flying colors." In City Hall Square the decoration that attracted the most attention was one in front of the Park Theater showing a straggling village of Brooklyn in 1746, the primitive ferry of 1814, then the completed bridge of 1883, and after that a view of the East River as it would look in 1983 with a hundred bridges spanning it. Joralemon Street on the Heights, Remsen, Montague, and Pierrepont, the streets running toward the river, were banked with flags and bunting. And in the little parks at the ends of these streets, at Columbia Heights, the trees were filled with Chinese lanterns and most of the biggest houses had lanterns strung all the way from basement to roof.

Every store window along Fulton, from the ferry to City Hall, on both sides of the street, and every doorway were decorated. A jeweler had made a miniature bridge with gold chain for the cables. A florist had made a bridge eight feet long, complete with bridge trains and boats passing below, all of flowers. Store windows carried framed portraits of the Chief Engineer and his father, Henry Cruse Murphy, Mayor Low, and General Slocum. And a sign in one window recalled something said a long time before: "Babylon had her hanging garden, Egypt her pyramid, Athens her Acropolis, Rome her Athenaeum; so Brooklyn has her Bridge."

On the Heights the two most elaborately decorated private homes were those of Seth Low and Washington Roebling, the two places where the President was to be a guest. Clusters of silk flags were in the mayor's windows and over his spacious doorway hung the flag of Brooklyn. Down the street, toward the bridge, the entire front of the Chief Engineer's house was covered with flags, shields, flowers, and the coat of arms of New York and Brooklyn. Over the street, suspended high enough for carriages to pass beneath, was one immense American flag.

The river in the distance below was probably the most arresting spectacle of all. The water was actually a bright blue and it looked that morning as though every variety of ship afloat had gathered in a great, elongated flotilla that extended from the bay to somewhere upstream beyond the bridge. Flags were flying from the masts of ships tied up at the wharves below and along the opposite shore. "It was as if the forest of masts had blossomed beneath the influence of the young spring sunshine into a thousand gorgeous dyes. Everywhere the eye glanced there floated from, and almost concealed the network of rigging, flags and banners and signals and streamers. . . . All the vessels anchored in the stream were likewise a mass of fluttering color above their dark hulls."

Sometime before noon the Atlantic Squadron came steaming up from the bay and into the river below the bridge, with the flagship, the *Tennessee*, anchoring about on a line with the Wall Street ferry. The others were strung out behind in a line reaching nearly as far as Governors Island and they too were covered with bunting and their crews of bluejackets could be seen quite plainly from the Brooklyn shore. One man later described how the gold trimmings on the officers' uniforms flashed in the sun.

The ferries kept churning back and forth to New York the whole morning and were packed with people. On the Fulton Ferry it was just about impossible to move an arm or leg. Hundreds of people, it seems, had decided that the best possible way to witness the day's events was to stick right on board and keep riding back and forth.

About noon there was a great surge toward Sands Street. Within half an hour at least ten thousand people had crammed into the narrow streets near the Brooklyn terminal, and a force of several hundred police, formed in a hollow square in front of the building, had all it could do to hold back the crowd. As in New York, vendors were everywhere, only here there seemed more of them, and along with pictures and commemorative medals, they were selling sheet music about the bridge and a variety of little facsimiles done in metal, wax, or confection. "On the whole it was a good-natured crowd," wrote one observer.

Brooklyn's part of the actual ceremonies got under way from City Hall at forty minutes past noon. The Twenty-third Regiment band in bright-red coats, followed by the Twenty-third Regiment in white helmets and blue coats, followed by a detachment of Fifth Artillery from Fort Hamilton and Marines from the Navy Yard, who in turn were followed by two hundred and some city officials, bridge trustees, and special guests, all in a body, led by the young mayor in a tall silk hat and followed by Mrs. Washington Roebling and her party in carriages, headed off down Remsen Street in the direction of the river, crossed Clinton, turned right at the next corner, onto Henry, and marched to the bridge. Their entire route was lined with crowds four and five deep. There were people looking down from rooftops and packed onto door stoops as mounted officers went clattering by and as one by one a great many familiar Brooklyn faces passed in review—ex-Comptroller Ludwig Semler . . . Judge McCue . . . Alfred Barnes . . . James Stranahan . . . William Kingsley . . . At Sands Street, where the police had cleared a path for them, all but a few of the civilians went directly into the station building, while the Twenty-third Regiment, Seth Low, William Kingsley, and a dozen others, at the command of "Route Step," marched out onto the bridge.

When the Erie Canal was opened in the autumn of 1825, there were four former Presidents of the United States present in New York City for the occasion—John Adams, Thomas Jefferson, James Madison, and James Monroe—as well as John Quincy Adams, then occupying the White House, and General Andrew Jackson, who would take his place. When the Brooklyn Bridge was opened on May 24, 1883, the main attraction was Chester A. Arthur.

Grover Cleveland, the portly new governor, was also there, and he, of course, would be the next President, but nobody knew that then and few even speculated on the prospect. In fact, if there was excitement about Cleveland's presence that May morning, it was mostly because people were anxious to see what the man looked like. The only other noteworthy figure to look for was Abram Hewitt, who was never exactly a crowd-pleaser.

But the strapping Arthur was considered a New Yorker and he looked like a President if any man ever did. When he stepped into the sunshine from the main entrance of the Fifth Avenue Hotel at twelve forty, the response from the crowd was overwhelming. On his arm was Mayor Edson, an erect, gray, scholarly-looking man in gold-rimmed spectacles. A few steps behind were Grover Cleveland and Henry Slocum, all smiles and arm in arm now.

Arthur was dressed in black frock coat, white tie, and a flat-brimmed black beaver hat that he kept taking off in response to the ovation. "The women in the crowd raised their hands above the heads of the men and waved their handkerchiefs," wrote one of the dozens of reporters covering the event, "and from the swarming windows on either hand similar feminine signals of hearty welcome met the Chief Magistrate's eye as he stepped into his open carriage." Cleveland went unrecognized for several minutes, but then he stood up in his carriage and lifted his hat, and the people, having concluded who he was, responded wholeheartedly.

The procession moved off, the "Dandy" Seventh Regiment and its band and a mounted police escort leading the way. Twenty-five carriages went rolling down Fifth to 14th Street, then east on 14th as far as Union Square, where they turned south again, down Broadway to City Hall. The greatest crowds anyone could remember seeing in New York lined the sidewalks and Chester A. Arthur, it would appear, was beloved by one and all. If he was not exactly a Jefferson or a Jackson nobody seemed to mind in the least.

It was one thirty when the procession wheeled into City Hall Park, where the press of people was almost beyond control. In another ten minutes Arthur and Cleveland had stepped down from their carriages and participated in a review of sorts. Then everyone formed up behind the Seventh Regiment, and with the band playing for all it was worth, everybody set off for the bridge on foot.

The historic ceremonial march to Brooklyn was made on the elevated promenade, but the roadways to either side of the promenade had already been opened to ticket holders and so thousands upon thousands of people lined the way. The band and soldiers went first, in route step as specified. Bayonets glinted in the sunshine and the music, whatever it was, had a decided effect on the dark-suited civilians. "It was a cheerful, jingling air which seemed to put life into the feet," wrote a spectator.

Up on the New York tower a lone photographer was busily at work under a black hood.

Arthur, it was said, "trod the pathway with an elastic step" and "looked with evident admiration at the structure opening up to view." (The next day a Broadway shoe merchant took space in the papers to announce that his "easy walking" shoes, made on patented lasts, had been "tested" at the opening of the bridge by President Arthur.) Cleveland was described as having a "wobbling gait," like a London alderman.

Just before the New York tower the Seventh Regiment halted, formed into two lines at the right of the promenade, and presented arms as the President passed by. William Kingsley, who had been waiting with a delegation of trustees in the shadow of the arches, removed his hat and stepped forward to grasp the President's hand. There was then a lively exchange of greetings and introductions, while the band played "Hail to the Chief" four times.

Guns were booming at Fort Hamilton and the Navy Yard as they all started out onto the main span. For Mayor Low and his Brooklyn reception committee, waiting beneath the opposite tower, only the heads of the lead men in the band could be seen as the procession approached—because of the gentle upward bow of the bridge. But as they came on, the heads gained shoulders and brass instruments. The oncoming figures not only grew larger as the distance narrowed, but more of each figure came steadily into view. They seemed to be coming up out of the planking of the promenade, the way approaching ships rise out of the horizon. Soon they had legs and feet and were on the downhill side of the span. Behind them, meanwhile, the solitary photographer on the New York tower had wheeled about 180 degrees and was back at work under his hood again. The band was playing full blast and the crowds on the carriageways to either side were cheering and waving as Arthur, "an Apollo in form," trod by overhead. "The President ran his eye around the horizon with the air of one appreciating the happy combination of the works of God and man. He filled his lungs with the refreshing breeze . . ."

Just before the Brooklyn tower the soldiers again parted ranks for the others to pass through. Seth Low made the official greeting for the City of Brooklyn, the Marines presented arms, a signal flag was dropped nearby and instantly there was a crash of a gun from the *Tennessee*. Then the whole fleet commenced firing. Steam whistles on every tug, steamboat, ferry, every factory along the river, began to scream. More cannon boomed. Bells rang, people were cheering wildly on every side. The band played "Hail to the Chief" maybe six or seven more times, and as the New York *Sun* reported, "the climax of fourteen years' suspense seemed to have been reached, since the President of the United States of America had walked dry shod to Brooklyn from New York."

Under the arched roof of the great iron terminal some six thousand people were waiting. Enormous American flags hanging overhead were the only decorations but shafts of sunshine slanted through the long banks of windows and fell on the crowd like floodlights. The President and the Governor were to sit on a raised platform along the west side of the building, while directly opposite was a section for the trustees, city officials, clergy, and speakers of the day. Everyone else was packed onto a temporary wooden floor between these sections.

At the sight of Arthur "the great multitude in the station arose and gave vent to the wildest enthusiasm." Handkerchiefs, parasols, and hats were waved in the air. The shouting even drowned out the band.

Presently James S. T. Stranahan began rapping for order with his cane and Bishop A. N. Littlejohn of Long Island stood up to offer a prayer.

All afternoon, as the speeches dragged on, thousands of men, women, and children went walking back and forth across the bridge, stopping now and then to exchange greetings with friends on neighboring housetops or to gaze down the smokestacks of the excursion steamers that floated slowly under the bridge with the outgoing tide. People were saying the ferries looked like water bugs from such a height. They waved to the crowds on the ferries and the crowds on the ferries waved back. Everybody seemed on a holiday. They joined arms. Some sang. Rooftops all along the river had been converted into "summer gardens," in the expression of the day, where thousands more spent the afternoon drinking beer, singing, and enjoying the glorious sunshine.

What was it all about? What was everyone celebrating? The speakers of the day had a number of ideas. The bridge was a "wonder of Science," an "astounding exhibition of the power of man to change the face of nature." It was a monument to "enterprise, skill, faith, endurance." It was also a monument to "public spirit," "the moral qualities of the human soul," and a great, everlasting symbol of "Peace." The words used most often were "Science," "Commerce," and "Courage," and some of the ideas expressed had the familiar ring of a Fourth of July oration. Still, everything considered, the speeches were quite appropriate on the whole and revealed much about the way people felt about the day and the bridge. The only real problem was that most of the audience never heard a word that was said, the big open-ended terminal being about the worst imaginable place to hold such a ceremony. There was no way to close off the din from outside and even under the best of circumstances the acoustics would have been miserable. For anybody sitting more than fifty feet from the speakers' stand, which meant nearly everyone, it was more like watching men go through the motions of making a speech.

Kingsley, the first speaker, got up very slowly, his long, rigid figure seeming to unfold like a telescope, as someone remarked. He looked deadly serious the whole time and kept his head down as he read his speech. But the audience had its attention fixed on the ceiling to his right, where a flagstaff had come loose at one end and was swinging to and fro, its shiny brass spear aimed straight down. The people directly below were packed in so tightly that nobody could move out of the way and it was impossible to reach the spear. So there it stayed swinging ever so gently and silently as one by one the speakers went through their pantomime orations.

Mayor Low was next after Kingsley and as he stepped to the rostrum one well-dressed woman sitting nearby was heard to exclaim, "Why he is no more than a boy!" The *Eagle* said later he was more like the valedictorian of the day and complimented him on his excellent voice, which apparently some people could actually hear. True or not, President Arthur was seen to yawn behind his fan, then whisper something to Secretary of State Frelinghuysen, which made them both laugh, and farther down the line, Secretary of the Treasury Folger appeared to be taking a nap.

Mayor Low was not very long about what he had to say and Mayor Edson, who held his speech in kid gloves, took even less time, which may have accounted for the enormous cheer when he sat down. At that point Stranahan was about to introduce Hewitt, the main speaker, apparently having forgotten that Jules Levy, a cornet player, was supposed to be next on the program. But the smiling Levy stepped forward all the same and as the *Times* reported "put the great multitude in a good humor" by playing "The

Star-Spangled Banner." Secretary Folger was seen to wake up and on the last note the crowd cheered so mightily that Levy did "Hail Columbia" as an encore. Again there was an ovation, but when Levy looked as though he was going to play still one more time, Stranahan seized him by the arm and led him off the platform. The audience was greatly disappointed by this and Levy looked furious. When Stranahan started to introduce Hewitt, Levy, from off stage, began playing "Yankee-Doodle." There was a roar of laughter and Chester A. Arthur appeared to be more pleased by this part of the program than any other. Levy took his time on the last notes, while Stranahan, a man with no music in his soul, according to one account, just stood glumly waiting.

About the time Stranahan finished introducing Hewitt, a shaft of sunlight had fallen on the Presidential head and neck, whereupon an Army officer appeared from somewhere with a lady's parasol, which he held over the portly Arthur until the close of the exercises. The *World*, now a Democratic paper, said he looked like an Asiatic potentate under the parasol.

It was also remarked that Abram Hewitt looked pale and rather "delicate" when he got up to speak, and by the time he finished, the audience had grown tired and extremely restless. But then the final speaker, the Reverend Dr. Storrs, was standing where Hewitt had been, a handsome, vibrant figure with flowing gray locks, who obviously felt at home before such a vast assembly and who punctuated every sentence with a nod, from the waist up, as if driving home each statement with his forehead. Storrs spoke for nearly an hour.

All told, the speeches, prayers, and cornet solos ran to nearly three hours and the Bridge Company's gold-embossed commemorative booklet containing the full text of everything said runs to 122 pages. Neither Chester Arthur nor Grover Cleveland said a word from the rostrum; they had not been asked to speak nor did anyone expect them to. They were there as honored guests only, to watch and listen and enjoy themselves. Nor did anyone make any public mention of the Queen's birthday.

Hewitt's address would be generally regarded in retrospect as the most successful of the day and was probably the finest he ever gave. He himself liked it so much that he had it published as a pamphlet three years later for his mayoralty campaign against Henry George and Theodore Roosevelt. But every speaker that afternoon seemed to be saying that the opening of the bridge was a national event, that it was a triumph of human effort, and that it somehow marked a turning point. It was the beginning of something new, and although none of them appeared very sure what was going to be, they were confident it would be an improvement over the past and present.

Henry Murphy, the assistant engineers, "the humblest workman," were all praised by one speaker or another and Kingsley, Seth Low, and Hewitt each in his way extolled the genius of the Roeblings. Hewitt compared John A. Roebling to Leonardo da Vinci but said Colonel Roebling was an even greater engineer than his father. Then he solemnly declared that the name of Emily Warren Roebling, a name he had not been quite sure of the week before, would be forever "inseparably associated with all that is admirable in human nature, and with all that is wonderful in the constructive world of art."

Hewitt said, too, that he could vouch for the manner in which all bridge business had been conducted, that no money had been stolen by Tweed, that the whole money raised had been "honestly expended," which was the part of his speech that drew the warmest response from those up front and on the platform. And disregarding, or perhaps misunderstanding, Roebling's skeptical remarks about progress since the Pyramids, he compared the $2.50 day's pay of the average bridge worker with the wage scale of ancient Egypt, which he figured at two cents a day in 1883 money. That in Hewitt's view was real progress. The bridge was a vindication, a heroic and monumental end result of modern industrialism, of labor and capital, of democracy, of new "methods, tools and laws of force"—of the nineteenth century. Even the *Times*, never an admirer of Abraham Hewitt, liked this part of his speech.

But it was the neatly combed little valedictorian, Seth Low, who came closer than anyone that day to expressing what was probably everyone's most deeply felt response to the bridge. "The beautiful and stately structure fulfills the fondest hope," he said. ". . . The impression upon the visitor is one of astonishment that grows with every visit. No one who has been upon it can ever forget it. . . . Not one shall see it and not feel prouder to be a man."

The Chief Engineer had sat alone at his window, his field glasses trained on the bridge, watching the procession until the last top-hatted figures at the tag end passed beneath the arches of the Brooklyn tower. Then he had stretched out on his bed for a rest. Sometime near four Emily had returned, having left the Sands Street terminal midway through the speeches. He put on a Prince Albert coat and went downstairs on her arm, to the front parlor, where they took a seat on the sofa and waited for the first guests to arrive. But it was nearly five thirty before President Arthur alighted from a carriage at the canvas canopy outside. The crowd in the street by then was such that the police were just able to keep a narrow path open to the door.

The house was decorated as if for a wedding. Both mantels in the drawing room were banked with red and white roses, wisteria, white lilacs, and in the center were clusters of calla lilies. On either side of the folding doors was a huge shield of roses. There were more roses and lilacs in gilt baskets and vases of cut flowers distributed through every room. And the balustrade on the stairway was trimmed with smilax all the way to the top floor.

There were busts of both the Chief Engineer and his father standing on one drawing-room mantel. On the elder Roebling's white marble head Emily had placed a wreath of immortelles, while the one of her husband wore a laurel wreath decorated with tiny American flags and a white satin ribbon on which she had had printed in red and blue: "Chief Engineer Washington A. Roebling, May 24, 1883. Brooklyn Bridge. Let him who has won it bear the palm."

A band was playing on a balcony above the drawing room, on the river side of the house, and through the doors beneath the balcony, out in the garden overlooking the river, stood a grand marquee and long tables of food and refreshments.

Emily and Washington Roebling stood side by side, just inside the parlor door, as the President and Seth Low entered the room together. "The engineer was pale, but he showed no excitement," one observer noted. She was dressed in heavy black silk, trimmed in crepe, with a knot of violets in her belt. She was described by the papers as beautiful and vivacious.

It was said the President warmly congratulated the engineer as they shook hands. After that people kept pressing through the door in great numbers. In all there were more than a thousand guests, including Grover Cleveland, the two mayors, all the speakers of the day (Abram Hewitt did make an appearance, after all), Mr. and Mrs. William C. Kingsley, General and Mrs. Henry Slocum, Stranahan and his wife, all the other trustees and wives, the assistant engineers, Ferdinand and Charles Roebling and their wives, Elvira Stewart, Professor and Mrs. Methfessel, Moses Beach from next door, Simeon Chittenden, Henry Pierrepont, A. S. Barnes, William Sellers of the Edge Moor Iron Company, Ludwig Semler, former Mayor Grace, Judge McCue, Hamilton Fish, William Evarts, Congressman Flower, and the Reverend and Mrs. Henry Ward Beecher.

Roebling remained standing only ten minutes or so, then went back to the sofa, where he sat, not saying much, Emily beside him. The President meanwhile gave all the appearance of having a splendid time. He tapped his foot to the band music, admired the flowers, went out into the garden, shook a great many hands, and stayed perhaps an hour in all. Once he was gone, Roebling excused himself and there was a burst of applause as he went slowly back up the stairs. The reception lasted another hour or more after that, but for Roebling his first and last ceremonial duty as Chief Engineer was over.

Everyone on both sides of the river was waiting for dark. Those whose job it was to describe the scene in words went to great lengths to do it justice. One reporter who was out on the bridge wrote that the innumerable boats and ships on the river looked like a sleeping city. Another man who was also on the bridge wrote this:

> As the sun went down the scene from the bridge was beautiful. It had been a perfect day. Up and down on either side of New York the bright blue water lay gently rippling, while to the south it merged into the great bay and disappeared toward the sea. The vast cities spread away on both sides. Beyond rolled the hilly country until it was lost in the mists of the sky. All up and down the harbor the shipping, piers, and buildings were still gaily decorated. On the housetops of both Brooklyn and New York were multitudes of people . . .
>
> The great buildings in New York loomed up black as ink against the brilliant background of the sky. The New York bridge pier looked somber and gloomy as night. But in Brooklyn the blaze of the dying sun bathed everything gold. The great building looked like burnished brass . . . In the west the sun sent its last tribute to the bridge in a series of great bars of golden light that shot up fanlike into the blue sky. Gradually the gold melted away, leaving the heavens cloudless. The sky was a light blue in the west, but

> grew darker as it rose, until it sank behind Brooklyn in a deep-sea blue.
>
> Slowly the extremities of the twin cities began to grow indistinct. . . . The towers of Brooklyn lost their golden hue. They seemed to sink slowly into the city itself. In New York the outlines of the huge buildings became wavering and indistinct.
>
> Then one by one the series of electric lights on the bridge leaped up until the chain was made from Brooklyn to New York. Dot by dot flashes of electric lights sprang up in the upper part of New York. The two great burners at Madison and Union Squares flared up, and the dome of the Post Office in New York set a circlet of diamonds out against the relief of the sky. The streets of the two cities sparkled into life like the jets on a limitless theatrical chandelier, and the windows of the houses popped into notice hundreds at a time. Long strings of lanterns were run over the rigging of the shipping in the harbor, and red and green port and starboard lights seemed numberless. The steamers sped to and fro on the water, leaving long ripples of white foam, which glistened in the light like silver.

In Brooklyn every public building was ablaze with gaslight. The Music Academy had a gas-jet rendition of the bridge out front. Houses draped with Chinese lanterns looked like Christmas trees. There were strings of lanterns over Montague Street and a block over, on Pierrepont, the Historical Society windows were lighted with hundreds of candles. Columbia Heights was nearly as bright as day with gaslights, lanterns, candles. Simeon Chittenden had a big sign in front of his house done in gas jets—"WELCOME TO BROOKLYN'S GUESTS"—but as the President was driven past later in the evening, a gust of wind blew out half the letters.

Every street on the Heights looked like a carnival. Indeed the crowds in both cities were far greater now than at any time earlier in the day. No traffic was moving anywhere near the river. Uptown New York and the inland sections of Brooklyn were all but deserted. Where there had been a hundred people watching by the river during the day, now there were a thousand, or at least so it seemed. The *Times* estimated there were 150,000 people just in the neighborhood of City Hall.

Suddenly a solitary rocket shot into the sky over Columbia Heights and burst into a spray of blue stars. It had come from the mayor's house, where the dinner for the President had been going on.

Almost instantly the lights on the bridge went out. For a moment not a thing could be seen of it. Then there was a long, distant hissing sound, a sudden roar, and fifty rockets exploded simultaneously high over the main span of the bridge, while at least twenty bombs burst higher still, from above the towers, and poured down great showers of gold and silver. "From an elevated point the city seemed to be in volcanic action, with the spouting crater on the suspension bridge."

At its final meeting on May 14, as its last official act, the Executive Committee of the Bridge Company had contracted with the New York firm of Detwiller & Street, Pyrotechnists, to put on a display of fireworks "worthy of the place and the occasion." In all, fourteen tons of fireworks—more than ten thousand pieces—were set off from the bridge.

It lasted a solid hour. There was not a moment's letup. One meteoric burst followed another. Rockets went off hundreds at a time and were seen from as far away as Montclair, New Jersey. Bombs exploded incessantly above the towers, bathing the bridge in red. In the strange light, firemen on the bridge could be seen in strong silhouette and the water from their hoses looked like molten silver. Meantime, innumerable gas balloons were being sent aloft. They were fifty feet in circumference and loaded with fireworks and as they swung into the sky, one by one, they scattered balls of colored fire over the river.

At each burst of a rocket a huge roar went up from the shores. Hundreds of thousands of people were watching—probably the biggest crowd ever gathered in New York until that time—and nobody, in all his days, had ever seen anything like this.

Nearly every boat on the water was making some sort of noise or display. Rockets and fireworks were shooting up from the middle of the river and down the bay. On one big excursion steamer, ablaze with lights, a calliope was shrieking out "America." Bands were playing on board other boats.

Rockets were going up all over New York meantime—and in Brooklyn. From the middle of the bridge now came great thunderclap reports as zinc balls, fired from mortars, burst five hundred feet up, fairly illuminating the two cities, like sustained lightning.

And finally, at nine, as the display on the bridge ended with one incredible barrage—five hundred rockets fired all at once—every whistle and horn on the river joined in. The rockets "broke into millions of stars and a shower of golden rain which descended upon the bridge and the river." Bells were rung, gongs were beaten, men and women yelled themselves hoarse, musicians blew themselves red in the face. And then when it was all over and nearly quiet again and the boats on the river were beginning to untangle themselves, there was one last memorable touch that not even Detwiller & Street, Pyrotechnists, could have arranged. "Hardly had the last falling spark died out," wrote an editor who had been watching from the top of the Tribune Building, "when the moon rose slowly over the further tower and sent a broad beam like a benediction across the river."

The grand reception for the President at the Music Academy, which began almost immediately after the fireworks, was considered a great success. The President, Grover Cleveland, William Kingsley, General Slocum, and twenty or thirty others stood on the stage, surrounded by a small forest of potted palms, while the people of Brooklyn were permitted to pass by and pay their respects. The procession lasted until ten. Arthur was in fine humor still, bowing, smiling, playing his part exactly as everybody would have wanted him to. And one member of his Cabinet was heard to remark, "Why I thought that Brooklyn had one hotel and a shipyard or two, but it's quite a town."

It had been announced in advance that the bridge would be thrown open to the public at the stroke of midnight and that anyone might cross upon payment of one cent. Enormous crowds had gathered at both ends.

At eleven twenty two young men with blond mustaches raised the windows of the Brooklyn toll booths and H. R. Van Keuren—a good Brooklyn name, reporters decided—was the first to put his money in the box. The first lady to cross was a Mrs. C. G. Peck of Baltic Street. The first vehicle from the Brooklyn side was an old-fashioned top-wagon drawn by a bony white horse whose large white hooves came down on the bridge floor with a noise like the discharge of musketry. The driver was a Charles Overton from Coney Island, who had been waiting at the gate for two hours and who also managed to make the first trip across from the New York side, since the gates there opened ten to fifteen minutes later than those in Brooklyn.

A fierce struggle ensued at the New York gates, characteristically, and perhaps it was altogether appropriate that the first man through was the Keeper of the City Hall. Once he was beyond the gate, there was an even more violent rush from behind. Police began swinging their clubs and several people had been rather roughly treated by the time the whole crowd—perhaps three thousand people—was strung out in a long orderly line.

People poured across the bridge through the entire night and were still coming with the first light in the sky. According to a count kept by the *Times*, the first beggar to cross came over from Brooklyn, as did the first drunk, the first policeman, the first hearse (which was empty), the first "dude," the first Negro, and the first musician, a Scottish bagpiper who marched over playing "The Campbells Are Coming."

How late the Chief Engineer and his wife stayed up watching from their window, who may have been with them during the evening, what they said to each other, or what reflections went unsaid as bombs and rockets burst over the bridge, can only be guessed at.

In another time and in what would seem another world, on a day when two young men were walking on the moon, a very old woman on Long Island would tell reporters that the public excitement over the feat was not so much compared to what she had seen "on the day they opened the Brooklyn Bridge."

Epilogue

FOR NEARLY fifty years after it was completed the Brooklyn Bridge reigned supreme as the most magnificent, if not technically the largest, suspension bridge on earth.

In its initial days as a public thoroughfare it was commonly referred to as "The Eighth Wonder of the World" and it was an even greater sensation than anyone had expected. On its first full day, May 25, 1883, a total of 150,300 people crossed on foot and 1,800 vehicles went over carrying an unknown number of others. The following day, a Saturday, the count was down. But on Sunday, May 27, a spectacular spring day, 163,500 people went "strolling" on the Great Bridge. One veteran New York policeman said he had never in all his experience seen such crowds. "It seems to me as if the people have got the bridge craze," he said.

And then, tragically, on Thursday, May 31, a week to the day after the bridge was opened, the very thing Roebling had warned against happened.

It was Memorial Day, a holiday in both cities, and the weather was ideal. There had been thousands on the bridge all morning. But C. C. Martin had allowed pedestrian traffic on both carriageways and that had taken pressure off the promenade until some time near three thirty when the crowds began to build rapidly. Probably twenty thousand people were on the bridge by four, or the approximate time of the panic. When it was all over, twelve people had been trampled to death.

The trouble began at the top of a narrow flight of stairs leading to the promenade at the end of the New York approach. A crowd pressing up the stairs was running head on into another crowd coming in the opposite direction, from the New York tower. There were fifteen steps in all, broken into two flights by a landing just seven feet wide. When the two oncoming throngs met there, it was virtually impossible for anyone to move either way and people approaching from behind, in both directions, kept trying to shove their way forward.

But from what several eyewitnesses said later, it seems some sort of order might have been restored had not a woman coming down the stairway lost her footing. Another woman began to scream at the top of her lungs and there was an immediate rush to see what was happening. Those who were packed onto the stairway tried desperately to hold back the crowd but it was impossible. In an instant three or four more lost their balance and fell. Meanwhile, the crowds farther back on the promenade kept advancing, nobody knowing what was going on up ahead, and in a moment the whole stairway was packed with dead and dying men, women, and children. People were shrieking and screaming and those who suddenly found themselves at the brink of the stairway and saw what was happening turned to shout for those behind to move back, but then they too lost their step and went over on top of the trampled bodies below. The most terrifying crush was on the promenade just back from the top of the stairs. Numerous people had their clothes torn off. In places, it was reported, people were jammed so tight that blood oozed from their noses and ears.

Hats, umbrellas, gloves, shoes, loose change, fell between the bridge train tracks and rained down on the housetops and streets below. Among a group of boys playing in the streets was Al Smith. "That was my first view of a great calamity," he said later. "I did not sleep for nights."

Other explanations would be given later. It would be said that somebody out in the middle of the bridge began to scream that it was falling. It would be said a gang of "roughs" from New York had started pushing and shoving people. Probably there is some truth to both accounts.

Lawsuits as a result of the accident added up to half a million dollars, but no negligence was proved. A coroner's jury reprimanded the Bridge Company for the narrowness of the stairway and for employing too few police. The Bridge Company blamed the newspapers for having created an "undefined feeling of insecurity" about the bridge, but promptly doubled the number of police on the promenade.

C. C. Martin remained in charge. He was officially named Chief Engineer on July 9, 1883, after Roebling had submitted his formal resignation, and he would hold that job until 1902, devoting, in all, thirty-three years of his life to the bridge.

Martin's full force for operating and maintaining the bridge was comparable in size to that needed for a large ship or fair-sized factory of the day. He had one assistant engineer, a chief mechanical engineer, who had charge of the steam engines and rope traction, three assistant mechanical engineers, three oilers, and three firemen. There were six locomotive engineers, six locomotive firemen, one master of transportation, forty-five conductors, a superintendent of tolls, nineteen collectors, one trainmaster, four train dispatchers, four yardmen, and five switchmen. A master machinist had charge of the machine shop and locomotives. There were two blacksmiths, a foreman of carpenters with "a force of men changing with the exigencies of the work," a foreman of car repairs and inspector of grips, a foreman of labor and general work, one captain of police, one sergeant, three roundsmen, and eighty-six policemen. Counting Martin, the grand total came to 201 full-time employees.

The bridge trains began running in September and worked to perfection. By the time it was a year old 37,000 people a day were using the bridge, or very nearly as many people as fourteen ferries were handling the year it was begun. In their first full year of service the bridge trains carried 9,234,690 passengers, but then the completion of the Brooklyn Elevated to Fulton Ferry more than doubled the patronage. In 1885 the bridge trains handled nearly twenty million passengers. The trains ran twenty-four hours a day and by 1888, just five years after the bridge was built, they were handling more than thirty million passengers a year. The terminals were expanded, more cars were put into service. Furthermore, the ferries were still in business, to the surprise of people, and would be for a long time to come. The last Brooklyn ferry, between Hamilton Avenue and the Battery, stopped running on June 30, 1942.

In May of 1884, P. T. Barnum, "in the interest of the dear public," took a herd of twenty-one elephants, including the famous "Jumbo," over the bridge to Brooklyn and thereupon declared that he, too, was now perfectly satisfied as to the solidity of the masterpiece.

And inevitably, perhaps, there were certain individuals who would see the bridge as a challenge to their manhood or as a means of doing away with themselves. The bridge was scarcely in full operating order before they began leaping from it, for glory or oblivion, and frequently with the unintended result.

The first to try for glory was Robert E. Odlum, a brawny swimming instructor from Washington, D.C. On May 19, 1885, to divert the bridge police who were waiting to stop him, Odlum sent a friend onto the bridge to go through the motions of jumping. Then he came riding up in a closed carriage, stepped out, climbed onto the railing, and, dressed in trunks and a bright-red swimming shirt, jumped to his death, with one arm thrust straight over his head, the other clamped firmly to his side.

Steve Brodie, the only man ever to become famous for jumping from the bridge, probably never did. He was a personable, unemployed Irishman in his early twenties, who, not long after Odlum's much publicized failure, began boasting that he would be the next to jump. Bets were made along the Bowery, but just when Brodie intended to jump remained a mystery. Then on July 23, 1886, it was announced he had done it and lived to tell the tale. Several friends said they had been witnesses, that they had watched him plummet straight into the river, where he was picked up by a passing barge. But nobody else had seen his jump and it was commonly said among the skeptics, of whom there were a great many, that a dummy had been dropped from the bridge and that Brodie merely swam out from shore in time to surface beside the passing barge.

Brodie was put in jail briefly for his supposed feat, then opened a saloon that became a favorite Bowery stop for sight-seers and slumming parties. In the main barroom hung a large oil painting of the bridge and there for all to see was Brodie plunging toward the water. For further historical documentation, there was a framed affidavit from the barge captain who claimed to have rescued the hero.

But Steve Brodie's jump from the Brooklyn Bridge would be fixed forever in the public imagination by a play called *On the Bowery*, which opened in 1894. Brodie was the star and his big scene was a leap from the bridge, done with all sorts of elaborate special effects, only this time it was to save the girl, who had been thrown off by the villain. The play was a smash hit and eventually toured the country. A bridge sweeper sang a moral ballad, Brodie sang "My Pearl's a Bowery Girl!" ("My Poil's a Bowery Goil!") and, for encores, a song called "The Bowery," written for an earlier production, which became a standard part of every performance.

Brodie became rich and famous, but died of diabetes at age thirty-six or thereabouts, in 1901. For years after his celebrated jump people kept asking him why he did not do it again, only this time with reliable witnesses. His answer was always the same: "I done it oncet."

Others kept on trying. Larry Donovan, a pressman at the *Police Gazette*, was the first to jump successfully. He went over the side wearing a red shirt like Odlum and a pair of baseball shoes. In 1887 James Martin, a painter's assistant on the bridge, fell off and lived and the following year a young man named Byrnes jumped to impress his girl friend and he too lived. In 1892 Francis McCarey jumped and was killed, but it seems that was what he wanted, so probably he ought to be considered the first suicide.

Then there was a man who jumped wearing a derby hat and was still wearing it when he surfaced in the river quite unharmed and another man who went off wearing immense canvas wings. He sailed a thousand feet upstream before landing safely on the water. But by the turn of the century the jumping craze had ended.

The bridge remained a subject of endless fascination for almost everybody who saw it. For the millions of immigrants arriving in New York through the 1880's and 1890's and on into the new century, it was one of the first things to be seen of the New World as they came up the bay. It was one of the landmarks they all looked for, the great world-famous symbol of the faith that was literally moving mountains. And the fact that it had been designed by an immigrant and built largely by immigrant workers did much naturally to enhance its appeal.

In truth there is really no end to the number of things the bridge meant to people. For whole generations growing up in New York and Brooklyn it was simply a large, dominant, and generally beloved part of the natural order of things. The river without the bridge or Brooklyn without the bridge would have been unthinkable and year after year people went to it on especially fine days, or at moments of personal stress or joy, the way people go to a mountain or walk beside the sea.

For countless people their first walk on the bridge would remain one of childhood's earliest memories. Countless others would tell how it was the place where they fell in love. No doubt it very often was. Al Smith was among those who loved to sing "Danny by My Side," the opening line of which runs "The Brooklyn Bridge on Sunday is known as lovers' lane."

In *A Tree Grows in Brooklyn*, the most popular of the many novels to be written with a Brooklyn setting, a young World War I soldier from Pennsylvania says, "I thought if ever I got to New York, I'd like to walk across the Brooklyn Bridge." It was something felt by whole generations of Americans before and since. They would come from every part of the country, take photographs of it and from it with one of the new Kodak cameras introduced not long after the bridge was finished, or buy some of the stereopticon views that sold by the millions. They would ride bicycles across, take honeymoon strolls by moonlight, carry newborn babies proudly down the promenade, or scatter the ashes of the dearly departed from the middle of the main span.

It was a place to go on stifling summer evenings, to take some exercise to and from work, to walk the baby, to watch the gulls, to find relief from the city. Its promenade was and would remain one of the most exhilarating walks on the continent. To be on the promenade of the Brooklyn Bridge on a fine day, about halfway between the two towers, looking over the harbor and the city skyline, was to be at one of the two or three most soul-stirring spots in America, like standing at the rim of the Grand Canyon.

Just why this bridge, more than all others, has had such a hold on people is very hard to pin down. But in the years since it opened it has been the subject of more paintings, engravings, etchings, lithographs, and photographs than any man-made structure in America. There are probably a thousand paintings and lithographs of the bridge by well-known artists alone.* It has been the setting for scenes in films, for Maxwell Anderson's *Winterset*, and for all

* Joseph Pennell, Joseph Stella, John Marin, Childe Hassam, Georgia O'Keeffe, O. Louis Guglielmi, Raoul Dufy, Ludwig Bemelmans, Lyonel Feininger, Albert Gleizes, and Max Weber are some of the artists who have taken the bridge as their subject. Several, such as Marin and Stella, have gone back to it many times. Stella's powerful abstraction *The Bridge* (1918) is probably the best known of all the paintings.

kinds of advertising. (It would seem that a whole chronological display of female fashions in America, since the advent of photography, could be assembled just from pictures posed on the bridge year after year.) It has been used repeatedly on postcards, Christmas cards, book jackets, posters, record jackets. It has been the symbol for a New York television network and for a popular Italian chewing gum.

There have been songs about the bridge, besides the one Al Smith liked, and a great many poems, nearly all of which have been less than memorable. The one notable exception is *The Bridge* by Hart Crane, who, in the 1920's, to identify as closely as possible with his subject, moved into Washington Roebling's old house at 110 Columbia Heights. In Crane's powerful but not altogether coherent masterpiece, the bridge is seen as a shining symbol of affirmation at the end of an epic search through the American past. It is the "Tall Vision-of-the Voyage," spare, "silver-paced," and all-redeeming.

The finest thing written at the time the bridge was opened appeared in *Harper's Weekly*. The author was a newspaperman named Montgomery Schuyler and his article, "The Bridge as a Monument," was not only the first critical review of the great work, but a bugle call, as Lewis Mumford would say, for serious architectural criticism in America. Schuyler did not think much of the bridge as a work of art. Still, everything considered, he judged it "one of the greatest and most characteristic" structures of his century. "It so happens," he wrote, "that the work which is likely to be our most durable monument, and to convey some knowledge of us to the most remote posterity, is a work of bare utility; not a shrine, not a fortress, not a palace, but a bridge."

The towers, he believed, would outlast everything else on either shore, and he asked his readers to imagine some future archaeologist surveying the ruins of New York, "a mastless river and a dispeopled land." The cables and roadway would have long since disintegrated, he said. The Roeblings would be as forgotten as the builders of the Pyramids. Only the towers of the Great Bridge would remain standing and the archaeologist would have "no other means of reconstructing our civilization." "What will his judgment of us be?"

Henry James, writing soon after the turn of the century, would see something darkly ominous in the looming silhouette of the bridge and its shuttling trains. New York for him had become a "steel-souled machine room," the end product of which was "merciless multiplications" and the bridge was a "monstrous organism," marking the beginning of a new age. For James the prospect was chilling.

By the 1920's, however, the bridge was a unique source of "joy and inspiration" for the critic Lewis Mumford.

> The stone plays against the steel; the heavy granite in compression, the spidery steel in tension. In this structure, the architecture of the past, massive and protective, meets the architecture of the future, light, aerial, open to sunlight, an architecture of voids rather than solids.

The bridge proved, he said, that industrialism need not be synonymous with ugliness. It was something done exceedingly well by Victorian America. "All that the age had just cause for pride in—its advances in science, its skill in handling iron, its personal heroism in the face of dangerous industrial processes, its willingness to attempt the untried and the impossible—came to a head in Brooklyn Bridge."

Others, later, would see it as a symbol of liberation, of release from the "howling chaos" on either shore. It would be said that at heart it was a monumental embodiment of the open road, the highway call, the abiding rootlessness that runs in the American grain—"not so much linking places as leaving them and shooting untrammeled across the sky." And an age that can no longer regard it as an engineering marvel has declared it a work of art. One prominent contemporary American architect has gone so far as to say it is one of the two works of architecture in New York of any real importance, the other one being Central Park.

It has also, of course, been taken quite for granted by millions who use it regularly and quite sentimentally by some. It can be seen as merely one of a number of different ways to get to or from Brooklyn or as the grandest sort of memento of a New York that was, a serene, aspiring emblem rising out of an exhilarating and confident age too often remembered solely for its corruption and gimcrackery. It can be seen as the beginning of modern New York—of monumental scale, of structural steel—or the end of old Brooklyn. It is all these. And possibly its enduring appeal may rest on its physical solidity and permanence, the very reverse of rootlessness. It says, perhaps, as does nothing else built by Americans before or since, that we had come to stay.

For Brooklyn, on a more practical level, it did everything its proponents had promised. It stimulated growth, raised property values, and provided a safe, reliable alternative to the ferries. It put Brooklyn on the map.

Rush hours at the terminals were like nothing ever witnessed before, not even at the old Fulton Ferry slip in Brooklyn, not even on the uptown platforms of New York's elevated trains. Certainly there was little semblance of the smooth, efficient transfer of humanity that John A. Roebling had pictured. But the bridge also withstood the Blizzard of 1888; it carried trolley cars, along with everything else, when they were installed on the carriageways and elevated trains when they replaced the cable cars. It accommodated ever greater numbers of people year by year. But it was not enough.

In 1903 the Williamsburg Bridge was completed upstream from the Navy Yard, from designs by an RPI man with the old Brooklyn name of Leffert Lefferts Buck. Heavy, ungainly-looking, built entirely of steel with a stiffening truss no less than forty feet deep, it was four and a half feet longer than the Brooklyn Bridge, which meant it was now the world's largest suspension bridge. One of the assistant engineers was C. C. Martin's son, Kingsley Martin, and the cables were of Roebling wire. Wilhelm Hildenbrand and Charles Roebling were in charge of the cable making.

Six years later, when the Brooklyn Bridge was handling half a million people a day, two more bridges were finished. The Manhattan, another suspension bridge, was built almost side by side with the Brooklyn Bridge, just upstream. The Queensboro Bridge, a cantilever, took a route John A. Roebling once considered, over Blackwell's (now Welfare) Island.

More than a dozen tunnels were built beneath the river for subways, railroads, water lines, and automobiles.* And for these reasons primarily Brooklyn changed beyond anyone's imagining.

In 1898 with a population of nearly a million people and still the third-largest city in the United States, Brooklyn had relinquished its independence to become a borough of New York. By 1930 Brooklyn's population was greater than that of Manhattan. Old

* The first subway, between Bowling Green, in Manhattan, and Joralemon Street, was completed in 1908; and the Brooklyn-Battery Tunnel, opened in 1950, is one of the longest underwater tunnels in the world.

Brooklyn families had become an infinitesimal minority, the Heights a tiny picturesque but inconsequential segment of a Brooklyn that spread over eighty-nine square miles, or four times the area of Manhattan. Even the name Brooklyn became synonymous with things never heard of before the turn of the century—the Dodgers, Murder Incorporated—and the butt of innumerable jokes. One favorite vaudeville remark about the bridge went, "All that trouble, just to get to Brooklyn."

In 1931 the George Washington Bridge was completed over the Hudson with a span more than twice that of the Brooklyn Bridge and six years later the Golden Gate Bridge, larger and still more awesome, was built at the opposite end of the continent. By contrast to such gleaming creations, the Brooklyn Bridge seemed an antique and there was even talk of tearing it down.

In 1944 the elevated trains stopped running over the bridge and the old iron terminal buildings were dismantled. A team of engineers began a painstaking examination of the entire structure to see what ought to be done about it. When they had concluded their studies two years later, it was announced that all the bridge needed was a new coat of paint.

Washington Roebling did not live to see the bridge eclipsed by the George Washington or Golden Gate Bridges, both of which were built with Roebling wire, but he came very close to it. Ironically—incredibly—the crippled, tormented legendary Chief Engineer lived on until 1926. He outlasted them all—Hewitt and Seth Low, each of whom became mayor of New York; Stranahan, who lived to be ninety and to see a statue of himself put up in Prospect Park; Kingsley, who died only a few years after the bridge was built, of a nervous stomach at age fifty-two; Eads, who died in 1887, while trying to enlist support for a fantastic ship railway across the Isthmus of Tehuantepec; Slocum, whose name would be remembered for one of the worst disasters in American history, the burning, in 1904, of the *General Slocum*, a New York excursion steamer; and every one of the assistant engineers, each of whom, except for Martin, went his own separate way professionally once the bridge was finished.

On the night Seth Low was defeated for re-election as mayor of New York, Roebling came up from Trenton to watch the returns in front of the *Herald* building and was very pleased by the results. But all he ever said for attribution concerning the old management of the bridge was that so far as he knew no money had ever been stolen—a decidedly different conclusion, of course, from what he said in his private notes years before. His exact statement, written at age seventy-eight, in a letter to an old friend then compiling a Trenton history, was this: "So far as I know not a dollar was stolen politically or otherwise. There was no thievery—no political robbery or peculation. In spite of the assertions of all the newspapers to the contrary. It is the unscrupulous press that makes most of the trouble."

He seems to have kept in touch with his former assistants over the years. McNulty named a son after him and Hildenbrand did a good deal of engineering work for John A. Roebling's Sons. Hildenbrand, in fact, had quite a life after he left Brooklyn. He built the Pikes Peak Railway in Colorado, directed the complete renovation of the Ohio River bridge at Cincinnati, and built a suspension bridge of his own, of about the same span as the Ohio River bridge, at Mapimi, Mexico. Not long after Hildenbrand died, in 1908, Roebling wrote, "Soon I will be the last leaf on the tree."

For four years after the completion of the bridge, he and Emily lived in Troy, in order to be near their son, who was then at Rensselaer. Roebling spent his time quietly. Often the old pains and cramps returned with a vengeance and he felt himself a "used-up man." Still the sustained separation from both the bridge and the wire business seemed to be what was needed. His health improved gradually but steadily.

In the spring of 1888, when his son was graduated, the three of them moved back to Trenton. John went to work at the mill and was married the following year. In the meantime, plans were drawn up, to Emily's specifications, for a "commodious mansion in the Tudor style," which Roebling had built for her on West State Street, on grounds sloping down to the Delaware River. The house took several years to finish. It was a great baronial affair of huge gables and towering brick chimneys, and on the street side there was a big stained-glass rendition of the Brooklyn Bridge, complete with clouds sailing by and ships passing below. At night, all lit up, this window was considered one of Trenton's "sights" and for years people went out of their way to see it.

They moved in in 1892. She bought the most expensive carriages and the finest horses, which she insisted always on driving herself, her coachman seated behind her. For a number of years she seems to have had quite a good time. She entertained often and beautifully to judge by some old clippings from the Trenton society columns. She became active in women's clubs. She studied law at New York University and received her degree. She worked on a book about Cold Spring.* She went to Europe twice. The first time they went together, but the second time, in 1896, she went without him and to Russia as well, where she was one of the few Americans present at the coronation of Tsar Nicholas and the Empress Alexandra. A formal photograph of Emily Roebling in the dress she wore to the coronation shows an erect, confident-looking woman in her early fifties, a little stout but rather regal herself in silvery white satin.

From Roebling's correspondence during these years, one gets a picture of him living in semiseclusion, privately very proud of her accomplishments, absorbed in all manner of interests—his mineral collection, his greenhouse, bird-watching, astronomy, the paleontological history of New Jersey. He also wrote a two-hundred-page biography of his father, which he misplaced and which was never found, most unfortunately. His one recreation was riding the trolleys, which he seems to have enjoyed enormously, riding out into the nearby countryside to look for wild flowers. And these he could identify and refer to by their Latin names as readily as a trained botanist.

When he felt well enough, he traveled with her—to the Columbian Exposition in Chicago in 1893, to Martha's Vineyard, which he liked, to Nantucket, which he did not, and to New York, where they went out on the bridge together unnoticed.

He read Schiller and Goethe, Carlyle's six-volume *Frederick the Great*, which he had read before as a young man, and Tolstoy, his favorite. He followed the stock market and made a very great deal of money at it.

He also kept an attentive eye on the wire business, in which he was again a major stockholder, but drew no salary. He liked to take his lunch at the office, which was the old family home long since so built over and swallowed up by the mills that it was barely identifiable. He was always available for his opinion and his opinion carried some weight apparently. Once when his brothers were all for

* With careful editing and numerous annotations she managed to turn a rather dry, colorless diary kept by a Putnam County preacher into an engaging chronicle. She also included an additional chapter on the Warren family. Titled *The Journal of the Reverend Silas Constant*, it was published in 1903.

selling out to U. S. Steel, he said no and that was that. But his relations with his brothers were always strained. Ferdinand was so bothered by the grand new house when it was first finished that he refused to enter it. For two years Ferdinand and he did not speak.

His one male confidant was his son, but communication with him had to be carried on by mail. John had lasted only a short time in the wire business. It is impossible to tell just what went wrong. The official explanation was that John had retired because of his health. But it was commonly said in Trenton that family affections and loyalties (if there were any) stopped at the door of the Roebling works. If a young man did not measure up inside, or if there were personality conflicts, then he departed rapidly, even if his name was Roebling.

Years before, Edmund Roebling, youngest of old John's sons, had been unable to get along with Ferdinand and was "kicked out," according to Washington Roebling. "He was now adrift with too much money, became a globe-trotter, and was somewhat dissipated." Edmund never returned to Trenton, never married, lived on in seclusion on the Upper West Side of New York.

Emily's health began to fail about the time the new century began. It would be said later that the strain of her experiences in Brooklyn was the principal cause. Her eyes gave her trouble at first and she took to wearing dark glasses. In the fall of 1902 Roebling went into Roosevelt Hospital in New York for some intestinal surgery and in early December, in Trenton, while he was still recuperating, she had a collapse of some sort. The doctors said she had stomach ulcers and were not encouraging. He kept going back and forth to Trenton from the hospital, a torturous trip for him, to be with her for a day or two at a time.

An additional night nurse was hired in early February. He was home to stay by then and joked with her about renaming the house Roebling Hospital. Though pitifully weak, she refused to give up and announced she would go to the mineral baths at Sharon Springs, New York. But she never did. She died of cancer of the stomach on February 28, 1903, and was buried at Cold Spring. He was too weak to make the trip.

In the file of Roebling's letters kept by his son there is an envelope marked "Undated notes, clippings, etc, found among W.A.R.'s papers after his death." Among the items in the envelope is a much-worn paper on which Roebling had copied in pencil an epitaph Mark Twain inscribed on the grave of his daughter:

Warm Summer Sun shine kindly here
Warm Summer Wind blow softly here
Green Sod above, lie light, lie light
Good night, Dear heart, good night, good night.

For five years after Emily's death he lived alone with the house and servants. The wire business in this time became the biggest in the world. The demand seemed unending—for telegraph wire, baling wire, electrical wire, for wire cloth, for bridge cable wire, for the construction of the Panama Canal, for wire rope for cable roads, coal mines, ships, oil rigs, logging machinery, tramways, elevators. The Otis Elevator Company was buying nearly all its cable from the Roeblings. The sons of John A. Roebling had become millionaires many times over. Washington Roebling's own estate would be approximately $29 million.

On fall afternoons in 1904, in Kinkora, New Jersey, a tiny village on the Delaware ten miles below Trenton, he watched the building of a new mill complex and an entire town of brick houses and broad streets that was to become known eventually as Roebling. His brothers had announced that the town was being built only out of "plain business necessity" and that there was to be nothing utopian about it. But to the surprise of very few, it turned out to be one of the best-planned industrial towns ever built in America, a model in every respect, as company towns went. In the view of many old admirers of the family, it was a fitting extension of ideas that had spurred John Roebling on at Saxonburg so many years before.

The memory of his father remained a looming presence for Roebling. Confusion over which one of them had built the Brooklyn Bridge became increasingly common as the years passed and for him a sore subject. Time was cheating him out of everything, he lamented, even his identity. When the family decided to erect a larger-than-life-size statue of the old man and the sculptor said the few photographs on hand were not suitable, Roebling agreed to pose. He sat several times, his head turned as though studying some distant horizon, a sheaf of plans on one knee. When the finished bronze was unveiled with much to-do in Trenton's Cadwalader Park in 1908, a great many people came up to tell him how much it looked like him.

To the surprise of almost everyone, he married again, that same year. She was a widow of about his own son's age, Mrs. Cornelia

Witsell Farrow of Charleston, South Carolina. How they happened to meet, when, or how long he had been contemplating marriage are all unclear. ". . . these relationships are those of the heart, not governed by reason or judgment," he had written to John. "A second marriage late in life cannot be judged by the standard of the first because its motives are usually quite different." John and the rest of the family heartily approved of the decision once he announced it, and of Mrs. Farrow, who, it was said, helped him "take a less gloomy view of things." The Colonel even "became at times almost jovial."

He who had weathered everything just lived on interminably, forever "bearing up," people said. His teeth were pulled, one by one, and in his letters to John he complained repeatedly of physical torment and in particular of excruciating pains in his jaw. He was seized with a terror of contracting tetanus and dying like his father. "And yet people say how well you look," he wrote, "I feel like killing them."

In April 1912 his nephew and namesake, Washington A. Roebling, III, Charles Roebling's son, went down on the *Titanic* and the family was news again. An editor of the New York *Times* wrote to ask if he would be good enough to explain, for the historical record, the part his mother had played in helping John A. Roebling build the Brooklyn Bridge. Roebling wrote back, explaining patiently that he was the one who had built the bridge, not his father, and that it was Emily, not his mother, who had been associated "for fourteen long years with the various phases of the work."

When the income tax came along in 1913, it was as though the country was coming apart at the seams. "It means 100,000 spies to snoop into everybody's business and affairs." When war broke out in Europe he shuddered at the fate of mankind. "It has come to this pass, that for an extra German to live, he must kill somebody else to make room for him. We can all play at that game. It means perpetual universal war."

And still he carried on, writing long, affectionate letters to John, taking solitary walks down West State Street. "War in the kitchen as usual," he reported to John in August of 1916. "The cook touched the laundress's smoothing iron. War to the Knife, peace impossible—damages 5 strands of hair, 4 aprons torn, 2 scratches. Starvation threatens!" Somebody was watering his whiskey.

His "oddities" became a favorite topic of conversation in Trenton. When he and Cornelia dined out at the homes of friends or one of his brothers, he would frequently proceed, without a word of explanation, to make himself comfortable on the nearest sofa and go fast asleep. He hated gloves and refused to wear them even in the coldest weather. He disliked automobiles intensely and refused ever to ride in one. Jigsaw puzzles became his "narcotic," but never satisfied with those to be found in stores, which he considered much too easy, he had his specially made, from large photographs or reproductions of paintings, each puzzle with a thousand to three thousand pieces.

He wrote his correspondence on anything at hand—a scrap of cheap note paper, the back of an old invitation, some stray bit of Emily's stationery found in a bottom drawer. His handwriting, again as perfect as copperplate, was so small that most of his aged friends were unable to read it without a magnifying glass.

When out on his walks he was known to step into a gateway or to appear suddenly fascinated by the contents of a store window if he thought he could avoid a conversation. The standard explanation locally was that the Roeblings were all a little odd that way, but the fact was that talking was often physically painful for him. Strangers were constantly stopping him on the street. Often a mother or father with a small boy in tow would ask if the boy might shake his hand and years later these same boys would remember him as "a nice, courteous old gentleman." Among beggars and other Trenton people interested in charity he was known as a soft touch.

There seems to have never been a day of his life in all the years following the bridge when he did not know some kind of physical discomfort or outright pain. Privately, like old men everywhere, he was preoccupied with his health, as well as material possessions he no longer had any use for.

Nature remained his solace. He had planted a grove of Siberian crab apples behind the house. "Four have agreed to bloom one year," he wrote, "and four the next year. How good they are." He had also acquired a new companion, a rather disreputable-looking Airedale, a stray he named "Billy Sunday." It became a common thing to see them come down the long drive as he set off on a walk, a small, fragile old gentleman in pinstripes and boater, advancing slowly, stiffly, the dog trailing at his heels. Or they would stand together in front of his tall iron gate waiting for the trolley. There was no regular stop there but the trolley stopped just the same and dog and master would climb aboard, everybody inside watching.

As was widely known, Billy Sunday was the one dog in Trenton with a special pass to ride free on the trolleys. Once Roebling was seated, Billy would slip between his legs and curl up under the seat.

Ferdinand Roebling died in 1917 and Charles the year after. Karl G. Roebling, Ferdinand's oldest son, was named head of John A. Roebling's Sons. But three years later Karl dropped dead on a golf course. Within days it was decided that there was only one person left who could possibly take charge of the vast industrial empire.

The New York and Brooklyn papers made much of the announcement. "A little old soldier of eighty-four, Col. Washington A. Roebling, the man who built Brooklyn Bridge and the son of the man who planned it, is fighting today his last fight," wrote the New York *World*, "is fighting to get his work done in spite of all his enemies—illness, debility, pain, loneliness, bereavement, the terrible depression of the man who has outlived his generation."

Roebling ran the company for the next five years and the business prospered exceedingly. "I claim a small part of this as the result of my management," he confided to his son. Others credited him with more than a small part.

He got up each morning at about seven thirty, had his breakfast, then, like the men in the mill, took the trolley to work, accompanied by his dog. His day was the full eight hours, the same as everyone. He had no secretary, preferring to handle his correspondence himself, which he wrote always in longhand. He was all but blind in one eye, almost totally deaf, and weighed perhaps 120 pounds. He looked so frail, so very old, like Lee in his final photographs, with the same snow-white beard and sunken eyes, that people wondered how in the world he could possibly manage, knowing, as most everybody did, what he had been through in his life. But the extraordinary thing is he did not simply manage. He was highly innovative, forceful, and seemed to know absolutely all there was to know about every facet of the business. He decided to change all the mills over to electric power, instead of steam, a momentous and costly move. An entirely new department for the electrolytic galvanizing of wire was set up under his direction and the contract for the cables of the Bear Mountain Bridge, over the Hudson River—among other bridges—was taken and completed during the time he was in charge.

In one interview he was asked how he was able to carry on. "Because it's all in my head," he answered. ". . . It's my job to carry the responsibility and you can't desert your job. You can't slink out of life or out of the work life lays on you."

In 1924 at the request of the Butler County Historical Association he sat down and wrote a detailed account of the early days of Saxonburg, and to a correspondent he wrote, "Long ago I ceased my endeavor to clear up the respective identities of myself and my father. Many people think I died in 1869."

The house next door was sold and torn down. Electric street lights were installed along West State Street. "The Great White Way in Trenton has come our way," he wrote in despair. "Every 50 ft. will be installed a huge arc lamp to light up the front of the house and keep us from sleeping." His own downward progress he described as accelerative, like gravity.

In the spring of 1926 it was obvious to every one that he was failing rapidly. By May he was down to less than a hundred pounds. "Think not that I am improving—growing weaker daily—body racks with pain—head bowed down in sheer apathy—bones crack when rolled over—fall down when I try to stand. Please leave me alone—and in peace," he wrote to John's wife. But then he added a P.S.: "A surprise: for several years—ten—a night-blooming cereus stalk has been knocked about in the greenhouse. Last night it suddenly bloomed, was brought to my bedside at 10 P.M. A delicate odor filled the room—a wonderful flower—much larger than a rose. A calyx filled with snow-white petals curved outward and oval-pointed. This morning it is gone—to sleep the sleep of ages again."

He lingered on for two more months. The only thing he had left, he said, was his brain and for that, he added, he was extremely grateful.

He died peacefully at age eighty-nine, on July 21, 1926, with his wife, son, and several others at his bedside. There is no record of any last words being said. The end came at three thirty in the afternoon.

All of the bridges built by John A. Roebling are gone now except two—the Cincinnati Bridge and an aqueduct over the Delaware built in 1848 above Port Jervis, New York, which has been converted into an automobile bridge and is the oldest suspension bridge in America. His house at Saxonburg still stands, however, as does the church he built there and a small shed in which the first reels of iron wire were stored. John A. Roebling's Sons has since been sold to the Colorado Fuel and Iron Company.

Washington Roebling's house on West State Street was offered to the state of New Jersey to be used as the governor's mansion, but the offer was declined because it was felt that the upkeep would be too costly. The house was torn down in 1946 to make room for a parking lot. The house at 110 Columbia Heights, Brooklyn has also been torn down. His mineral collection, which numbered some sixteen thousand pieces and included all but four of the known minerals on earth, was given by his son to the Smithsonian Institution.

As he had requested, Roebling was buried at Cold Spring, beside Emily. No statues were put up in his honor. The graves were very plainly marked.

In 1948 D. B. Steinman and his New York engineering firm were retained by the city to prepare plans to increase the highway capacity of the Brooklyn Bridge. With some fifty men assigned to the project, an extensive remodeling was carried out over a number of years. The trolley and el tracks were removed, the roadways were widened to three lanes in each direction, and additional trusswork was built. The changes, which cost more than nine million dollars, altered the over-all appearance of the bridge very little.

In 1964 the bridge was officially declared a National Historic Landmark. It now carries more than 121,000 trucks and automobiles a day and on the average Sunday, in good weather, more than a thousand people go walking or bicycling on the promenade, which is still the only one of its kind. There are bronze plaques on both towers, beside the promenade, listing the names of John A. and Washington A. Roebling, the trustees, the assistant engineers, and the master mechanic. The plaques were put up when the bridge was first completed. In the time since, two more plaques, one for each tower, have been added to honor Emily Roebling.

The towers themselves, though long since dwarfed by the skyline of downtown Manhattan, remain unique. Nothing to compare to them has been built in America. Since the towers of the mammoth suspension bridges built in the twentieth century are of steel, the towers of the Brooklyn Bridge are both the first and the last monumental stone gateways on the North American continent.

A combined force of some thirty men looks after the bridge. It gets a new coat of paint every five years or so and according to the engineers at the New York Department of Public Works, of all the bridges on the East River, it is the one that gives them the least trouble. With normal maintenance, say the engineers, the bridge will last another hundred years. If parts are replaced from time to time—even entire cables if necessary, which would be perfectly possible—then, "As far as we are concerned, it will last forever." Perhaps it will.

Appendix

Brooklyn Bridge Vital Statistics*

Length of river span	1,595 feet 6 inches
Length of each land span	930 feet
Length of the Brooklyn approach	971 feet
Length of the New York approach	1,562 feet 6 inches
Total length of the bridge	5,989 feet
Full width of the bridge floor	85 feet
Number of cables	4
Diameter of each cable	15¾ inches
Length of each cable	3,578 feet 6 inches
Number of wires in each cable	5,434
Total length of wire in each cable	3,515 miles
Miles of wrapping wire on each cable	243 miles 943 feet
Weight of each cable	1,732,086 pounds
Ultimate strength of each cable	24,621,780 pounds
Number of suspenders from each cable, main span	208
Number of suspenders from each cable, land span	86
Strength of each suspender	70 tons
Greatest weight on a single suspender	10 tons

* All figures are based on the bridge as it was when completed in 1883.

Greatest weight on a single cable	3,000 tons
Depth of Brooklyn foundations below high water	44 feet 6 inches
Depth of New York foundations below high water	78 feet 6 inches
Size of Brooklyn caisson	168 feet x 102 feet
Size of New York caisson	172 feet x 102 feet
Launching weight, Brooklyn caisson	3,000 tons
Launching weight, New York caisson	3,250 tons
Height of Brooklyn caisson when launched	14 feet 6 inches
Height of Brooklyn caisson when completed	21 feet 6 inches
Height of New York caisson when launched	14 feet 6 inches
Height of New York caisson when completed	31 feet 6 inches
Size of each tower at high-water mark	140 feet x 59 feet
Size of each tower at top	136 feet x 53 feet
Total height of each tower above high water	276 feet 6 inches
Height of roadway at towers	119 feet
Height of arches above roadway	117 feet
Height of towers above roadway	159 feet
Width of openings through towers	33 feet 9 inches
Total masonry in Brooklyn tower	38,214 cubic yards
Total masonry in New York tower	46,945 cubic yards
Size of each anchorage at base	129 feet x 119 feet
Size of each anchorage at top	117 feet x 104 feet
Height of anchorages in front	89 feet
Height of anchorages in rear	85 feet
Weight of each anchorage	60,000 tons
Total number of anchor plates	8
Weight of each anchor plate	23 tons
Weight of suspended superstructure from anchorage to anchorage, 3,400 feet	6,620 tons
Total weight of the bridge (exclusive of masonry)	14,680 tons
Grade of roadway	3¼ feet in 100 feet

Chronology of Construction

Center-line surveys begun	June 1869
Ground broken for Brooklyn tower foundation	January 3, 1870
Brooklyn caisson launched	March 19, 1870
Brooklyn caisson towed to position	May 3 and 4, 1870
Work commenced inside Brooklyn caisson	May 21, 1870
First stone laid on Brooklyn foundation	June 15, 1870
Fire discovered in Brooklyn caisson	December 1, 1870
Brooklyn foundation completed	March 11, 1871

New York caisson launched	May 8, 1871
New York caisson towed to position	September 11, 1871
Roebling stops descent of New York caisson	May 18, 1872
New York tower foundation completed	July 12, 1872
Brooklyn anchorage begun	February 1873
New York anchorage begun	May 1875
Brooklyn tower completed	June 1875
Brooklyn anchorage completed	November 1875
New York tower completed	July 1876
New York anchorage completed	July 1876
First wire for cable making stretched	August 14, 1876
E. F. Farrington makes first crossing	August 25, 1876
First cable wire run out	May 29, 1877
Cable making begun	June 11, 1877
Cable spinning completed	October 5, 1878
Understructure for bridge floor completed	December 1881
Trusswork and promenade completed	April 1883
Bridge opened	May 24, 1883

Notes

The following abbreviations have been used throughout these notes:

JAR:	John A. Roebling
WAR:	Washington A. Roebling
EWR:	Emily Warren Roebling
JAR II:	John A. Roebling II
HCM:	Henry Cruse Murphy
LER:	*Laws and Engineer's Reports*
RUL:	Rutgers University Library
RPI:	Rensselaer Polytechnic Institute
LIH:	Long Island Historical Society
ASCE:	American Society of Civil Engineers

For further details on the books cited, the reader is referred to the Bibliography.

PART ONE

1 The Plan

Page 21 "The shapes arise!": Whitman, *Sound of the Broad-Axe.*

Page 21 Meetings with the consultants: Minutes kept by WAR. RPI.

Page 22 Biographical sketches of the consultants: *National Cyclopedia of American Biography*, *Dictionary of American Biography*, and memorial tributes published by the ASCE.

Page 24 "If there is to be a bridge": Barnard, "The Brooklyn Bridge."

Page 24 1867 charter: *An Act to incorporate the New York Bridge Company.* Chapter 399. Passed April 16, 1867. LER, pp. 3–7.

Page 25 JAR's formal proposal of 1867 was officially titled *Report of John A. Roebling, C.E., to the President and Directors of the New York Bridge Company, on the Proposed East River Bridge.* LER. Most of the descriptive material concerning the proposed bridge has been drawn from this report, which was published in 1870 but which appeared first in the *Eagle*, September 10, 1867.

Page 25 Brooklyn interest in the bridge: Virtually every Brooklyn publication had something good to say for the bridge. The new and short-lived *Brooklyn Monthly*, for example, was nearly as enthusiastic as Roebling, saying in its issue for May 1869, "When it is finished, the East River Bridge will, without comparison, be the grandest monument of its kind on this continent, if not in the world . . ."

Page 26 "As the great flow of civilization": JAR, *Report of John A. Roebling, C.E.* LER.

Page 27 "Lines of steamers, such as the world never saw before": *Ibid.*

Page 27 "Lo, Soul, seest thou not God's purpose": Whitman, *Passage to India.*

Page 27 "Singing my days": *Ibid.*

Page 27 "The completed work, when constructed in accordance with my designs": JAR in a covering letter for his report, addressed to the President and Directors of the New York Bridge Company, September 1, 1867. LER.

Page 28 Navy engineer's plan for East River dam: Brooklyn *Union*, January 7, 1869.

Page 28 New York Polytechnic Society sessions: Reported in various issues of the *Eagle*, February 1869.

Page 28 Another bridge and a tunnel besides: Brooklyn *Union*, December 22, 1869.

Page 30 "A force at rest": *Scientific American*, Vol. XII, 1865.

Page 35 Congressional legislation: *An Act to establish a bridge across the East River.* Public, No. 53. Approved by Congress March 3, 1869. LER.

Page 35 The make-up of the Bridge Party: Thomas Kinsella in the *Eagle*, April 16, 1869; also a reminder book kept by WAR, RUL.

2 Man of Iron

Page 39 "We may affirm absolutely . . . without passion": Bartlett, *Familiar Quotations.*

Page 40 "A wet bandage around the neck"; "*A full cold bath every day*": Two small JAR notebooks on the water cure, dated 1852. RPI.

Page 40 Man of iron . . . poised . . . confident: Various obituaries and contemporary biographical sketches; also a speech delivered by Henry D. Estabrook at the unveiling of JAR's statue, June 30, 1908, RUL.

Page 41 ". . . Never known to give in": *Eagle*, July 26, 1869.

Page 41 "One of his strongest moral traits": Stuart, *Lives and Works of Civil and Military Engineers of America*, p. 325.

Page 41 "Sir, you are keeping me waiting": *Ibid.*, p. 81.

Page 42 Christoph Polycarpus and Friederike Dorothea Roebling: *Ibid.*, p. 9.

Page 42 Hegel's favorite pupil: *Ibid.*, p. 12.

Page 42 "It is a land of hope": Hegel, *Lectures on the Philosophy of History*, London, 1890.

Page 42 ". . . the heart of him into cold storage": Henry D. Estabrook at the unveiling of JAR's statue, June 30, 1908. RUL. Estabrook was uncommonly candid about his long-deceased subject. His source appears to have been WAR.

Page 43 Nothing could be accomplished without an army of functionaries: JAR, *Diary of My Journey from Muehlhausen in Thuringia via Bremen to the United States of America in the Year 1831*, p. 113.

Page 43 Cash gift for Mühlhausen: Schuyler, *The Roeblings*, p. 15.

Page 44 The description for JAR's voyage to America is drawn entirely from his *Diary of My Journey*.

Page 44 Six thousand dollars in cash: WAR, *Early History of Saxonburg*, p. 7.

Page 44 Trunkful of books: Most of these volumes are in the RPI collection.

Page 45 "If one earnestly desires it": JAR, *Diary of My Journey*, pp. 18–19.

Page 45 ". . . a cheerful carefree disposition": *Ibid.*, p. 54.

Page 45 ". . . then one perceives in the foam": *Ibid.*, p. 57.

Page 46 The founding of Saxonburg: WAR, *Early History of Saxonburg;* JAR, "Letters to Ferdinand Baehr, 1831."

Page 46 Saxonburg as "the future center of the universe": WAR to JAR II, winter 1893–94. RUL.

Page 46 "My father would have made a good advertising agent": WAR, *Early History of Saxonburg*, p. 11.

Page 46 ". . . no unbearable taxes": JAR "Letters to Ferdinand Baehr, 1831."

Page 47 ". . . valuable attribute of industry . . . They have made good farmers": *History of Butler County, Pennsylvania*, p. 289.

Page 47 "I cannot reconcile myself": Schuyler, *The Roeblings*, p. 44.

Page 47 "So he took to engineering again": WAR to JAR II, winter 1893–94. RUL.

Page 48 "The iron ore on Laurel Hill": JAR, RUL; also quoted in Steinman, *The Builders of the Bridge*, p. 53.

Page 49 German periodical the source of the wire rope idea: Schuyler, *The Roeblings*, p. 50.

Page 49 "His ambition now became boundless": WAR to JAR II, winter 1893–94. RUL.

Page 49 ". . . farmers were metamorphosed into mechanics": Schuyler, *The Roeblings*, p. 60.

Page 49 WAR's description of wire rope making: WAR, *Early History of Saxonburg*, pp. 13–14.

Page 49 ". . . benefactors to mankind who employ science": Quoted in Steinman, *The Builders of the Bridge*, p. 73.

Page 50 "As this work is the first of the kind": Craig, *The Olden Time*, Vol. 1, pp. 45–48.

Page 51 "The progress of the fire": Pittsburgh *Gazette* reporter quoted in Lorant, *Pittsburgh; The Story of an American City*, p. 110.

Page 51 "Great Central Railroad" speech: *American Railroad Journal*, Special Edition, 1847; also quoted in part in Schuyler, *The Roeblings*, pp. 65–71.

Page 52 Never home in springtime: Elvira Roebling to JAR, March 14, 1860. RUL.

Page 52 JAR's letters to Charles Swan are in the RUL; also quoted at length in Schuyler, *The Roeblings*, pp. 93–114.

Page 53 "I for my part wish the blacks all good fortune": JAR, *Diary of My Journey*, p. 118.

Page 53 ". . . legs under my mahogany long enough": Schuyler, *The Roeblings*, p. 189.

Page 53 "When a whole nation . . . steeped for a whole century in sins"; "We cannot close our eyes to the appalling fact": JAR philosophical papers. RUL.

Page 54 "A pure-hearted woman or one gifted with warmer affections": WAR to EWR, August 2, 1864. RUL.

Page 54 "My dearly beloved wife, Johanna": Bible page reproduced in Schuyler, *The Roeblings*, opposite p. 99.

Page 55 Prayed he would never have to read Roebling's philosophy: *Ibid.*, p. 13.

Page 55 "We are born to work and study"; "True life is not only active"; "It is a want of my intellectual nature"; "Human reason is the work of God": JAR philosophical papers. RUL.

Page 56 Davis plan proposed to Horace Greeley: The letter in the RUL collection is undated but refers to the "recent foreign war," meaning the war with Mexico no doubt, so it was probably written between 1848 and 1850. There is no indication whether the letter ever appeared in the *Tribune*.

Page 57 The incident involving young Edmund Roebling, as well as Edmund's subsequent life, is described by WAR in a private memorandum written March 16, 1922. RUL.

Page 58 "A man may be content with the success of an enterprise": JAR philsophical papers. RUL.
Page 58 "The latest sensation we have had here are spiritual communications": Ferdinand Roebling to WAR, November 12, 1867. RUL.
Page 59 Séances: From original questions and notes made by JAR. RPI.
Page 62 Light topcoat and soft felt hat: A rare photograph taken at Niagara of the engineers in the Bridge Party shows both JAR and WAR. It is the only known photograph of father and son together and reveals how remarkably alike they looked. RPI.

3 The Genuine Language of America

Page 63 The description of the Bridge Party's tour has been drawn almost entirely from three long articles by Thomas Kinsella that appeared in the *Eagle*, April 16, 17, and 26. Interestingly, the local papers in Pittsburgh, Cincinnati, and Niagara Falls did little more than mention the arrival of the "visitors from the East."
Page 63 James Finley is a fascinating but somewhat shadowy figure. He is given only passing mention in most histories of civil engineering and is referred to as a justice of the peace or judge, but in the classic work *The Planting of Civilization in Western Pennsylvania* by Solon J. and Elizabeth Hawthorn Buck, Finley is an itinerant preacher, who earlier in his career had been sent into the wilderness of western Pennsylvania to put down a burgeoning new-state movement—a mission he accomplished with amazing skill and speed. His patented chain bridge was first described by Thomas Pope in *A Treatise on Bridge Architecture*, published in 1811.
Page 64 Smithfield Street Bridge: *American Railroad Journal*, February 21, 1846; also in Craig, *The Olden Time*, Vol. I, pp. 286–288.
Page 64 Allegheny River Bridge: White and von Bernewitz, *The Bridges of Pittsburgh.*
Page 65 "The bridge will be beautiful": JAR to Charles Swan, June 21, 1859, RUL; also quoted in Schuyler, *The Roeblings*, p. 108.
Page 65 "Washington is about the work": JAR to Charles Swan, RUL; also quoted in Steinman, *The Builders of the Bridge*, p. 206.
Page 66 Cincinnati Bridge: JAR, "The Cincinnati Suspension Bridge," *Engineering* (London), Vol. 40, pp. 22–23, 49, 74–76, 98–99, 140–141; JAR, *Report of John A. Roebling, C.E., to the President and Directors of the New York Bridge Company, on the Proposed East River Bridge*, LER; Schuyler, *The Roeblings*, pp. 125–128; Farrington, *A Full and Complete Description of the Covington and Cincinnati Suspension Bridge with Dimensions and Details of Construction.*
Page 69 "The Germans about here are mostly loyal": JAR to Charles Swan, spring of 1863, RUL; also quoted in Schuyler, *The Roeblings*, p. 110.
Page 69 "The size and magnitude of this work far surpass any expectations": WAR to Charles Swan, March 16, 1865, RUL; also in Schuyler, *The Roeblings*, p. 234.

Page 70 "Leave bridgebuilding to younger folks": JAR to Charles Swan, April 1865, RUL; also quoted in Schuyler, *The Roeblings*, p. 114.

Page 71 "You drive over to Suspension Bridge": Quoted in Gies, *Bridges and Men*, p. 188.

Page 71 Niagara Bridge: Schuyler, *The Roeblings*, pp. 118–124; Steinman, *The Builders of the Bridge*, pp. 157–193; Stuart, *Lives and Works of Civil and Military Engineers of America;* Kirby and Laurson, *The Early Years of Modern Civil Engineering*, pp. 155–156. There is also a superb scale model of the bridge on display in the Museum of History and Technology at the Smithsonian Institution.

Page 72 *Maid of the Mist* shoots the rapids: The best description is in Anthony Trollope's *North America.*

Page 74 Early suspension bridges: Of the numerous histories of bridges the most readable and reliable is *Bridges and Men* by Joseph Gies. See also *Bridges and Their Builders* by David B. Steinman and Sara Ruth Watson.

Page 75 Charles Ellet: Stuart's biographical sketch in *Lives and Works of Civil and Military Engineers of America*, pp. 257–285, miscellaneous newspaper clippings, RPI.

Page 76 Roebling aspires to be Ellet's assistant: Letter quoted in Schuyler, *The Roeblings*, pp. 54–55.

Page 76 Homer Walsh: Steinman, *The Builders of the Bridge*, p. 163.

Page 77 Ellet drew up cannon: WAR to F. M. Colby of Dodd, Mead & Co., February 1907. RUL.

Page 78 "Before entering upon any important work": Stuart, *Lives and Works of Civil and Military Engineers of America*, p. 325.

Page 78 "The only real difficulty of the task": JAR, *Report on the Niagara Bridge*, Buffalo, 1852. RUL.

Page 78 JAR not the innovator of stiff roadway, anchor stays, or the first to spin cables in place: Steinman mistakenly credits Roebling with all three, either directly or by implication, in *The Builders of the Bridge*, pp. 81, 172.

Page 78 "In the anxiety to obtain a light roadway": ASCE, *Transactions*, 1868–71, a paper by Edward P. North, March 4, 1868, which contains one of the very best accounts of the evolution of the suspension bridge and its refinements.

Page 79 JAR's disdain for English bridgebuilders: JAR letter quoted by Stuart, *Lives and Works of Civil and Military Engineers of America*, pp. 306–308.

Page 80 Eyewitness account of Wheeling Bridge failure: Wheeling *Intelligencer*, May 18, 1854; also quoted in Steinman, *The Builders of the Bridge*, p. 171.

Page 80 ". . . there are no safer bridges": Steinman and Watson, *Bridges and Their Builders*, p. 209.

Page 81 JAR's explanation of the Wheeling failure: Steinman, *The Builders of the Bridge*, pp. 182–183.

Page 81 Ellet rebuilt the bridge himself: Research by C. M. Lewis, S.J., of Wheeling College, West Virginia, reported in ASCE, *Civil Engineering*, September 1969.

Page 81 "My bridge is the admiration of everybody"; "We had a tremendous gale"; "No one is afraid to cross": JAR to Charles Swan. RUL.
Page 83 Slocum toast: *Eagle*, July 26, 1869.
Page 83 ". . . the great achievements of the present": Whitman, *Passage to India.*
Page 84 "one of the victories of peace": *Harper's Weekly*, May 29, 1869.
Page 84 "The chief engineers became his heroes": Sullivan, *The Autobiography of an Idea*, pp. 247–249.

4 Father and Son

Page 85 "Nothing lasts forever": WAR to JAR II, March 6, 1894. RUL.
Page 85 Job applicants and JAR's comments: JAR's address book, 1869. RUL.
Page 86 WAR's notes and diagrams for the center line: Black leather notebook kept by WAR, 1869. RPI.
Page 87 "Your Turkish Bath tickets came today": WAR to JAR, May 21, 1869. RUL.
Page 88 Meetings with Rawlins: Described by WAR in several letters to JAR, June 1869. RUL.
Page 88 Consultants' approval published: *Report of the Board of Consulting Engineers to the Directors of the New York Bridge Company.*
Page 89 Revisions in design as a result of War Department directive: WAR.
Page 90 "Introductory Remarks," *Pneumatic Tower Foundations of the East River Suspension Bridge.* LER.
Page 90 "This bridge is to be built": *The New York Times*, July 23, 1869.
Page 90 "He felt at his age he could ill afford to lose any time": WAR in an "Introduction" to JAR's *Long and Short Span Railway Bridges.*
Page 90 This description of the accident is drawn largely from an account in the *Eagle*, July 22, 1869.
Page 91 "There is no such thing as chance": Steinman, *The Builders of the Bridge*, p. 320.
Page 93 Death of JAR: Various items in the *Eagle* in the days that followed; later remarks made by WAR (RUL and RPI); Schuyler, *The Roeblings*, pp. 139–140; description of tetanus in *The Merck Manual of Diagnosis and Therapy.*
Page 93 Instructions to Ed Riedel: Schuyler, *The Roeblings*, p. 140.
Page 93 "He who loses his life from injuries": *Eagle*, July 22, 1869.
Page 95 "The name of John A. Roebling": EWR to JAR, January 6, 1868. RUL.
Page 96 Gifts for Elvira: From purchases listed in JAR's Private Cash Account, 1867–69. RPI.
Page 96 Wedding gifts for the second Mrs. JAR: *Ibid.*
Page 97 Contents of will: JAR will dated September 14, 1867, RUL; also covered in some detail in Schuyler, *The Roeblings*, pp. 145–146.
Page 97 Funeral: Both the *Eagle* and the Trenton *Daily State Gazette* for July 26, 1869, carried long descriptive accounts.

Page 99 "With its inspiration gone": Steinman and Watson, *Bridges and Their Builders*, p. 236; Steinman, *The Builders of the Bridge*, p. 323.
Page 100 "Not long since, before the accident": *Eagle*, July 22, 1869.
Page 100 "First—I was the only living man": WAR to James Rusling, January 23, 1916. RUL.
Page 102 ". . . At the time of his death he was already arranging": WAR to William Couper, July 26, 1907. RUL.
Page 102 "The great boast of this land . . . jabbering and wrangling politicians": *Eagle*, July 27, 1869.

5 Brooklyn

Page 103 "transformed . . . from insignificance": *The City of Brooklyn*, a guidebook.
Page 104 Third-largest city: Syrett, *The City of Brooklyn, 1865–1898*, p. 12.
Page 104 Types of manufacturing: *Ibid.*, pp. 14–15.
Page 104 "an enigma to the respectable": *Ibid.*, p. 29.
Page 105 East River shipyards and virtues as a harbor: Albion, *The Rise of New York Port (1815–1860)*.
Page 106 More ships than New York and Hoboken combined: Syrett, *The City of Brooklyn*, p. 139.
Page 106 Salt air "pure and bracing . . .": Stiles, *A History of the City of Brooklyn*, Vol. II, p. 504.
Page 106 "the most majestic views of land and ocean": Attributed to James S. T. Stranahan in *The City of Brooklyn.*
Page 107 Banquet on board *City of Brooklyn: Eagle*, April 15, 1869.
Page 107 Hezekiah Pierrepont and the development of Brooklyn Heights: Stiles, *A History of the City of Brooklyn.*
Page 108 "Almost everybody appears to have built his house": *Eagle*, June 22, 1872.
Page 108 "elegant equipages, well-dressed grooms": *Old Brooklyn Heights*, pp. 33–34.
Page 108 "His knowledge of fish": *National Cyclopedia of American Biography.*
Page 109 Henry Ward Beecher: Smith, *Sunshine and Shadow in New York*, pp. 86–100; McCabe, *Lights and Shadows of New York Life*, pp. 655–657; Rourke, *Trumpets of Jubilee;* a long profile in the *Eagle*, March 10, 1869.
Page 110 "He went marching up and down the stage": Kaplan, *Mr. Clemens and Mark Twain*, pp. 23–24.
Page 110 "Our institutions live in him": *Eagle*, March 10, 1869.
Page 111 "A more intelligent body": From the Springfield (Mass.) *Republican*, quoted in the *Eagle*, January 1872.
Page 111 Charles Dickens on Brooklyn: Quoted in Still, *Mirror for Gotham*, p. 204.
Page 112 Brooklyn slums: According to *The New York Times*, June 30, 1866, "dirt and filth and poverty reign triumphant . . . Here homeless and vagabond children, ragged and dirty, wander about . . . de-

caying garbage, dead animals, filth and unclean privies, with crowds of unwashed human beings [are] packed together . . ."

Page 113 The Kingsley-McCue-Murphy meeting is reported in the *Eagle*, May 24, 1883; also in Steinman, *The Builders of the Bridge*, pp. 302–303.

Page 113 General Johnson's opposition to a bridge: Long Island *Star*, February 13, 1834; Trachtenberg, *Brooklyn Bridge; Fact and Symbol*, pp. 35–36.

Page 114 HCM's Mansion House speech: A commemorative booklet on the farewell dinner, LIH; also quoted in the *Eagle*, December 2, 1882.

Page 114 William C. Kingsley: Obituaries in *The New York Times*, New York *World*, and *Eagle*, February 21, 1885; in memoriam booklet, *W. C. Kingsley*, LIH; Green, *A Complete History of the New York and Brooklyn Bridge from its Conception in 1866 to its Completion in 1883;* Stiles, *The Civil, Political, Professional and Ecclesiastical History of the County of Kings and the City of Brooklyn*, Vol. I, pp. 463–464; *Eagle History of Brooklyn*, *Eagle*, May 24, 1883; scrapbooks in LIH collection; Syrett, *The City of Brooklyn*, pp. 74–76.

Page 114 Henry C. Murphy: Obituaries in *The New York Times*, New York *World*, and *Eagle*, December 2, 1882; scrapbooks in LIH collection; Green, *A Complete History of the New York and Brooklyn Bridge;* Stiles, *The Civil, Political, Professional and Ecclesiastical History of the County of Kings and the City of Brooklyn*, Vol. I, pp. 360–366; *Eagle*, May 24, 1883; Stiles, *A History of the City of Brooklyn*, Vol. II, pp. 266–270.

Page 115 McLaughlin the first to be called "Boss": Syrett, *The City of Brooklyn*, p. 71.

Page 116 "very earnest in manner": Stiles, *A History of the City of Brooklyn*, p. 269.

Page 117 "It was not a change for the better": *Eagle*, December 2, 1882.

Page 118 "Mr. Murphy only failed as a politician": Stiles, *The Civil, Political, Professional and Ecclesiastical History of the County of Kings and the City of Brooklyn*, Vol. I, p. 364.

Page 119 WAR's private remarks on the role played by Julius Adams: Personal notebook, entry dated January 6, 1880. RPI.

Page 120 HCM named president: *New York and Brooklyn Bridge Proceedings, 1867–1884*, p. 319.

Page 120 The name Roebling "invaluable": Kingsley, in a speech given on the opening of the bridge, May 24, 1883.

Page 120 "Confidence on the part of the public": *New York and Brooklyn Bridge Proceedings*, p. 320.

6 The Proper Person to See

Page 122 "Who owns the City of New York today?": Quoted in Syrett, *The City of Brooklyn, 1865–1898*, p. 19.

Page 122 Tweed's prior interest in the Brooklyn ferry lines: Lynch, *"Boss" Tweed*, pp. 70–75.

Page 123 New York in 1869: Still, *Mirror for Gotham;* McCabe, *Lights and Shadows of New York Life;* Smith, *Sunshine and Shadow in New York; Harper's Weekly* for 1869; Crapsey, *The Nether Side of New York.*

Page 124 ". . . a rich field for clever money lovers": Olof Olson to his brother, September 11, 1869, quoted in *Land That Our Fathers Plowed*, David Greenberg, ed., University of Oklahoma Press, 1969.

Page 125 William M. Tweed and his cohorts: Werner, *Tammany Hall;* Lynch, "*Boss*" *Tweed;* Callow, *The Tweed Ring; Harper's Weekly;* Bryce, *The American Commonwealth*, Vol. II; McCabe, *Lights and Shadows of New York Life;* Smith, *Sunshine and Shadow in New York; Dictionary of American Biography.*

Page 126 "I don't care a straw for your newspaper articles": Callow, *The Tweed Ring*, p. 254.

Page 127 "Tweed had an abounding vitality": Bryce, *The American Commonwealth*, Vol. II, p. 383.

Page 127 Tweed and the first session of the Executive Committee: *New York and Brooklyn Bridge Proceedings, 1867–1884*, p. 526.

Page 130 Tweed's testimony: As recorded before a committee of the Common Council of the City of New York, September 18, 1877; quoted also in *Testimony in the Miller Suit to Remove the East River Bridge*, "Exhibit A," pp. 58–63.

Page 131 "a strong combination made against the measure": Kingsley to JAR, April 16, 1868. RUL.

Page 133 Chambers Street courthouse: The best account of this incredible story is in Callow, *The Tweed Ring*, the chapter titled "The House That Tweed Built," pp. 198–206, which also appeared in *American Heritage*, October 1965.

Page 135 Bridge Company stockholders as of autumn 1869: *New York and Brooklyn Bridge Proceedings*, "Exhibit C," Part I, p. 167.

Page 137 ". . . therefore he was the proper person to see": *Eagle*, September 19, 1877.

Page 139 Beach tunnel: *Scientific American*, February 19, 1870; "Alfred Ely Beach and His Wonderful Pneumatic Underground Railway" by Robert Daley, *American Heritage*, June 1961.

Page 141 Black Friday: Swanberg, *Jim Fisk: The Career of an Improbable Rascal*, pp. 149–153.

Page 142 Cardiff Giant: *Harper's Weekly*, October 1869; Franco, "The Cardiff Giant: A Hundred-Year-Old Hoax." The Giant itself is still drawing crowds at the Farmers' Museum, Cooperstown, New York. Once having seen the Giant, most twentieth-century onlookers find it hard to believe anyone ever took it seriously.

Page 143 "all were disgusted": Adams, *The Education of Henry Adams*, p. 273.

7 The Chief Engineer

Page 145 Assistant engineers: Various memoirs published by the ASCE; *National Cyclopedia of American Biography;* biographical sketches in the *Eagle*, May 24, 1883; odd notes made by WAR, RPI.

Page 147 Claims of Samuel Barnes B. Nolan: *Scientific American*, August 7, 1869.
Page 148 "the details not having been considered": New York *Tribune*, May 23, 1883.
Page 148 "very versatile attainments": From an unpublished biographical sketch of WAR by EWR. RPI.
Page 149 "rather indifferent to matters of courtesy": EWR to JAR, January 6, 1868. RUL.
Page 149 "History teaches us that no man can be great unless a certain amount of vanity enters into his composition": WAR considered his brother Ferdinand the perfect example of such vanity. The quote is from WAR's draft of an obituary for Ferdinand, April 15, 1917, RUL; also quoted somewhat differently in Schuyler, *The Roeblings*, p. 307.
Page 149 "a peculiarity of the Roebling mind": WAR to JAR II, May 24, 1896. RUL.
Page 149 "It might be argued if a man inherits everything": WAR's obituary for his brother Charles, October 1918, RUL; also in Schuyler, *The Roeblings*, pp. 324–325.
Page 150 WAR's passport: RUL.
Page 150 "Roebling is a character": *Lyman*, *Meade's Headquarters*, p. 240; also quoted in Schuyler, *The Roeblings*, p. 195.
Page 151 "reverently chose . . . the name that most inspired him": Steinman, *The Builders of the Bridge*, p. 41.
Page 151 WAR named for Washington Gill: WAR to JAR II, July 4, 1904. RUL.
Page 151 Baptized by postmaster Shilly: WAR, *Early History of Saxonburg*, p. 12.
Page 151 "well-built, sturdy, quiet boy": JAR to his brother Christel, undated. RUL.
Page 151 ". . . a black bear walked down Main Street": WAR, *Early History of Saxonburg*, p. 9.
Page 151 Saxonburg social life: *Ibid.*, p. 17.
Page 151 Ferdinand Baehr and Waterloo stories: *Ibid.*, p. 18.
Page 152 WAR's love of Saxonburg and disappointing return visit: WAR to JAR II, January 5, 1926. RUL.
Page 152 "Being the 'Roebling boy' ": WAR, *Early History of Saxonburg*, p. 20.
Page 152 Story of Massy Harbison: "The Touching Narrative of Massy Harbison," from *Our Western Border*, Charles McKnight, Philadelphia, 1875, pp. 685–695.
Page 152 Pigeons, thunderstorms, and the great comet of 1843: WAR, *Early History of Saxonburg*, p. 21.
Page 153 "In regard to the mustache you covet so": Laura Roebling to WAR, December 7, 1856. RUL.
Page 154 Courses at RPI: Greene, *The Rensselaer Polytechnic Institute;* also Steinman, *The Builders of the Bridge*, pp. 196–197.
Page 154 "Under such a curriculum the average college boy of today": Steinman, *The Builders of the Bridge*, p. 197.
Page 154 "that terrible treadmill of forcing an avalanche of figures . . .

unusable knowledge that I could only memorize, not really digest": Schuyler, *The Roeblings*, pp. 173–174.

Page 155 "My candle is certainly bewitched . . . no woman had sense enough to understand his love": WAR to EWR, about April 14, 1864. RUL.

Page 155 "Our temperaments are so very different": RUL.

Page 155 Letter written Thanksgiving Day: *Ibid.*

Page 156 "left the school as mental wrecks": Quoted in Schuyler, *The Roeblings*, p. 174.

Page 156 "Pittsburgh is getting along quite smart": WAR to Charles Swan, April 11, 1859, RUL; also quoted in Schuyler, *The Roeblings*, p. 182.

Page 156 Penn Street boardinghouse: WAR to Charles Swan, May 2, 1858. RUL.

Page 157 "There is a perfect mania here for improvements": WAR to Charles Swan, April 11, 1859, RUL; also quoted in Schuyler, *The Roeblings*, p. 182.

Page 157 "dark, cloudy, smoky afternoons": WAR to Charles Swan, November 13, 1858, RUL; also quoted in Schuyler, *The Roeblings*, p. 179.

Page 157 "This is my first letter to you in 1860": WAR to Charles Swan, January 23, 1860, RUL; also quoted in Schuyler, *The Roeblings*, pp. 184–185.

Page 157 "My enlistment was rather sudden": WAR to James Rusling, February 18, 1916. RUL.

Page 158 "Loafing in the camp": Undated letter. RUL.

Page 158 "This is a mean little town": WAR to Elvira Roebling, July 19, 1861. RUL.

Page 158 "This artillery business": WAR to Charles Swan, July 31, 1861. RUL.

Page 158 "could make a violin talk": Letter of condolence written to the second Mrs. WAR by George R. Brown, president of the Eastchester Savings Bank, Mount Vernon, New York, August 2, 1926. RUL.

Page 158 "My father being too old to rough it": WAR to James Rusling, February 18, 1916. RUL.

Page 159 Swims the Shenandoah with tape in his mouth: WAR to Ferdinand, June 8, 1892. RUL.

Page 159 Surprised Jeb Stuart at his breakfast: WAR to JAR, August 24, 1862. RUL.

Page 159 Describes bridge: WAR to Charles Swan, August 3, 1862. RUL.

Page 159 Fate of Harpers Ferry bridge: WAR to James Rusling, February 18, 1916. RUL.

Page 160 Incident with the statue of Washington's mother: Schuyler, *The Roeblings*, pp. 193–194.

Page 160 With Hooker at Chancellorsville: WAR to James Rusling, February 18, 1916. RUL.

Page 160 Reconnaissance from a balloon: *Ibid.;* also Schuyler, *The Roeblings*, p. 191.

Page 160 Trip home for maps: WAR to Oliver W. Norton, July 13, 1915. RUL.

Page 161 WAR's account of his day on Little Round Top: Letter to a Colonel Smith of New York, July 5, 1913. RUL.

Page 161 "Roebling was on my staff": Schuyler, *The Roeblings*, p. 193.

Page 161 "I was the first man on Little Round Top": WAR to James Rusling, February 18, 1916. RUL.

Page 162 WAR and Warren before the Battle of the Crater: *Ibid.*

Page 162 ". . . I was in the Civil War for four years and saw Lincoln on two occasions": WAR to I. E. Boos, June 19, 1921, RUL; also quoted in Schuyler, *The Roeblings*, pp. 196–197.

Page 162 "They must put fresh steam on the man factories . . . the rest think it is about played out to stand up and get shot": WAR to EWR, June 23, 1864. RUL.

Page 163 ". . . the conduct of the Southern people": WAR to EWR, July 7, 1864. RUL.

Page 163 Description of meeting Emily: WAR to Elvira Roebling, February 26, 1863. RUL.

Page 163 JAR's letter on the engagement: JAR to WAR, March 30, 1864. RUL.

Page 163 "I like her very much": JAR to WAR, November 17, 1864. RUL.

Page 164 "I dare say you could not sleep": WAR to EWR, August 14, 1864. RUL.

Page 164 "This day might be signalized": WAR to EWR, November 16, 1864. RUL.

Page 165 "The town is horribly dull": WAR to EWR, August 6, 1864. RUL.

Page 165 "I have now more lasting memories": WAR to EWR, September 10, 1864. RUL.

Page 165 "I have been solacing myself": WAR to EWR, April 11, 1864. RUL.

Page 165 ". . . the greatest giver of us all [is] gone": WAR to EWR, December 25, 1864. RUL.

Page 166 Trip to Europe: Described in numerous lengthy letters from WAR to JAR, in both the RPI and RUL collections.

Page 167 Letter to JAR describing Keystone Bridge works: WAR to JAR, October 11, 1868. RUL.

Page 169 Family differences over Edmund: WAR, private memorandums dated July 20, 1898, and March 16, 1922. RUL.

Page 169 Reminders and comments on stone: WAR's personal notebook, 1869. RPI.

PART TWO

8 All According to Plan

Page 173 "The foundations for the support": JAR, *Report of John A. Roebling, C.E., to the President and Directors of the New York Bridge Company, on the Proposed East River Bridge*, p. 20. LER.

Page 174 Dimensions of the Brooklyn caisson, as well as all other descriptive data: WAR, *First Annual Report of the Chief Engineer of the East River Bridge*, LER; WAR, *Pneumatic Tower Foundations of the East River Suspension Bridge*, LER.

Page 176 Barometer analogy: *Harper's Weekly*, December 17, 1870.

Page 177 "The extreme rise and fall": WAR, *First Annual Report of the Chief Engineer*, pp. 8–9. LER.

Page 178 Webb & Bell contract: Kingsley, *First Annual Report of the General Superintendent of the East River Bridge*, p. 23. LER.

Page 179 "A pile which was sixteen inches in diameter": WAR, *First Annual Report of the Chief Engineer*, p. 11. LER.

Page 179 "The character of this material": *Ibid.*, pp. 11–12.

Page 180 James B. Eads: There is no real biography of the remarkable Eads. The following have been used as general biographical background: Dorsey, *Road to the Sea;* Woodward, *A History of the St. Louis Bridge;* Gies, *Bridges and Men; Dictionary of American Biography.*

Page 180 "Eads's Turtles": Catton, *Grant Moves South*, pp. 102–103.

Page 181 JAR calls St. Louis people fools: JAR to WAR, November 10, 1867. RUL.

Page 182 Carnegie, Linville, and the Keystone Bridge Company: Carnegie, *Autobiography*, pp. 119–121.

Page 183 "an achievement out of all proportion": Kirby and Laurson, *The Early Years of Modern Civil Engineering*, p. 162.

Page 184 Material on early use of compressed air and resulting cases of caisson sickness is from *The Effects of High Atmospheric Pressure, Including the Caisson Disease* by Andrew H. Smith, M.D., pp. 4–10. LER.

Page 186 "A workman walking about with difficult step": Woodward, *A History of the St. Louis Bridge.*

Page 186 "The fatigue of ascent added not a little": *Ibid.*

Page 186 Eads's views on the problem of caisson sickness are contained in a long article in *Scientific American*, December 24, 1870.

Page 187 The launching of the Brooklyn caisson was described in considerable detail by all of the following: *Eagle*, March 19, 1870; *Engineering* (London), June 10, 1870; *Scientific American*, July 9, 1870; Collingwood, *A Few Facts about the Caissons of the East River Bridge*, LER.

Page 187 "more like a huge war leviathan": *Eagle*, March 19, 1870.

Page 188 The only known reference to Roebling's visit to St. Louis and his sessions with Eads is an exchange of letters in *Engineering* (London) in the issues for May 16, June 27, and September 5, 1873.

Page 189 "I do not want any news carried between myself and Mr. Ellet": JAR to Charles Swan, April 21, 1849, RUL; also quoted in Schuyler, *The Roeblings*, p. 82.

Page 191 ". . . one of the wonders of the nineteenth century"; "hidden from the gaze of mortal eyes"; "as placidly as a swan": *Eagle*, May 3, 1870.

Page 191 ". . . they had been upon the monster": *Ibid.*, May 4, 1870.

Page 191 Roebling, Paine, and Collingwood go down for first time on May 10: WAR, *Pneumatic Tower Foundations of the East River Suspension Bridge*, p. 24. LER.

9 Down in the Caisson

Page 195 The descent of the Brooklyn caisson and the work that went on inside it were the subjects of many articles in newspapers and technical publications in the year 1870. Of particular interest were those in the following: *Eagle*, June 20; *Scientific American*, July 9; *Van Nostrand's Eclectic Engineering Magazine*, October; *Journal of the Franklin Institute*, October; and *Harper's Weekly*, December 17. But nearly all of this chapter has been drawn from a paper read before the ASCE by Francis Collingwood on June 21, from Master Mechanic E. F. Farrington's *Concise Description of the East River Bridge*, and from WAR's own annual report to the directors of the Bridge Company. An excellent scale model of the caisson can be seen on display at the Smithsonian Institution.

Page 195 "We have no precedent just like this bridge": WAR, *Report of the Chief Engineer of the New York and Brooklyn Bridge, January 1, 1877*, p. 5. LER.

Page 195 "The material now became sufficiently exposed": WAR, *Report of the Chief Engineer to the Board of Directors of the New York Bridge Company, June 5, 1871*, p. 4. LER.

Page 196 "Inside the caisson everything wore an unreal, weird appearance": Farrington, *Concise Description of the East River Bridge*, pp. 27–28.

Page 197 "An unearthly and deafening screech": *Scientific American*, July 9, 1870.

Page 197 Use of limelights: WAR, *Report of the Chief Engineer, June 5, 1871*, pp. 35–37. LER.

Page 199 Varieties of rock uncovered: *Ibid.*, pp. 4–5.

Page 200 "Moreover, a settling of the caisson of six inches": *Ibid.*, p. 6.

Page 201 "The noise made by splitting blocks": *Ibid.*, p. 23.

Page 201 "Levels were taken every morning": Collingwood, *A Few Facts about the Caissons of the East River Bridge*. LER.

Page 201 Techniques for removing boulders from under the shoe: WAR, *Report of the Chief Engineer, June 5, 1871*, pp. 8–10. LER.

Page 202 "five months of incessant toil . . . we were almost tempted to throw the buckets overboard": *Ibid.*, pp. 15–17.

Page 203 "When the lungs are filled with compressed air": *Ibid.*, p. 15.
Page 204 Side friction: Collingwood, *A Few Facts about the Caissons of the East River Bridge*. LER.
Page 205 WAR "conspicuous for his presence and exertions": Kingsley, *Report of the General Superintendent, New York Bridge Company*, p. 54. LER.
Page 205 Lowering of air pressure gives added twelve hundred tons: Collingwood, *A Few Facts about the Caissons of the East River Bridge*. LER.
Page 205 Apprehensions about blasting: WAR, *Report of the Chief Engineer, June 5, 1871*, pp. 11–12. LER.
Page 206 WAR uses revolver: *Ibid.*, p. 12.
Page 207 "For night is turned into day": New York *Herald*, December 3, 1870.
Page 207 Work schedule and work force: WAR, *Report of the Chief Engineer, June 5, 1871*, pp. 38–39; Kingsley, *Report of the General Superintendent, New York Bridge Company*, p. 52. LER.
Page 209 Pneumatic water closet: Collingwood, *A Few Facts about the Caissons of the East River Bridge*. LER.
Page 210 Roebling follows Eads's system, convinced increased oxygen intake is the heart of the problem: WAR, *Report of the Chief Engineer, June 5, 1871*, pp. 39–40. LER.
Page 210 Steam coils in air locks: *Ibid.*, p. 40.
Page 211 Great Blowout: *Ibid.*, pp. 20–21; Farrington, *Concise Description of the East River Bridge*, pp. 20–21.
Page 212 Weight variation in columns of water: Collingwood, *A Few Facts about the Caissons of the East River Bridge*. LER.
Page 213 "To say that this occurrence was an accident": WAR, *Report of the Chief Engineer, June 5, 1871*, p. 20. LER.

10 Fire

Page 231 "When the perfected East River bridge": *Eagle*, June 22, 1872.
Page 231 Modifications in New York caisson: WAR, *Report of the Chief Engineer to the Board of Directors of the New York Bridge Company, June 5, 1871*, pp. 45–49. LER.
Page 232 "This bold and peculiarly American design": *Harper's Weekly*, November 19, 1870.
Page 232 "the rapidity with which the work has proceeded": *Scientific American*, November 12, 1870.
Page 232 Cause of the fire and description of the fire itself have been drawn from the following: WAR, *Report of the Chief Engineer, June 5, 1871*, pp. 29–35, LER; *Eagle*, December 2, December 3, December 5, 1870; Farrington, *Concise Description of the East River Bridge*, pp. 22–24; *Engineering* (London), December 30, 1870; *Journal of the Franklin Institute*, February 1871.
Page 233 Attempts to extinguish fire: WAR, *Report of the Chief Engineer, June 5, 1871*, pp. 29–30. LER.
Page 234 Boring into the roof: *Ibid.*, p. 31.

Page 234 WAR's efforts "almost superhuman": Kingsley, *Report of the General Superintendent, New York Bridge Company*, p. 54. LER.
Page 235 WAR's collapse: *Ibid.;* Farrington, *Concise Description of the East River Bridge; Eagle*, December 3–5, 1870.
Page 236 Discover mass of living coals: WAR, *Report of the Chief Engineer, June 5, 1871*, p. 31. LER.
Page 237 "He appeared calm and collected": *Eagle*, December 2, 1870.
Page 238 Damage estimated at $250,000: New York *Herald*, December 3, 1870.
Page 238 *World* charges sabotage: December 2, 1870.
Page 238 Fire marshal's hearing: *Eagle*, December 5, 1870.
Page 239 Begin filling work chambers with concrete: *Eagle*, December 23, 1870.
Page 240 Blowout of supply shaft: WAR, *Report of the Chief Engineer, June 5, 1871*, pp. 24–27. LER.
Page 242 Repairing the fire damage: *Ibid.*, pp. 32–35; Farrington, *Concise Description of the East River Bridge*, pp. 22–24.
Page 245 Fresh-water springs: WAR, *Report of the Chief Engineer, June 5, 1871*, p. 28. LER.

11 The Past Catches Up

Page 248 Launching of the New York caisson: *Eagle*, May 8–9, 1871.
Page 250 Tweed's daughter's wedding: Werner, *Tammany Hall*, pp. 190–193; Lynch, *"Boss" Tweed*, pp. 359–360; New York *Sun*, June 1, 1871.
Page 252 Activities of Matthew J. O'Rourke: Lynch, *"Boss" Tweed*, pp. 354, 361.
Page 252 Watson the nerve center of the Ring: Werner, *Tammany Hall*, p. 209.
Page 253 "You must do just as Jimmy tells you": *Ibid.*, p. 161.
Page 253 O'Rourke's estimate of Ring thefts: *Ibid.*, p. 160.
Page 253 Attempt to bribe Jones: *Ibid.*, p. 210.
Page 253 Attempt to bribe Nast: *Ibid.*, pp. 211–212.
Page 254 *Times* attack: Callow, *The Tweed Ring*, pp. 256–261.
Page 254 Orange riot: Lynch, *"Boss" Tweed*, pp. 367–369; Swanberg, *Jim Fisk: The Career of an Improbable Rascal*, pp. 234–240; Strong, *The Diary of George Templeton Strong*, entries for July 1871.
Page 255 Cooper Union meeting of September 4: Lynch, *"Boss" Tweed*, pp. 370–371; Werner, *Tammany Hall*, pp. 217–218.
Page 256 Cartoon of Tweed in the shadow of the gallows: *Harper's Weekly*, October 21, 1871.
Page 257 "At home again amidst the haunts of my childhood": Lynch, *"Boss" Tweed*, pp. 377–378.
Page 257 George Templeton Strong on the "Boss of New York": Strong, *Diary*, entry for January 27, 1871.
Page 258 Elections in Brooklyn: Syrett, *The City of Brooklyn, 1865–1898*, pp. 56–60.

Page 258 Accident: C. C. Martin interview published in the *Eagle*, May 24, 1883.

Page 260 "This has been the case from the first": Kingsley, *Report of the General Superintendent of the New York Bridge Company*, p. 32. LER.

Page 261 Six thousand illegal votes: Syrett, *The City of Brooklyn*, p. 59.

Page 261 Kingsley interview in the *World:* Quoted in the *Eagle*, December 15, 1871.

Page 262 Kingsley's name a football: Beecher at Kingsley's funeral, published in memorial book, *W. C. Kingsley*. LIH.

Page 263 Tweed's appearances at the meetings of the Executive Committee: "Exhibit J; A Full Synopsis of the Minutes of the Respective Executive Committees Thereof, From September 1869 to June 1st, 1883," *New York and Brooklyn Bridge Proceedings, 1867–1884*, pp. 526–566.

Page 264 "*Resolved*, That fifteen per centum on the amount of expenditure": *Ibid.*, p. 552.

Page 265 "I had no understanding with him, sir": *Testimony in the Miller Suit to Remove the East River Bridge*, "Exhibit A," February 15, 1879, p. 62.

Page 265 Kingsley's "claim . . . liquidated": *New York and Brooklyn Bridge Proceedings*, p. 572.

Page 265 Erasure made in the records: "Exhibit No. 4," *Minority Report* by Demas Barnes, December 16, 1872, *New York and Brooklyn Bridge Proceedings*, p. 96.

Page 266 Tweed indicted and arrested: Werner, *Tammany Hall*, p. 233.

Page 267 Death of Fisk: Swanberg, *Jim Fisk*, pp. 271–278.

12 How Natural, Right, and Proper

Page 269 "Although the bridge from every element of its use": "Exhibit No. 4," *Minority Report* by Demas Barnes, *New York and Brooklyn Bridge Proceedings*, p. 100.

Page 269 "It is true that Tweed, Connolly, and Sweeny are among the subscribers": *Eagle*, April 10, 1872.

Page 270 Kingsley's letter to the *Eagle* and *Union* appeared April 17, 1872.

Page 271 Committee of Fifty's letter in answer to Kingsley: *Eagle*, April 22, 1872.

Page 271 Kingsley's second letter: *Eagle*, April 29, 1872.

Page 272 Replacements for Tammany quartet: "Synopsis of the Minutes of Proceedings of the Corporators, Directors, and Stockholders of the New York Bridge Company and Also of the Trustees of the New York and Brooklyn Bridge, Comprehending a Period of 16 Years, Viz.; From May 13th, 1867, to June 1st, 1883" ("Exhibit I"), *New York and Brooklyn Bridge Proceedings*, p. 332.

Page 272 Hewitt swings into action: "The New York Bridge Company and the Trustees of the New York and Brooklyn Bridge, A Full Synopsis of the Meetings of the Executive Committees Thereof, From September 17th, 1869, to June 1st, 1883" ("Exhibit J"), *New York and Brooklyn Bridge Proceedings*, pp. 575–577.

Page 273 WAR's report: *Report of the Chief Engineer on Prices of Materials, and Estimated Cost of the Structure, East River Bridge, June 28, 1872* ("Exhibit No. 2"), *New York and Brooklyn Bridge Proceedings*, pp. 74–83.

Page 275 Barnes called an ass and a quack: *Eagle*, June 22, 1872.

Page 275 Kinsella faces down scandal: Syrett, *The City of Brooklyn, 1865–1898*, p. 95.

Page 276 ". . . He is the thinker who acts": *Eagle*, June 22, 1872.

Page 277 Predict bridge to cost forty million dollars: *Scientific American*, July 15, 1872.

Page 277 Kingsley's "agreement" at an end: Directors' Meeting, November 4, 1872, *New York and Brooklyn Bridge Proceedings*, p. 334.

Page 277 Beecher scandal breaks: Johnston, *Mrs. Satan*, pp. 159–178; Shaplen, "The Beecher-Tilton Case," Part II.

Page 278 Majority report: "Exhibit No. 3," *New York and Brooklyn Bridge Proceedings*, pp. 84–89.

Page 279 Barnes's minority report: "Exhibit No. 4," *Ibid.*, pp. 90–101.

Page 281 Executive Committee report: "Exhibit No. 5," *Ibid.*, pp. 102–131.

Page 283 "This Company was chartered as a private company": *Ibid.*, p. 109.

Page 283 Kingsley back at ten-thousand-dollar salary: *Ibid.*, p. 336.

Page 284 Kingsley takes leave of absence: *Ibid.*, p. 586.

13 The Mysterious Disorder

Page 289 "Knowing from the reports of other similar works": Kingsley, *Report of the General Superintendent of the New York Bridge Company*, p. 33. LER.

Page 289 "To such of the general public": WAR, *Third Annual Report of the Chief Engineer, June 1, 1872*, p. 7. LER.

Page 291 "Considerable risk and some degree of uncertainty": *Ibid.*, p. 8.

Page 292 Depth of bedrock: *Ibid.*, p. 9.

Page 293 "The great timber foundation was now complete!": *Ibid.*, p. 13.

Page 293 Paine's mechanical signaling system: Collingwood, *Further Notes on the Caissons of the East River Bridge*. LER.

Page 295 Caisson sinking six to eleven inches a day: *Ibid.*

Page 295 Sand pipes: WAR, *Third Annual Report of the Chief Engineer*, pp. 18–20, LER; Farrington, *Concise Description of the East River Bridge*, pp. 25–26; Collingwood, *Further Notes on the Caissons of the East River Bridge*. LER.

Page 297 "The downward movement of the caisson": WAR, *Third Annual Report of the Chief Engineer*, p. 26. LER.

Page 298 Change of work shifts: Collingwood, *Further Notes on the Caissons of the East River Bridge*. LER.

Page 299 Dr. Smith's nine rules: Smith, *The Effects of High Atmospheric Pressure, Including the Caisson Disease*, p. 13. LER.

Page 300 "The habits of many of the men": *Ibid.*, p. 14.
Page 300 "The utmost efforts of the expiratory muscles": *Ibid.*, p. 15.
Page 300 "Hence, the pulse is small": *Ibid.*, p. 16.
Page 301 Experiment with pigeons: *Ibid.*, p. 20.
Page 301 Experiment with dog: *Ibid.*, p. 28.
Page 303 Sample case histories: *Ibid.*, pp. 35–37.
Page 304 "When it is severe, local numbness": Sodeman, *Pathologic Physiology*, p. 238.
Page 305 Remedies employed: *Ibid.*, pp. 32–33.
Page 305 Walter Reed at Brooklyn City Hospital: *Ibid.*, p. 39.
Page 306 "Indeed, it is altogether probable": *Ibid.*, p. 30.
Page 306 "Experience teaches": *Ibid.*, p. 7.
Page 306 Smith rules more time in the lock: *Ibid.*, p. 30.
Page 307 "The natural impatience of the men": *Ibid.*, p. 30.
Page 307 Theory of "special predisposition": *Ibid.*, p. 29.
Page 308 "The testimony of all observers": *Ibid.*, p. 27.

14 The Heroic Mode

Page 309 Smith's explanation, "overpowering physical force," blood "retreats," etc.: Smith, *The Effects of High Atmospheric Pressure, Including the Caisson Disease*, pp. 25–26. LER.
Page 311 Prior discovery by Paul Bert: *Ibid.*, p. 27.
Page 312 "It frequently happened under my observation": *Ibid.*, p. 34.
Page 312 "by applying the heroic mode": WAR, *Third Annual Report of the Chief Engineer, June 1, 1872*, p. 24. LER.
Page 312 Difficulty of taking patient into the caisson: Smith, *The Effects of High Atmospheric Pressure*, p. 34. LER.
Page 313 Death of John Myers: *Ibid.*, p. 41.
Page 314 Death of Patrick McKay: *Ibid.*, p. 40.
Page 314 "Perhaps if they had known": Josephson, *Al Smith, Hero of the Cities*, p. 20.
Page 314 Caisson workers strike: Kingsley, *Report of the General Superintendent of the New York Bridge Company*, pp. 34–35, LER; *Eagle*, May 8, 1872.
Page 315 "The surface was evidently very irregular": WAR, *Third Annual Report of the Chief Engineer*, p. 21. LER.
Page 316 WAR estimates a hundred lives to go to bedrock: EWR, unpublished biographical sketch of WAR. RPI.
Page 316 Strata undisturbed since time of deposit: WAR, *Third Annual Report of the Chief Engineer*, p. 22. LER.
Page 316 A time of "intense anxiety": EWR, unpublished biographical sketch of WAR. RPI.
Page 316 First spur of bedrock described: WAR, *Third Annual Report of the Chief Engineer*, p. 23. LER.
Page 317 Death of Reardon: Smith, *The Effects of High Atmospheric Pressure*, p. 40. LER.
Page 317 Differences of level at the extreme corners: Collingwood, *Further Notes on the Caissons of the East River Bridge*. LER.

Page 317 "The labor below is always attended with a certain amount of risk": WAR, *Third Annual Report of the Chief Engineer*, p. 29. LER.

Page 318 "Relief from the excruciating pain": WAR, *Pneumatic Tower Foundations of the East River Suspension Bridge*, p. 88, *fn*. LER.

Page 319 Cholera epidemic at Niagara Falls: JAR to Charles Swan, July 29, 1854, RUL; also quoted in Schuyler, *The Roeblings*, p. 95.

Page 319 "He determined not to have it": *Beecher's Magazine*, January, 1871; also quoted in Schuyler, *The Roeblings*, p. 96.

Page 319 Business carried on by WAR in the fall of 1872: *New York and Brooklyn Bridge Proceedings, 1867–1884*, pp. 579–583.

Page 320 WAR's efforts the winter of 1872–73; EWR, unpublished biographical sketch of WAR, RPI; also WAR notes, letters, specifications, etc., RPI.

Page 321 Requests leave of absence: Meeting of the Board of Directors, April 21, 1873, *New York and Brooklyn Bridge Proceedings*, p. 339.

Page 322 "My plan would be as follows": Smith, *The Effects of High Atmospheric Pressure*, p. 34. LER.

PART THREE

15 At the Halfway Mark

Page 325 "Everything has been built to endure": Francis Collingwood in a speech before the First Annual Meeting of the Alumni of RPI, New York, February 18, 1881.

Page 326 "The love of praise is, I believe,": Dorsey, *Road to the Sea*, p. 163.

Page 326 Tweed escapes: Werner, *Tammany Hall*, p. 244.

Page 327 Beecher on trial: Shaplen, "The Beecher-Tilton Case," Part II.

Page 328 ". . . probably no great work was ever conducted": EWR, unpublished biographical sketch of WAR. RPI.

Page 330 Granite and gravity: WAR, *Report of the Chief Engineer of the New York and Brooklyn Bridge, January 1, 1877*, p. 6. LER.

Page 330 Limestone in anchorages: Collingwood, *Notes on the Masonry of the East River Bridge*. LER.

Page 331 Arrangement of the anchor plates and anchor bars: WAR, *Report of the Chief Engineer, January 1, 1877*, pp. 6–8, LER; *Specifications for Anchor Plates, New York Anchorage, East River Bridge, 1875*, LER; *Specifications for Iron Anchor Bars, New York Anchorage, East River Bridge, April, 1875*, LER; "Up Among the Spiders; or How the Great Bridge Is Built," *Appleton's Journal*, January 1878; Conant, "The Brooklyn Bridge."

Page 331 Work on the approaches: WAR, *Report of the Chief Engineer, January 1, 1877*, pp. 23–32. LER.

Page 332 Model of the bridge: Brooklyn *Union*, May 25, 1878.

Page 333 Tower work: WAR, *Report of the Chief Engineer, January 1, 1877*, pp. 4–5, LER; *Scientific American*, August 10, 1872; Collingwood, *Notes on the Masonry of the East River Bridge*, LER.

Page 334 "There are times when standing alone on this spot": Farrington, *Concise Description of the East River Bridge*, pp. 57–59.

Page 335 Deaths from tower and freak accidents: *Eagle*, May 18, 1876; interview with C. C. Martin, *Eagle*, May 24, 1883.

Page 336 The bridge as an obstruction to navigation: *Iron Age*, April 27, 1876; *Scientific American*, May 6, 1876. The hearings were reported in detail in the *Eagle*, April 24 and May 21, 1876.

Page 337 Charter amended: *An Act to amend an act* . . . Chapter 601. Passed June 5, 1874. LER.

Page 337 New York Bridge Company dissolved: *An Act providing that the bridge in the course of construction over the East River* . . . Chapter 300: Passed May 14, 1875. LER.

Page 338 "Before winter shall drive the workmen": *Eagle*, July 11, 1876.

Page 338 "One thing is certain": *Ibid.*

16 Spirits of '76

Page 340 Specifications: *Specifications for Granite Face-stone and Arch-stone, Required for the New York Tower, East River Bridge, April, 1875*, LER; original copy, RPI.

Page 341 Correspondence with Brooklyn: WAR and EWR, letter books. RPI.

Page 341 "It is one thing to sit in your office": WAR to JAR II, August 20, 1907. RUL.

Page 342 "I would further add, *now* is the time": WAR to HCM, February 25, 1875. LER.

Page 342 Physical discomforts: WAR to JAR II, May 5, 1894.

Page 342 "There is a popular impression": EWR, unpublished biographical sketch of WAR. RPI.

Page 343 Could neither read nor write: WAR to James Rusling, February 18, 1916. RUL.

Page 343 "Regarding your health": WAR to Francis Collingwood, undated. RPI.

Page 344 Note with check for minerals: RPI.

Page 344 WAR to HCM concerning Keystone Bridge rumor: December 6, 1875. RPI.

Page 344 Eads lawsuit: Papers on file at RPI; exchange of letters between Eads and WAR, *Engineering* (London), May 16, June 27, September 5, 1873.

Page 344 "Its perusal has left only the one prominent impression"; "My actual experience in the St. Louis caisson"; "You might as well patent contrivances in a ship's rigging"; "In conclusion I beg to assure Captain Eads": *Engineering* (London), June 27, 1873.

Page 346 G. K. Warren and the St. Louis Bridge: Gilbert and Billington, "The Eads Bridge and Nineteenth-Century River Politics."

Page 347 "I am willing to accede to the proposition": WAR to William Paine, May 10, 1876. RPI.

Page 347 "My health has become of late so precarious": WAR to HCM, December 1875. RPI.

Page 348 WAR on his brother Charles: WAR, "Memorial to Charles Roebling," October 1918, RUL; also quoted in Schuyler, *The Roeblings*, pp. 324–326.

Page 349 "He lost no opportunity": WAR, notes for what was apparently to be an autobiographical sketch, written July 1898. RUL.

Page 349 Feelings of indignation: *Ibid.*

Page 349 Personal expenses: WAR, notebooks. RPI.

Page 350 "Their grounds cover fourteen acres": *Eagle*, August 8, 1876.

Page 351 Roebling Centennial display: *Ibid.;* photograph, RPI. The section of cable made up for the Centennial Exhibition is now on display at the Smithsonian Institution, in the Museum of History and Technology.

Page 351 Wire: *Specifications for Steel Cable Wire, for the East River Suspension Bridge—1876*, original copy, RPI; also LER.

Page 351 Machinery Hall: Brown, *The Year of the Century: 1876*, pp. 112–137.

Page 352 Starting of the Corliss engine: *Scientific American*, May 20, 1876.

Page 352 "It was a scene to be remembered": *Ibid.*

Page 352 "The engineer sits reading his newspaper": *The Atlantic Monthly*, July 1876.

Page 353 WAR to return to Brooklyn: WAR to William Paine, undated. RPI.

Page 354 "He is a man of great resource": WAR to HCM, May 6, 1876. RPI.

Page 354 Telegrams: Originals in scrapbook kept by EWR. RPI.

17 A Perfect Pandemonium

Page 355 The description of hanging the first rope is drawn from the following: *Eagle*, August 14 and 15, 1876; New York *Herald*, August 15, 1876; New York *Tribune*, August 15, 1876; *Scientific American*, September 2, 1876; *Van Nostrand's Eclectic Engineering Magazine*, October 1876; Farrington to WAR, December 30, 1876, LER; Farrington, *Concise Description of the East River Bridge*, pp. 28–30.

Page 357 "In a few seconds the rope began to move": Farrington, *Concise Description of the East River Bridge*, p. 30.

Page 358 "When it is considered that one has to climb": New York *Herald*, August 15, 1876.

Page 361 Farrington's ride: *Eagle*, August 25, 1876; New York *Herald*, New York *Tribune*, New York *Times*, Brooklyn *Argus*, August 26, 1876; Farrington to WAR, December 30, 1876, LER; Conant, "The Brooklyn Bridge."

Page 361 Ten thousand spectators: New York *Tribune*, August 26, 1876.

Page 364 "The ride gave me a magnificent view": Farrington, *Concise Description of the East River Bridge*, p. 36.

Page 365 Farrington complains of notoriety: Farrington to WAR, December 30, 1876. LER.

Page 365 "He does most of the brain work": Unidentified clipping in a scrapbook kept by EWR. RPI.

Page 366 High-wire acrobatics on Saturday, August 26: *Eagle*, same day; New York *Herald*, *Tribune*, *Sun*, and *World* for August 28.

Page 369 "Mr. Harry Supple was all that could be desired": William Paine to WAR, December 31, 1876. LER.

Page 369 Second day of acrobatics, Monday, August 28: *Eagle*, same day; New York *Herald*, *Tribune*, *Sun*, and *World* for August 29.

Page 371 "I have carried out your instructions": Farrington to WAR, December 30, 1876. LER.

18 Number 8, Birmingham Gauge

Page 372 Hewitt and Tilden: Nevins, *Abram S. Hewitt*, pp. 305–310.

Page 373 "Hewitt was as true a patriot": *Ibid.*, p. 317.

Page 373 "who played the game for ambition": Adams, *The Education of Henry Adams*, p. 373.

Page 373 "the best-equipped, the most active-minded": *Ibid.*, pp. 294–295.

Page 374 Hewitt resolution: Meeting of the Trustees, New York and Brooklyn Bridge, September 7, 1876, *New York and Brooklyn Bridge Proceedings, 1867–1884*, pp. 383–384; *Eagle*, September 7 and 8; New York *Tribune*, September 8, 1876.

Page 374 "I am very strongly opposed": *Tribune*, September 8, 1876.

Page 375 Tweed arrested in Spain: Werner, *Tammany Hall*, pp. 247–251.

Page 375 Letter of resignation: WAR to HCM, September 8, 1876. RPI.

Page 376 WAR to the *Eagle:* Undated. RPI.

Page 377 "I was publicly and specifically singled out": WAR to HCM, September 11, 1876. RPI.

Page 378 Haigh's matrimonial adventures: *Eagle*, January 6, 1880.

Page 379 WAR returns to New York City by barge: Unidentified clipping in a scrapbook kept by EWR. RPI.

Page 379 "There is something colossal in the look of the East River piers": *Ibid.*

Page 379 HCM notified of stock sale: WAR to HCM, November 2, 1876. RPI.

Page 379 Aspinwall proposal: *New York and Brooklyn Bridge Proceedings*, pp. 384–386; *Eagle*, November 14, 1876.

Page 380 Presidency stolen: Nevins, *Abram S. Hewitt*, p. 320 *ff*.

Page 380 Men to be trained: WAR to HCM, November 6, 1876. RPI.

Page 381 Oil kettles, sample ferrule, iron and steel rope: WAR to Farrington, November 16, 1876. RPI.

Page 381 "Man is after all a very finite being": WAR to James S. T. Stranahan, November 20, 1876. RPI.

Page 381 Technical instructions to Trenton: WAR to Ferdinand Roebling, undated. RPI.

Page 382 Requirements for Number 8, Birmingham Gauge: *Specifications for Steel Cable Wire, for the East River Suspension Bridge—1876*, original copy, RPI; also LER.

Page 382 Opening and contents of the bids: Meeting of the Trustees, November 4, 1876, *New York and Brooklyn Bridge Proceedings*, pp. 387–389; also, Meeting of the Executive Committee, December 6, 1876, *Proceedings*, pp. 643–645.

Page 383 Reporter sees Martin and HCM: New York *Herald*, December 16, 1876.

Page 383 "If one man's samples": WAR to HCM, December 15, 1876. LER.

Page 384 Hill's computations: New York *Herald*, December 16, 1876.

Page 385 Aspinwall and Kinsella comment: New York *Herald*, December 20, 1876.

Page 385 Hill's answer: New York *Herald*, December 21, 1876.

Page 386 Executive Committee Meeting of December 23, 1876: *New York and Brooklyn Bridge Proceedings*, pp. 645–646.

Page 386 WAR's report on tests: WAR to HCM, December 18, 1876. LER.

Page 387 Board of Directors' Meeting of December 28, 1876: *New York and Brooklyn Bridge Proceedings*, p. 389.

Page 387 "Unquestionably Bessemer steel wire is the cheapest": *Eagle*, January 10, 1877.

Page 388 Model of cable and Hildenbrand drawing: *Eagle*, December 26, 1876.

Page 388 Trustees' Meeting of January 11, 1877: *New York and Brooklyn Bridge Proceedings*, pp. 389–391.

Page 388 "The assurance of the correct performance": WAR, *Report of the Chief Engineer of the New York and Brooklyn Bridge, January 1, 1877*, p. 18. LER.

Page 388 Slocum requests Army engineers: *New York and Brooklyn Bridge Proceedings*, pp. 389–391.

Page 389 Hewitt letter: *Eagle*, January 12, 1877.

Page 390 Brooklyn Theater fire: New York *Times*, December 6, 1876.

Page 390 Ashtabula disaster: Gies, *Bridges and Men*, pp. 125–130. Footnote: *Ibid.*, p. 130.

Page 391 Trustees' response to Hewitt's letter: *New York and Brooklyn Bridge Proceedings*, pp. 389–391; *Eagle*, January 12, 1877.

Page 391 Kinsella's comments on decision: *Eagle*, January 16, 1877.

Page 393 "They can help us and the public": *Union*, January 16, 1877.

Page 394 "My attention has been called": *Eagle*, January 23, 1877.

Page 395 "It has become the deepest of mysteries": *Union*, January 18, 1877.

Page 396 "In laying this plan": WAR, private notes, undated. RPI.

19 The Gigantic Spinning Machine

Page 397 "I never saw better days for bridge work": *Eagle*, clipping in a scrapbook kept by EWR, no date. RPI.

Page 397 ". . . no man can be a bridge builder": Unidentified clipping, dated February 12, 1877, in a scrapbook kept by EWR. RPI.

Page 398 "The undulating of the bridge": New York *Tribune*, February 19, 1877.

Page 398 WAR's sign at the footbridge entrance: From a photograph.

Page 399 "Trinity Church steeple was fencing": New York *Tribune*, February 12, 1877.

Page 399 Farrington discloses imagined report of crossing (*fn.*): Farrington, *Concise Description of the East River Bridge*.

Page 399 "While Revs. Drs. Storrs and Buddington": *Eagle*, February 22, 1877.

Page 400 Lengthy descriptions of the wire spinning and of the array of apparatus involved were published in the *Eagle*, June 1 and July 7, 1877, and in *Appleton's Journal*, January 1878; "The Gigantic Spinning Machine": *Eagle*, July 6, 1877.

Page 405 Report of WAR's return to Brooklyn: *Eagle*, May 20, 1877.

Page 406 HCM and the footbridge craze: *New York Illustrated Times*, August 18, 1877.

Page 408 "I started to go once": *Ibid.*

Page 408 Seaman's epileptic fit: *The New York Times*, September 20, 1877.

Page 409 *Eagle*'s comments on suicide: October 19, 1877.

Page 409 "It is as brittle as glass": WAR to Paine, December 3, 1877. RPI.

Page 410 "This is what Mr. Kinsella is pleased to call the best": WAR to HCM, December 3, 1877. RPI.

Page 410 Kinsella says cost no issue: *Eagle*, December 4, 1877.

Page 410 "All of which is bosh": *Union and Argus*, December 4, 1877.

Page 411 Accident at the Brooklyn anchorage: *Eagle*, December 23, 1877; *Union and Argus*, December 24.

Page 412 "The brick arch fell because it had a right to fall": WAR to HCM, December 31, 1877. RPI.

Page 413 "There are so many points to be considered": WAR to Hildenbrand, January 9, 1878. RPI.

Page 414 "I want you to help me get out a specification": WAR to Farrington, February 9, 1878. RPI.

Page 415 January 8 meeting of the Executive Committee: *New York and Brooklyn Bridge Proceedings*, pp. 667–668.

Page 416 "Of course more or less legal information is required": *Union and Argus*, January 8, 1878.

20 Wire Fraud

Page 434 "Yet the existence of evil in human life": JAR, "Life and Creation," 1864. RUL.

Page 434 Storm of January 31, 1878: WAR, *Communication from Chief Engineer W. A. Roebling, In Regard to the Method of Steam Transit Over the East River Bridge*, p. 8, LER; *Eagle*, January 31, 1878.

Page 435 Murphy predicts 1880 completion: *Eagle*, February 5, 1878.

Page 435 WAR plans for bridge trains: *Eagle*, March 4, 1878.
Page 436 "An ingenious arrangement": WAR, *Communication from Chief Engineer W. A. Roebling*, p. 6, *fn.*
Page 436 "Neither, must we overlook the effect": *Ibid.*, p. 5.
Page 436 *Minnesota* clips a cable: *Eagle*, March 4, 1878.
Page 437 Death of Tweed: Werner, *Tammany Hall*, pp. 257–258.
Page 437 "He never thought of angels": *Ibid.*
Page 437 "If he had died in 1870": Callow, *The Tweed Ring*, p. 298.
Page 437 "Alas! Alas! young men": *Ibid.*, quoted, p. 297.
Page 438 "A villain of more brains": Quoted in Werner, *Tammany Hall*, p. 263.
Page 438 "Well, the Brooklyn people have no right": New York *Sun*, clipping in a scrapbook kept by EWR. RPI.
Page 440 Virtually every paper on both sides of the river carried a long account of the breaking of the cable. This description has been drawn chiefly from the following: Brooklyn *Union and Argus*, June 14 and 15, 1878; New York *Herald*, June 15, 1878; *Eagle*, June 14 and 15, 1878; New York *Times*, June 15 and 16, 1878; New York *World*, June 16.
Page 441 "It will not sway from side to side": New York *World*, June 30, 1878.
Page 442 HCM cuts back the work: *Eagle*, August 12, 1878.
Page 445 The exchange of letters between WAR and HCM concerning the Haigh wire deception is contained in "Exhibit No. 6," *New York and Brooklyn Bridge Proceedings*, pp. 132–138. They include: WAR to HCM, July 9 and 22, 1878; HCM to WAR, July 25, 1878; WAR to HCM, July 28 and August 6, 1878.
Page 446 Trustees' meeting of August 5, 1878: *New York and Brooklyn Bridge Proceedings*, p. 441.
Page 447 Trustees' meeting of August 7, 1878: *New York and Brooklyn Bridge Proceedings*, p. 441.
Page 447 WAR's private notes on Haigh: RPI.
Page 447 "We have brought machinery to a pitch": George, *Social Problems*, p. 19.
Page 448 "The thousands who daily cross": *Eagle*, August 8, 1878.
Page 449 "It has pleased the average penny-a-liner": EWR, unpublished biographical sketch of WAR. RPI.
Page 449 "Each must hang in its own peculiar length": *Appleton's Journal*, January 1878.
Page 450 Close call on the buggy: *Eagle*, January 5, 1879.
Page 451 Wrapping wire contract changed: Meeting of the Executive Committee, September 12, 1878, *New York and Brooklyn Bridge Proceedings*, p. 682.
Page 451 "The end, then, is near": *Eagle*, October 5, 1878.

21 Emily

Page 452 "At first I thought I would succumb": WAR, sometime in the spring of 1903. RUL.

Page 454 "Mrs. Roebling is a tall and handsome woman": Trenton *Gazette*, April 15, 1894.

Page 454 "I would send you a little tintype": WAR to Elvira Roebling, March 5, 1864. RUL.

Page 454 "You know, darling, that your presence": WAR to EWR, April 1, 1864. RUL.

Page 455 "This full moon evening": WAR to EWR, April 15, 1864. RUL.

Page 455 "After all, dear Emmie": WAR to EWR, April 4, 1864. RUL.

Page 455 "Look for a big thief next winter": WAR to EWR, July 4, 1864. RUL.

Page 455 "Does the *Mary Powell* run": WAR to EWR, August 1, 1864. RUL.

Page 456 "Your letter describing the visit": WAR to EWR, June 19, 1864. RUL.

Page 456 Ferdinand's reaction to EWR: WAR to EWR, September 1, 1864. RUL.

Page 456 "When the two hopefuls": WAR to EWR, September 25, 1864. RUL.

Page 456 "I still entertain a lively remembrance": *Ibid.*

Page 457 "aspired to no higher distinction": Blake, *History of Putnam County, New York.*

Page 457 Parrott guns bombard Storm King: Pelletreau, *History of Putnam County, New York.*

Page 458 Career of G. K. Warren prior to the Civil War: Taylor, *Gouverneur Kemble Warren.*

Page 458 The distressing thing about Indian fighting: Catton, *A Stillness at Appomattox*, pp. 51–52.

Page 458 *Picnic on the Hudson*: The painting hangs in the Julia Butterfield Memorial Library, Cold Spring, New York.

Page 459 "I think we will be a pair of lovers": WAR to EWR, November 18, 1864. RUL.

Page 460 Warren and Sheridan at Five Forks: Catton, *A Stillness at Appomattox*, pp. 348–357.

Page 460 WAR's view of Five Forks: WAR to James Rusling, February 18, 1916. RUL.

Page 461 *Effie Afton* case: Gies, *Bridges and Men*, p. 151; Sandburg, *Abraham Lincoln*, pp. 124–125.

Page 461 "I have heard men like Humphreys": WAR, private memorandum, written sometime in 1914. RUL.

Page 463 "It is whispered among the knowing ones": New York *Star*, December 17, 1879.

Page 463 Edge Moor Iron official writes directly to EWR: W. H. Francis to WAR, October 28, 1879. The letter is contained in the scrapbook kept by EWR from May 1878 to October 1882. RPI.

Page 465 Secretary of State Evarts retained as council: *Eagle*, January 16, 1879.

Page 465 Decision of Supreme Court of New York: *Ibid.*

Page 465 Decision of Court of Appeals: *Eagle*, March 25, 1879.

Page 465 Miller suit: New York *Herald* and *Eagle*, February 28, 1879;

Eagle, March 7; New York *Herald* and *Eagle*, March 31; *Eagle*, March 23, 1879; *Testimony in the Miller Suit to Remove the East River Bridge.*

Page 466 Slocum charges that the engineers are taking bribes: *Union and Argus*, May 3, 1879; New York *World* and New York *Sun*, May 4, 1879.

Page 466 "And I want to say right here": *Eagle*, May 6, 1879.

Page 467 Davidson and Ferdinand Roebling testify: New York *Sun*, New York *Star*, New York *World*, May 7, 1879; *Union and Argus* and *Eagle*, May 8, 1879.

Page 467 "I hope I have heard for the last time": WAR to Slocum, May 6, 1879. RPI.

Page 468 Engineers exonerated: *Eagle*, New York *Sun*, New York *Herald*, May 28, 1879.

Page 468 Kinsella declines to serve again, new faces on the board: *Eagle* and *Union and Argus*, June 9, 1879.

Page 469 Steinmetz attacks Kingsley: *Eagle*, June 10, 1879.

Page 469 Murphy appears to be out: *Ibid.*

Page 469 Murphy in again: *Eagle*, June 25, 1879.

Page 470 Tay Bridge disaster: Gies, *Bridges and Men*, pp. 134–146.

Page 470 "WILL THE TAY DISASTER BE REPEATED": New York *Herald*, January 11, 1880.

Page 471 De Lesseps illustration: EWR's scrapbook for May 1878 to October 1882, RPI; the illustration is from New York *Daily Graphic*, February 28, 1880.

Page 471 *Frank Leslie's Illustrated Newspaper's* view of cable work: November 15, 1879.

Page 472 J. Lloyd Haigh at Sing Sing: Unidentified clipping in EWR's scrapbook. RPI.

Page 473 Kingsley-Steinmetz scene: The incident was widely reported and the dialogue differs somewhat from one account to another. This version is a composite from what appeared in the *Eagle*, the New York *Star*, the *Sun*, the *Herald* and the *Tribune* on October 12, 1880, all of which were carefully entered in EWR's scrapbook.

Page 475 RPI alumni dinner: *Eagle*, February 19, 1881; *Engineering News*, February 26, 1881.

Page 475 "The men who have come from the Institute": *Engineering News*, February 26, 1881.

Page 475 Rossiter Raymond at the RPI dinner for 1882 (*fn.*): Unidentified clippings. RPI.

Page 477 There is no known description of the view from WAR's window written at the time. This one has been derived from contemporary photographs of New York taken from the Brooklyn side of the river.

Page 478 EWR leads the first walk over the bridge: New York *Star*, *Eagle*, and the *Union and Argus*, December 13, 1881.

22 The Man in the Window

Page 479 "The best way to secure rapid and effective work": New York *Star*, August 23, 1882.

Page 479 Trustees' meeting of December 12, 1881: *New York and Brooklyn Bridge Proceedings, 1867–1884*, pp. 461–462; also New York *Star*, Brooklyn *Union and Argus*, and the *Eagle*, for December 13, 1881.

Page 479 Stranahan's customary method: *Eagle*, December 13, 1881.

Page 481 Total expenditures January 1, 1882; also HCM's estimate: *Eagle*, January 10, 1882.

Page 481 Meeting of the trustees, October 13, 1881: *New York and Brooklyn Bridge Proceedings*, pp. 457–458.

Page 481 "When I consented to make this change": WAR to HCM, January 9, 1882, p. 11. LER.

Page 482 Cables could uproot the anchorages: *Ibid.*, p. 12.

Page 483 Seth Low at his first trustees' meeting: *Eagle*, January 10, 1882.

Page 483 "the first scholar in college": *Dictionary of American Biography.*

Page 483 Low's campaign for mayor: Syrett, *The City of Brooklyn, 1865–1898.*

Page 484 Meeting of the trustees, June 12, 1882: *New York and Brooklyn Bridge Proceedings*, p. 468.

Page 484 Robert Roosevelt's letter to Mayor Grace: New York *Herald*, June 14, 1882.

Page 484 HCM talks to the press: New York *Sun*, June 16, 1882.

Page 485 "His plans and diagrams are all about him": *Ibid.*

Page 485 Sellers of Edge Moor ridiculed: *Eagle*, New York *Sun*, New York *Herald*, June 27, 1882.

Page 485 Meeting of the trustees, June 26, 1882: *New York and Brooklyn Bridge Proceedings*, pp. 469–470.

Page 486 Slocum's remarks: *Eagle*, New York *Sun*, June 27, 1882.

Page 486 "unsubstantial fabric of a dream": Quoted in Syrett, *The City of Brooklyn*, p. 153.

Page 486 The best roundup of rumors concerning the health and mental decline of WAR: New York *Sun*, July 31, 1882.

Page 486 WAR's letter of explanation: *New York and Brooklyn Bridge Proceedings*, pp. 473–474.

Page 487 WAR claims he is powerless to push Sellers: WAR to HCM, July 19, 1882, RPI; also New York *Sun*, August 17, 1882.

Page 487 "Newport has never looked more attractive": *Eagle*, July 3, 1882.

Page 487 WAR's "cottage" at Newport: The house still stands; it is now a Catholic convalescent home and is located, ironically, beside the Newport end of the gigantic new suspension bridge over Narragansett Bay.

Page 489 WAR will not "dance attendance on the Trustees": Draft of a long letter to Comptroller Campbell, undated. RPI. It is not known whether the letter was sent.

Page 489 WAR will not be "dragged into the board and put on exhibition": Draft of a letter to the New York *Sun*, probably written in July 1882; probably never sent. RPI.

Page 489 "no less than one hundred and twenty politicians": *Ibid.*

Page 489 "This is the same General Slocum": *Ibid.*

Page 490 Kingsley overpaid by $175,000: *Ibid.*

Page 490 "I have always had bitter enemies in the Board": Draft of letter, WAR to Comptroller Campbell, undated. RPI.

Page 490 "I have over and over again been interviewed": *Ibid.*

Page 490 Low reported to be out of town briefly: New York *Herald*, August 3, 1882.

Page 491 "Mr. Roebling, I am going to remove you because it pleases me": WAR, undated notes, written sometime in late August 1882. RPI.

Page 491 Death of G. K. Warren and decision of the military court: Taylor, *Gouverneur Kemble Warren.*

Page 491 "Please make it convenient to be present": *Eagle*, August 17, 1882.

Page 491 Trustees' meeting of August 22: *Eagle*, New York *Sun*, New York *Star*, New York *Evening Post*, New York *World*, New York *Herald*, New York *Tribune*, August 23, 1882.

Page 491 Low's comments on WAR at the meeting of the trustees, August 22, 1882: *Eagle*, same date; New York *World*, New York *Sun*, New York *Star*, New York *Herald*, August 23.

Page 492 "WHEREAS, The Chief Engineer of this Bridge": *New York and Brooklyn Bridge Proceedings*, pp. 477–478.

Page 493 Editorial comments in New York papers: All for issues of August 23, 1882.

Page 494 *Iron Age* comment: Issue of August 31, 1882; Newport *Daily News:* August 24, 1883; Trenton *Daily State Gazette* and the *Eagle:* August 23, 1882.

Page 496 "I take the liberty of writing to express to you my heartfelt gratitude": EWR to Ludwig Semler, undated. RPI.

Page 496 "Nobody should be convicted before he is tried": *Eagle*, September 5, 1882.

Page 497 WAR would as soon "be out of the bridge" if Kingsley is to decide his fate: Letter to William Paine, September 10, 1882. RPI.

Page 497 Semler reports on visit to WAR: *Eagle*, September 7, 1882.

Page 498 Visit of the *World* reporter to Newport: Described by EWR in her letter to William Marshall, undated. RPI.

Page 499 Meeting of the trustees, September 11, 1882: The entire session was heavily reported by all of the following, from which this account has been drawn, New York *Evening Post*, New York *Sun*, New York *Times*, New York *Star*, New York *Herald*, New York *World*, New York *Tribune*, Brooklyn *Union and Argus*, and the *Eagle*, all for September 12, 1882; also *New York and Brooklyn Bridge Proceedings*, pp. 478–481.

Page 504 Low tells reporters he is pleased with outcome: *Eagle*, September 12, 1882.

Page 504 "I actually believe that all that ails him": *Ibid.*

23 And Yet the Bridge Is Beautiful

Page 505 "And yet the bridge is beautiful in itself": *Scientific American*, September 22, 1883.

Page 506 The *Times* on Mrs. Vanderbilt's party: March 27, 1883.

Page 507 The *World* on the "Bridge Frauds": The first of a long series of articles appeared on September 18, 1882, under a headline, "THE BRIDGE RING, OVERWHELMING PROOFS OF SYSTEMATIC JOBBERY AND OFFICIAL CORRUPTION."

Page 508 Kinsella interview: *World*, September 19, 1882.

Page 508 The seating of the Tammany delegates: The best account of the Syracuse convention is in Nevins, *Grover Cleveland; A Study in Courage*.

Page 508 The *Times* pinpoints Slocum's association with the bridge as the chief cause of his failure to get the nomination: September 22, 1882.

Page 509 Mayors Low and Grace appoint accountants to examine the Bridge Company's books: "Report of the Committee Appointed by the Board of Trustees," *New York and Brooklyn Bridge Proceedings, 1867–1884*, pp. 1–6.

Page 509 Report of the accountants: *Ibid.*, pp. 7–64.

Page 510 The bridge as a memorial to HCM: Special meeting of the trustees, December 2, 1882, *New York and Brooklyn Bridge Proceedings*, pp. 483–484.

Page 512 "What a relief it will be": Barnard, "The Brooklyn Bridge."

Page 513 *Scientific American* editor describes the bridge as seen from the river: Issue of September 22, 1883.

Page 515 United States Illuminating Company gets contract for arc lamps: Meeting of the Executive Committee, February 12, 1883, *New York and Brooklyn Bridge Proceedings*, pp. 745–747; "Report of the Committee on Lighting the Bridge, April 9, 1883," *Proceedings*, pp. 161–164.

Page 516 "The scene suggested the subterranean laboratory of a magician": Quoted in Frederick L. Collins, *Consolidated Gas Company of New York*, 1934, pp. 268–270.

Page 516 EWR explains how superstructure should be made: *Times*, May 23, 1883.

Page 517 EWR's ride over the bridge: Only passing mention of the event was made in the papers and then weeks after it happened (*Times*, May 23, 1883).

Page 517 The fact that the rooster went along turns up only in the Trenton papers many years later, when the bird, stuffed and mounted, was a conversation piece in the Roebling home.

Page 517 The interview with the *Union* reporter appeared May 16, 1883.

Page 519 Full accounts of the preparations for "The People's Day" appeared in just about every paper, off and on, throughout the preceding week.

Page 521 The *World* now favors the bridge: Swanberg, *Pulitzer*, p. 74.

Page 521 WAR's concern about fireworks display: Letter to Stranahan, dated May 5, 1883. RPI.

Page 522 Hewitt writes for wages of bridge workers, etc.: May 3, 1883, RPI; WAR's answer to Hewitt, RPI.

Page 523 "I wish you would make one of my party of ladies": EWR to Mrs. William G. Wilson, May 17, 1883. New York Historical Society.

Page 524 Hewitt letter of May 18, 1883: RPI.

24 The People's Day

Page 525 Every newspaper on both sides of the East River went to great lengths to describe the opening of the Great Bridge. The sources used most here were the *Eagle*, the *Union*, New York *Times*, *Sun*, *World*, and *Tribune*, all for May 25, 1883.

Page 525 Estimates on crowds pouring into New York: *Times*.

Page 526 "One moment they were clambering clumsily": *Sun*.

Page 529 "It was as if the forest of masts had blossomed": *Tribune*.

Page 531 "The women in the crowd raised their hands": *Sun*.

Page 532 Arthur trods the bridge "with an elastic step": *Sun*.

Page 532 Arthur "an Apollo in form": *Ibid*.

Page 535 Large portions of the Opening Day addresses were carried in the papers of May 25, but the complete text is contained in a commemorative book, *Opening Ceremonies of the New York and Brooklyn Bridge*, published by the Bridge Company.

Page 537 WAR's day and the reception at 110 Columbia Heights were also covered in most newspaper accounts of the day's events. See in particular the *Tribune* and the *Sun*.

Page 538 "As the sun went down the scene from the bridge was beautiful": *Sun*.

Page 541 "Why I thought Brooklyn had one hotel": *Ibid*.

Page 542 The most interesting account of the first crowds to cross the bridge was provided by the *Times*.

Epilogue

Page 543 Attendance figures for first three days: New York *Tribune*, May 26, 28, 1883.

Page 543 The Memorial Day tragedy: New York *Times* and *Tribune*, June 1, 1883.

Page 544 "That was my first view of a great calamity": Josephson, *Al Smith, Hero of the Cities*, p. 24.

Page 545 Martin's force: Martin, *Report of the Chief Engineer and Superintendent of the New York and Brooklyn Bridge, June 1, 1884*. LER.

Page 546 Barnum's elephants: *Times*, May 17, 1884.

Page 546 Brodie: *Times*, July 23, 24, 1886; Botkin, *New York City Folklore*, pp. 218–223.

Page 547 Other jumpers: "This Alluring Roadway," *The New Yorker*, May 17, 1952.

Page 549 Montgomery Schuyler's article appeared in *Harper's Weekly*, the issue of May 26, 1883; also included in Mumford, *Roots of American Architecture*, pp. 159–168.

Page 549 James: See his *American Scene;* for the best analysis of how he and other American intellectuals and artists responded to the bridge, see Trachtenberg, *Brooklyn Bridge; Fact and Symbol*, pp. 129–165.

Page 550 "The stone plays against the steel" plus other Mumford comments on the bridge: *Sticks and Stones*, pp. 114–117; *The Brown Decades*, pp. 96–106.

Page 550 "not so much linking places as leaving them": Scully, *Modern Architecture*, p. 17.

Page 550 The prominent American architect is Philip Johnson.

Page 552 All that was needed was a new coat of paint: "This Alluring Roadway," *The New Yorker*, May 17, 1952.

Page 552 Roebling's pleasure over Low's defeat: WAR to JAR II, November 3–6, 1903. RUL.

Page 553 "So far as I know not a dollar was stolen": WAR to James Rusling, January 23, 1916. RUL.

Page 553 Hildenbrand's post-Brooklyn career: ASCE, *Transactions*, Vol. 77, 1914.

Page 553 "Soon I will be the last leaf on the tree": WAR to JAR II, April 26, 1909. RUL.

Page 554 EWR at the coronation of Tsar Nicholas: WAR to JAR II, May 24, 1896. RUL.

Page 555 WAR blocks sale to U.S. Steel: *Ibid.*, p. 352.

Page 555 "He was now adrift": WAR in a private memorandum dated March 16, 1922. RUL.

Page 555 The founding of Roebling, New Jersey: Schuyler, *The Roeblings*, pp. 359–361, 367–373.

Page 556 Size of WAR's estate: Schuyler, *The Roeblings*, p. 278.

Page 557 ". . . these relationships are those of the heart": WAR to JAR II, March 21, 1908. RUL.

Page 557 WAR becomes "almost jovial": Schuyler, *The Roeblings*, p. 269.

Page 557 "And yet people say how well you look": WAR to JAR II, October 19, 1893. RUL.

Page 557 *Times* letter: RUL.

Page 557 "It means 100,000 spies": WAR to JAR II, September 15, 1913. RUL.

Page 557 "It has come to this pass": WAR to JAR II, October 9, 1914. RUL.

Page 557 "War in the kitchen": WAR to JAR II, August 11, 1916. RUL.

Page 557 WAR's "oddities": Schuyler, *The Roeblings*, pp. 263, 274.

Page 558 "a nice, courteous old gentleman": Author's interview with W. H. Pearson, formerly of Trenton.

Page 558 WAR known as a soft touch: Schuyler, *The Roeblings*, p. 262.

Page 558 "Billy Sunday": *Ibid.*, p. 274.

Page 559 "I claim a small part of this": WAR to JAR II, June 10, 1922. RUL.

Page 560 "It's my job to carry the responsibility": New York *World* interview quoted in the Trenton *Times*, June 13, 1921.

Page 560 "Think not that I am improving": WAR to Mrs. JAR II, May 14, 1926. RUL.

Page 562 "As far as we are concerned, it will last forever": Jack Schiff, city engineer in charge of all East River bridges, in an interview with the author, March 16, 1971.

Picture Credits

The Bettmann Archive, Inc.—pp. 426, 427, 428
Brown Bros.—pp. 229 (right), 230
Culver Pictures, Inc.—pp. 425 (top), 429 (top), 430–31
Mrs. James L. Elston—p. 215
Harper's New Monthly Magazine—pp. 220, 221 (bottom), 417, 418
Harper's Weekly—pp. 226–27, 419, 422
Library of Congress—endpapers
Museum of the City of New York—pp. 221 (top), 222 (top), 224–25, 421 (bottom), 424, 429 (bottom), 432–33
New-York Historical Society—pp. 420, 423, 425 (bottom)
New York *Illustrated Times*—p. 421 (top)
New York Public Library—pp. 222 (bottom), 223, 228, 229 (three engravings at left)
Rensselaer Polytechnic Institute—p. 218
Rutgers University—pp. 216, 219
Smithsonian Institution, Division of Mechanical and Civil Engineering—p. 217

Bibliography

Manuscript Sources

There are two collections of Roebling manuscript papers: the Roebling Collections in the Library of Rensselaer Polytechnic Institute, Troy, New York, and in the Special Collections of the Library of Rutgers University, New Brunswick, New Jersey. Both are of vast scope and value and have been almost totally ignored by all but two or three scholars.

The latter collection contains numerous notebooks, ledgers, diaries, and other documents belonging to John A. Roebling, in addition to his philosophical papers, patents, numerous drawings, sketches, and the early papers and records of John A. Roebling's Sons. But the most important part of the collection, so far as the telling of this story, is the file of Washington Roebling's correspondence. The letters cover a span of nearly seventy years and include, for example, all of his war letters to Emily, plus those written to his son in the years after the completion of the bridge. This correspondence has been carefully arranged by the late Clarence E. Case, a prominent New Jersey attorney and friend of the Roebling family. Like everything else in the collection the letters are readily accessible. Interested scholars ought to be warned, however, that in editing the letters for a typed transcription, Mr. Case cut a great deal that he considered of too personal or too technical a nature.

The RPI collection is the larger of the two and contains far more concerning the Brooklyn Bridge. It includes hundreds of letters, notebooks, reports, cashbooks, and personal memorandums relating to the careers of both John A. and Washington Roebling. It includes drawings of all of John A. Roebling's bridges, his various preparatory schemes for the towers of the Brooklyn Bridge, numerous plans and blueprints of the bridge, photographs, and his private library. It includes the notes he kept during the spiritualist séances of 1867, Washington Roebling's letters from Europe that same year, and two large scrapbooks kept by Emily Roebling from April 1876 to October 1882. Most important of all, it includes the letter books and private notes kept by Washington and Emily Roebling during the years the bridge was being built. Recently the entire collection was classified and catalogued for the first time by Robert M. Vogel of the Smithsonian Institution, with a grant from the American Society of Civil Engineers.

Newspapers, Magazines, and Technical Journals

Use was made of numerous newspapers, magazines, and technical journals. Many of these were in the form of clippings included in the scrapbooks kept by Emily Roebling; the rest were consulted in various libraries. Of the papers consulted the most valuable by far was the Brooklyn *Eagle*.

Newspapers: Boston *Post*, Brooklyn *Argus*, Brooklyn *Eagle*, Brooklyn *Leader*, Brooklyn *Union*, Brooklyn *Union and Argus*, Cincinnati *Daily Gazette*, Cold Spring *Recorder*, Coney Island *Sun*, Long Island *Star*, New York *Commercial Advertiser*, New York *Daily Graphic*, New York *Daily Witness*, New York *Evening Express*, New York *Evening Mail*, New York *Evening Post*, New York *Evening Telegram*, New York *Herald*, New York *Independent*, New York *Mail and Express*, New York *Mercury*, New York *Star*, New York *Sun*, New York *Times*, New York *Tribune*, New York *World*, Newport *Daily News*, Niagara Falls *Gazette*, Pittsburgh *Gazette*, Trenton *Daily State Gazette*, Troy *Record*.

Magazines and technical journals: *American Heritage*, *American Railroad Journal*, *Appleton's Journal*, *Architects and Mechanics' Journal*, *Beecher's Magazine*, *Brooklyn Monthly*, *Civil Engineering*, *Engineering* (London), *Engineering News*, *Frank Leslie's Illustrated Newspaper*, *Harper's New Monthly Magazine*, *Harper's Weekly*, *The Iron Age*, *Journal of the Franklin Institute*, *Mechanics* (New York), *The New Yorker*, *Puck*, *The Railroad Gazette*, *St. Nicholas Magazine*, *Scientific American*, *Transactions* (American Society of Civil Engineers), *Van Nostrand's Eclectic Engineering Magazine*, *Western Pennsylvania Historical Magazine*.

Works Relating Directly to the Brooklyn Bridge

Barnard, Charles, "The Brooklyn Bridge." *St. Nicholas Magazine*, July 1883.

Barnes, A. C., *The New York and Brooklyn Bridge*. (Pamphlet) Brooklyn, 1883.

Brooklyn Bridge: 1883–1933. Published by the City of New York Department of Plant and Structures, 1933.

Conant, William C., "The Brooklyn Bridge." *Harper's New Monthly Magazine*, May 1883.

East River Bridge, Laws and Engineer's Reports, 1868–1884. Brooklyn, 1885. (This very rare volume contains all of the following, most of which were published separately during the time the bridge was being built.)

An Act to amend an act entitled "An Act to incorporate the New York Bridge Company, for the purpose of constructing and maintaining a bridge over the East River, between the cities of New York and Brooklyn," passed April sixteenth, eighteen hundred and sixty-seven, and to provide for the speedy construction of said bridge. Chapter 601. Passed June 5, 1874.

An Act to establish a bridge across the East River, between the cities of Brooklyn and New York, in the State of New York, a post road. Public, No. 53. Approved by Congress March 3, 1869.

An Act to incorporate the New York Bridge Company, for the purpose of constructing and maintaining a bridge over the East River, between the cities of New York and Brooklyn. Chapter 399. Passed April 16, 1867.

An Act providing that the bridge in the course of construction over the East River, between the cities of New York and Brooklyn, by the New York Bridge Company, shall be a public work of the cities of New York and Brooklyn, and for the dissolution of said Company, and the completion and management of the said bridge by the said cities. Chapter 300. Passed May 14, 1875.

Collingwood, Francis, *A Few Facts about the Caissons of the East River Bridge.* Paper read at the third annual convention of the American Society of Civil Engineers, June 21, 1871; printed originally in ASCE *Transactions;* also *Engineering* (London), February 16 and 23, 1872.

——— *The Foundations for the Brooklyn Anchorage of the East River Bridge.* Paper read before the American Society of Civil Engineers, June 10, 1874; in *Transactions.*

——— *Further Notes on the Caissons of the East River Bridge.* Paper read at the fourth annual convention of the American Society of Civil Engineers, June 5–6, 1872; in *Transactions;* also *Engineering* (London), October 18 and 25, 1872.

——— *Notes on the Masonry of the East River Bridge.* Paper read before the American Society of Civil Engineers, November 1, 1876; in *Transactions.*

——— *Progress of Work at the East River Bridge.* Paper read before the American Society of Civil Engineers, June 17, 1879; in *Transactions.*

Kingsley, William C., *First Annual Report of the General Superintendent of the East River Bridge.* Eagle Book and Job Printing Department, Brooklyn, 1870.

——— *Report of the General Superintendent, New York Bridge Company.* Eagle Book and Job Printing Department, Brooklyn, 1871.

——— *Report of the General Superintendent of the New York Bridge Company*. Eagle Book and Job Printing Department, Brooklyn, 1872.

Martin, C. C., *Report of the Chief Engineer and Superintendent of the New York and Brooklyn Bridge, June 1, 1884*. Eagle Book and Job Printing Department, Brooklyn, 1884.

Report of the Executive Committee of the New York Bridge Company, June 1, 1872. Eagle Book and Job Printing Department, Brooklyn, 1872.

Report of the Officers of the New York Bridge Company to the Board of Directors, February, 1875. Eagle Print, Brooklyn, 1875.

Reports of Assistant Engineers and Master Mechanic, 1875–1876.

Roebling, John A., *Report of John A. Roebling, C.E., to the President and Directors of the New York Bridge Company, on the Proposed East River Bridge*. Eagle Book and Job Printing Department, Brooklyn, 1870.

Roebling, Washington A., *Communication from Chief Engineer W. A. Roebling, In Regard to the Method of Steam Transit Over the East River Bridge*. March 4, 1878.

——— *First Annual Report of the Chief Engineer of The East River Bridge*. Eagle Book and Job Printing Department, Brooklyn, 1870.

——— *Pneumatic Tower Foundations of the East River Suspension Bridge*. Eagle Book and Job Printing Department, Brooklyn, 1872.

——— *Report of the Chief Engineer to the Board of Directors of the New York Bridge Company, June 5, 1871*. Eagle Book and Job Printing Department, Brooklyn, 1871.

——— *Report of the Chief Engineer of the East River Bridge on Prices of Materials and Estimated Cost of the Structure*. Eagle Book and Job Printing Department, June 28, 1872.

——— *Report of the Chief Engineer of the New York Bridge Company, 1874*. Eagle Book and Job Printing Department, Brooklyn, 1874.

——— *Report of the Chief Engineer of the New York and Brooklyn Bridge, January 1, 1877*. Eagle Print, Brooklyn, 1877.

——— *Report of the Chief Engineer on the Strength of the Cables and Suspended Superstructure of the Bridge, Made to the Board of Trustees, January 9, 1882*.

——— *Report of the Chief Engineer on the Tests of the Samples of Wire, 1876.*

——— *Third Annual Report of the Chief Engineer, June 1, 1872.* Eagle Book and Job Printing Department, Brooklyn, 1872.

Smith, Andrew H., M.D., *The Effects of High Atmospheric Pressure, Including the Caisson Disease.* Eagle Book and Job Printing Department, Brooklyn, 1873.

Specifications for Anchor Plates, New York Anchorage, East River Bridge, 1875.

Specifications for Corners, Facing and Archstone, of Granite, Required for the New York Anchorage, East River Bridge, 1875.

Specifications for Cut Face-stone, Backing and Archstone of Limestone, Required for the New York Anchorage, East River Bridge, 1875.

Specifications for Cut Face-stone and Backing, Limestone and Granite, Required for the New York Anchorage, East River Bridge, 1875.

Specifications for Granite Cut Stone, Required for the Parapets at the roadway, Brooklyn and New York Towers, East River Bridge, 1876.

Specifications for Granite Face-stone and Archstone, Required for the New York Tower, East River Bridge, April, 1875.

Specifications for Iron Anchor Bars, New York Anchorage, East River Bridge, April, 1875.

Specifications for Saddles and Saddle-Plates for the Brooklyn and New York Towers, East River Bridge, 1874.

Specifications for Steel Cable Wire, for the East River Suspension Bridge—1876.

Specifications for Wire Ropes for the East River Bridge.

Farrington, E. F., *Concise Description of the East River Bridge, with Full Details of the Construction.* (Pamphlet) C. D. Wynkoop, New York, 1881. Republished in 1969 by Boro Book Store, Brooklyn.

Green, S. W., *A Complete History of the New York and Brooklyn Bridge from its Conception in 1866 to its Completion in 1883, with Sketches of the Lives of J. A. Roebling, W. A. Roebling, and H. C. Murphy.* (Pamphlet) New York, 1883.

An Illustrated Description of the New York and Brooklyn Bridge. (Pamphlet) Published by John A. Roebling's Sons, Trenton, no date.

New York and Brooklyn Bridge Proceedings, 1867–1884. Brooklyn, 1885. (This thick invaluable volume contains the only records kept of the meetings of the directors—later the trustees—of the Bridge Company, the meetings of the Executive Committee, lists of the stockholders at various periods during the construction of the bridge, lists of real estate purchased, plus all other reports, letters, etc., emanating from the Bridge Company during the time from May 1867 to June 1884.)

Opening Ceremonies of the New York and Brooklyn Bridge. Brooklyn, 1883.

Report of the Board of Consulting Engineers to the Directors of the New York Bridge Company. The Standard Press Print, Brooklyn, 1869.

Schuyler, Montgomery, "The Bridge as a Monument." *Harper's Weekly,* May 24, 1883. Also included in Lewis Mumford's *Roots of Contemporary American Architecture,* Reinhold Publishing Corporation, New York, 1952; paperback edition, Grove Press, New York, 1959.

Testimony in the Miller Suit to Remove the East River Bridge. Albany, 1879.

Trachtenberg, Alan, *Brooklyn Bridge; Fact and Symbol.* (An excellent analysis of the bridge as a cultural symbol in America.) Oxford University Press, New York, 1965.

Works Relating to John A. and Washington A. Roebling

There is no first-rate biography of either John A. or Washington A. Roebling. The closest thing to it and the one reliable source of family history is *The Roeblings; A Century of Engineers, Bridgebuilders and Industrialists* by Hamilton Schuyler, Princeton University Press, 1931. Other works consulted were these:

Farrington, E. F., *A Full and Complete Description of the Covington and Cincinnati Suspension Bridge with Dimensions and Details of Construction.* Cincinnati, 1867.

John A. Roebling. An Account of the Ceremonies at the Unveiling of a Monument to His Memory; Address by H. Estabrook. Roebling Press, Trenton, 1908.

Roebling, John A., *Diary of My Journey from Muehlhausen in Thuringia via Bremen to the United States of North America in the Year 1831,* trans. by Edward Underwood. Roebling Press, Trenton, 1931.

——— *Final Report of John A. Roebling, Civil Engineer, to the President and Directors of the Niagara Falls Suspension and Niagara Falls International Bridge Companies.* Rochester, N. Y., 1855

——— "The Great Central Railroad from Philadelphia to St. Louis." *American Railroad Journal,* Special Edition, 1847.

——— "Letters to Ferdinand Baehr, 1831." *Western Pennsylvania Historical Magazine,* June 1935.

——— *Long and Short Span Railway Bridges.* (Includes an introduction by Washington Roebling.) D. Van Nostrand, New York, 1869.

——— *Report to the President and Board of Directors of the Covington and Cincinnati Bridge Company.* Trenton, 1867.

Roebling, Washington A., *Early History of Saxonburg.* Trenton, 1924.

Steinman, D. B., *The Builders of the Bridge.* Harcourt, Brace and Company, New York, 1945. (The author was a famous bridgebuilder himself and was long considered the authority on John A. Roebling. His book, however, was based on superficial research and contains many inaccuracies.)

Vogel, Robert M., *Roebling's Delaware and Hudson Canal Aqueducts.* Smithsonian Institution Press, Washington, D. C., 1971.

Books on Bridges and Bridgebuilders

Dorsey, Florence L., *Road to the Sea.* Rinehart & Company, New York, 1947.

Gies, Joseph, *Bridges and Men.* Doubleday and Company, Garden City, N. Y., 1963.

Gilbert, Ralph W., Jr., and Billington, David P., "The Eads Bridge and Nineteenth-Century River Politics." Paper read at the First National Conference on Civil Engineering: History, Heritage and the Humanities, Princeton University, 1970.

Jacobs, David, and Neville, Anthony E., *Bridges, Canals and Tunnels.* American Heritage Publishing Company, New York, 1968.

Jakkula, A. A., *A History of Suspension Bridges in Bibliographical Form.* Agricultural and Mechanical College of Texas, College Station, Texas, 1941.

Kirby, Richard Shelton, and Laurson, Philip Gustave, *The Early Years of Modern Civil Engineering.* Yale University Press, New Haven, 1932.

Pope, Thomas, *A Treatise on Bridge Architecture, in which the Superior Advantages of the Flying Pendent Lever Bridge are Fully Proved.* New York, 1811.

Steinman, David B., and Watson, Sara Ruth, *Bridges and Their Builders*, 2nd rev. ed. Dover Publications, New York, 1957.

Stuart, C. B., *Lives and Works of Civil and Military Engineers of America.* D. Van Nostrand, New York, 1871.

Vose, George L., *Bridge Disasters in America.* Boston, 1887.

White, Joseph, and Von Bernewitz, M. W., *The Bridges of Pittsburgh.* Cramer Printing and Publishing Company, Pittsburgh, 1928.

Woodward, C. M., *A History of the St. Louis Bridge.* G. I. Jones and Company, St. Louis, 1881.

General Works

An Account of the Dinner by the Hamilton Club to Honor James S. T. Stranahan, December 13, 1888. Brooklyn, 1889. Collection of the Long Island Historical Society.

Adams, Henry, *The Education of Henry Adams.* Houghton Mifflin Company, Sentry Edition, Boston, 1961.

Albion, Robert Greenhalgh, *The Rise of New York Port (1815–1860).* Charles Scribner's Sons, New York, 1939.

Andrist, Ralph K., ed., *The Confident Years.* American Heritage Publishing Company, New York, 1969.

Blake, William, *History of Putnam County, New York.* New York, 1849.

Botkin, B. A., ed., *New York City Folklore.* Random House, New York, 1956.

The Brooklyn City and Business Directory, 1869, 1870.

Brown, Dee, *The Year of the Century: 1876.* Charles Scribner's Sons, New York, 1966.

Bryce, James, *The American Commonwealth.* Macmillan and Company, New York and London, 1895.

Buck, Solon J. and Elizabeth Hawthorn, *The Planting of Civilization in Western Pennsylvania.* University of Pittsburgh Press, 1939.

Butterfield, Roger, *The American Past.* Simon and Schuster, New York, 1966.

Callender, James H., *Yesterdays on Brooklyn Heights.* The Dorland Press, New York, 1927.

Callow, Alexander B., Jr., *The Tweed Ring.* Oxford University Press, New York, 1966.

Carnegie, Andrew, *Autobiography.* Houghton Mifflin Company, Boston, 1920.

Catton, Bruce, *Grant Moves South.* Little, Brown and Company, Boston, 1960.

——— *A Stillness at Appomattox.* Doubleday and Company, Garden City, N. Y., 1954.

The City of Brooklyn. (A guidebook, probably written by Henry R. Stiles.) New York, 1871.

Condit, Carl W., *American Building.* University of Chicago Press, 1968.

Craig, Neville B., *The Olden Time*, Vol. I. Pittsburgh, 1846.

Davis, Andrew Jackson, *Free Thoughts Concerning Religion.* Boston, 1854.

——— *The Magic Staff, An Autobiography.* New York, 1857.

Ellis, Edward Robb, *The Epic of New York.* Coward-McCann, New York, 1966.

Fiftieth Anniversary Program. Putnam County Historical Society.

Fiske, Stephen, *Off-Hand Portraits of Prominent New Yorkers.* George R. Lockwood and Son, New York, 1884.

Foster, L. H., *Newport Guide.* 1876.

Franco, Barbara, "The Cardiff Giant: A Hundred-Year-Old Hoax." *New York History*, October 1969.

Genung, Abram, *The Frauds of the N.Y.C. Government Exposed.* New York, 1871.

George, Henry, *Social Problems.* New York, 1883.

Hansen, Marcus Lee, *The Immigrant in American History.* Harvard University Press, 1940.

Historical Sketch of the Fulton Ferry. Brooklyn, 1879.

History of Butler County, Pennsylvania. Waterman, Watkins and Company, Chicago, 1883.

Howard, Henry W. B., ed., *History of The City of Brooklyn.* The Brooklyn Daily Eagle, Brooklyn, 1893.

Johnston, Johanna, *Mrs. Satan.* G. P. Putnam's Sons, New York, 1967.

Josephson, Matthew, *Edison.* McGraw-Hill, New York, 1959.

Josephson, Matthew and Hannah, *Al Smith, Hero of the Cities.* Houghton Mifflin Company, Boston, 1969.

Kaplan, Justin, *Mr. Clemens and Mark Twain.* Simon and Schuster, New York, 1966.

Keller, Morton, *The Art and Politics of Thomas Nast.* Oxford University Press, New York, 1968.

W. C. Kingsley. Privately published, Brooklyn, 1885. Collection of the Long Island Historical Society.

Kouwenhoven, John A., *The Columbia Historical Portrait of New York.* Doubleday and Company, Garden City, N. Y., 1953.

Lancaster, Clay, *Old Brooklyn Heights.* Charles E. Tuttle Company, Rutland, Vt., 1961.

Lorant, Stefan, ed., *Pittsburgh; The Story of an American City.* Doubleday and Company, Garden City, N. Y., 1964.

Lyman, Colonel Theodore, *Meade's Headquarters.* Little, Brown and Company, Boston, 1922.

Lynch, Denis Tilden, *"Boss" Tweed.* Boni and Liveright, New York, 1927.

McCabe, James B., Jr., *Lights and Shadows of New York Life*, National Publishing Company, Philadelphia, 1872.

Mumford, John Kimberly, *Outspinning the Spider.* New York, 1921.

Mumford, Lewis, *Sticks and Stones.* Boni and Liveright, New York, 1924.

——— *The Brown Decades.* Harcourt, Brace and Company, New York, 1931.

Murphy, Mary Ellen, and others, eds., *A Treasury of Brooklyn.* William Sloane Associates, New York, 1949.

Nevins, Allan, *Abram S. Hewitt.* Harper and Brothers, New York, 1935.

——— *Grover Cleveland; A Study in Courage.* Dodd, Mead and Company, New York, 1932.

New York at Gettysburg, Vol. III. J. B. Lyon Company, Albany, 1900.

The Newport Directory, 1882.

Old Brooklyn Heights. Brooklyn Savings Bank, 1927.

Overton, Jacqueline, *Long Island's Story.* Doubleday, Doran and Company, Garden City, N. Y., 1929.

Pelletreau, William S., *History of Putnam County, New York.* Philadelphia, 1886.

Roebling, Emily Warren, *The Journal of the Reverend Silas Constant.* J. B. Lippincott Company, Philadelphia, 1903.

Rourke, Constance Mayfield, *Trumpets of Jubilee.* Harcourt, Brace and Company, New York, 1927.

Sandburg, Carl, *Abraham Lincoln, The Prairie Years and The War Years.* 1-vol. ed. Harcourt, Brace and Company, New York, 1954.

Scully, Vincent, Jr., *Modern Architecture.* George Braziller, New York, 1961.

Shaplen, Robert, "The Beecher-Tilton Case." *The New Yorker*, June 5 and 12, 1954.

Smith, Matthew Hale, *Sunshine and Shadow in New York*, J. B. Burr and Company, Hartford, 1868.

Stiles, Henry R., *The Civil, Political, Professional and Ecclesiastical History and Commercial and Industrial Record of the County of Kings and the City of Brooklyn, N. Y., from 1683 to 1884.* Munsell and Company, New York, 1884.

——— *A History of the City of Brooklyn: Including the Old Town and Village of Brooklyn, the Town of Bushwick, and the Village and City of Williamsburg.* 3 vols. Brooklyn, 1867–1870.

Still, Bayrd, *Mirror for Gotham*. New York University Press, 1956.
Strong, George Templeton, *The Diary of George Templeton Strong*, Allan Nevins and Milton Halsey Thomas, eds. The Macmillan Company, New York, 1952.
Stryker, Roy, and Seidenberg, Mel, *A Pittsburgh Album, 1758–1958*. Pittsburgh Post-Gazette, Pittsburgh, 1959.
Sullivan, Louis, *The Autobiography of an Idea*. Dover Publications, New York, 1956.
Swanberg, W. A., *Jim Fisk: The Career of an Improbable Rascal*. Charles Scribner's Sons, New York, 1959.
——— *Pulitzer*. Charles Scribner's Sons, New York, 1967.
Syrett, Harold Coffin, *The City of Brooklyn, 1865–1898*. Columbia University Studies in History, Economics and Public Law, No. 512, New York, 1944.
Taylor, Emerson Gifford, *Gouverneur Kemble Warren*. Houghton Mifflin Company, Boston, 1932.
Trollope, Anthony, *North America*. Alfred A. Knopf, New York, 1951.
Twain, Mark, and Warner, Charles Dudley, *The Gilded Age*. American Publishing Company, Hartford, 1873.
The Uncollected Poetry and Prose of Walt Whitman, Emory Holloway, ed. New York, 1932.
Wall, Joseph Frazier, *Andrew Carnegie*. Oxford University Press, New York, 1970.
Weld, Ralph Foster, *Brooklyn Village, 1816–1834*. Columbia University Press, New York, 1938.
Werner, M. R., *Tammany Hall*. Doubleday, Doran and Company, Garden City, N. Y., 1928.
Whitman, Walt, *The Portable Walt Whitman*. The Viking Press, New York, 1945.

Reference Works

Appleton's Cyclopedia of American Biography. New York, 1888.
Bartlett, John, *Familiar Quotations*. Thirteenth and Centennial Edition, Little, Brown and Company, Boston, 1955.
Biographical Directory of the American Congress, 1774–1961. U. S. Government Printing Office, Washington, D. C., 1961.
Dictionary of American Biography. Charles Scribner's Sons, New York, 1937.
Encyclopedia Americana.
Keller, Helen Rex, *The Dictionary of Dates*. The Macmillan Company, New York, 1934.
The Merck Manual of Diagnosis and Therapy. Merck, Sharp, and Dohme Research Laboratories, 1966.
National Cyclopedia of American Biography.
Sodeman, William A., M.D., and Sodeman, William A., Jr., M.D., *Pathologic Physiology*. W. B. Saunders Company, Philadelphia, 1967.
White, Norval, and Willensky, Elliot, eds., *AIA Guide to New York City*. The Macmillan Company, New York, 1968.

Index

[NOTE: References to illustrations are in *italics*.]